Preventing Occupational Disease and Injury Second Edition

Edited by

Barry S. Levy, M.D., M.P.H.
Gregory R. Wagner, M.D.
Kathleen M. Rest, Ph.D., M.P.A.
James L. Weeks, Sc.D.

American Public Health Association
800 I Street NW
Washington, DC 20001-3710

American Public Health Association
800 I Street, NW
Washington, DC 20001–3710
www.apha.org

Cover photographs by Earl Dotter illustrate airborne, ergonomic, safety, and physical hazards at work, all of which are preventable.

Georges C. Benjamin, MD, FACP
Executive Director

Hugh W. McKinnon, MD, MPH
APHA Publications Board Liaison

Printed and bound in the United States of America
Set In: Palatino and Helvetica Condensed
Interior Design and Typesetting: Terence Mulligan
Cover Design: Michele Pryor
Printing and Binding by Automated Graphic Systems, Inc., White Plains, MD

ISBN 0-87553-043-5
2M 11/04

Contents

Part I: Strategies for Prevention

Part II: Occupational Diseases and Injuries

Part III: Special Topics

TABLE OF CONTENTS FOR PART II, BY SUBJECT AREA*

Some chapters appear under multiple categories.

Hematologic and Lymphatic Disorders

Infectious Diseases

Kidney and Bladder Disorders

Liver and Other Gastrointestinal Disorders

Musculoskeletal Disorders

Neurologic Disorders

Other Disorders

Physical Hazards Disorders

Reproductive and Endocrine Disorders

Respiratory Disorders

Skin Disorders

Stress Disorders

Toxic Effects

Preface

Occupational health is a field within public health devoted to the prevention of occupational disease and injury. Public health professionals and others with responsibilities or interest in controlling or eliminating occupational disease and injury need appropriate tools to act effectively. Prevention depends on four fundamental tasks: (1) anticipation of the potential for disease or injury; (2) recognition of occupational disease and injury; (3) evaluation of relevant data; and (4) design and implementation of control measures. This book, thoroughly updated from its first edition published in 1991, provides information to assist in the performance of these tasks.

Part I presents an integrated and multidisciplinary approach to prevention, describing anticipation, recognition, evaluation, and control in the workplace. Public health practice requires a technical understanding of problems as well as an appreciation of the social, economic, and political context in which problems occur. Therefore, we consider the context of work in addition to the technical aspects of diseases and their prevention.

Part II is a compendium of many of the adverse health outcomes caused in whole or in part by work. Part II is organized alphabetically and presents in a consistent format for each disease or injury entity: a description of the disease or injury, occurrence data, causes, pathophysiology, and preventive measures. Other pertinent issues are discussed and further readings are listed. Where possible, disease entries are classified by the International Classification of Diseases, 10th revision (ICD-10). Sentinel Health Events/Occupational (SHE/Os), diseases that should always trigger public health investigations, are so identified.

We have attempted to provide useful data on the occurrence of specific occupational diseases and injuries in Part II. In the United States, as elsewhere, sufficient knowledge of the magnitude of occupational disease and injury is lacking. Data on the incidence of occupational injuries are often not reliable, and estimates of occupational disease rates are often rough approximations.

Part III presents information on taking an occupational history, work organization, populations at high risk, preventing occupational disease and injury in developing countries, resources for disaster preparedness, and organizations and websites that can be other sources of information.

This book is not designed to be used as a guide to the diagnosis and clinical management of occupational disease and injury. A number of texts devoted to that purpose are widely available. Instead, this book is designed to provide public health practitioners, interested and informed lay readers, and others guidance in identifying and preventing occupational disease and injury.

Topics discussed are drawn from those that are documented in the scientific literature. However, many causes of occupational disease and injury are unknown or poorly documented. There are many more occupational diseases and injuries than one could cover in this book. Our list of topics, therefore, is inevitably incomplete.

Given the close relationship between occupational disease and injury and environmental disease and injury, this book is relevant to the prevention of disease and injury caused by nonoccupational hazards. Because exposures are often more intense in the workplace than in the nonoccupational environment, investigation of workers' health and safety may help identify diseases, injuries, and other adverse health effects that can be caused by exposures outside the workplace.

We do not address factors in disease and injury that are largely accepted as being matters of personal choice, such as smoking, drug use, alcohol consumption, food intake, and exercise separately. Such risk factors are discussed in relation to particular outcomes, such as smoking and chronic lung diseases, and alcohol and traumatic injuries. Similarly, we do not title diseases that have both occupational and nonoccupational causes as "occupational," because all conditions described in the book are, at times, occupational. Thus, for example, we list "asthma," rather than "occupational asthma."

The occupational environment is unique for the practice of public health. This uniqueness arises from the nature of both work and occupational disease and injury. Both are therefore discussed as they relate to public health practice. Government intervention is a regular feature of public health practice and is one important form of intervention for workplace health and safety. Public health agencies at federal, state, and local levels have the power, for example, to impose quarantines, close swimming pools, and withdraw contaminated food, consumer items, or drugs from the market in order to reduce the risk of disease or injury. It is a logical extension of public health practice for government agencies to have similar powers to enter the workplace to prevent occupational disease and injury. This handbook can provide tools for public health consultation and intervention.

In occupational health, as in other fields of public health, some of the most difficult dilemmas concern whether, how, and when to intervene. Prudent public health practice requires action to prevent disease and injury, even in the absence of "definitive" data. Delay can result in more harm. Arguments for implementing preventive measures generally depend on assessment of the magnitude and severity of the problem, the quality of data on which this assessment is based, and the feasibility of effective controls.

Public health in the workplace holds out a significant promise: Because the workplace is a wholly human invention, virtually everything that happens there can be changed. It is therefore possible to eliminate or to control conditions in the workplace that are harmful—to a degree unmatched in other settings—thus significantly reducing the occurrence of occupational disease and injury.

We intend that this second edition of *Preventing Occupational Disease and Injury* continue to contribute to this effort.

—*The Editors*

Acknowledgments

We are grateful to the many people who contributed to the development of this book.

We acknowledge the ongoing support and sage advice of Ellen Meyer, the Director of Publications at the American Public Health Association, and the excellent work of Terence Mulligan in production of the book.

Our advisory committee played an important role in shaping the book at a critical stage in its development. We thank its members: William B. Bunn, III, Jane Lipscomb, Melissa McDiarmid, James A. Merchant, David Michaels, Margaret Quinn, and Luz Maritza Tennassee.

We acknowledge Heather Merrell for her outstanding work in preparing the manuscript and coordinating communication among many authors, reviewers, and editors.

Peer reviewers of chapters made many helpful comments and suggestions. We thank: Harriett Ammann, Henry Anderson, Dean Baker, John Balmes, William Beckett, Robert Castellan, Richard Clapp, Scott Deitchman, Michael Ellenbecker, Brad Evanoff, Nancy Fiedler, Marilyn Fingerhut, Howard Frumkin, Tee Guidotti, William Halperin, Howard Hu, Lynn Jenkins, Augustine Kposowa, David Kriebel, Paul Landsbergis, Boris Lushniak, Melissa McDiarmid, Vernon McDougall, James Melius, Carolyn Murray, Ben Nemery, Raymond Neutra, Stephen Olenchock, David Parkinson, Lee Petsonk, Glenn Pransky, Randy Rabinowitz, Carol Rao, Beth Rosenberg, Jonathan Rutchik, Nancy Sahakian, Steven Sauter, Dennis Schusterman, Gordon Smith, Lorraine Stallones, Naomi Swanson, Sam Telford, Luz Maritza Tennassee, Paul Vinger, and Richard Wedeen.

This book is designed in a similar format as the APHA handbook *Control of Communicable Diseases Manual*. We acknowledge those who edited that publication through 18 editions.

Finally, we thank our families for their support, encouragement, and understanding.

—*The Editors*

List of Contributors

Reza Alaghehbandan, M.D.
Research Analyst, Research and Development Division
Newfoundland and Labrador Centre for Health Information
1 Crosbie Place
St. John's, NL A1B 3Y8
Canada
709-757-2420
709-757-2411
<rezaa@nlchi.nf.ca>

Michael D. Attfield, Ph.D.
Surveillance Branch Chief
Division of Respiratory Disease Studies
National Institute for Occupational Safety and Health
MS G900.2
1095 Willowdale Road
Morgantown, WV 26505-2888
304-285-5737
<mda1@cdc.gov>

Robin Baker, M.P.H.
Director, Labor Occupational Health Program
Center for Occupational and Environmental Health
School of Public Health, UC Berkeley
2223 Fulton St. - 4th floor
Berkeley, CA 94720-5120
510-643-8900
510-643-5698 (fax)
<rbaker@uclink4.berkeley.edu>

Sherry L. Baron, M.D., M.P.H.
Coordinator, Priority Populations
National Institute for Occupational Safety and Health
4676 Columbia Parkway, R-10
Cincinnati, OH 45226
513-458-7159
513-458-7105
<SBaron@cdc.gov>

Michael E. Bigby, M.D.
Associate Professor of Dermatology
Harvard Medical School
Beth Israel Deaconess Medical Center
330 Brookline Avenue
Boston, MA 02215
617-667-4995
617-496-0560 (fax)
<mbigby@bidmc.harvard.edu>

A. Yvonne Boudreau, M.D., MSPH
Medical Epidemiologist
Centers for Disease Control and Prevention
National Institute for Occupational Safety and Health
Denver Federal Center, Building 53, Room I-1405
P.O. Box 25226
Denver, CO 80225-0226
303-236-5945
303-236-6072 (fax)
<ABoudreau@cdc.gov>

Sean M. Bryant, M.D.
Assistant Professor
University of Cincinnati College of Medicine
Department of Emergency Medicine
Division of Toxicology
231 Albert Sabin Way
ML 0769, P.O. Box 670769
Cincinnati, OH 45267-0769
513-558-7821
513-558-5791 (fax)
<bryantsm@ucmail.uc.edu>

Susan N. Buchanan, M.D., M.P.H.
Occupational Medicine Research Fellow
Department of Occupational Medicine
University of Illinois at Chicago
835 S. Wolcott MC 685
Chicago, IL 60612
312-413-0369
312-413-8485 (fax)
<sbucha@uic.edu>

David C. Christiani, M.D., M.P.H., M.S.
Professor of Occupational Medicine and Epidemiology
Harvard School of Public Health
Harvard Medical School
665 Huntington Avenue, I-1407
Boston, MA 02115
617-432-1260
617-432-3441 (fax)
<dchristi@hsph.harvard.edu>

Richard W. Clapp, D.Sc., M.P.H.
Professor
Boston University School of Public Health
715 Albany Street
Boston, MA 02118
617-638-4620
617-638-4857 (fax)
<rclapp@bu.edu>

Marc Croteau, M.D., M.P.H.
Division of Occupational and Environmental Medicine
University of Connecticut School of Medicine
263 Farmington Avenue
Farmington, CT 06030-4587
860-679-4035
860-679-4587 (fax)
<croteau@uchc.edu>

Mark R. Cullen, M.D.
Professor of Medicine and Public Health
Director, Occupational and Environmental Medicine
Yale University School of Medicine
135 College Street, 3rd Floor
New Haven, CT 06510
203-785-5885
203-785-7391 (fax)
<mrcullen@aol.com>

Letitia K. Davis, Sc.D., Ed.M.
Director, Occupational Health Surveillance Program
Massachusetts Department of Public Health
2 Boylston Street, 6th Floor
Boston, MA 02116
617-988-3335
617-624-5696 (fax)
<Letitia.Davis@state.ma.us>

Carl-G. Elinder, Ph.D., M.D.
Professor and Head of Department of Renal Medicine at Huddinge
Karolinska University Hospital and Karolinska Institutet
S-141 86 Stockholm, Sweden
46-8-585-82527
46-8-711 4742 (fax)
<carl.elinder@klinvet.ki.se>

William L. Eschenbacher, M.D.
Chief, Surveillance Branch
Division of Surveillance, Hazard Evaluation and Field Studies
National Institute for Occupational Safety and Health
4676 Columbia Parkway, M/S R17
Cincinnati, OH 45226
513-841-4184
513-841-4489 (fax)
<wae7@cdc.gov>

Robert G. Feldman, M.D. *(Deceased)*
Former Professor and Chairman
Department of Neurology
Boston University Schools of Medicine and Public Health
Boston, MA 02118

Marilyn A. Fingerhut, Ph.D.
International Coordinator
National Institute for Occupational Safety and Health
200 Independence Avenue SW, Room 715H
Washington DC 20201
202-205-8898
202-260-4464
<mfingerhut@cdc.gov>

C. Michael Fored, M.D., P.D.
Researcher, Department of Medicine, Karolinska Institutet
Clinical Epidemiology Unit
Karolinska University Hospital M9:01
SE-171 76 Stockholm
Sweden
46 8 517 79181
46 8 517 79304 (fax)
<Michael.Fored@medks.ki.se>

Judith E. Gold, Sc.D.
Professor, Department of Work Environment
University of Massachusetts Lowell
One University Avenue
Lowell, MA 01854
978-934-4378
978-452-5711 (fax)
<judith.gold@uml.edu>

Rose H. Goldman, M.D., M.P.H.
Chief, Occupational and Environmental Medicine
Cambridge Health Alliance
Associate Professor of Medicine
Harvard Medical School
Associate Professor of Medicine
Harvard School of Public Health
Cambridge Hospital
1493 Cambridge Street
Cambridge, MA 02139
617-665-1580
617-665-1672 (fax)
<rgoldman@challiance.org>

Michael R. Grey, M.D., M.P.H.
Professor of Medicine
Chief, Division of Occupational/Environmental Medicine
University of Connecticut School of Medicine
263 Farmington Avenue
Farmington, CT 06030-6210
860-679-4576
860-679-4587 (fax)
<grey@nso.uchc.edu>

Andrew S. Gurwood, O.D., FAAO, Diplomate
Associate Professor of Clinical Sciences
The Eye Institute of PA College of Optometry
Staff, Department of Ophthalmology
Albert Einstein Medical Center
1200 W. Godfrey Avenue
Philadelphia, PA 19141
215-276-6134
<Agurwood@pco.org>

Rana A. Hajjeh, M.D.
Medical Epidemiologist
CDC, National Center for Infectious Diseases
Atlanta, GA
and
Head, Disease Surveillance Program
Navy Medical Research Unit 3 (NAMRU-3)
Cairo, Egypt
Postal address: NAMRU-3
PSC 452, Box 130
FPO AE 09835
011-20-2-342-8933 (phone and fax)
<HajjehR@namru3.org>

Steven C. Hatch, M.D.
Resident
New England Medical Center
21 Cabot Street
Waltham, MA 02453
781-209-2824
<shatch@tufts-nemc.org>

Michael Hodgson, M.D., M.P.H.
Director, Occupational Health Program
VHA (136)
810 Vermont Avenue, NW
Washington, DC 20420
202-273-8353
<muh7@mail.va.gov>

Linden T. Hu, M.D.
Associate Professor of Medicine
Tufts-New England Medical Center
750 Washington Street
Boston, MA 02111
617-636-8498
617-636-3216 (fax)
<Lhu@tufts-nemc.org>

Joseph J. Hurrell, Ph.D.
Associate Director for Science
Division of Surveillance, Hazard Evaluation, and Field
 Studies
National Institute for Occupational Safety and Health
555 Ridge Avenue
Cincinnati, OH 45213
513-841-4403
513-841-4493 (fax)
<joseph.hurrell@cdc.hhs.gov>

Janice M. Huy, M.S.
Deputy Director
Office of Research and Technology Transfer
National Institute for Occupational Safety and Health
4676 Columbia Parkway, MS R6
Cincinnati, OH 45226
513-841-4245
513-841-4483 (fax)
<jmh4@cdc.gov>

Patricia A. Janulewicz
Department of Neurology
Boston University School of Medicine
715 Albany Street, Suite C-329
Boston, MA 02118

E. Lynn Jenkins, M.A.
Branch Chief
Division of Safety Research
National Institute for Occupational Safety and Health
1095 Willowdale Road
Morgantown, WV 26505-2888
304-285-5913
304-285-6235 (fax)
<lynn.jenkins@cdc.hhs.gov>

Jeffrey V. Johnson, Ph.D.
Professor
Department of Family and Community Health
University of Maryland
655 Lombard Street, Room 655B
Baltimore, MD 21201-1579
410-706-0799
410-706-4914 (fax)
<jjohn012@son.umaryland.edu>

Evan D. Kanter, M.D., Ph.D.
Staff Psychiatrist
VA Puget Sound Health Care System
Acting Assistant Professor
Department of Psychiatry and Behavioral Sciences
University of Washington School of Medicine
1660 South Columbian Way - 116MHC
Seattle, WA 98108
206-764-2007
206-764-2572 (fax)
<ekanter@u.washington.edu>

James P. Keogh, M.D. *(Deceased)*
Former Professor and Director
Occupational Health Project
University of Maryland
School of Medicine
Baltimore, MD 21201

Howard Kipen, M.D., M.P.H.
Director and Professor of Occupational Health
Environmental & Occupational Health Sciences Institute
UMDNJ-Robert Wood Johnson Medical School
170 Frelinghuysen Road
Piscataway, NJ 08854
732-445-0123 x 629
732-445-3644 (fax)
<kipen@eohsi.rutgers.edu>

William H. Kojola, M.S.
Industrial Hygienist
Department of Occupational Safety and Health
AFL-CIO
815 Sixteenth Street, NW
Washington, DC 20006
202-637-5003
202-508-6978 (fax)
<bkojola@aflcio.org>

Kathleen Kreiss, M.D.
Field Studies Branch Chief, Division of Respiratory Disease Studies
NIOSH
1095 Willowdale Road
Morgantown, WV 26505
304-285-5751
304-285-5820 (fax)
<kxk2@cdc.gov>

James Leigh, MB, BS, Ph.D., M.D., M.A., MSc, BLegS, C Eng,
 FAFOM, FAFPHM
Director, Centre for Occupational and Environmental Health
Senior Lecturer, School of Public Health
University of Sydney, Australia
61 2 9416 4577
<jleigh@bigpond.com>

Nancy Lessin, B.A., M.S.
Health and Safety Coordinator
Massachusetts AFL-CIO
389 Main Street
Malden, MA 02148
781-324-8230
781-324-8225 (fax)
<nlessin@massaflcio.org>

Barry S. Levy, M.D., M.P.H.
Adjunct Professor
Department of Family Medicine and Community Health
Tufts University School of Medicine
P.O. Box 1230
20 North Main Street, Suite 200
Sherborn, MA 01770
508-650-1039
508-655-4811 (fax)
<blevy@igc.org>

Jane A. Lipscomb, R.N., Ph.D., FAAN
Associate Professor
University of Maryland School of Nursing
655 W. Lombard Street
Baltimore, MD 21201
410-706-7647
410-706-0295 (fax)
<lipscomb@son.umaryland.edu>

Donald K. Milton, M.D., DrPH
Lecturer on Occupational and Environmental Health
Exposure, Epidemiology, and Risk Program
Harvard School of Public Health
Landmark Center
401 Park Drive
P.O. Box 15677
Boston, MA 02215
617-998-1036
617-384-8859 (fax)
<dmilton@hsph.harvard.edu>

Seth Nelson
Department of Neurology
Boston University School of Medicine
715 Albany Street, Suite C-329
Boston, MA 02118

Robert J. Nordness, M.D., M.P.H.
Officer in Charge
Naval Operational Medical Institute
Detachment Naval Undersea Medical Institute
144 Dunns Corner Road
Westerly, RI 02891
401-952-0120
<RJNordness@US.MED.NAVY.MIL>

Patrick J. O'Connor-Marer, Ph.D.
Pesticide Safety Education Program Coordinator
University of California Statewide IPM Program
One Shields Avenue
Davis, CA 95616
530-752-7694
530-752-9336 (fax)
<pjmarer@ucdavis.edu>

William B. Patterson, M.D., M.P.H., FACOEM
Assistant Professor of Environmental Health
Boston University School of Public Health
Chief Medical Officer
Occupational Health + Rehabilitation
175 Derby Street, Suite 36
Hingham, MA 02043
781-741-5175
781-741-5499 (fax)
<bpatterson@ohplus.com>

Edward L. Petsonk, M.D.
Senior Medical Officer, Surveillance Branch
Division of Respiratory Disease Studies
NIOSH
1095 Willowdale Road, MS G900.2
Morgantown, WV 26505-2888
304-285-5754
304-285-6111 (fax)
<elp2@cdc.gov>

Laura Punnett, Sc.D.
Professor, Department of Work Environment
University of Massachusetts Lowell
One University Avenue
Lowell, MA 01854
978-934-3269
978-452-5711 (fax)
<laura_punnett@uml.edu>

Marcia H. Ratner, Ph.D.
Departments of Neurology and Pharmacology
Boston University School of Medicine
715 Albany Street, Suite C-329
Boston, MA 02118
617-638-4440
617-638-5354 (fax)
<marcia@bu.edu>

Kathleen M. Rest, Ph.D., M.P.A.
Executive Director
Union of Concerned Scientists
2 Brattle Square
Cambridge, MA 02135
617-547-5552
617-864-9405 (fax)
<krest@ucsusa.org>

Leon Robertson
(current information not available)

Cecile S. Rose, M.D., M.P.H.
Director, Occupational Medicine Program
Division of Environmental and Occupational Health Sciences
National Jewish Medical and Research Center
1400 Jackson Street
Denver, CO 80206
303-398-1867
303-398-1452 (fax)
<rosec@njc.org>

Kenneth D. Rosenman, M.D.
Professor of Medicine
Chief, Division of Occupational and Environmental Medicine
Michigan State University
117 West Fee
East Lansing, MI 48864-1315
517-353-1846
517-432-3606
<Rosenman@msu.edu>

Annette MacKay Rossignol, Sc.D.
Professor, Department of Public Health
Oregon State University
322 Waldo Hall
Corvallis, OR 97330-6406
541-737-3840
541-737-4001 (fax)
<Anne.Rossignol@oregonstate.edu>

William A. Rutala, Ph.D., M.P.H.
Professor of Medicine
University of North Carolina School of Medicine
CB# 7030, Bioinformatics Building
130 Mason Farm Road, 4th Floor
Chapel Hill, NC 27599-7030
919-966-3432
<brutala@unch.unc.edu>

Mohammed A. Saleem, M.D., M.P.H.
Attending Physician
Employee Health Service
Division of Occupational and Environmental Medicine
John H. Stroger, Jr. Hospital of Cook County
230 N. Cadinal Avenue
Addison, IL 60101
630-833-9740
<altaf63@yahoo.com>

Marc B. Schenker, M.D., M.P.H.
Professor and Chair
Department of Epidemiology and Preventive Medicine
University of California at Davis
One Shields Avenue
Davis, CA 95616-8638
530-752-5676
530-752-3239 (fax)
<mbschenker@ucdavis.edu>

Michael A. Silverstein, M.D., M.P.H.
Assistant Director
Washington State Department of Labor and Industries
P.O. Box 44600
Olympia, WA 98504-4600
360-902-5495
360-902-5619 (fax)
<silm235@lni.wa.gov>

Rosemary K. Sokas, M.D., M.O.H.
Director, Environmental and Occupational Health Sciences
M/C 922 UIC
UIC School of Public Health
2121 W. Taylor Street
Chicago, IL 60612
312-355-4497
312-413-9898 (fax)
<sokas@uic.edu>

Emily A. Spieler, J.D.
Dean and Edwin Hadley Professor of Law
Northeastern University School of Law
400 Huntington Avenue
Boston, MA 02115
617-373-3307
617-373-8793 (fax)
<e.spieler@neu.edu>

Mark R. Stephenson, Ph.D.
Senior Research Audiologist
National Institute for Occupational Safety and Health
4676 Columbia Parkway, Mail Stop C-27
Cincinnati, OH 45226
513-533-8144
513-533-8139 (fax)
<mark.stephenson@cdc.hhs.gov.>

Arthur C. Upton, M.D.
Clinical Professor Environmental and Community Medicine
University of Medicine and Dentistry of New Jersey
Robert Wood Johnson Medical School
Room N112
675 Hoe's Lane
Piscataway, NJ 08854
732-235-3460
732-235-9607 (fax)
<acupton@eohsi.rutgers.edu>

Satya B. Verma, O.D., FAAO, Diplomate
Director, Community Eye Care Programs
Pennsylvania College of Optometry
8360 Old York Road
Elkins Park, PA 19027
215-780-1345
215-780-1327 (fax)
<satya@pco.edu>

Gregory R. Wagner, M.D.
Senior Advisor
National Institute for Occupational Safety and Health
Washington, DC
and
Visiting Professor
Harvard School of Public Health
665 Huntington Avenue
Boston, MA 02115
617-432-6434
<grw3@cdc.gov>

Elizabeth M. Ward, Ph.D.
Director, Surveillance Research
Department of Epidemiology and Surveillance Research
American Cancer Society
1599 Clifton Road, NE
Atlanta, GA 30329-4251
404-327-6552
404-327-6450 (fax)
<elizabeth.ward@cancer.org>

David J. Weber, M.D., M.P.H.
Professor of Medicine, Pediatrics and Epidemiology
University of North Carolina at Chapel Hill
Medical Director, Occupational Health and Hospital Epidemiology
University of North Carolina Health Care System
CB#7030 Burnett-Womack
Bioinformatics Building
UNC at Chapel Hill
Chapel Hill, NC 27599-7030
919-966-2536
919-966-1451 (fax)
<DWeber@unch.unc.edu>

James L. Weeks, Sc.D., CIH
Senior Scientist
ATL International, Inc.
20010 Century Blvd., Suite 500
Germantown, MD 20874
240-364-6009
301-972-6904 (fax)
<jweeks@atlintl.com>

David N. Weissman, M.D.
Senior Medical Officer, Health Effects Laboratory Division
National Institute for Occupational Safety and Health
1095 Willowdale Road, MS L-4020
Morgantown, WV 26505
304-285-6261
304-285-6126 (fax)
<DWeissman@cdc.gov>

Laura S. Welch, M.D.
Medical Director
The Center to Protect Workers' Rights
Adjunct Professor
School of Public Health and Health Sciences
George Washington University
8484 Georgia Avenue
Silver Spring, MD 20910
301-578-8500
301-578-5782 (fax)
<lwelch@cpwr.com>

Acronyms and Abbreviations

ACGIH	American Conference of Governmental Industrial Hygienists
ACLS	advanced cardiac life support
AOEC	Association of Occupational and Environmental Health Clinics
ANSI	American National Standards Institute
ATSDR	Agency for Toxic Substances and Disease Registry (DHHS)
BEI	Biological Exposure Index
BLS	Bureau of Labor Statistics (DOL)
CDC	Centers for Disease Control and Prevention (USPHS)
CT	computerized tomography
dB	decibel
dB(A)	decibel (A scale)
DHHS	Department of Health and Human Services (U.S.)
DOL	Department of Labor (U.S.)
DOT	Department of Transportation (U.S.)
EPA	Environmental Protection Agency (U.S.)
FAA	Federal Aviation Administration (DOT)
FAO	Food and Agricultural Organization (U.N.)
FDA	Food and Drug Administration (U.S.)
FIFRA	Federal Insecticide, Fungicide, and Rodenticide Act
FRA	Federal Railroad Administration (DOT)
FRC	Federal Railroad Commission (DOT)
IDLH	immediately dangerous to life or health
ILO	International Labor Organization
ISO	International Standards Organization
MRI	magnetic resonance imaging
MSDS	material safety data sheet
MSHA	Mine Safety and Health Administration (DOL)
MSHAct	Mine Safety and Health Act (Mine Act)
NCHS	National Center for Health Statistics (DHHS)
NEISS	National Electronic Injury Surveillance System
NHTSA	National Highway Transportation Safety Administration (DOT)

NIEHS	National Institute of Environmental Health Sciences (DHHS)
NIOSH	National Institute for Occupational Safety and Health (DHHS)
NSC	National Safety Council
OSHA	Occupational Safety and Health Administration (DOL)
OSHAct	Occupational Safety and Health Act
PEL	Permissible Exposure Limit
ppb	parts per billion
ppm	parts per million
REL	Recommended Exposure Limit
SARA	Superfund Amendments and Reauthorization Act
SEER	Surveillance, Epidemiology and End Results program (National Cancer Institute)
STEL	Short-Term Exposure Limit
TLV	Threshold Limit Value
TWA	Time-Weighted Average
USPHS	U.S. Public Health Service (DHHS)
WHO	World Health Organization (U.N.)

Foreword

The hope and promise of the Occupational Safety and Health Act of 1970 was to assure "every working man and woman in the United States safe and healthful working conditions." While clear progress has been made in reducing occupational illnesses and injuries, at the end of the 20th century 16 working men and women in the United States were still dying on the job daily (6,000 annually) and over 13,000 were being injured daily (over 5 million annually). From 1970 to 2000, however, dramatic changes occurred in the working environment, the organization of work, the demographics of the workforce, the provision of occupational health care, and the regulatory environment—all of which have provided new challenges for occupational disease and injury prevention.

The workplace of the 21st century will relate more to provision of services and will be more dispersed—away from large manufacturing plants to smaller employers in more rural locations. Workers will be more diverse and more will work with disabilities. New computer-based and other technology, workplace design, globalization, and off-site work will continue to change the organization of work. Occupational health services will be more often contracted out to local providers and the need for many more types of occupational health professionals will be extended to ergonomists, psychologists and behavioral health specialists, physician assistants and nurse practitioners, lawyers with training in public health law, and health and safety managers and workers with health and safety responsibilities.

Education and training of the occupational safety and health workforce must therefore extend from traditional occupational health and safety professionals—nurses, physicians, industrial hygienists, and safety specialists—to include a range of allied occupational health professionals, health and safety managers, and workers with health and safety responsibilities. Indeed, occupational safety and health education must extend to the general public, with basic concepts of prevention of occupational diseases and injuries introduced as elements of primary and secondary education.

New educational materials designed for this extended occupational health and safety workforce are therefore needed. This second edition of *Preventing Occupational Disease and Injury*, which has been written for both health and safety professionals and informed lay readers, meets this need extremely well.

James A. Merchant, M.D., Dr. P.H.
Dean of the University of Iowa College of Public Health

Luz Maritza Tennassee, M.D., M.P.H., M.H.Sc.
Head of Occupational Health at Pan American Health Organization

PART I

Strategies for Prevention

Synopsis of Part I

An integrated, multidisciplinary strategy for preventing occupational disease and injury, comprises four fundamental tasks: anticipation, recognition, evaluation, and control (Figure 1 on page 8). Primary prevention seeks to eliminate hazardous work exposures and working conditions, thus preventing their associated adverse health outcomes.

Anticipation

Anticipation involves imagining adverse events that might occur in the future and taking steps to prevent and forestall them. Anticipation is most important when new facilities are being built, new equipment and technology are being purchased, work processes and practices are being redesigned, and new workplace policies are being considered. A comprehensive hazards inventory can help anticipate and recognize risks.

Recognition

Surveillance, screening, and inspections of the work environment are fundamental tools of public health for preventing occupational disease and injury.

Surveillance is the ongoing gathering, analysis, and dissemination of data on the occurrence of disease and injury in a population in order to guide further investigation, evaluation, and control. Surveillance systems can be workplace- or community-based. They can rely on existing data sources, such as death certificates, workers' compensation, or hospital records, or create new ones. Surveillance can *actively* recruit participants or identify workplace hazards, or *passively* rely on existing data collected for other reasons. Passive surveillance usually detects only symptomatic disease. If surveillance detects a cluster or a change in the frequency of disease or injury, additional investigation and/or intervention is warranted.

Exposure and hazard surveillance are usually performed through workplace inspections, either regularly to identify and monitor potential hazards, or when a problem is suspected or a disease, injury, accident, or "near-miss" has occurred. Inspections may be required by statutes or regulations, insurance carriers, or labor-management contracts.

Medical screening attempts to detect disease at an early treatable stage. It depends on the availability of suitable screening tests and effective treatment or other interventions.

Biological monitoring is a form of surveillance that detects and measures toxic substances or their metabolites in bodily fluids or the dysfunction that they cause. It is a form of exposure monitoring that supplements environmental monitoring.

Evaluation

Information obtained must be evaluated using some combination of epidemiology, toxicology, exposure assessment, clinical assessment, and risk assessment.

Epidemiology is the study of the distribution and determinants of disease and injury in populations. Its ultimate purpose is to identify causes of disease and injury for which intervention would reduce risk. Chance, bias, and confounding are important factors to consider in interpreting individual studies. For helping to determine if a causal relationship exists, epidemiologic principles can be applied in the interpretation of a body of literature. They include strength of association, consistency, temporality, biologic gradient (dose-response), coherence, and plausibility.

Toxicological studies in vivo with non-human animals or in vitro, such as with cell cultures, permit control and measurement of dose or exposure as well as response. This helps to identify and describe pathophysiology, metabolic pathways, and dose-response relationships, but often at doses much higher than humans would encounter. For results to be relevant to humans, they need to be extrapolated from high to low dose and from animals to humans.

Exposure and hazard assessment is used to measure and evaluate exposures, hazards, and control measures and to determine compliance with exposure limits. The basic parameters of relevance when measuring exposure to chemical and physical hazards are concentration or intensity of exposure; duration, frequency, and latency; and determinants of exposure.

Exposure limits are useful benchmarks for evaluating exposure. Some limits are legal requirements, others are not. For many substances, there are none. Exposure that exceeds a legal limit may result in a penalty. Exposure that exceeds a recommended limit may identify a public health problem. Measuring exposure can be expensive, time-consuming, and sometimes unnecessary or not feasible. A less-rigorous exposure assessment method is control banding, which organizes chemicals by hazard classification, determines the amount of each chemical in use, assesses their volatility/dustiness, and considers the number of workers potentially exposed. This information may be sufficient for identifying the need for workplace controls, and may be especially useful in developing countries.

Clinical assessment of individual workers, combined with information about exposure, enables health care providers to evaluate illnesses and injuries and their work-relatedness. It can also help to identify problems early, enabling secondary prevention and creating a useful benchmark for later assessment.

Risk assessment helps to translate scientific knowledge of hazards into rational, evidence-based decisions and policies. By creating a common metric (the probability of illness or injury), it can help to set priorities among hazards and to define objectives for controlling a specific hazard. However, it is difficult to implement in practice. There is a range of opinion concerning which mathematical models are appropriate, what levels of risk are acceptable, and what level is significant.

Control

Decisions concerning control must often be made in the face of uncertainty. The precautionary principle is a recent articulation of the basic public health practice to prevent illness and injury even in the absence of thoroughly documented hazards. It is based on the following principles:

- Preventive action is appropriate even when causality or magnitude of effect are uncertain.
- The burden of proof should be shifted from having to prove an activity is harmful to having to prove that it is safe.
- A wide range of solutions should be considered.
- All interested parties should be involved in solving problems.

Upon establishing the need for intervention, one must design and implement specific hazard controls. A hierarchy of controls is standard practice in industrial hygiene. From most to least efficient, it is comprised of:

- *Inherently safe production* reduces hazards by using materials and methods that are inherently safer, such as: water-soluble solvents, which are less neurotoxic than organic solvents; bolts, which are quieter to install than rivets; and non-silica abrasives, which are less hazardous than sand.

- *Engineering controls* remove hazards from the work environment by such measures as local exhaust ventilation, enclosures, and shielding.

- *Personal protective equipment* aims to protect individual workers with such devices as respirators, earplugs and earmuffs, gloves, and aprons. These devices, while inexpensive, may be inconvenient to wear and vary in the degree of protection they confer. They must be designed and maintained for each worker.

While these generic types of controls differ for chemical, physical, biological, and psychosocial hazards, the principles upon which they are based are broadly applicable. Other aspects of prevention are discussed below.

Occupational Health and Safety Law

Occupational health and safety law must strike a balance between the rights of workers to a safe work environment and the rights of employers to manage their own affairs. (See Chapter 2.) Direct regulatory intervention by OSHA, MSHA, and state agencies evolved as the means of protecting workers' health and safety because alternative methods, relying on market forces, liability, contracts, and property rights, did not create adequate incentives for employers to control risks of injury or disease.

Effective regulation requires:

- Laws that define principles for regulation, obligations of employers, and the authority of government;
- An agency with appropriate expertise for identifying, evaluating, and controlling hazards;
- A mechanism and criteria for deciding what hazards warrant intervention;
- A basis for establishing a minimally acceptable level of protection;
- Dissemination of information to employers and employees about regulation;
- Enforcement methods to ensure compliance; and
- Protection of workers who participate in enforcement activities.

Regulations resulting from multiple laws may work in concert to promote prevention.

Economics considerations and cost allocation can provide additional incentives for prevention. A negligence-based tort system requires employers—or, more likely, third parties such as manufacturers of goods that caused injury—to pay for damages if it can be shown that they were negligent. A social insurance system spreads liability among a group of employers and covers costs of a larger proportion of injured workers, but at a lower benefit level. Most countries combine elements of both approaches.

Education and Training

Although behavior change is difficult and occupational diseases and injuries arise from many causes, education and training can provide workers and others with knowledge and skills needed to recognize, evaluate, and control hazards. (See Chapter 3.) In any specific workplace, education consists of identifying workers' specific needs, gaining support from potential participants, identifying educational objectives, establishing educational methods, and implementing and evaluating education.

Occupational Health and Safety Programs at Work

Health and safety on the job is a management responsibility that requires formulating a plan with specific objectives and a strategy for achieving them. (See Chapter 4.) Integral to this plan are both management commitment (manifest in a written program, a person responsible, and the commitment of resources) and employee involvement (manifest in meaningful participation in evaluating problems and making decisions about how to prevent disease and injury). Other aspects of a health and safety program include:

- Hazard identification through surveillance, screening, and incident investigation;
- Hazard control, with an emphasis on engineering control of health hazards and passive control of safety hazards;
- Employee education and training; and
- Recordkeeping, analysis, and evaluation.

Occupational health and safety programs can control hazards and reduce the frequency of injuries and illnesses. Critical to their success is commitment of time, attention, and resources appropriate to a given workplace and its actual and potential hazards.

State and Local Public Health Departments

The conventional responsibilities of state and local public health departments include conducting disease and injury surveillance, addressing the needs of vulnerable populations, investigating disease outbreaks, and developing and implementing intervention measures. (See Chapter 5.) In many instances, conventional concerns of public health departments—such as licensing of health care facilities, ensuring adequate sanitary conditions in restaurants, and assuring good indoor air quality in schools—target environments where public health and occupational health needs overlap.

Public health departments can also complement the activities of other agencies, such as OSHA, in the surveillance of diseases and injury—in part, because of state laws mandating reporting by health care providers of lead poisoning, pesticide poisoning, and silicosis. In addition, disease-specific registries and poison control centers can identify occupational diseases not identified otherwise.

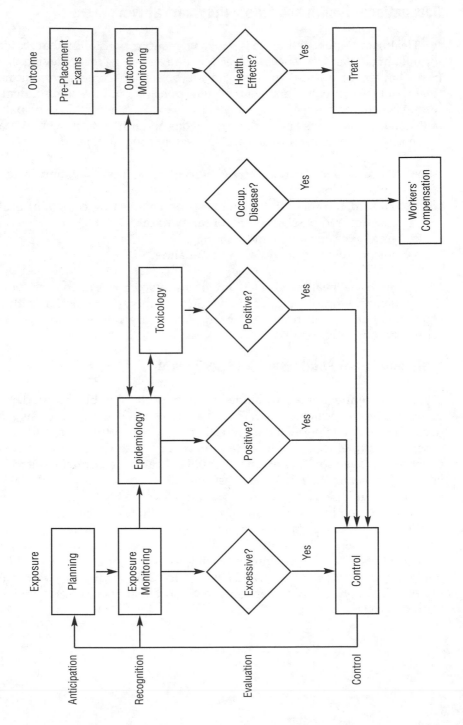

Figure 1: An integrated strategy for preventing occupational disease and injury. This figure depicts activities and events in rectangular boxes and analytic tasks in diamonds. The solid lines with arrows represent a logical (though not always chronological) sequence of events. From left to right, this diagram proceeds from exposure to outcome. Thus, industrial hygiene and safety engineering activities concerned with recognition, evaluation, and control of hazardous exposures are on the left; medical activities concerned with early detection, diagnosis of conditions, and treatment of people are on the right; and epidemiologic and toxicological activities, sciences that consider exposure and outcome together, are in the center.

From top to bottom, the diagram starts with anticipation, proceeds through data gathering and analysis. All paths converge on hazard controls, the ultimate aim of occupational health and safety practice.

Chapter 1
A Public Health Approach to Preventing Occupational Diseases and Injuries

James L. Weeks, Gregory R. Wagner, Kathleen M. Rest, and Barry S. Levy

Each year in the United States, 5000 to 6000 workers die from acute traumatic occupational injuries. The same industries report the highest rates of fatal injury every year—underground mining, construction, agriculture-forestry-fishing, and transportation. The construction, manufacturing, and agriculture sectors report the highest rates of nonfatal injury. Consistently, the most common causes of occupational injuries are motor vehicle crashes (on and off the road), violent acts and homicide, and falls from heights. Strains and sprains are the most common types of reported injuries, with back injuries accounting for the most lost time. There is, in short, sufficient knowledge to anticipate the potential for and types of many fatal and nonfatal occupational injuries in order to direct preventive efforts.

Similarly, the causes of, and methods to prevent, many occupational diseases are well known. The causes of some diseases, such as lead poisoning and silicosis, have been known since antiquity; successful prevention and control methods have been available for many decades. Dermatitis, musculoskeletal injuries, and noise-induced hearing loss continue to occur, despite many advances in our understanding of their causes and how to prevent them. With the development of more sophisticated means to measure exposures, identify subclinical effects, and process large amounts of information, more occupational causes of disease are being identified, including the occupation-attributable fraction of such common ailments as cancer, cardiovascular disease, and stress-related conditions.

Work can be hazardous. More importantly, most hazards can be anticipated. Knowledge about hazards and the methods to control them exist in many places: the scientific literature, regulatory agencies, workers' compensation organizations, the collective experience of workers and their employers, insurance companies, industry and trade organizations, labor unions, health and safety professionals, and elsewhere. At times, this knowledge is acquired and applied only after injuries, illnesses, or even catastrophes have occurred.

The following represents a systematic public health approach to preventing occupational diseases and injuries. (See Figure 1 on page 8.)

Anticipation

Anticipation is the cornerstone of the public health approach to controlling occupational disease and injury. It is a concept that merges two important elements: prediction and action. According to the *New Oxford American Dictionary* (2001), to anticipate is "...to expect or predict;...to take action in order to be prepared." In occupational health, anticipation involves an active expectation that hazards will occur, and an obligation to take action to eliminate or control them before harm occurs. Thus, occupational health and safety professionals have critical tasks to perform. They must (1) acquire the knowledge necessary to prevent disease and injury, (2) use it to anticipate problems, and (3) intervene to prevent their occurrence and reoccurrence. Occupational disease and injury are not an inevitable part of work.

Anticipation can take many forms. These include factoring worker health and safety into the design of work, workplaces, and work processes and practices and into the selection of safest/least hazardous materials and methods. Training and educating workers, managers, and health and safety employees about aspects of workplace technologies are also key. Preplacement and return-to-work medical examinations can play a role in anticipation and prevention. When coupled with the necessary and specific information about the workplace and demands of a job, an understanding of a worker's abilities and limitations can help predict and prevent future problems. Other aspects of anticipation include promoting good labor-management relations, forming active worksite health and safety committees, and eliminating payment and incentive mechanisms that encourage supervisors and workers to cut corners. Worksite-specific policies as well as overarching public policies and regulations can stimulate and support these efforts.

When employers and other decision-makers create or modify workplaces, they routinely anticipate and consider many factors, such as the price and accessibility of materials, the availability and training of the workforce, the most appropriate and efficient work methods and technologies, proximity to markets, and transportation. Health and safety hazards must find their place among these factors. Decision-makers should incorporate hazard control into their planning and decision–making processes. Then, steps to protect worker health and safety can be taken early in the design phase, before commitments are made that will be difficult and expensive to change—literally before they are cast in concrete.

Hazards Inventory

A comprehensive hazards inventory can be a useful tool in the anticipation and recognition of risk. Hazards can be inventoried by their form and route of exposure (Table 1) and can be evaluated with estimates or measurements of exposure combined with an assessment of potential harmful effects. This haz-

Table 1. Hazards Inventory

I. **Physical Hazards**
 A. Noise and vibration
 B. Extremes of heat and cold
 C. Radiation
 D. Barometric pressure
 E. Other

II. **Ergonomic**
 A. Repetitive motion
 B. Excessive force
 C. Awkward posture
 D. Other and aggravating conditions

III. **Chemical Hazards**
 A. Inhalation
  1. Particulate matter
  2. Gases and vapors
 B. Skin Absorption
 C. Ingestion
 D. Other routes

IV. **Biological Hazards**
 A. Infectious microorganisms
 B. Chemical hazards of biological origin
 C. Animals and plants

V. **Psychosocial Hazards**
 A. Work load; speed and hours of work
 B. Control over work
 C. Social isolation
 D. Work organization
 E. Abusive social environment

ards inventory involves a systematic enumeration of physical, ergonomic, chemical, biological, and psychosocial hazards present in the workplace; the routes of exposures; and an estimation of the frequency and intensity of potential exposure.

Physical hazards are different forms of energy. These include noise, vibration, ionizing and non-ionizing radiation, extremes of temperature and of barometric pressure, and rapid change in barometric pressure. In general, risk of injury increases with the energy level. Specific hazards are described in more detail in Part II (see, for example, Hearing Loss, Noise-Induced; Radiation, Ionizing and Nonionizing; Hyperbaric Injury; Injuries, Fatal and Nonfatal; and Heat Stress). Circumstances that can result in acute traumatic injury, such as working at heights or on busy highways, are also physical hazards.

Ergonomic hazards are physical motions or positions that may cause acute or chronic injury. These include repetitive or awkward motion, excessive force, flexion or extension, or awkward or static posture. Effects of these hazards are aggravated by stress, temperature extremes, work organization, and other environmental factors. (See, for example, Part II chapters on Upper–Extremity Musculoskeletal Disorders; Carpal Tunnel Syndrome; Tendonitis and Tenosynovitis; Peripheral Nerve Entrapment; and Low Back Pain Syndrome; and the Part III chapter on Work Organization.)

Chemical hazards are classified as solids, liquids, or gases that most commonly enter the body by inhalation, ingestion, or absorption through the skin. Harmful effects depend on the nature of substances, the magnitude of exposure and dose, and the duration of exposure. Inhalation is the most common route of entry for chemical hazards, although chemicals may be ingested if they contaminate food, drink, or smoking materials or are coughed up and swallowed. Evaluating the potential health effects of airborne particles (dusts, mists, and fumes) requires knowledge of their identity and concentration, as well as information about their size (diameter). Particle diameter determines the site of their deposition in the lung which, in turn, determines the site of injury and whether the particle is absorbed systemically (see Box). Some explosive and flammable gases and vapors, organic particles, such as coal, grain, and sugar, and some metal aerosols, such as magnesium and aluminum, may also create risk of fire or explosion.

Biological hazards include infectious microorganisms, plant or animal toxins, and animals. Microorganisms may (a) cause frank disease, such as viral hepatitis or Lyme disease; (b) cause allergic reactions, such as those associated with molds; (c) deplete oxygen; or (d) produce toxic gases. Plants may produce toxins. Animals may attack and transmit infections to zookeepers, veterinarians, postal delivery workers, and other workers. These hazards are described in Part II chapters on such topics as Human Immunodeficiency Virus (HIV) Infection; Tuberculosis; Hepatitis, Viral; and Zoonoses.

Psychosocial hazards result from a complex interplay of job demands, skills, decision-making latitude, personal control of work, work organization, and social interactions. (See Part II chapters on Stress, Collective Stress Disorder, Depression, Post-Traumatic Stress Disorder, Homicide and Assault, and Suicide, and the Part III chapter on Work Organization.)

Recognition

Hazard inventories coupled with knowledge and experience related to the workplace, work processes, and workforce are valuable tools for anticipating

---

Box

Aerosols

- Particles less than 5–10 μm in diameter are called respirable. They are likely deposited in the alveoli and terminal brochcioles where gas exchange occurs.
- The relative solubility of aerosols, gases, and vapors also affects the site of deposition and injury or absorption, with the more highly water soluble likely to be absorbed in the upper airways.
- Respirable particles are more likely to be retained and absorbed than large particles and thus cause lung injury or systemic disease.
- Repetitive deposition of aerosols in the airways may result in chronic airway irritation.
- Extremely small particles may behave differently and may be absorbed and distributed systemically.

---

health and safety risks and implementing prevention strategies. However, ongoing monitoring of hazards and of the health of workers is needed to identify and respond to changing or unanticipated risks to health and safety in both stable and unstable work environments. Surveillance, or tracking, of workplace hazards and worker health can help in the recognition of risk and adverse effects.

Surveillance

Surveillance, as defined in the *Dictionary of Epidemiology*, is "the ongoing scrutiny [of the occurrence of disease and injury], generally using methods distinguished by their practicality, uniformity, and frequently their rapidity, rather than by complete accuracy. Its main purpose is to detect changes in trends or distributions in order to initiate investigative or control measures." Surveillance is a fundamental part of public health practice. A surveillance system for occupational disease and injury control should (a) acquire information about hazardous exposures and diseases and injuries (outcomes), (b) analyze this information, and (c) disseminate and interpret it to those who need it. Mere information-gathering is not sufficient; the point of surveillance is to prevent disease and injury, not only to document its occurrence. Thus, a surveillance system must be linked with the capability to investigate further and to intervene to prevent disease or injury.

Surveillance strategies differ for acute and chronic health outcomes. For acute conditions, in which the time between exposure and outcome is short and/or the relationship of outcome to work is apparent, exposures and conditions that have caused disease or injury are more easily identified and controlled. Specific causes are more difficult to identify for chronic conditions, which are often multifactorial in nature but may result from long-term and/or low-level exposure or appear after many years (latency). The exposures that caused the disease may have changed or disappeared by the time the disease becomes clinically apparent. Even if the exposures that caused the disease remain in the workplace, the disease may not occur until late—well after the worker leaves the workplace for another job or to retire. Moreover, for irreversible conditions, control of hazards is effective at preventing disease or injury only prospectively and cannot correct harm that has already occurred. Therefore, surveillance for some chronic or long-latency health effects should persist even after hazards are controlled.

Identification of hazards that cause chronic effects often requires more knowledge than is available from a single workplace or population of workers. Information gleaned from the literature or from professional sources knowledgeable of problems in other settings may establish the need for controls. It is not prudent to wait for the development of chronic disease in any single workplace before reducing exposure that has caused problems elsewhere.

For both acute and chronic conditions, surveillance of exposures and of outcomes can be practiced both inside and outside specific workplaces. At the level of individual workplaces, surveillance involves systematic workplace inspections, measurement and evaluation of exposure, examination of workers, recordkeeping, and reporting of health effects and exposures. Surveillance at the workplace is thus an essential ingredient for managing a disease and injury prevention program. When practiced in settings apart from individual workplaces—for example, at local, state, or national agencies, hospitals, or disease or injury registries—surveillance can involve acquiring and analyzing data from a wide variety of sources, including employer reports, workers' compensation claims, hospital records, police reports, disease registries, and poison control centers. Regardless of the source of surveillance data, however, most information arises from workplaces, and intervention must eventually focus on workers, employers, and workplaces.

Hazard Surveillance

Workplace inspections should occur regularly to identify new problems and ensure that existing controls and prevention strategies are adequate and maintained. They also should be conducted (a) immediately after injuries, accidents, or near-misses occur to identify manifest and root causes; and (b) when someone on the job suspects a problem and requests an inspection. Inspections may be required by statute or regulation, by some insurance carriers, or by a labor-management contract.

Inspections can be conducted by health and safety committees, workers at the job site, health and safety professionals, engineers, or inspectors from outside the workplace, such as those from regulatory agencies, insurance carriers, parent corporate offices, or labor unions. Workers have regular and sustained experience with the workplace and are essential witnesses to circumstances surrounding specific incidents. Both employers and employees should be represented to ensure a balanced and full assessment of the hazards. Safety engineers bring needed expertise and experience to any inspection. Inspections by people not in daily contact with a job, such as government inspectors, insurance agents, or experts from parent corporations or unions, bring fresh perspectives and knowledge of pertinent regulations and guidelines. They may also bring needed incentives in the form of citations, changes in insurance rates, and other penalties or rewards.

Regardless of the reason prompting them, inspections are an important source of data for use in later analysis. Therefore, information should be documented in a consistent form. Periodic inspections not related to particular incidents should be systematic and custom-made for each workplace or industry. Checklists and a plan to visit every job site are useful. A walk-through survey should follow the flow of work from start to finish, accounting for uses, storage, waste, byproducts, disposal of all materials, and maintenance operations. Results should be documented and discussed by a work-site committee. Records of prior inspections, committee meetings and actions, and injuries should be available to monitor performance. The frequency of inspections depends on the degree of hazard. In the high-hazard underground coal mining industry, for example, certain inspections are required by statute prior to each work shift.

Employers are legally required to record work-related injuries that result in medical treatment, time away from work, and restricted activity. These records may be obtained by the Occupational Safety and Health Administration (OSHA) or by workers and their representatives. At a specific workplace, these records are useful surveillance tools. Although intended to cover both diseases and injuries, these records are inherently more likely to reflect acute conditions, such as traumatic injuries, and certain acute diseases, such as contact dermatitis and acute poisoning, than chronic conditions. Under the Mine Safety and Health Administration (MSHA), records of injuries and accidents are required to be reported (not merely recorded) and are available from MSHA.

Monitoring and measuring some occupational hazards are required by law. Employers under OSHA jurisdiction are required to maintain exposure records and make them available to OSHA or to workers or their representatives. When analyzed collectively, surveillance data can be used to identify potential hazards and, in some instances, to estimate exposure for individuals or populations over the period the regulations have been in effect. Exposure monitoring data for the mining industry are available from MSHA.

Medical (Health) Surveillance (Biological Monitoring of Effects)

The purpose of medical surveillance is to promote prevention by identifying the distribution and trends in the occurrence of disease and injury in populations. Outcome measures may range from individual signs and symptoms to well-defined diseases. Most often, medical surveillance results in secondary rather than primary prevention, because it can only identify individuals already affected by occupational exposures. When combined with hazard surveillance, however, analysis of medical surveillance data can complement primary prevention efforts by also identifying hazards. Surveillance is distinguished from screening (described below) by its concern with a target population; screening is primarily concerned with individuals.

Medical surveillance may be conducted at specific workplaces or in community settings. Employer medical departments usually have easy access to workers and to workplaces and thus are well situated to detect acute conditions, monitor active workers' health regularly, link medical with exposure data, and implement programs for the early detection and prevention of occupational conditions. Conditions caused by multiple exposures at different workplaces and chronic conditions that may not appear until after retirement are harder to detect by workplace-based surveillance programs. Moreover, workplace-based medical departments serve only a small minority of the working population—usually those situated in large or exceptionally high-risk workplaces.

Surveillance efforts that are based on data from the community, clinical settings, and registries complement workplace surveillance programs. Such efforts may be designed and implemented by government agencies, hospitals, or clinics, or they may be based on networks of health care providers. For example, the Sentinel Event Notification System for Occupational Risk (SENSOR) program developed by the National Institute for Occupational Safety and Health (NIOSH) and implemented by some states is a small-scale model of disease surveillance for selected disease outcomes with well-defined clinical features. The states may use several data sources, including case reports from a select group of health care providers, hospital discharge data, and workers' compensation data. A state agency collects and analyzes the information and reports and follows up to help activate preventive measures.

Medical surveillance may be active, in which populations of workers are selected, recruited, and examined, or it may be passive, relying on existing data collected at medical facilities for other reasons. Passive surveillance usually detects only symptomatic disease and cannot be relied on to uncover conditions earlier in their natural history. It also requires that health professionals be able to recognize the effects of occupational exposures in individuals in clinical settings. Because many occupational and nonoccupational diseases resemble each other and because occupational diseases are only suspected when an exposure history is obtained, passive surveillance cannot be relied on to detect many work-related diseases without a complementary effort to assess occupational exposures.

Active surveillance for work-related conditions requires assessment of exposure prior to conducting surveillance in order to define and select the appropriate population. Workers selected for active surveillance are usually at high risk for disease or injury. Selection criteria include assessment of current or past exposure based on measurements (if possible), employment history, or similar parameters. Because health effects depend on the identity of hazards, exposure assessment is also required prior to selecting medical testing procedures. For example, workers exposed to lead should receive regular laboratory tests for blood lead levels; workers exposed to silica, chest x-rays; and workers exposed to noise, audiometric examinations.

The occupational contribution to illness often cannot be recognized when individuals are considered in isolation from similarly exposed workers. Thus, results of surveillance should be analyzed in populations classified by exposure. Basic epidemiological methods are used to analyze such data, so that when aggregate findings are linked with assessment of occupational exposure, the results can be used to identify, evaluate, and control hazards.

Medical Screening

The purpose of medical screening is the early detection of disease or conditions for which treatment can successfully affect morbidity or mortality. Screening is a form of secondary prevention. Medical screening programs at work can provide the data for ongoing population health surveillance efforts. Screening usually consists of performing physical examinations and specific tests for the purpose of detecting disease at an early treatable or remediable stage. Ideally, screening should be designed and administered within an overall program that identifies people at risk, educates and informs them about the screening program, implements the screening program, and appropriately follows up with diagnostic tests on those who screen positive and with treatment for those who have disease.

Various types of screening tests are relevant to occupational health, ranging from pulmonary function tests that help detect respiratory impairment associated with work-related lung disease to tests and procedures for the early detection of various types of cancer. Important considerations in a screening program include the sensitivity of a test (the degree to which it correctly identifies those with the disease or condition), the specificity of a test (the degree to which it correctly identifies those without the disease or condition), and predictive value positive (the likelihood that a person who screens positive for a disease or condition actually has that disease or condition). There is often a trade-off between sensitivity and specificity in choosing among various screening tests and in determining a cut-off for abnormality in a given screening test; that is, a highly sensitive screening test is likely to be less specific, and a highly specific screening test is likely to be less sensitive. While screening is considered secondary prevention, it can indirectly identify hazardous workplace situations and lead to effective primary prevention.

Biological Monitoring of Exposure

Biological monitoring is a form of surveillance to estimate exposure based on biological assays of material, such as urine, blood, and exhaled air, collected from workers. Data from such examinations may be an important complement to industrial hygiene and can provide data for formal epidemiologic investigations. However, biological monitoring should never be a substitute for effective environmental monitoring of toxic exposures, and, since the difference between a biomarker of exposure and a sign of injury may not be clear, the potential ethical implications of biomonitoring should always be considered.

Sentinel Health Events

Intervention is an essential part of surveillance. If surveillance reveals the presence of a Sentinel Health Event/Occupational (SHE/O), identifies sick or injured people, or identifies people with preclinical signs or symptoms or other effects, multiple interventions should be considered: (a) referral of affected individuals to appropriate diagnostic and treatment services, (b) investigation of the workplace for identification of possible additional cases and causes, (c) implementation of measures to control the causative or excessive exposure, (d) provision of information about state workers' compensation programs and benefits, and (e) reporting of findings to the affected group of individuals and to the relevant authorities. Conditions that are considered to be SHE/Os are indicated as such at the start of some chapters in Part II.

Evaluation

Systematic analysis will help determine if intervention to control hazards is needed and will help inform the intervention effort. As illustrated in Figure 1 (see page 8), control is required (a) when there is excessive exposure, (b) when epidemiological or toxicological analysis demonstrates a positive relationship between exposure and health outcome, or (c) when occupational diseases are manifest. Several disciplines and tools can be critical aids in analyzing and evaluating data and other information obtained through surveillance and screening efforts. Below, we describe selected issues in epidemiological and toxicological investigations and in exposure measurement.

Epidemiology

Epidemiology is the study of the occurrence, distribution, and determinants of disease and injury in populations. It is an observational rather than an experimental science. Events and associations are explored in the actual contexts in which they occur rather than in experimental settings. The value of epidemiology comes at the expense of not being able to control variables directly and with the constraint of having to conduct studies on available populations.

Epidemiological investigations of occupational disease and injury are used for decision-making in much the same way as they are in other aspects of public health practice. The purposes of occupational epidemiology are to (a) identify and assess causes of disease and injury and, thereby, (b) identify opportunities for prevention, (c) evaluate or determine exposure limits, and (d) evaluate control measures.

The principal tasks of epidemiologists are formulating hypotheses to investigate, developing suitable study designs, selecting study and comparison groups (with respect to size and composition), determining and analyzing outcome data in relation to various factors, addressing potential sources of bias and confounding, and drawing conclusions from studies. The aim is to increase the probability that observations of associations (or the lack of associations) between exposures and health effects are valid. This is done by controlling, to the extent feasible, selection, observation, and random bias, and by analyzing data for confounding.

Solving problems of bias and chance does not, by itself, directly address the question of whether an outcome is, in fact, caused by a particular exposure or condition. In evaluating the relevant medical and scientific literature to determine whether exposure to a specific chemical or other factor likely caused a disease or medical condition, one should consider (a) the soundness and relevance of each individual study (methodology, results, and conclusions); and (b) the overall body of applicable scientific information. To help determine causality, guidelines or principles have been developed, including:

(a) <u>Temporality</u>: The disease or injury occurred after the exposure.
(b) <u>Strength of Association</u>: The magnitude of risk, such as the relative risk, associated with the purported cause is large.
(c) <u>Consistency</u>: There is general consistency of the results among relevant studies.
(d) <u>Biologic Gradient</u>: There is a positive dose–response or exposure–response relationship.
(e) <u>Plausibility</u>: Given current scientific information, the association is biologically plausible.

(f) <u>Coherence</u>: A cause–and–effect interpretation for an association does not conflict with what is known of the natural history and biology of the disease.

Only the first of these is essential in establishing a cause–and–effect relationship.

A conclusion that a demonstrated association reflects a cause-and-effect relationship is usually based on some, but not all of these factors, and a sound integrative assessment of the strengths and weaknesses of the available data. Conversely, the adage "Absence of proof is not proof of absence" is critically important in determining whether to continue to attend to suspected hazards when available epidemiologic studies do not support a causal association or appear "negative." These issues are not unique to occupational health studies; they are frequently encountered throughout public health and are discussed extensively in public health and epidemiology textbooks.

An important factor in occupational epidemiology involves the select nature of working populations. Participation in the labor force is restricted for some people who are disabled and/or chronically ill or injured; other people are selected out when they become ill or injured. In the United States, access to health care is often determined by work status, and people with better health care access may have better health. These selection processes result in the "healthy worker effect," in which workers as a group are somewhat healthier than those not working. Since adverse health effects can appear after departure from the workforce or may cause early departure from hazardous jobs, it is often necessary to select comparison populations, also composed of workers, and to find former or absent workers and determine their health status.

Working populations that are studied are also select groups. Because of practical limits on the availability of sufficiently large populations, the workers and the occupational hazards that epidemiologists study are often in larger workplaces or companies that have low turnover and are accessible. Epidemiologists are less likely to investigate workplaces that are small and widely dispersed, where labor turnover is high, or where there are significant obstacles to their accessibility. Such workplaces present significant logistical problems. Part-time, undocumented, and self-employed workers are unlikely to be studied.

Toxicology and Biochemistry

Toxicology and biochemistry are also useful disciplines for evaluating hazards. The toxic effects of many exposures are known primarily by their effects on animals or through biochemical investigations using bacteria or other cells. Information from such investigations can help evaluate biological plausibility and disease mechanisms. Dose–response relationships can also be evaluated. Risk to humans can be estimated by considering differences in physiology and by comparing the biochemical or toxicological potency of some substances with that of other substances whose effects on humans are better understood.

The clear advantage of toxicological and biochemical investigations are that they are experimental. Dose (as distinguished from exposure) and bias can be controlled, and outcome can be assessed to a degree unmatched in epidemiological investigations. The disadvantages are that investigators must extrapolate from one species to another and often from high to low dose, or they must make inferences from a biochemical reaction in cells to effects in whole organisms. In making extrapolations, investigators must compare humans with other animals concerning disease mechanisms and routes of exposure (such as whether the substance was inhaled, ingested, or injected). To aid in assessing risk, NIOSH and regulatory agencies such as the Environmental Protection Agency (EPA) and OSHA have developed methods of extrapolation.

Exposure Assessment

Exposure Measurement: Exposure measurement provides a more precise assessment of the hazard. Measurement of exposure is appropriate when hazards are suspected or reasonably predictable based on a hazards inventory combined with an assessment of work practices. Measuring exposure is done to (a) identify hazards in order to implement controls, (b) evaluate controls, (c) determine compliance with standards, and (d) assess exposure for epidemiologic research.

The basic parameters for measuring exposure to chemical and physical hazards are

- the concentration or intensity of exposure (measured volumetrically [ppm] or gravimetrically [mg/m^3], or by energy units [such as dBA and mSv]);
- the duration, frequency, and latency of exposure; and
- the determinants of exposure.

The determinants of exposure are those variables inherent in any occupational setting that affect exposure and are roughly classified as the conditions of use and the labor process. These concepts are applicable to jobs in either goods-producing or service sectors. The conditions of use include the level of production, the nature and form of raw materials, products and byproducts, maintenance processes, and use of industrial hygiene controls. The potential for accidental exposures from spills or breakdowns can be estimated with the assistance of people on the job: workers, supervisors, maintenance personnel, and others.

The labor process includes what workers do that affects exposure. This can include, for example, where and how they sit or stand, when and how they do certain parts of their jobs, whether they use personal protective equipment, the circumstances under which they might take a break, and so on. Some experienced workers may know certain "skills of the trade" that enable them to reduce exposure and these may be passed on to less experienced workers.

Measurement methods depend on the particular hazards and circumstances under which exposure occurs. Airborne hazards are typically measured by analyzing a sample of air at a particular point in time (a "grab sample") or averaged over an appropriate time period (a time-weighted average, or TWA), such as a work shift. Most exposure limits are linked to 8-hour work shifts. Work shifts, however, are becoming increasingly varied in length requiring more interpretation of exposure measurements. Standard methods in the United States have been developed by NIOSH in its *Manual of Analytical Methods* (available at the NIOSH website).

Some novel exposure measuring methods have been developed for measuring exposure associated with tasks (rather than work shifts) and for identifying controls. Task-based exposure assessment must be designed for the duration of each task. This type of exposure measurement can then link exposure and the need for controls to specific tasks. Real–time monitoring—linking a real–time instrument with a video camera and recording a task while simultaneously measuring exposure—is a useful tool for identifying specific sources of exposure and thus for developing controls.

Exposure varies. The major sources of variation are associated with the source itself as the determinants of exposure change. Sampling and analytical error, inherent in any measurement, is typically a smaller source of variation. Any sampling strategy should be able to account for variability by an appropriate selection of the type, number, and timing of samples and a written report should identify the sources and include measurements of variability.

Standards and Exposure Limits: Measurements are often evaluated by comparison with exposure limits and standards. Some standards are legal requirements; others are not. Legal standards are set by federal, state, or local governmental regulatory agencies. Setting standards is governed by the nearly identical language of the OSHAct and the MSHAct. (See next chapter on Occupational Health and Safety Law.) If exposure exceeds a legal standard, the employer is obligated to reduce it. If exposure exceeds a standard that is not a legal requirement, a public health need may dictate reducing exposure, but legal resources for compelling reduction may be lacking. The principal federal regulatory agencies involved in setting workplace health and safety standards are OSHA and MSHA, both in the Department of Labor. Both agencies set legally enforceable exposure limits for toxic substances or other hazards called permissible exposure limits (PELs). These are limits based on a

TWA exposure; they also set some short-term exposure limits (STELs or ceilings), and, for some hazards, conditions that might result in skin absorption. PELs exist for about 600 chemical substances commonly found at workplaces.

Exposure limits proposed by the American Conference of Governmental Industrial Hygienists (ACGIH) (threshold limit values, or TLVs, and biological exposure indices, or BEIs) and by NIOSH (recommended exposure limits, or RELs) do not have the force of law. Although most are identical with PELs, there are many differences. For example, only the ACGIH has proposed BEIs, some TLVs cover topics not covered by OSHA or MSHA (such as heat stress and ultraviolet radiation), and many TLVs and RELs for airborne hazards are more stringent than PELs. RELs are not required to meet the "feasibility" test of PELs.

There are limits in the scope, completeness, and enforcement of workplace health and safety standards. Inevitably, therefore, health effects associated with work can occur with little or no evidence of exceeding an exposure limit. In these situations, it may be necessary to investigate the work environment in detail to identify the potential causes of workers' ill health. Federal, state, or local public health agencies may conduct small- or large-scale epidemiological investigations; both the OSHAct and the Mine Act provide for health hazard evaluations (HHEs), which generally are more focused investigations of health problems in a specific workplace. These investigations are conducted by NIOSH and can be requested by employers or by workers or their representatives. The names of requestors may be kept confidential. To request an HHE, contact NIOSH.

Control Banding: Measuring exposure can be expensive and time-consuming and, in some cases, it may be unnecessary, impractical, or impossible, especially for small enterprises and companies in developing countries. A recently developed approach to assessing chemical exposures to guide decisions about control is called control banding. This method is relatively simple and intuitively appealing. It:

- organizes chemicals into exposure classes based on their hazard classification according to international criteria,
- determines the amount of each chemical in use,
- assesses its volatility/dustiness, and
- considers the number of workers potentially exposed.

This information is then used to apply conventional industrial hygiene control methods (see discussion of exposure control, below) or to seek specialist advice.

The control banding concept has been used in specific circumstances by a variety of organizations. The Health and Safety Executive (HSE) of the United Kingdom and the International Labor Organization (ILO) have developed tools that assist safety officials and employers in applying control banding techniques in the workplace (see http://www.coshh-essentials.org.uk and http://www.ilo.org/public/english/protection/safework/ctrl_banding/index.htm).

Such an assessment provides useful and sometimes sufficient information for making rational decisions about the need for intervention or the adequacy of existing controls. It may also assist in making decisions about the need for more formal and rigorous evaluations, including exposure measurements.

Clinical Assessment

Clinical evaluation of an individual worker or group of workers enables physicians and other health care providers to determine the nature of worker illnesses and injuries and their degree of work-relatedness. Information for the assessments is collected through medical and occupational histories, physical examinations, and/or various laboratory, imaging, and other clinical tests. Information obtained clinically needs to be reviewed in the context of, and integrated with, information obtained on exposures and working conditions, along with relevant medical and scientific literature regarding possible causal associations between workplace exposures and adverse health effects. Clinical assessment can also identify problems early, thus helping to assess future health risks that a worker may be facing and ideally stimulating the implementation of preventive measures to reduce these risks.

Risk Assessment

Risk assessment is a process that we use intuitively on a regular basis. We use it when we suspect or identify a problem. Generally, we try to characterize the problem, assess its magnitude, estimate the likelihood of harm, communicate our thinking to others if needed, and use all this to determine a course of action. Individuals use this basic process in many arenas—from making decisions about health and lifestyle issues to deciding how to handle finances. Health care professionals assess the risks and benefits when providing medical care to patients. Employers, workers, and occupational health professionals do this in the context of assessing and controlling workplace health and safety hazards.

In the context of public policy, risk assessment is a more formalized and rigorous process. Regulatory agencies increasingly conduct risk assessments in the course of making regulatory or policy decisions about occupational and environmental exposures. Often, the final product of a formal risk assessment is a statement such as, "The lifetime risk of this illness increases by 10^{-5} (one per 100,000 population) with each increase of each 1 mg-year/m^3 of exposure."

The appeal of risk assessment is the creation of a common metric—the probability of illness or injury—that can be used to address two important aspects of decision-making: (a) setting priorities among hazards; and (b) considering options for controlling a particular hazard, such as the setting of an exposure limit. Given inevitable resource limitations, risk information is a valuable asset to decision-making.

Nevertheless, despite its scientific trappings, risk assessment is also a value-laden process. Its outcomes are heavily dependent on the models and methods chosen and the variables used in the calculations. Uncertainty is handled differently by different models, and the inputs selected influence the outputs. Options and decisions that flow from the process may make assumptions about "acceptable risk," which can vary among the many stakeholders affected by the policy or regulation. The values and assumptions underlying conclusions based on risk assessment may not be clearly articulated.

Risk assessment is generally distinguished from risk management and risk communication. There is a large and rich body of literature on all of these activities. Readers involved in communication and decision-making related to occupational safety and health are encouraged to acquaint themselves with these critical disciplines.

Control

The previously described tasks of occupational health practice—anticipating, recognizing, and evaluating hazards—produce information used to identify and assess workplace hazards and the circumstances that produce them. But prevention of work-related disease and injury requires action: selecting and implementing the strategies needed to eliminate or reduce hazardous workplace conditions. Decisions are guided by the knowledge generated by the previous steps, by professional experience and codes of practice, and by relevant statutory mandates and regulatory requirements. A host of prevention and control strategies are available, and the precautionary principle can help guide decisions in the face of uncertainty.

The Precautionary Principle

In public health, as in most other professions, it is usually necessary to take action based on imperfect or missing information. Research might produce answers or reduce uncertainty, but the scientific enterprise takes time and resources and thus could tolerate the very hazards we are committed to control. Often, decisions cannot wait. The basis for taking preventive actions is the "best available evidence"—the same legal requirement embodied in both the OSHAct and the Mine Act.

The precautionary principle embraces the fundamental precepts of public health; John Snow in the 1850s in London removed the handle from the

Broad Street pump without a full understanding of an epidemic of diarrheal illness and thereby interrupted a large outbreak of cholera. In recent years, this practice has been articulated more formally and proposed as a principle to guide public health policy for recognizing and controlling environmental health hazards. A consensus statement was developed in 1988 to address the following situation: "When an activity raises threats of harm to human health or the environment, precautionary measures should be taken even if some cause–and–effect relationships are not fully established scientifically." Four propositions were articulated:

(1) Preventive action should be taken even under conditions of uncertainty.
(2) The burden of proof should be shifted from having to prove an activity is harmful to having to prove that it is safe.
(3) A wide range of possible solutions should be considered.
(4) Everyone with an interest should be encouraged to participate in making decisions about what to do.

The precautionary principle tolerates uncertainty and elevates health and safety as the primary goal. Precautionary action is consistent with the "anticipation" aspect of the conventional formulation of the practice of industrial hygiene. Shifting the burden of proof also shifts the burden of uncertainty. It suggests erring on the side of caution.

The principle encourages consideration of a wide variety of solutions. This is consistent with a full understanding of the hierarchy of controls described below and the many types of interventions that can be employed in the workplace. In any particular situation, the workers, employers, and safety and health professionals involved can often identify a wide range of options for prevention and control. Solutions that appeal to common interests and are broadly understood are more likely to be supported and implemented. A wide base of participation in problem-solving that includes persons with a stake in the outcome will often result in better, more effective, and more efficient decisions.

Strategies for Action

Upon establishing the need for intervention—based on exposure levels, epidemiological or toxicological evidence, manifest cases of occupational disease or injury, and the occurrence of "near misses," and guided by the precautionary principle, it is necessary to design and implement specific hazard controls. In the work environment, primary prevention involves controlling the exposure or hazard at its source. This often requires engineering solutions, which employ basic methods of industrial hygiene and safety engineering. Other strategies, such as training, education, and administrative actions, can also be useful, especially when engineering interventions are not feasible or affordable. The following discussion briefly describes the essentials of industrial hygiene control methods for primary prevention. Detailed information about specific control methods can be found in textbooks and is also discussed in chapters in Part II.

Hazard Controls

In the workplace, primary prevention of disease and injury is largely an engineering problem, using basic concepts of industrial hygiene and injury prevention for the purpose of eliminating or controlling hazards. Below, we discuss methods of controlling health hazards and preventing disease; later, we discuss methods of controlling safety hazards and preventing injury.

Hierarchy of Controls

An elementary principle of reducing the risk of illness and injury from exposure to health hazards is that one reduces risk by reducing exposure. The most efficient way to reduce exposure is to eliminate the hazard altogether; this common sense concept is the foundation for a basic principle of industrial hygiene. There is a hierarchy of controls: The most efficient control is to eliminate the hazard by preventing its generation at the source. The next most efficient control is to prevent its dispersal into the environment with engineering controls. The least efficient control is to protect the individual worker with appropriate personal protective equipment. This translates into inherently safe production; environmental or engineering controls that prevent releasing hazards into the work environment and protect against unintended contact with hazardous equipment; or, least preferred, the use of personal protective equipment to protect each exposed worker.

Inherently safe production uses materials that are less toxic and methods that are less hazardous and whose safety features are an inseparable part of the process. This concept can and should be built into the design of machines or jobs just as are other features such as their efficiency or payload or the amount of power that is consumed. Inherently safe production methods are implemented best and most easily when new jobs or processes are being developed and before commitments are made that might have to be undone, sometimes at significant cost. In many instances, development of inherently safer methods will require research and development and will require a shift in perspective—from an emphasis on assessing and managing risks to an emphasis on developing inherently safer technological alternatives. This is the basis for emphasizing anticipation as a key ingredient in preventing occupational disease and injury.

The concept of inherently safe production may also be used to revise an existing job, process, or industry. The opportunity to devise safe production is not lost if the need becomes apparent only after hazards appear. Revisions in jobs are routine in industry, and inherently safe production methods should take their place alongside other revisions for other reasons. Given the opportunity and support, effective and efficient solutions can be developed with industrial hygienists, engineers, workers, and supervisors who are familiar with a particular job.

There are as many examples as there are work processes. Examples include substituting water for organic solvents in paints, glues, and parts-cleaning operations; substituting impact for pneumatic hammers; reducing the weight of manually handled objects (such as boxes or dry–wall panels); changing pay schedules to reduce incentives to cut corners; working at night to avoid heat stress; using non-silica materials, such as abrasive blasting agents; employing closed systems when using volatile chemicals; and isolating and controlling access to high-hazard operations.

Engineering controls and personal protective equipment are discussed below in relation to chemical and biological hazards and in relation to physical hazards. When chemical or biological hazards are released into the environment, workers may absorb them by any of several routes. Engineering controls and personal protective equipment prevent absorption by blocking these routes and, consequently, we organize the discussion below by the several potential routes of absorption.

Chemical and Biological Hazards

Airborne chemical hazards—gases, vapors, particulate matter, and microorganisms—are common. Inhalation is a frequent route of absorption. And ventilation is a frequent engineering control. Ventilation can control not only toxic substances but also oxygen deficiency, air temperature, and humidity. In general, there are two types of ventilation for hazard control: local exhaust and dilution. Local exhaust ventilation removes contaminated air from as close to its source as possible, removing it from the worker's breathing zone, cleaning it by means appropriate to the hazard (for example, dust particles may be removed by a filter and organic vapors with an absorbent material), and releasing it outside the workplace. In designing local exhaust ventilation systems, it is common practice to design controls that would enclose the source of the hazard completely and provide access as necessary for work and maintenance. This allows fewer opportunities for hazards to escape into the worker's breathing zone and reduces the need to remove and treat large quantities of air. There are standard designs of local exhaust systems for a wide variety of industrial processes, so practitioners need not redesign systems from first principles for each installation. (See the ACGIH *Industrial Ventilation Manual of Recommended Practice*.) Dilution ventilation is less efficient for controlling hazards and is used in circumstances when a hazard of relatively low toxicity is not released from discrete locations.

Control of infectious occupational diseases is based on standard methods of infectious disease control (see Heymann DL [Ed.]. *Control of Communicable*

Diseases Manual [18th ed.]. Washington, DC: APHA, 2004). Although exposures to some biological hazards are most common in the health care industry, they may also occur in agriculture, among emergency response workers, and in other industries and occupations. Indoor air quality of offices and schools may also be affected by bioaerosols (see Building-Related Illness).

Respirators appear to be a simple solution to protecting workers from airborne hazards and they do provide protection but not until several problems are addressed. Respirators leak. They make breathing more difficult (by both increasing the resistance to breathing and by increasing the physiological dead space (where air is inspired but no exchange occurs). They are uncomfortable. They interfere with communication and sometimes with vision. Some, especially air-supplied respirators, are burdensome. Some workers cannot wear respirators due to other medical problems, such as lung or heart disease or claustrophobia. And different hazards and different tasks require different kinds of respirators. Because of these problems, if respirators are used, they should be part of a respiratory protection program.

OSHA and MSHA both describe such programs (29 CFR 1910.134; 30 CFR 56.5005, 57.5005, 70D) based, in part, on a consensus standard developed by the American National Standards Institute (ANSI Z-88). These regulations and standards state that there are only two generic circumstances under which respirators may be used: (a) as temporary measures (during emergencies or while other controls are being implemented), or (b) when engineering controls are not feasible.

When respirators are used, they should be introduced with a respiratory protection program. This should include:

- hazard evaluation;
- task evaluation;
- selection of the appropriate respirator for the hazard;
- worker training and education about the hazard, its warning properties, and proper use and limitations of the respirator;
- fit-testing to prevent leakage;
- medical evaluation of workers to determine their ability to use respirators;
- monitoring of hazards and stress associated with the use of respirators;
- monitoring air quality for air-supplied respirators;
- maintenance and care of respirators; and
- proper storage and maintenance.

Certain tasks, such as strenuous work, needs to communicate, and work in close quarters, can be impaired by the use of the respirators and should therefore be evaluated. If the workplace is a confined space, additional precautions are needed to provide for emergency escape and rescue.

Communication among workers who have to wear respirators should be organized before work is begun in a hazardous environment.

Dermal absorption usually results from direct contact with liquids, although some vapors may be absorbed through the skin. Factors determining absorption are the concentration of the liquid at the surface of the skin, the surface area exposed, and the duration of exposure. Thus, dermal absorption can be eliminated or reduced by preventing contact, reducing the concentration of the chemical in contact with the skin, or reducing the surface area or duration of exposure.

Contact with chemical hazards usually results from specific work practices, housekeeping problems, or accidental spills. Each of these causes can be modified to reduce risk of exposure. Therefore, the work process itself should be evaluated to reduce the risk of contact and to change materials or work practices.

In some instances, gloves, aprons, or other forms of personal protective equipment are appropriate. However, different chemicals require different kinds of barrier fabric to be effective. Selection is facilitated by use of the *NIOSH Pocket Guide to Chemical Hazards*.

Ingestion of hazards may occur if food, drink, or smoking materials become contaminated, if workers place contaminated fingers or implements into their mouths, or if splashes occur. Chemical hazards may also be ingested secondary to inhalation, coughing, and swallowing; preventing ingestion by this route is the same as preventing inhalation. Otherwise, preventing ingestion requires, for example, preventing eating, drinking, or smoking in areas where contamination might occur and by providing a clean place to eat and facilities (and time) for workers to clean up before eating.

Other routes of exposure to chemical or biological hazards are by injection (by hypodermic needles or paint guns) or by transplacental transport. Preventing exposure by these routes requires more specialized means.

Physical Hazards

Noise, ionizing and nonionizing radiation, and radiant heat are all forms of radiant energy and because of this similarity, exposure can be controlled with certain generic means. Risk of injury from these hazards is proportional to the amount of energy generated, released, and absorbed. Thus, one can reduce the risk of injury by:

- reducing the amount of energy generated or released,
- shielding or enclosing the source or the worker (in a cab or booth),
- increasing the worker's distance from the source,
- reducing the duration of exposure, or
- providing the worker with personal protective equipment.

Shielding the source is possible because radiant energy can be reflected or absorbed with appropriate shielding material. Increasing the distance between the worker and the source is useful because radiant energy attenuates by the inverse square rule (its intensity declines with the inverse square of the distance from the source). Reducing the time of exposure is useful because harmful effects are proportional to the duration of exposure. For more detail, see Part II chapters on Hearing Loss, Noise-Induced; Radiation, Ionizing and Nonionizing; and Heat Stress.

Evaluation and control of exposure to ionizing radiation is complex, and readers are advised to consult references listed in the chapter on this subject in Part II. Energy emitted from lasers, a unique form of radiant energy, is monochromatic, synchronized, and highly focused; consequently, it does not follow the inverse square rule. The risk of injury from a laser beam depends on the strength of the beam and the probability of intercepting the beam itself.

Vibration is mechanical energy transmitted directly to a worker's body by vibrating tools or machines or vehicles. Depending on its frequency and where on the body it is transmitted, vibration can affect limbs, organs, or the whole body. It is best controlled by eliminating the source, such as by using impact as opposed to pneumatic hammers or by isolating the worker from the source. Low-frequency whole-body vibration, such as from moving vehicles, can be controlled by ensuring proper vehicle suspension, maintaining smooth roadways, and providing adequate seats (see Hand-Arm Vibration Syndrome and Low–Back Pain Syndrome in Part II).

Personal protective equipment to protect workers from physical hazards depends on the nature of the hazard. Thus, hearing protection (ear muffs or plugs) may be needed to protect workers from noise, eye shades to protect from ultraviolet (UV) radiation or flying objects, and reflective clothing to protect against radiant heat. In each case, however, these forms of personal protective equipment come with their own problems. Hearing protection masks other important sounds (such as back-up alarms and noise of oncoming equipment), interferes with communication, and is often uncomfortable. (For more details and for a discussion of a comprehensive hearing conservation program, see the Part II chapter on Hearing Loss, Noise-Induced.) Protective clothing may protect workers, but it also may be burdensome and may increase heat stress.

Ergonomic Hazards

As discussed in the chapters on Musculoskeletal Disorders, Carpal Tunnel Syndrome, Tendonitis and Tenosynovitis, and Low Back Pain Syndrome, controlling the hazard through engineering controls (substituting mechanical for physical energy) is not only the preferred method, but may be the only effective preventive measure. In some instances, personal protective equipment, such as splints and back belts, has been used for primary or secondary prevention, but with little effect. Administrative controls, such as work breaks for keyboard operators and others engaged in highly repetitive tasks, have also been used.

Ongoing Monitoring

Workplaces and the people who work in specific workplaces change over time. Materials used in the production process change. Work processes are modified. Workers and managers change jobs. Some workplaces, such as most construction worksites, are dynamic. In others, change comes more slowly. Therefore, an effective and comprehensive program for prevention of occupational disease and injury requires ongoing monitoring of hazards and health in order to assure the adequacy of controls in place and prevention strategies in use, and to assist in the recognition of new problems before they become widespread. The intensity of the monitoring should be proportionate to the level of risk and the stability of the work process. Workplace hazard surveillance and worker health surveillance, along with periodic updating of a comprehensive hazard inventory, are useful approaches for ongoing monitoring.

Further Reading

Adams RM (Ed.). *Occupational Skin Disease* (3rd ed). Philadelphia: W.B. Saunders Company, 1999.

American Conference of Governmental Industrial Hygienists. *Threshold Limit Values and Biological Exposure Indices.* Cincinnati: ACGIH; published annually.

American Conference of Governmental Industrial Hygienists. *Industrial Ventilation: A Manual of Recommended Practice* (25th ed.). Cincinnati: ACGIH, 2004.

Ashford NA. Industrial safety: The neglected issue in industrial ecology. *Journal of Cleaner Production* 1997; 5: 46-52.

Baxter PJ, Adams PH, Aw TC, et al. (Eds.). *Hunter's Diseases of Occupations* (9th ed.). London: Arnold Publishers, 2000.

Blanc PD, Rempel D, Maizlish N, et al. Occupational illness: case detection by poison control surveillance. *Annals of Internal Medicine* 1989; 111: 238–244.

Bollinger N, Schutz RH. *NIOSH Guide to Industrial Respiratory Protection.* Washington, DC: U.S. Public Health Service, National Institute for Occupational Safety and Health, 1987. (DHHS [NIOSH] Publication, No. 87–116).

Burgess WA. *Recognition of Health Hazards in Industry: A Review of Materials and Processes* (2nd ed.). New York: John Wiley & Sons, 1995.

Burgess WA, Ellenbecker MJ, Treitman RD. *Ventilation for Control of the Work Environment* (2nd ed). New York: John Wiley & Sons, 2004.

Centers for Disease Control and Prevention. Guidelines for investigating clusters of health events. *MMWR* 1990; 39 (RR–11).

Checkoway H, Dement JM, Fowler DP, et al. Industrial hygiene involvement in occupational epidemiology. *Journal of the American Industrial Hygiene Association* 1987; 48: 515–523.

Checkoway H, Pearce NE, Crawford–Brown DJ. *Research Methods in Occupational Epidemiology,* (2nd ed.). New York: Oxford University Press, 2004.

Couturier AJ (Ed.). *Occupational and Environmental Infectious Diseases: Epidemiology, Prevention, and Clinical Management.* Beverly Farms, MA: OEM Press, 2000.

Ellenbecker MJ. Engineering controls as an intervention to reduce worker exposure.

American Journal of Industrial Medicine 1996; 29: 303–307.

Feldman RG. *Occupational & Environmental Neurotoxicology.* Philadelphia: Lippincott–Raven Publishers, 1999.

First M. Engineering control of occupational health hazards. *Journal of the American Industrial Hygiene Association* 1983; 44: 621–626.

Frazier LM, Hage ML. *Reproductive Hazards of the Workplace.* New York: John Wiley & Sons, 1998.

Freund E, Seligman PJ, Chorba TL, et al. Mandatory reporting of occupational diseases by clinicians. *MMWR* 1990; 39:19–28.

Graham JD, Green LC, Roberts MJ. *In Search of Safety: Chemicals and Cancer Risk.* Cambridge, MA: Harvard University Press, 1989.

Greenland S. Randomization, chance, and causal inference. *Epidemiology* 1990; 1:421–429.

Griefe A. Hazard surveillance: Its role in primary prevention of occupational disease and injury. *Applied Occupational and Environmental Hygiene* 1995; 10:737–742.

Haddon W Jr. The changing approach to the epidemiology, prevention, and amelioration of trauma: the transition to approaches etiologically rather than descriptively based. *American Journal of Public Health* 1968; 58: 1431-1438.

Haddon W Jr. A logical framework for categorizing highway safety phenomena and activity. *Journal of Trauma* 1973; 12:193–207.

Halperin W, Baker EL Jr. (Eds.). *Public Health Surveillance.* New York: Van Nostrand Reinhold, 1992.

Harber P, Schenker MB, Balmes JR (Eds.). *Occupational and Environmental Respiratory Diseases.* St. Louis: Mosby–Year Book, Inc., 1996.

Hathaway GJ, Proctor NH, Hughes JP (Eds.). *Chemical Hazards of the Workplace* (4th ed.). New York: Van Nostrand Reinhold, 1996.

Heymann DL. (Ed.). *Control of Communicable Diseases Manual* (18th ed.). Washington, DC: American Public Health Association, 2004.

Hill AB. *A Short Textbook of Medical Statistics.* London: Hodder & Stoughton, 1984.

Karvonen M, Mikheev MI (Eds.). *Epidemiology of Occupational Health.* WHO Regional Publications, European Series No. 20. Copenhagen: World Health Organization Regional Office for Europe, 1986.

Klaassen CD (Ed.). *Casarett & Doull's Toxicology: The Basic Science of Poisons* (6th ed.). New York: McGraw–Hill, 2001.

Kriebel D. Occupational injuries: Factors associated with frequency and severity. *International Archives of Occupational and Environmental Health* 1982; 50:209–218.

LaDou J. *Occupational & Environmental Medicine* (2nd ed.). Stamford, CT: Appleton & Lange, 1997.

Last JM. *Public Health and Human Ecology.* East Norwalk, CT: Appleton & Lange, 1987.

Last JM (Ed.). *A Dictionary of Epidemiology* (Fourth Edition). New York: Oxford University Press, 2001.

Lauwerys RR, Hoet P. *Industrial Chemical Exposure: Guidelines for Biological Monitoring* (2nd ed.). Boca Raton, FL: Lewis Publishers, 1993.

Levy BS, Wegman DH (Eds.). *Occupational Health: Recognizing and Preventing Work-Related Disease and Injury* (4th ed.). Philadelphia: Lippincott Williams & Wilkins, 2000.

Lewis RL. *Sax's Dangerous Properties of Industrial Materials (11th Ed.) (3 Vols.).* Hoboken, NJ: Wiley Interscience, 2004.

MacKenzie EJ, Shapiro S, Smith RT, et al. Factors influencing return to work following hospitalization for traumatic injury. *American Journal of Public Health* 1987; 77:329–334.

Maizlish NA (Ed.). *Workplace Health Surveillance: An Action-Oriented Approach*. New York: Oxford University Press, 2000.

McCunney RJ (Ed.). *A Practical Approach to Occupational and Environmental Medicine* (3rd ed.). Philadelphia: Lippincott Williams & Wilkins, 2003.

Monson RR. *Occupational Epidemiology* (2nd ed.). Boca Raton, FL: CRC Press, 1990.

Morrison AS. *Screening in Chronic Disease*. New York: Oxford University Press, 1985.

Mulhausen JR, Damiano J. *A Strategy for Assessing and Managing Occupational Exposures* (2nd ed.). Fairfax, VA: American Industrial Hygiene Association, 1998.

NIOSH Pocket Guide to Chemical Hazards and Other Databases (DHS [NIOSH] Publication No. 2004-103). Washington, DC: NIOSH, 2004. (Available as a CD from NIOSH: 1-800-35-NIOSH.)

Paul M (Ed.). *Occupational and Environmental Reproductive Hazards: A Guide for Clinicians*. Baltimore, MD: Williams & Wilkins, 1993.

Plog BA (Ed.). *Fundamentals of Industrial Hygiene* (5th ed.). Chicago: National Safety Council, 2004.

Pollack ES, Keimig DG (Eds.). *Counting Injuries and Illnesses in the Workplace—Proposals for a Better System*. Report of the Panel on Occupational Safety and Health Statistics, National Research Council. Washington, DC: National Academy Press, 1987.

Rempel D (Ed.). Medical Surveillance in the Workplace. *Occupational Medicine: State of the Art Reviews* 1990; 5(3):435–652.

Robertson LS. *Injury Epidemiology. Research and Control Strategies*, (2nd ed.). New York: Oxford University Press, 1998.

Robertson LS. (Weight x speed) + instability = hazardous trucks (Editorial). *American Journal of Public Health* 1988; 5&7:486–487.

Roelofs CR, Barbeau EM, Ellenbecker MJ, et al. Prevention strategies in industrial hygiene: A critical literature review. *American Industrial Hygiene Association Journal* 2003; 64: 62–67.

Rom WN (Ed.). *Environmental & Occupational Medicine* (3rd ed.). Philadelphia: Lippincott-Raven, 1998.

Rosenstock L, Cullen MR, Brodkin C, Redlich C (Eds.). *Textbook of Clinical Occupational and Environmental Medicine (2ⁿᵈ Ed.)*. St. Louis, MO: Elsevier Science, 2004.

Rothman KJ, Greenland S (Eds.). *Modern Epidemiology* (2nd ed.). Philadelphia: Lippincott-Raven Publishers, 1998.

Rothstein MA. *Medical Screening of Workers*. Washington, DC: Bureau of National Affairs, 1984.

Shapiro J. *Radiation Protection* (2nd ed.). Cambridge, MA: Harvard University Press, 2002.

Smith GS, Kraus JF. Alcohol and residential, recreational, and occupational injuries: A review of the epidemiologic evidence. *Annual Review of Public Health* 1988; 9:99–121.

U.S. Department of Labor, Bureau of Labor Statistics. *Annual Survey of Occupational Injury and Illnesses*. Washington, DC: U.S. Government Printing Office, published annually.

Waller J. *Injury Control: A Guide to the Causes and Prevention of Trauma*. Lexington, MA: Lexington Books, 1985.

Weeks JL, Fox MB. Fatality rates and regulatory policies in underground bituminous coal mining: United States, 1959–1981. *American Journal of Public Health* 1983; 73:1278–1280.

World Health Organization. *Epidemiology of Work-Related Diseases and Accidents.* Technical Report Series 777. Geneva: World Health Organization, 1989.

Zimmerman HJ. *Hepatotoxicity: The Adverse Effects of Drugs and Other Chemicals on the Liver* (2nd ed.). Philadelphia: Lippincott Williams & Wilkins, 1999.

Zwetsloot, GIJM, Ashford NA. The feasibility of encouraging inherently safer production in industrial firms. *Safety Science* 2003; 41 (2/3): 210-240.

Chapter 2
Occupational Health and Safety Law

Emily Spieler

The purpose of this section is to describe the role of the law in promoting improved occupational safety and health and in preventing occupational disease and injuries.

The law is more central to the successful prevention of occupational diseases and injuries than it is in most other public health arenas. Because occupational disease and injury often occur in the private sector, achievement of public health goals requires direct legal intervention in core economic activities. Not surprisingly, political and legal barriers limit the government's ability to intrude into these activities.

The law balances individual and corporate rights against the best public health outcome. In this sense, the law serves both as an important tool for achieving public health objectives and, at times, as an obstacle. In the case of occupational safety and health, this underlying tension is exacerbated by the fact that the right of workers to a safe working environment may be compromised by competing rights extended to their employers who are themselves entitled to due process and protection from excessive government intrusion.

There are many manifestations of the legal tensions between the public health goals of the Occupational Safety and Health Act [29 U.S.C. §651 et seq.)]and other principles that may be seen as limiting the promotion of public health. For example, at its heart, the Act broadly mandates that every employer "shall furnish to each of his employees employment and a place of employment which are free from recognized hazards that are causing or likely to cause death or serious physical harm to his employees." [29 U.S.C.A. §654(a)(1)]. In furtherance of this clear public health goal, the Act then authorizes the government to "enter without delay . . . and to inspect..." workplaces [29 U.S.C.A. §657(a)(1,2)]. But this broad right of entry was limited by a decision of the U.S. Supreme Court that balanced this public health principle against the Fourth Amendment right of employers to demand that the government seek a search warrant before searching premises. [See Marshall v. Barlow's Inc., 436 U.S. 207 (1978)]. In another example of this kind of balancing of interests, the OSHAct itself requires OSHA to get a court order before closing a work operation, even if the inspector finds an imminent (that is, "about to happen" [OED]) danger, resulting in critical delays to regulatory intervention [29 U.S.C.A. §662(a)(b)]. These limitations on the power of the law may vary, depending on specific circumstances. For example, the Mine Safety and Health Act is more protective of public health because of the history of the mining industry and public recognition of the high level of risk. Thus, mining inspectors may close an imminently dangerous mine without seeking court approval.

Legal Issues in the Prevention of Occupational Injury and Disease

In the sections below I describe aspects of the legal environment that affect the prevention of occupational disease and injury. In each section, I describe possible legal interventions and the particular model that is currently utilized in the United States.

Direct Government Regulatory Intervention

Governments can set enforceable rules to govern health and safety conditions in workplaces. Regulation is an essential tool to achieve reduction of workplace risks because alternative methods (legal regulation of contracts, property, and liability for injuries) generally have failed to create adequate incentives for employers to assess and eliminate risks. For example, much of the cost associated with death and disabilities resulting from occupational exposures in the United States is externalized from workplaces and therefore does not create sufficient incentives for employers to reduce risks.

Effective regulatory intervention requires all of the following:

* A law that sets out the principles for regulation.

In the United States, general industry is governed by the Occupational Safety and Health Act (OSHAct) [29 U.S.C. §651 et seq.] and the mining industry is governed by the Mine Safety and Health Act (Mine Act) [30 U.S.C. §801 et seq.]. These laws have similar provisions but differ in their enforcement powers. For instance, most inspections under the OSHAct are discretionary, while most inspections under the Mine Act are mandatory.

The OSHAct covers the vast majority of private establishments in general industry, but excludes all public sector employers. Federal employees are covered by a broad presidential Executive Order (Exec. Order No. 12196 [February 26, 1980]) that extends the provisions of OSHA to the federal employment sector. Many states have state laws with similar provisions that cover state and local public employees. A variety of less well known laws provide some health and safety coverage to workers in specified industries such as transportation and longshoring, or protection against particular exposures, such as use of radioactive nuclides.

The stated purpose of the OSHAct is to "assure as far as possible every working man and woman in the Nation safe and healthful working conditions" [29 U.S.C.A. §651(b)]. Employers are obligated to comply both with standards and with the "general duty clause" that requires every employer to "furnish to each of his employees employment and a place of employment which are free from recognized hazards that are causing or are likely to cause death or serious physical harm to his employees" [29 U.S.C. §654(a)(1)].

- An agency with appropriate expertise.

To be effective in its regulatory role, a government agency must have expertise that will:

- identify significant hazards;
- evaluate the risks of illness or injury caused by these hazards;
- design appropriate regulations;
- inform affected employers, employees, and employee representatives of requirements;
- enforce the regulations through workplace inspections and imposition of penalties; and
- establish appropriate review processes for rule-making and enforcement activities.

For example, the OSHAct established the Occupational Safety and Health Administration (OSHA) and the Occupational Safety and Health Review Commission. Similarly, the Mine Act established the Mine Safety and Health Administration (MSHA) and the Mine Safety and Health Review Commission. Both agencies have all of these responsibilities and powers. Both U.S. laws set out standard administrative procedures for rule-making and review of enforcement activities. Both laws also set out internal appellate procedures and establish the federal courts as the judicial appellate bodies for review of agency decisions.

- A mechanism to decide what hazards warrant intervention.

Regulations generally target hazards that create an unacceptable risk of injury or disease for workers. Both the decision to regulate a significant hazard, and the decision regarding the specific regulatory requirements, involve an assessment of whether the risk is excessive. This determination will vary based upon who is making the determination and upon political and economic concerns.

Under the OSHAct, OSHA must show that any rule (called a standard in the statute) is "reasonably necessary or appropriate to provide safe or healthful employment and places of employment" [29 U.S.C. §652(8)]. In interpreting this language, the U.S. Supreme Court held that OSHA demonstrate that any standard is "reasonably necessary and appropriate to remedy a significant risk of material health impairment." Industrial Union Dept., AFL-CIO v. Marshall ("the benzene case") [448 U.S. 607 (1980)]. The Court also said that OSHA must "make a threshold finding that a place of employment is unsafe in the sense that significant risks are present and can be eliminated or lessened by a change in practices" [Id]. In making this finding, OSHA must be prepared to estimate the number of injuries or illnesses that a regulation would prevent. That is, the Court said that OSHA had to assess risk of injuries or illnesses, determine that the risk was significant, and determine that a proposed standard would reduce that risk to an acceptable level. The Court specifically rejected the notion that the OSHAct required the creation of risk-free workplaces.

- A defined basis for establishing the minimal level of protection that is to be required by any regulation.

In every regulatory system, there must be a decision regarding the level of protection that is required. For example, feasibility is one measure of the appropriate level of minimum protection, and this is the measure adopted in the OSHAct. But even if feasibility is the measure, there may not be general consensus as to the definition of feasibility. Should feasibility be governed by currently available technology? In the alternative, regulations can encourage the development of new and more protective technologies. Should economic feasibility be a consideration for an industry or for any particular employer? What trade-offs are appropriate to consider? For example, are cost-benefit analyses appropriate in the designing of health and safety regulations, and, if so, how does one measure the benefits or the costs of illness, injury, disability, and death? Are there differences in the appropriate approach to safety risks (that may result in immediate death or injury) and health risks (where cumulative exposure over time may produce adverse health effects)? How does one balance feasibility against the level of protection due to workers?

These questions are answered under the OSHAct as follows. Under the Act, regulation of all risks is subject to the "reasonably necessary or appropriate" language in Section 3(8). In regulating a health hazard, OSHA is specifically required to set a standard that "most adequately assures, to the extent feasible, on the basis of the best available evidence, that no employee will suffer material impairment of health or functional capacity even if the employee has regular exposure to the hazard dealt with by such standard for the period of his or her working life" [29 U.S.C. §655(b)(5)]. In addition to the attainment of the highest degree of health and safety protection for the employee, other factors are consideration of the latest available scientific data in the field, the feasibility of the standards, and experience gained under this and other health and safety laws [Id].

In interpreting these requirements, the U.S. Supreme Court endorsed the language that requires OSHA to demonstrate that compliance with a standard is feasible (that is, "capable of being done") and rejected the argument that OSHA had to show that the benefits to be derived from a regulation must outweigh the costs of complying with it. The court said, "Congress itself defined the basic relationship between costs and benefits, by placing the 'benefit' of worker health above all other considerations save those making attainment of this 'benefit' unachievable . . . [C]ost-benefit analysis is not required by the statute because feasibility analysis is" [American Textile Mfrs. Inst. v. Donovan, 452 U.S. 490, 508-09 (1981) ("cotton dust case")]. Since this decision, courts that have reviewed OSHA standards have required OSHA to make detailed findings regarding feasibility of compliance with a standard. Technological feasibility is therefore based upon the existence of currently available (not hoped-for) technology.

OSHA (in response to a remand from a court) has set out principles that require a determination of feasibility for safety standards as well. Under the regulations, safety standards must also set the most cost-effective means for achieving the goal [58 Fed. Reg. 16,612-23 (1993)]. Explicit cost-benefit analysis is not required, however.

The specific requirements of health and safety rules can vary. Health standards generally set permissible exposure levels (based upon time-weighted averages, short-term [15-minute] exposure limits, and ceiling levels), establish monitoring and reporting requirements, and provide for medical surveillance. Safety standards can either set specification requirements (setting the precise requirements for compliance) or performance standards (allowing employers to design their specific compliance within guidelines).

- Dissemination of information to affected employers and employees regarding the regulatory requirements.

In order to comply with a rule, people must be aware of it and understand it. In the occupational safety and health field, both employers and employees need information. In the United States, employers receive information directly from the regulating agencies, as well as through various trade and commercial associations.

Four OSHA requirements provide access to information for employees.

(1) Employers must post information regarding the law in an accessible and visible place.
(2) The Hazard Communication Standard requires training, warnings, and communication of information to workers who use hazardous chemicals [29 CFR §1910.1200]. Generally, employees must request the material safety data sheets (MSDSs) to which they have legal access under the Hazard Communication Standard.
(3) The Access to Employee Exposure and Medical Records Standard

gives employees access to records that are maintained by employers regarding their exposure to toxic substances or their own medical condition [29 CFR 1910.1020]. This rule requires only that the employer provide the records that are maintained; it does not establish any record keeping requirements. Both individual employees and unions may request information under this standard. Unions can obtain access to individualized medical information only with the consent of the individual employee.

(4) Employees are allowed to view employers' record of injuries and ill-nesses that employers must maintain under the regulations.

In addition, unionized employees have access to information from employers as part of the collective bargaining process.

- Enforcement methods to ensure adequate compliance to both general and specific regulatory requirements.

Regulatory intervention is based upon the assumption that employers (and, in some instances, employees) will comply with the required rules without the ongoing presence of an enforcement agent. This may function simply as a normative principle: that is, people will sometimes comply with known rules without further action. But effective enforcement of regulatory requirements substantially increases the likelihood of compliance with rules, particularly unpopular or costly rules, by creating incentives (in the form of inspections, oversight, and penalties).

Once a rule is established, enforcement is designed to ensure compliance with the rule, not to reassess its feasibility. The rules are designed to reduce risks and therefore to prevent disease and injury. While risk assessment occurs during the development of regulatory standards, enforcement involves risk management.

Because rules can be ignored if they are not enforced, effective enforcement is at the heart of effective health and safety regulation. Enforcement is difficult because it requires both substantial resources and the political will to intrude into private workplaces.

Enforcement activities need to encourage compliance at all times, not only at the time of inspection. This is particularly important (and difficult) in the context of chronic health hazards, such as respirable dust or noise. For these hazards, conditions can vary substantially from one day to another and risk of disease is the result of exposure over time, but inspections can only identify brief instances of continually changing conditions.

Effective enforcement requires:

(1) Unambiguous communication of expectations and rules to employers and employees.

(2) On-site presence of the regulatory agency. In general, this is achieved

through inspections performed by health and safety professionals employed by the regulating agency. The utility of on-site inspections can be enhanced by empowering workers and unions to request and participate in them. Both OSHA and MSHA employ professionals who perform on-site inspections, but they differ markedly in the breadth and depth of their powers. In relation to the number of workers and workplaces under its jurisdiction, OSHA employs relatively few compliance officers and there are no statutory requirements for regular inspections of any workplaces. The agency must set priorities for inspections, but the majority of workplaces are never inspected. Inspections are only required after serious injuries or fatalities. In contrast, the Mine Act requires quarterly on-site inspections of underground mines and biennial inspections of surface mines, in addition to other inspections in response to specific concerns. Evidence suggests that the MSHA model creates a more effective tool to encourage regulatory compliance to reduce workplace risks.

(3) Specified powers for the on-site inspector or compliance officer. How extensive is the inspection? Can the inspector close down an operation or take other immediate on-site action? Can the employees participate in the inspection? The possible variations in on-site powers are illustrated by some of the key differences between the Mine Act and the OSHAct. MSHA inspectors can perform full inspections of an entire workplace; they can close down an operation if it poses an imminent hazard; and miners are guaranteed the right to participate in the inspection and to be paid for their time. In contrast, OSHA inspectors are often limited in the scope of an on-site inspection; they can only close down an operation by seeking intervention from a federal court (a cumbersome process); workers' participation in inspections is not compensated; and inspections cannot be challenged based upon a failure to involve the workers at the site.

(4) Penalties for noncompliance. The incentive to comply with the regulatory requirements rises with both the amount of the penalty and the likelihood that a substantial penalty will be imposed. OSHA sets penalties for serious, nonserious, and willful violations of both the rules and the general duty clause. OSHA penalties are often relatively small and therefore provide little financial incentive, on their own, for risk reduction. Additional economic incentive for compliance may derive from costs imposed by the market on employers who fail to comply, including the costs of OSHA appeal proceedings, higher insurance premiums, loss of contracts, or greater difficulty in raising capital.

(5) Requirements for abatement of noncomplying conditions. How aggressively does the regulatory agency require the employer to remove unlawful hazards that are discovered? To what extent can the

employer delay or eliminate a requirement for hazard removal through use of appeal procedures? Can employers obtain "variances" that allow noncompliance without demonstrating the effectiveness of alternative strategies to reduce risk? Whenever OSHA issues a citation, it sets a period of abatement within which the employer is required to correct the hazard. Under OSHA, however, this abatement period can be appealed.

(6) Appeal processes by which agency enforcement actions are reviewed. Are these review procedures efficient? Is the review conducted by individuals with expertise in health and safety matters? Can an employer delay abatement through use of these procedures? Do appeals often result in substantially reduced fines, through negotiation or decisions? These variables can dramatically impact the effectiveness of all components of enforcement activities. Under OSHA, employers can challenge the citation itself, the amount of the fine, and the period required for abatement. During the period of the appeal, the employer will not be required to abate the hazard, despite the continued exposure of workers to the regulated risk. If the employer appeals, employees or their representatives can become parties to the proceedings. If the employer does not appeal, however, employees may only challenge the period of abatement.

- Protection of workers who initiate and participate in enforcement activities.

Employees and unions provide critical information regarding health and safety conditions in workplaces and have vested interests in reducing workplace risks in order to maintain their own health and the viability of their continued employment. Effective protections require both substantive rights against retaliation and procedures that are speedy and accessible for employees. The ability of workers to participate effectively in preventive activities depends upon their ability (without retaliation) to seek information from their employers in order to investigate health concerns; to inform the regulatory agency regarding hazards; to participate in inspections or post-inspection proceedings, thus ensuring the inspections are thorough; and to refuse imminently dangerous work.

In the United States, these protections are theoretically offered under the OSHAct and the Mine Act, as well as under the general labor laws (National Labor Relations Act), and under collective bargaining agreements (union contracts). The OSHAct's specific anti-retaliation provisions are quite weak: workers only have 30 days to complain; the agency must agree to file the legal complaint in federal court; the remedy available to workers is limited to back pay and reinstatement. Few cases are ever filed. While both MSHA and OSHA guarantee to workers the right to refuse imminently dangerous work, the enforcement provisions under OSHA make this a difficult right to enforce. The general state of the law regarding job security is summarized in the next section.

Nonregulatory Legal Structures to Promote Prevention in the Workplace

Other than direct regulatory intervention, what other legal interventions might increase the likelihood that employers would reduce workplace risks? At least two other critical aspects of the law will affect the likelihood that employers will work to reduce injuries. The first involves the extent to which workers who are injured at work or who are active in health and safety organizing can maintain job security. The second issue relates to the allocation of the costs of morbidity and mortality caused by occupational hazards.

• Job security for employees.

The basic (generally unwritten) contract of employment between an employer and an employee can substantially affect an employer's incentive to provide a safe workplace. If a worker can be discharged for engaging in health and safety activities, or if a worker can be discharged without cost when he or she becomes medically unfit to perform a job, the employment relationship provides little incentive for the employer to prevent injuries. Conversely, if the employer must retain workers who raise concerns about health and safety and must pay the full costs associated with injury and disability, substantial incentives may exist to reduce health and safety hazards. Thus, a lifetime contract irrespective of work-caused impairment would create the highest level of incentive to reduce the risks. The least incentive exists when an employer can easily terminate employees for activism around health and safety or for inability to perform the pre-injury job at pre-injury performance levels.

Unlike most industrialized countries, private sector employees in the United States are generally governed by the "employment-at-will" doctrine. Under this rule, employers can terminate employees without justification and without offering an explanation. The rule also presumes that an employer is free to change the conditions of employment unilaterally: if employees continue to work, they are deemed to have accepted the new terms of work. This basic principle applies to health and safety conditions, subject to regulatory interventions. Union contracts generally amend the general "at will" rule to require "just cause" for discharge and to give unions the right to bargain over changed conditions. For a variety of complex reasons, however, the extent of unionization in the private sector workforce has fallen to under 9% in recent years in the United States. Public sector workers have greater basic legal protections from civil service systems, and are more likely to be unionized. States have developed laws that provide some protection for nonunion workers who are discharged in violation of "public policy," and these protections have been extended in a few states to include protection against retaliation for raising health and safety concerns. But these protections are limited.

There is no effective protection for U.S. workers who are medically unfit to continue to work. Despite provisions of the Americans with Disabilities Act (ADA), state anti-discrimination laws governing disability, limited protections for medical leave under the Family and Medical Leave Act, and some provisions of union contracts, most workers who are unable to work due to occupational injuries are not guaranteed continued employment by law. The U.S. Supreme Court has interpreted the provisions of the ADA in ways that limit the application of this Act to individuals with work-related disabilities. For example, in Toyota Motor Mfg., Kentucky, Inc. v. Williams, 534 U.S. 184 (2002), the court was reluctant to view a worker with carpal tunnel syndrome as disabled within the meaning of the ADA. The court wrote:

[T]o be substantially limited in performing manual tasks [a requirement for ADA protection], an individual must have an impairment that prevents or severely restricts the individual from doing activities that are of central importance to most people's daily lives.... the manual tasks unique to any particular job are not necessarily important parts of most people's lives. As a result, occupation specific tasks may have only limited relevance to the manual task inquiry. In this case, "repetitive work with hands and arms extended at or above shoulder levels for extended periods of time," the manual task on which the Court of Appeals relied, is not an important part of most people's daily lives. The court, therefore, should not have considered respondent's inability to do such manual work in her specialized assembly line job as sufficient proof that she was substantially limited in performing manual tasks....

The U.S. Supreme Court has also endorsed the principle that an employer can discharge an employee at risk for disease in order to achieve a safe workplace. In a case involving an employee with hepatitis C, the court relied on the employer's obligation to provide a safe workplace under the OSHAct to uphold the employer's right to terminate an employee because his liver disease would be exacerbated by continued exposure to toxins in an oil refinery [Chevron U.S.A. Inc. v. Echazabal, 536 U.S. 73 (2002)]. While this decision endorses the general principle that employers have an obligation to provide a safe and healthy working environment for each and every worker, it also underscores the fact that public policy does not guarantee job security to individual workers with underlying chronic disease who are at risk at work.

- Allocating the costs associated with occupational injuries and illnesses.

The cost of injury, illness, and death can be borne by workers and their families, by employers, or by social insurance systems that spread the costs in a variety of ways. Employers can be required to provide some form of benefits through public or private insurance schemes to cover medical costs, wage loss, and other economic and noneconomic losses.

Alternatively, through a negligence-based tort system, employers may have to pay the costs of injuries if they should have prevented the injury and failed to do so. This system provides both compensation for the employee who is injured and financial incentives to the employers, who would have to pay the costs to the employee. But these systems are also uncertain and fail to provide compensation to large numbers of injured workers who may not be able to provide the necessary legal proof regarding negligence or whose injuries are not sufficiently severe to warrant litigation. If insurance can be purchased by the employer to cover the costs, the costs are spread among a group, reducing the individual incentive to reduce risks. Individually-based experience-rating schemes correct for this to some extent.

Social insurance systems attempt to provide compensation more broadly and more efficiently. The amount paid to individual workers is generally reduced, but the worker is not expected to provide proof of negligence. The administrative structure in these systems is intended to be less complicated and provide speedier justice. These systems tend to spread the costs more, dulling even further the specific employer incentives.

Most countries utilize systems that combine elements of these two approaches. In the United States, most employees with work-related injuries are eligible for workers' compensation benefits. These benefits are established under state and some federal laws. In the United States alone, there are more than 55 different systems of compensation, requiring different proof, with different procedures, and providing differing levels of benefits. All of these systems provide medical benefits for the specific work-related injury, some replacement of lost wages, and some system for compensating for permanent impairments. All of these systems provide some form of insurance system that spreads the costs of injuries among a class or group of employers (with a limited amount of adjustment of insurance rates based upon employer-specific experience). Significantly, all of the state compensation systems are viewed as the exclusive remedy for workers who are injured at work. They therefore replace tort liability, protecting the employer from lawsuits involving the injuries in almost all situations. Workers' compensation is therefore both a system of compensation for workers and an insurance system to protect employers from liability for negligently caused injuries. If levels of compensation for workers in these programs are reduced, financial incentives for employers to reduce hazards are likewise reduced.

For a variety of reasons, workers' compensation programs generally are much more effective at providing compensation for traumatic injuries (particularly those that do not result in long-term impairment) than for occupational diseases that result from exposure to health hazards over time. To the extent that these systems provide some incentive for employers to reduce risks, these incentives therefore tend to be relatively less effective in encouraging the reduction of health hazards.

The amount of compensation provided by workers' compensation programs does not replace all of the economic losses of the worker, and it does not compensate the worker for any noneconomic losses. The costs that are not covered are borne by workers and their families and by other, more general, social insurance and social welfare programs. For example, a worker with a significant injury who is unable to continue working may apply for benefits under the Social Security Disability Insurance program. This program provides lifetime benefits to all individuals who have worked for a required period and who are permanently disabled from continuing to work, irrespective of the cause of their disability.

Although most occupationally injured employees in the United States cannot bring tort litigation directly against their employers, they can bring lawsuits against "third party" manufacturers of equipment or substances. These manufacturers generally have legal duties to warn users of their products of hazards and to eliminate known defects. These manufacturers are not protected by the exclusive remedy provisions of workers' compensation. Notably, however, this liability will not increase employers' incentives to reduce risks in workplaces.

Chapter 3
Occupational Health and Safety Education

Robin Baker

Training is an essential element of any successful workplace hazard control program. Yet training alone is not an adequate prevention strategy.

At the beginning of the 20th century, the field of occupational health was founded with the assumption that occupational diseases and injuries were the result of worker ignorance and carelessness. According to these early theories, developing safety consciousness in workers was a primary solution to workplace injuries. This notion remains all too popular, with ongoing debates as to what percentage of accidents is due to unsafe acts as opposed to unsafe conditions. Yet modern research points to a multiple causation theory of injuries and illness. This theory suggests the need for multifaceted prevention strategies, in which training is but one component. Major elements of an effective prevention program in the workplace include:

- management commitment,
- worker participation,
- workplace analysis,
- hazard prevention and control, and
- safety and health training.

Training is most effective when placed in this context of a comprehensive prevention strategy. The literature shows that training can be quite effective at the level of increasing hazard awareness and adoption of safe work practices. However, training programs that actually result in lower injury rates are those that are coupled with other interventions, such as engineering controls and changes in management practices.

Purpose of Training

In its narrowest form, training can provide the skills and motivation for workers to use personal protective equipment properly and follow safety procedures. Safety and health training should be an integral part of basic job skills training. But, in its broadest form, training can go beyond affecting individual compliance. It can also prepare workers to take an active role in promoting health and safety on the job.

Workers have an invaluable role to play in preventing injury and illness in the workplace. Technical experts who do not know a job from daily experience cannot anticipate the full range of potential hazards or potential solutions. Good training can unleash the wealth of knowledge held by workers—the practical experts—and provide the opportunity for them to share their knowledge with the technical experts, ideally before problems occur.

Informed and active workers not only protect themselves but benefit management as well, both by reducing lost time and compensation costs and by increasing productivity. They also protect society as a whole. Costly government inspections and enforcement programs have a much smaller role to play in a society in which effective cooperation is taking place between informed labor and management. And, of course, workplace health and safety problems do not stop at the plant gate: Toxic substances that are not properly controlled in the workplace can become a community problem.

Unfortunately, much of the training offered in the workplace is not intended to empower workers and use their expertise. Some training even encourages passivity in workers by seeking compliance with work rules without allowing for worker input. Worker passivity, however, is dangerous because it tolerates hazards rather than controlling them.

Training to overcome passivity should include how to identify and control hazards as well as how to get problems corrected (using company procedures or legal enforcement). A comprehensive injury and illness prevention policy would be one that requires this type of action-oriented training and in turn uses workers as full participants in the process of making the workplace safe (through the health and safety committees, union safety representatives, protection for whistle blowers, and other measures). (See discussion of workplace safety and health committees at the end of Chapter 4.) Model worker participation programs such as this exist in several countries, including Sweden, Australia, and Canada.

A Quality Training Process

Despite the lack of any such policy in the United States, it is possible to design and provide high-quality training that empowers workers and encourages participation. The first step toward designing successful training is to understand that training is a continuing process, not an event. It is a process that requires careful and skillful planning through each major stage.

Step One: Assess Needs. A thorough needs assessment forms the foundation for the entire planning process. This includes both a hazards assessment—identifying high–priority problems to be addressed—and a profile of the target population. Who can most benefit from training? What training has the target population already received? What knowledge and experience will the trainees bring to the process? What is the ethnic and gender makeup of the work force? What is the literacy level of the workers and what languages do they speak? Whom do they respect and whom do they mistrust? Needs assessment can be based on questionnaires, reviews of documents, observation in the workplace, and interviews with workers and their union representatives.

Step Two: Gain Support. Successful training also relies on identifying and involving key actors. The target population of training must be involved in the planning process; it is difficult to gain their trust without having sought their input. Who are the other key actors? Is it high–level management who must make the necessary resources available? Is it the union, that must provide access to the workforce or give the training credibility? Can a government or community organization provide the training, support, or follow-up activities? Whoever has a key role to play must be involved in the process through co-sponsorship, participation on a planning committee, or other means.

Step Three: Establish Training Objectives. Using information from the needs assessment, the planning team can identify appropriate objectives. A common mistake is to assume that the objective of training is to present information. What is presented matters less than what the target population receives. In addition, one must consider not just what the learner should know (knowledge objectives), but also what the learner will believe (attitude objectives) and do (behavior objectives) as a result of training.

There is a hierarchy of these objectives (Table 1). Providing information and teaching specific skills are the first level (though not always as simple as they sound). Attitude objectives are more challenging and behavior objectives are the hardest. For example, it is possible to communicate the risks of asbestos to steamfitters and instruct them on proper work practices. It is more difficult to change what steamfitters believe, such as to convince them that they and their fellow workers are at risk and that something can be done about it. A real challenge is to change individual behavior—to get them to actually use safe work practices and protective equipment. But the highest–level goal is for workers to engage together in making work safer, for example, by insisting on having the necessary tools and equipment and adequate time to do a job safely.

Table 1. Hierarchy of training objectives

Social Action

Individual Behavior

Attitudes/Beliefs

Knowledge and Skills

To summarize before going on to the actual design of training sessions: (a) The goal of training must be to empower workers to act intelligently, not to control them. (b) Workers must be involved in planning their own training. (c) Training should be aimed at providing what workers most need to know, believe, or do. (d) For both ethical and practical reasons, it is essential that learners agree with the learning objectives.

Step Four: Select Training Methods. It is important to select the right methods for the chosen objectives. In general, the more ambitious the objectives, the more intensive the methods must be. Selection of appropriate methods should be based on basic principles of adult education, which include the following:

- People learn best by building on what they already know, by incorporating new ideas into their already vast reservoir of learning. Adults wish to be respected for their experience in life. Therefore, effective methods are those that draw on participants' own knowledge.
- People learn in different ways. Each person has a particular learning style. In a group, some will learn best by reading, some by listening, and some by practicing. It is important to offer training in more than one way. Variety not only ensures that each cognitive style is addressed but also provides repetition to reinforce learning and, of course, combats boredom.
- People learn better through active, participatory methods than through passive measures. Lectures and written materials have their place in a full repertoire of methods. But case studies, role-plays, hands-on simulations, and other small-group activities that allow each individual to be involved are more likely to result in the retention and application of new learning. Participatory methods require more training time, smaller groups, and perhaps different skills than those many trainers currently possess. But to increase the impact of training, participation is essential.

Whatever methods are selected, the profile of the workforce must be considered. For example, the trainer may need to address language and literacy issues (see Box 1). The approach for young workers just entering the workforce may well be different than for more experienced workers (see Box 2).

Step Five: Implement Training. Actually conducting a well-planned training session becomes the easiest part of the process; the trainer simply needs to carry out the plan. The trainer is a facilitator who takes the learners through a series of activities designed to (a) explore new ideas or skills, (b) share their own thoughts and abilities, and (c) combine the two. If the session is well planned, including logistics (such as a comfortable facility and good outreach to get people there), everything should run itself.

Box 1

Literacy and Language Issues in Training

Almost half of all American adults have trouble with basic reading and math (literacy) skills. Approximately one-third of American adults read at or below an eighth grade level. More than 40% of new immigrants to the United States report that they do not speak English "well." Yet, the large majority of reading materials used in U.S. workplaces are between the 9th and 12th grade reading level. Material safety data sheets (MSDSs), used to convey information about hazardous substances on the job, are typically written at a college reading level. Many workers cannot comprehend critical health and safety information in the form they receive it. Trainers can address these problems by modifying both training activities and materials:

- Design training activities to rely less on reading of materials and more on discussion and hands-on, practical activities that allow participants to demonstrate their comprehension.
- Adapt training materials by using the simplest terms possible to convey information and by avoiding jargon. Short sentences and use of clear illustrations can make a big difference. Avoid loading a page with text and stay away from abstract graphs.
- Offer training programs and materials in the languages needed by a diverse workforce. There is a pressing need for high–quality, multilingual materials, as well as for skilled bilingual trainers and/or simultaneous translation of training sessions.

Often those with limited literacy skills will be highly skilled in other areas, so that their reading problems are not evident. Avoid making assumptions about the literacy level of your audience. Concentrate on making training accessible to everyone, rather than on identifying which workers have special needs. Even those with high-level literacy and English language skills prefer more active training and less dense, more graphic written materials.

Sources: (a) Szudy A. *The Right to Understand: Linking Literacy to Health and Safety Training*. Berkeley, CA: Labor Occupational Health Program, University of California at Berkeley, 1994; (b) National Research Council. *Safety is Seguridad*. Washington, DC: National Academies Press, 2003.

Box 2

Reaching Young Workers

Most young people in the United States enter the workforce before they finish high school. They are at high risk of injury because of their inexperience, lack of training, and hesitation to speak up or ask questions. An estimated 200,000 teens aged 14–17 are injured on the job each year. They are injured at a higher rate than adult workers, even though youths are legally prohibited from holding the most dangerous types of jobs. (See Chapter 8 on Populations at High Risk.)

As a society, we have done a poor job of preparing our youth for protecting themselves at work, and for contributing to injury and illness prevention in the workplace. In recent years, various efforts have been made to correct this problem, including the following:

High School Curricula. Several programs around the United States have developed and disseminated curriculum materials that allow teachers to introduce information about job hazards and workers' rights into core academic subjects from English to science, as well as career exploration classes.

Job Training. Many young people get their first job through a summer jobs program or other type of job training and placement service. Youth employment specialists are beginning to incorporate health and safety training into their programs to prepare teens for work.

Peer Education. Teens can be very effective trainers of other teens. This is the finding of pilot programs around the United States that looked at a variety of approaches for effectively reaching teens with occupational safety and health information. They found that teen peer trainers bring energy and enthusiasm to their teaching, speak the language of their peers, serve as role models to other youth, and provide fresh perspectives.

Public Information. A variety of new programs by the U.S. Department of Labor, as well as state coalitions such as the California Partnership for Young Worker Health and Safety, now reach out to teens through public awareness campaigns. They use websites, teen poster contests, school journalism contests, and other creative approaches.

Sources: (a) <www.youngworkers.org>; (b) National Institute for Occupational Safety and Health. *Promoting Safe Work for Young Workers: A Community-Based Approach* (DHHS [NIOSH] Publication No. 99-141). Washington, DC: NIOSH, 1999.

For many trainers, the biggest challenge is making the switch from being a good presenter or lecturer, to being an effective facilitator of participatory learning activities. In order to maximize group learning, it can be a good idea to utilize someone other than a technical expert or professional trainer. Several unions in the United States have been successfully training workers to serve as health and safety trainers for their peers.

Step Six: Evaluate and Follow Up. Evaluation is often the forgotten step in the training process, but it serves several essential purposes. It allows the learner to judge his or her progress toward a new knowledge, attitudes, or abilities; it allows the trainer to judge the effectiveness of the training to decide how to make continuous improvements; and it can document the success of training to justify future expenditures of resources. However, selecting evaluation measures can be tricky. Some training programs rely solely on reaction surveys that tell you what participants did or did not like, but not much more. On the other hand, training programs often are pressured to offer big results such as reduced injuries and compensation costs. However, since training alone does not change dangerous work environments, it is more appropriate to select measures that relate directly to realistic training objectives. What new knowledge, skills, attitudes and/or behaviors will result, and how will these contribute toward a safer work?

Follow-up is also critical. It involves planning how to reinforce new learning and how to support the application of new knowledge, inspiration, or skills resulting from training. For example, follow-up sessions to explore the application of learning can be scheduled; individual or small group consultations may be offered. Workers frequently report that the one thing which would most enhance their ability to apply training is having supervisors who also are trained and committed to safety.

Making a Commitment

A thorough step-by-step training process requires a significant commitment of time and resources. All too often, training is conducted in the most expedient manner to meet minimal legal requirements (see Box 3) rather than to educate the work force effectively. Some steps that could be taken to improve the quality of worker health and safety training include the following:

- Educate the management community as to the need for worker training in its broadest form.
- Enhance government regulations, including a generic training standard that applies to all workers, as well as a right for workers to participate fully in workplace health and safety programs.
- Support innovative programs to reach underserved workers with training, for example, through schools and community-based organizations.

Box 3

Required Training

Under OSHA:
HazCom
Specific toxic and hazardous substances
HAZWOPER
Special industries
Machinery and machine guarding
Welding, cutting, and brazing
Farm equipment
Construction safety orders
Machine training requirements
Federal employee programs

Under MSHA:
40 hours training in mine safety (general and mine–specific)
Annual refresher training
Certified persons for fire control and for dust sampling
Qualified persons for ventilation, electrical work, work as a hoisting
    engineer

Further Reading

American National Standards Institute (ANSI). *Criteria for Accepted Practices in Safety, Health, and Environmental Training.* 2001.

Colligan M (Ed.). Occupational safety and health training. *Occupational Medicine: State of the Art Reviews* 1994; 9.

National Institute for Occupational Safety and Health. *Assessing Occupational Safety and Health Training: A Literature Review* (DHHS [NIOSH] Publication No. 98-145). Washington, DC: NIOSH, 1998.

National Institute for Occupational Safety and Health. *Promoting Safe Work for Young Workers:A Community-Based Approach* (DHHS [NIOSH] Publication No. 99-141). Washington, DC: NIOSH, 1999.

National Research Council. *Safety is Seguridad.* Washington, DC: National Academies Press, 2003.

Occupational Safety and Health Administration. *Training Requirements in OSHA Standards and Training Guidelines* (U.S. Department of Labor Publication OSHA2254). Washington, DC: OSHA, 1998.

Szudy A. *The Right to Understand: Linking Literacy to Health and Safety Training.* Berkeley, CA: Labor Occupational Health Program, University of California at Berkeley, 1994.

Chapter 4
Occupational Safety and Health Management Programs in the Workplace

Michael Silverstein

A spects of an integrated plan to prevent occupational disease and injury described in Chapter 1 are intended to identify problems and to suggest effective solutions. It is not enough, however, merely to describe solutions; to be effective, to implement solutions in a rational and efficient manner, it is useful to have a systematic safety and health management programs in each workplace. While there are differing views about the substantive and operational details necessary for fully effective programs, there is considerable agreement about their basic elements. Guidelines were published by OSHA in 1989 (54 Fed. Reg. 3904-16, January 26, 1989). Similar versions have been proposed by counterpart agencies in the states of New Jersey, California, and Washington, and by the American Industrial Hygiene Association (AIHA). The description we offer closely follows the OSHA proposal. The usefulness of such programs is that there is, as observed by OSHA, "a strong correlation between the application of sound management practices in the operation of safety and health programs and a low incidence of occupational injuries and illnesses."

Major Elements and Recommended Actions

Management Commitment

Each workplace should:

- have a written policy on safe and healthful working conditions,
- communicate clear goals to all members of the organization,
- identify responsible individuals, and
- provide visible top management involvement in implementation.

Moreover, in order to ensure that objectives are actually met, it is necessary to assign responsibilities for all aspects of the program, provide adequate authority and resources to responsible persons, and then to hold these persons accountable. Also, programs' operations should be reviewed at least annually in order to identify and correct deficiencies.

Employee Involvement

Employees should be involved in the structure and operation of the program and the decisions that affect their safety and health. This involvement can take many forms, depending on the workplace. Many workplaces with union contracts, for example, have procedures for handling safety matters and may have a health and safety committee for handling safety issues.

Hazard Analysis and Accident Investigation

As described above, comprehensive baseline and periodic worksite surveys, including job hazard analyses, should be conducted for the purpose of identifying all hazards. Also, injury and illness trends should be analyzed over time to identify patterns with common causes that can be prevented. There should be scheduled periodic inspections whenever new substances, processes, or equipment are introduced that may represent new hazards, or whenever the employer is made aware of a previously unrecognized hazard.

In addition to routine surveys and analyses, employees should be encouraged, without fear of reprisal, to report conditions that appear hazardous. A system should be established to provide employees timely and appropriate responses to their reports. Finally, incidents and "close calls" should be investigated so their causes may be identified and controlled.

Hazard Prevention and Controls

Where feasible, hazards should be eliminated by design of the job or worksite. Where they cannot be eliminated, exposure to hazards should be reduced using engineering techniques, safe work practices, personal protective equipment, and administrative controls. Identified hazards should be corrected and documented in a timely manner. The concept of a preferred hierarchy of controls is discussed in Chapter 1 (see page 28).

Safety and Health Training

Safety and health training should be incorporated into other training about performance and job practices. It should be designed to ensure all employees' skill, awareness, and competency for understanding hazards and how to prevent harm from them, including the importance of accepting and following established protective procedures. Training should also be designed to ensure that managers and supervisors understand their responsibilities and the importance of carrying them out effectively. Training should be reinforced through continual performance feedback and enforcement of safe work practices. Additional health and safety training is needed—along with job-training—when new operations are implemented or when employees are given new job assignments. Training is discussed in more detail in Chapter 3.

Record Keeping and Program Evaluation

Employers are required by OSHA and MSHA to record accidents and illnesses. These records should be used to identify injury hazards and evaluate program effectiveness. (As noted elsewhere, such records are less useful at identifying chronic diseases. Records of exposure to chronic hazards, however, can identify risk of chronic disease.) To be effective in this task, workers and supervisors need to be educated about the importance of reporting injuries and how to do it. It is also useful to supplement injury records with data from personnel and workers' compensation records for the injury record systems. These records are useful for calculating injury rates and duration of disability for workers of interest, such as inexperienced high-risk workers, those in jobs at elevated workplaces, drivers, and other workers. Such measures are essential for evaluating performance in reducing rates.

Depending on the specific needs and contexts of each workplace, other possible program elements are:

- Facility and equipment maintenance, including training of maintenance workers;
- Emergency planning and response; and
- A medical program with first aid available on site and full medical care nearby.

AIHA, serving as the Secretariat for Accredited Standards Committee Z10, is developing a standard for occupational health and safety systems, built on a continuous improvement model. This model tentatively involves a continuous cycle of management planning, implementation and operation, checking and correcting, and review.

Special Considerations

Collective Bargaining Agreements

Unionized workplaces often have provisions negotiated into their collective bargaining agreements that supplement or modify the employer's safety and health program and provide additional tools for solving problems. These agreements vary greatly, but some common provisions include:

- Safety and health grievance or complaint procedures;
- Joint union-management safety and health committees;
- Positions for full-time or part-time union safety and health representatives;
- Special safety and health investigations or research projects;
- Jointly developed and delivered training programs;

- Pre-introduction safety and health review of proposed new chemicals, processes, materials, or equipment;
- Additional rights, beyond those provided by law, to information about chemicals, processes, materials, or equipment that may be hazardous; and
- The right to refuse unsafe work.

Safety and Health Committees

Many employers recognize the value of employee participation in safety and health programs. A common method for securing this participation is through an employer-employee safety and health committee. Such committees exist today in workplaces all over the country. In some cases, they are required by state safety and health rules or by collective bargaining agreements, but in most cases they exist as a matter of employer choice and discretion. Employers setting up these committees on their own discretion may inadvertently violate provisions of the National Labor Relations Act (NLRA) that prohibit management domination of union activities. Section 8(a)(2) of the National Labor Relations Act says it is an unfair labor practice ". . . to dominate or interfere with the formation or administration of any labor organization..." The National Labor Relations Board has interpreted this section to apply to workplace safety committees.

However, joint employer-employee safety and health committees can be set up in a manner consistent with the protections of the NLRA if their activities and operating procedures are carefully limited. For example, a labor-management committee might, by majority vote (not by consensus), provide ideas to management for its consideration. Even so, safety committees in nonunion workplaces are subject to legal challenge for unfair labor practices. Thus, it would be prudent to establish safety and health committees in nonunion workplaces with great care and attention to the requirements of the NLRA.

Chapter 5
Roles of State and Local Health Departments

*Letitia Davis**

Since the passage of the Occupational Health and Safety Act in 1970, OSHA and state labor departments that enforce workplace health and safety standards have come to be seen by the public as the government agencies with primary responsibility for worker safety and health at the state and local levels. These regulatory agencies play a central and essential function in protecting worker health. However, enforcement of workplace safety and health regulations is only one component of a comprehensive public health approach to workplace health and safety (Table 1). State public health agencies have a critical, complementary role to play in conducting surveillance of work-related diseases and injuries, investigating occupational health problems in the community, and implementing prevention activities to protect workers' health.

There are many opportunities for integrating occupational health into mainstream public health practice at the state and local levels. Public health has always focused on addressing the health concerns of those most in need. Many public health programs target specific underserved populations such as women, adolescents, minorities, immigrants, migrants, and those with disabilities. These are the same populations that comprise special worker populations whose occupational health needs have not been well addressed through conventional approaches to occupational safety and health. Consequently, the public health infrastructure provides numerous points of access for reaching special worker populations (Figure 1). Because public health agencies have regulatory responsibilities in the health care and food service industries, they also serve as a point of access to workers and employers in these industries.

There are also numerous points of convergence in public health practice, where the health concerns of workers and the general public intersect. Many health hazards, such as poor indoor air in schools and latex exposures in hospitals, threaten the health of workers as well as the general public and require solutions that protect all those at risk. For numerous contemporary public

*The author acknowledges Janet Johnston and Elise Pechter, who contributed substantially to the work on which this chapter is based.

Figure 1. Examples of Public Health Programs Serving Special Worker Populations[a], Massachusetts Department of Public Health (MDPH) , 2001

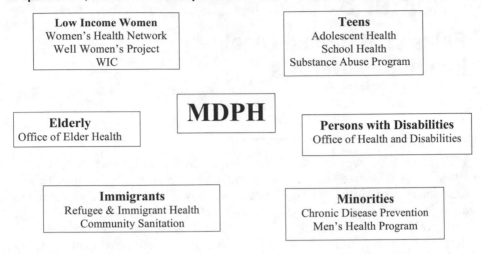

[a] National Institute for Occupational Safety and Health.[NIOSH] *National Occupational Research Agenda.* Washington, DC: U.S Department of Health and Human Services, Centers for Disease Control and Prevention, NIOSH, April 1996.

health problems, such as asthma, cardiovascular disease, and violence, occupational risk factors are among multiple contributing factors. Comprehensive approaches to these problems require collaborations between occupational health practitioners and colleagues in other fields of public health. Likewise many policy issues regarding emerging infectious diseases, such as SARS, and biological and chemical terrorism preparedness are also worker health issues, and decision-makers need the expertise of a public health team that includes experts in occupational health and worker representatives.

State public health agencies vary widely in their ability to identify and address occupational health concerns. While relatively few state public health agencies have comprehensive occupational health programs, a growing number are working to build occupational health capacity, many with support from NIOSH. The local public health infrastructure differs markedly among states and communities; local boards of health have varying degrees of responsibility for addressing occupational health concerns and usually do so within the context of broader public health mandates.

Surveillance of Work-Related Diseases and Injuries

Public health surveillance (public health tracking) is the ongoing systematic collection, analysis, and interpretation of health data coupled with the timely dissemination of information to those who need to know. It is the cornerstone of public health practice and essential for planning, implementation, and evaluation of public health activities. State public health agencies, which are vested with legal authority to require disease reporting and to collect health data, play a central role in surveillance. Although historically state health agencies focused on surveillance of communicable diseases, today they conduct surveillance of a full range of health outcomes and their determinants, including, for example, chronic diseases, injuries, and health behaviors.

The most widely cited statistics on occupational injuries and illness are generated largely outside of the public health infrastructure and are dependent on data recorded by employers, specifically on OHSA logs and in workers' compensation records. Yet state public health agencies, which have access to a wide variety of health data systems and epidemiological capacity, have an important complementary role to play in surveillance of occupational diseases and injuries, and their determinants. They are in a unique position to provide critically needed data on occupational diseases, which are known to be substantially undercounted in data based on OSHA logs and workers' compensation records. They can also use public health data sources, such as hospital discharge records and newly developing emergency department data systems, to generate information to augment and evaluate the conventional sources of occupational injury data. Public health population surveys, such as the Behavioral Risk Factor Surveillance System, can be used to collect information not only about occupational health conditions, but also the prevalence of workplace risk factors in the state.

As shown in Table 2, state health agencies are using a variety of public health data sets for tracking work-related diseases and injuries. A number of state health departments conduct case-based surveillance of occupational diseases, such as silicosis, work-related asthma, and pesticide-related disease, based on mandatory reporting by health care providers, hospitals, and, for some conditions, clinical laboratories. The mandatory reporting requirements vary from state to state. These case-based surveillance systems, like those used to track communicable diseases, involve collecting personally identifiable data on individuals that allow the state health agencies to conduct timely case follow-up and intervene in specific workplaces to protect other workers at risk. States also use existing population-based data sets, such as death certificates, to monitor occupational injury and illness trends by variables such as occupation, industry, age, and locale. Population-based cancer registries have also been used to generate hypotheses about possible links between occupation and cancer, leading to further etiologic research. Many state occupational health surveillance systems, such as those developed for adult lead poisoning, both allow for individual case follow-up and generate summary data that can be used to guide the development of broad-based prevention activities focusing on particular industries, hazards, or populations.

Surveillance of Occupational Lead Poisoning

Over 30 state health departments (and some state departments of labor and other agencies) have adult lead registries based on mandatory reporting of blood lead test results among adults by clinical laboratories. Over 80% of individuals reported to these registries have been exposed to lead at work. State health departments conduct follow-up of individual cases with high blood lead levels to obtain more information about sources of exposure and may intervene directly or refer workplaces to other agencies to reduce lead hazards at specific worksites. Summary data are used to identify high-risk industries, occupations, and populations. For example, Hispanic workers have been found to be overrepresented among those with elevated blood lead levels in some states, prompting outreach to the Spanish-speaking community about lead hazards in the workplace. Findings from these state lead registries are reported to the Adult Blood Lead Evaluation and Surveillance (ABLES) System at NIOSH and reported periodically in the CDC's *Morbidity and Mortality Weekly Report*.

Surveillance, to be meaningful, must be linked to intervention, which is, in large measure, local. Intervention involves interacting with individuals, businesses, and organizations in the community. State health agencies are well poised to actively link surveillance findings with intervention and prevention activities at the state and local levels.

Investigating Occupational Health Problems in the Community

Worksite Investigations

The occurrence of an occupational disease or injury can be considered a sentinel health event that serves as a warning sign that the prevention system has failed and other workers may be at risk. Many public health agencies have working relationships with federal or state OSHA offices and mechanisms in place for referring hazardous worksites identified through surveillance or public complaints to these agencies for investigation. Some state public health agencies also use their own in-house experts or partner with local health agencies to conduct nonregulatory investigations of workplaces where affected workers were exposed to hazards. In situations where there is no imminent or life-threatening danger, state health agencies often take a nonregulatory approach as a first step. The seriousness of the potential hazard, statutory authority of the agency, patient confidentiality, and health care provider recommendations are among the complex set of issues considered in determining the most appropriate worksite follow-up.

Research-oriented investigations conducted by health agencies go beyond assessing compliance with existing health and safety standards to identify other contributing factors, such as poorly designed equipment and work processes as well as lack of worker health and safety training. The health agencies not only provide feedback to specific employers and workers involved. They also can take the lessons learned in individual workplaces (or across similar workplaces) and translate them into educational materials to be disseminated broadly throughout the community—to employers, unions, community organizations, safety experts, health care providers, and local boards of health that serve the affected worker populations. Findings from these case follow-up investigations can also be used to generate anonymous cases studies to complement surveillance statistics. The combination of local statistics and compelling case studies that put faces on the numbers can be a powerful tool in influencing state and local decision-makers.

Investigating Fatal Occupational Injuries

In 2004, 15 states participated in the NIOSH Fatality Assessment Control Evaluation program (FACE). Through FACE, state safety experts conduct research-oriented investigations of targeted workplace fatalities to identify factors leading to the incidents. Examples of the types of occupational fatalities that have been targeted by FACE include machine-related deaths, falls, electrocutions, and deaths of youth under age 18. The FACE investigator visits the incident site and conducts interviews to collect information about what happened just before, during, and right after the fatal injury. For each investigation, a FACE report is prepared that describes the incident and includes recommendations to prevent similar deaths. These reports are distributed to those involved in the incident and others in the state who are in a position to act on the recommendations—employers in the industry, unions, advocacy organizations, equipment manufacturers, and policymakers. Summary information from multiple FACE investigations is also used to identify research needs and gaps in existing health and safety standards.

Investigating Disease Clusters in the Workplace

State epidemiologists have responsibility for conducting investigations of infectious disease outbreaks in the workplace. Occupational health experts may work together with the infectious disease epidemiologists in these investigations. State public health epidemiologists may also investigate unusual clusters of chronic diseases in the workplace, although there is considerable controversy about how many state resources should be devoted to cluster investigations involving small numbers of cases and limited historical exposure information. Some states have well-established protocols for responding to community concerns about unusual disease patterns that take into account the feasibility of conducting meaningful research. The NIOSH Health Hazard Evaluation (HHE) program is an important resource for state health agencies trying to resolve concerns about unusual disease patterns in specific workplaces. Partnerships with experts in local academic institutions can also enhance expertise available to state health agencies in conducting these investigations.

Preventing Occupational Disease and Injury

State public health agencies also have an important role to play in carrying out broad-based prevention efforts to protect worker health. Surveillance can identify specific industries, occupations, and populations at risk, and in some cases, local communities, where occupational health and safety problems need to be addressed. There are always competing prevention needs. Among the constellation of factors that ultimately shape prevention priorities are the seriousness and magnitude of the identified problem, feasibility of intervention, vulnerability of the population affected and its access to other occupational health resources, public concern, and political will. The full gamut of public health prevention strategies are available to the occupational health practitioner. These include, for example, information dissemination and education that empower people to take action, development of policies and plans that promote occupational health, and mobilizing new partnerships to solve state and local occupational health problems. State health agencies can also play an active part in assuring access to clinical occupational health services.

Occupational health practitioners in public health agencies can maximize limited prevention resources by building on the existing public health infrastructure in carrying out prevention efforts. Existing public health networks and programs serve as important points of access for reaching special populations of workers with information about health and safety risks, control strategies, occupational health services in the state, and legal rights. These access points are two-way streets that can be used not only to provide information, but also to obtain information from community members about their health and safety needs and experiences. Following are several examples of how occupational health practitioners in state health agencies have partnered with colleagues in other public health programs in conducting outreach to workers and employers. Occupational health programs have:

- Worked through school-based health and adolescent health programs to provide young workers with information about workplace health and safety and child labor laws;
- Collaborated with state-funded community health centers (CHCs) to document occupational health concerns of low-income, immigrant, and minority workers and to provide health and safety information to the CHC patient population;
- Enlisted local food safety specialists who conduct restaurant inspections in disseminating information about burn prevention strategies

to employers in the restaurant industry;
- Partnered with the child care safety programs to obtain information about latex glove use in day care centers; and
- Worked with public health bureaus responsible for hospital licensing and state epidemiologists to educate employee heath and infection control staff in hospitals about sharps injury surveillance and prevention.

Occupational health professionals working within the public health infrastructure also have a valuable opportunity to learn from colleagues experienced in working with underserved populations, who are familiar with issues of linguistic and cultural competency and passionate about serving their constituencies.

Public health agencies also serve as a critical point of access to the health care community, including both health care facilities and individual health care providers. Public health agencies are responsible for licensing health care facilities and assuring quality of patient care. While quality assurance activities are focused on the patient, they can have implications for health care workers as well. Massachusetts, for example, has passed requirements for hospitals to report sharps injuries among hospital workers to the state public health department, under its hospital licensing authority. State health agencies routinely interface with health care providers directly and through state professional organizations, and can play an active role in educating health care providers about diagnosis and treatment of occupational diseases. Many state health departments do so by participating in local meetings of health professionals, occupational health newsletters and alerts, and other educational outreach efforts. Several state health agencies have state mandates to provide clinical occupational health service and manage networks of occupational health clinics.

Many public health hazards are shared hazards, affecting workers and general public alike; policies to protect the public often have direct implications for workers' health. For example, a state public health policy to ban latex gloves in food services, driven by concern for consumers, affords protection to those working in the food service industry. Efforts to improve air quality in schools, prompted by concern for students, also impact favorably on school personnel. Occupational health practitioners often have important technical expertise that can contribute to development of public health policies to protect the public at large, with secondary gains for worker safety and health. In states with OSHA state plans (where the state is responsible for setting and enforcing workplace health and safety standards), public health agencies can have a direct influence on the development of workplace health and safety policies.

Public health practice needs to move beyond the "silo mentality," prompted, in part, by categorical funding, to develop comprehensive approaches to public health problems that bring together experts from various public health disciplines and include stakeholders in the process. Occupational health practitioners must be engaged in these efforts. Comprehensive public health programs to address health concerns, such as asthma, cardiovascular disease, violence, and bioterrorism, require collaboration among occupational health experts and colleagues in other public health domains. It is not possible, for example, to have a comprehensive state plan to reduce the burden of asthma without addressing work-related asthma, which is believed to account for 15% to 25% of asthma cases among adults. Many state health agencies today are working, with joint support from NIOSH and the National Center for Environmental Health, to include work-related asthma in their comprehensive asthma prevention plans. Similar collaborations are needed in other public health domains.

Conclusion

Most adults and many adolescents spend much of their waking lives at work. It is essential to consider the impact of work on health in the overall effort to protect the health of the public. At the turn of the 19th century, the health of workers was a central concern of the social reform movement to improve public health. Working conditions and other determinants of health, such as housing and sanitation, were seen as inextricably linked. In the 20th century, occupational health fell off the public health agency agenda in many states. The passage of the federal Occupational Safety and Health Act in 1970 was a public health victory, but regulation and enforcement is only one component of a comprehensive public health approach to worker safety and health. In the 21st century, we must integrate occupational health into mainstream public health practice at the state and local levels, by building occupational health capacity within state agencies, as well as new partnerships among health agencies and occupational health experts and advocates in the community.

Further Reading

Council of State and Territorial Epidemiologists. *The Role of the States in a Nationwide, Comprehensive Surveillance System for Work-Related Diseases, Injuries and Hazards.* A Report to NIOSH from the NIOSH States Surveillance Planning Work Group, 2001. Washington, DC: CSTE, 2001.

Maizlish N (Ed.). *Workplace Health Surveillance, An Action-Oriented Approach.* New York: Oxford University Press, 2000.

Stanbury M, Rosenman KD, Anderson HA. *Guidelines for Minimum and Comprehensive State-Based Activities in Occupational Safety and Health* (DHHS [NIOSH] Publication No. 95-10). Washington, DC: NIOSH, 1995.

Table 1. Public Health Core Functions[a]
and Essential Public Health Services[b]

Public Health Core Function	Essential Public Health Services
Assessment	Monitor health status to identify community problems
	Diagnose and investigate health problems and health hazards in the community
	Evaluate effectiveness, accessibility, and quality of health services
Policy Development	Mobilize community partnerships and action to solve health problems
	Develop polices and plans that support individual and community health efforts
	Research new insights and innovative solutions to health
Assurance	Inform, educate, and empower people about health issues
	Enforce laws and regulations that protect health and ensure safety
	Link people to needed personal health services and ensure the provision of health care when otherwise unavailable
	Assure an expert public health workforce

[a] Institute of Medicine. *The Future of Public Health.* Washington, DC: National Academy Press, 1988.
[b] Harrell JA, Baker EL. The essential services of public health. *Leadership in Public Health* 1994; 3:27-31.

Table 2. State Health Data Sources Used for Occupational Health Surveillance [*]

Data Source	Asthma	Silicosis	Teen work injuries	Amputations	illnesses	loss	poisoning	Dermatitis	Carpal tunnel syndrome	Burns	Agricultural injuries	Cancers	Lead/cadmium exposures
Case Reporting Sources													
Physician and nurse reports	X	X			X	X	X	X	X		X		
Emergency department logs			X		X		X				X		
Hyperbaric chamber reports							X						
Poison control reports					X								
Department of Agriculture reports					X						X		
State burn registry reports			X							X			
Data Systems													
Hospital discharge data	X	X	X				X			X			
Outpatient observation data									X	X			
Emergency department data	X		X							X			
Workers' compensation records	X	X	X	X		X		X	X	X			
State trauma registry				X									
Clinical laboratory reports													X
Death certificates		X	X		X						X	X	
Behavioral Risk Factor Surveillance	X								X				
Cancer registry												X	

*Partial listing. Table adapted from CSTE. *The Role of the States in a Nationwide Comprehensive Surveillance System for Work-Related Diseases, Injuries and Hazards* (see Further Reading)

PART II

Occupational Diseases and Injuries

Acne and Other Pilosebaceous Unit Disorders

ICD-10 L70-L73

Michael E. Bigby

Acne vulgaris is characterized by open and closed comedones (blackheads and whiteheads), erythematous papules and nodules, pustules, and large cysts. Acne vulgaris may be aggravated by occupational factors, such as exposure to cutting oils and machine grease. Other pilosebaceous unit and follicular abnormalities appear as (1) follicular papules (round elevated lesions <1 cm in diameter around hair follicles), pustules, and nodules; or (2) chloracne—comedones (follicular impactions of keratin and lipids), hyperpigmentation, and oil cysts.

Occurrence

The percentage of these disorders that is occupationally related is not known. The epidemiology of pilosebaceous unit abnormalities depends on the particular agent responsible and the resultant disorder. Machine-tool operators who are exposed to heavy oils, such as insoluble cutting oils and greases, may be at risk for acne or for developing oil folliculitis. Workers at risk for chloracne include those exposed to chlorinated hydrocarbons, including workers in herbicide manufacture and workers exposed to dioxin.

Causes

Both acne and folliculitis can be caused or aggravated by oil blockage of follicular orifices.

Pathophysiology

The blockage of follicular orifices may be due to the physical presence of the oil or grease within the follicular orifice or to changes in the epidermal cells lining the follicular orifices. Follicular blockage may cause bacterial overgrowth, accumulation of sebum within the follicle, follicular rupture, and inflammation.

Prevention

Exposure to the responsible agents may be minimized or avoided by using gloves, protective clothing, adequate ventilation, and closed systems, as well as educating and informing workers about hazards and what can be done to reduce them.

Other Issues

Although chloracne is not itself a disabling illness, it is important evidence of percutaneous absorption of chloracnogens, which has been associated with excess mortality from cardiovascular and respiratory diseases, an excess of diabetes cases, and an increased occurrence of cancer of the gastrointestinal, lymphatic, and hematopoietic systems.

Further Reading

Adams RM. *Occupational Skin Disease* (3rd ed.). Philadelphia: W.B. Saunders, 1999, pp. 135-140 and pp. 562-564.

Bertazzi PA, Bernucci I, Brambilla G, et al. The Seveso studies on early and long-term effects of dioxin exposure: A review. *Environ. Health Perspectives* 1998; 106 (Suppl. 2): 625-633.

Bigby M, Arndt KA, Coopman SA. Skin disorders. In Levy BS, Wegman DH (Eds.) *Occupational Health: Recognizing and Preventing Work-Related Disease and Injury* (4th ed.). Philadelphia: Lippincott Williams & Wilkins, 2000, pp. 537-552.

Suskind RR. Chloracne, the hallmark of dioxin intoxication. *Scandinavian Journal of Work, Environment and Health* 1985; 11: 165-171.

Acute Respiratory Infections, Including Influenza

ICD-10 J00-J06

Donald K. Milton

Acute upper respiratory infections (URIs) are infectious diseases of the nasopharynx and upper intra-thoracic airways caused by a variety of organisms including viruses and atypical bacteria. Mild to moderately symptomatic infections are most frequent and often referred to as the "common cold." Many of the agents responsible for these infections, however, can cause illnesses with a wide range of severity from completely asymptomatic to incapacitating generalized symptoms, asthma exacerbation, and involvement of other organs including pneumonia and rarely myopericarditis. Acute URIs are the major cause of asthma exacerbations, and thus indirectly the major cause of hospitalization among school-age children. These infections are primarily due to organisms transmitted from person-to-person. The route of transmission may be either by direct contact, by contaminated surfaces, via large droplets over short distances, or by airborne droplet nuclei. The relative importance of each of these routes is controversial, and may vary with age and environment as well as by organism.

Occurrence

Acute upper respiratory infections are the most common infections in the United States. They are the leading cause of acute illness, physician visits, and of school and work absences in the U.S. Estimates from the National Health Interview Survey, a cross-sectional study, suggest that there are about 0.55 to 0.65 URI cases per person per year among persons aged 18-64 with a total of about 52 million cases of influenza and common cold each year in the U.S. Longitudinal studies from the 1960s suggested that incidence rates are actually much higher: adult females had between 2.6 and 3.2 cases, and adult males had between 1.8 and 2.2 cases per person per year. There are few data on occupational risks for acute URI. The high rate among women is often attributed to exposure to young children. The seasonal occurrence of URIs throughout communities has been closely linked with the opening of schools after vacation. Thus, children appear to be an important link in the chain of transmission to the adult working population. Data from Japan suggest that school children are an important factor in the spread of influenza in the community. There is a small amount of data implicating working conditions in offices as factors in the transmission of URIs, with sharing an office or low rates of outdoor or otherwise sanitized air supply being the primary risk factors. A longitudinal study of office workers at a large insurance company suggested that in families with children, working adults were responsible for bringing home 32% of the infections occurring within families.

Causes

Most URIs are caused by three large groups of viruses: Paramyxoviridae (parainfluenza viruses 1-3 and respiratory syncytial virus), Picornaviridae (rhinoviruses, coxsackieviruses A and B, other enteroviruses, and echoviruses), and Coronaviridae (coronaviruses OC43 and 229E). Orthomyxoviridae (influenza A and B) and Adenoviridae (adenovirus) can also cause "cold" symptoms but more often are associated with more severe illness and pneumonia. Reoviridae cause common infections that have also been implicated. In addition to viruses, acute URIs can be caused by *Chlamydia pneumoniae* and *Mycoplasma pneumoniae*, although these infections like those with influenza and adenoviruses are also associated with more severe infections, particularly pneumonia, in older persons. The sore throat of beta-hemolytic streptococcal infections cannot reliably be clinically discriminated from that accompanying the common cold caused by viruses. Approximately 25% of URIs are due to unidentified agents. The new human coronavirus shown to cause severe acute respiratory syndrome (SARS) is notable for causing severe disease in adults, but primarily causing asymptomatic or mild upper respiratory infections in children.

Rhinoviruses are associated with approximately 40% and 65% of "common colds" in adults and children, respectively, and up to 80-92% of "colds" occurring during outbreaks. Experimental studies of rhinovirus infection transmission primarily demonstrate that the agent is not easily transmissible. Effective transmission with "normal" contact requires prolonged exposure to infected persons. The method of rhinoviral transmission remains controversial. Some experiments suggest that airborne transmission does not occur, while several others suggest that it is the primary mode of transmission among adults, and that transmission via fomites does not occur. Recent analysis of these experiments suggests that the apparently differing results regarding airborne transmission may be due to differences in ventilation rates, and relatively low exposures to airborne virus in the negative experiments.

Recent information on the agents responsible for URIs come from polymerase chain reaction (PCR)-based methods, which are more sensitive than culture for detection of infection with viruses and atypical bacteria. For example, in a study in the United Kingdom of children aged 9 to 11 years old followed longitudinally for symptoms of respiratory infection, nasal aspirate samples were collected both during symptomatic periods and at other times during follow-up. Samples analyzed by PCR detected one or more agents in 78% of symptomatic and 19% of asymptomatic samples. The detection rates in symptomatic samples were rhinovirus (31%), enterovirus (16%), respiratory syncytial virus (19%), coronaviruses OC43 and 229E (12%), influenza viruses A and B (3%), parainfluenza viruses 1-3 (2%), adenovirus (<1%), *M. pneumoniae* (15%) and *C. pneumoniae* (5%). Two or more viruses were detected in 20% of symptomatic and 3.4% of asymptomatic samples.

Influenza virus infection is among the most severe infections in this group. Influenza is thought to be transmitted primarily by small particle aerosols because (a) large numbers of viral particles are present in respiratory secretions of infected persons; (b) it is stable in air; and (c) experimental infection with small particle aerosols is more effective and representative of natural illness than nasal inoculation. The ability of influenza virus to infect many people nearly simultaneously suggests that the rate of infectious particle generation during infection is very high compared with, for instance, the tubercle bacillus or rhinovirus.

Pathophysiology

The agents causing acute URIs are primarily obligate intracellular pathogens. The mechanism of infection of nasopharyngeal mucosal and airway epithelial cells by most of these agents has not been well studied. In the case of rhinovirus, there is little pathologic alteration of the nasal mucosa. However, infected cells release a number of inflammatory mediators and these play important roles in the development of the cold syndrome. Neurologic reflexes also likely play a role in causing symptoms in response to rhinovirus infection. Recent studies show that the infection with rhinovirus and the ensuing inflammatory response is not limited to the nose.

Rather, the intra-thoracic airways are also infected and may be a particularly important site of inflammatory response in asthmatics with exacerbation of the underlying typical asthmatic inflammatory infiltrate. Similarly, coronavirus infection has been associated with exacerbations of chronic obstructive lung disease. Some of the agents, particularly the enteroviruses associated with the cold syndrome, also cause a viremia by which route they reach other organs and can cause other syndromes such as myopericarditis.

The picornaviruses have many serotypes, with rhinoviruses alone having 100 identified distinct serotypes. Immunity against reinfection with a particular rhinovirus serotype is probably long lasting, but may be overcome with a large infectious dose. However, there appears to be little protection afforded by previous infection with a different rhinovirus strain. Coronavirus, on the other hand, appears not to produce long-lasting immunity and reinfection is common.

Influenza infection produces long-lasting immunity. However, influenza is well known for its antigenic drift over time, producing new epidemic strains with occasional large antigenic shifts occurring in influenza A that result in pandemic strains. A new pandemic strain with severity similar to that of the 1918 pandemic is expected to occur eventually. Influenza A arises from a reservoir in birds, where it is transmitted by the oral-fecal route. Transmission to humans may occur primarily after passage though pigs, with recombination of genetic elements and acquisition of the ability to infect humans via the airborne route. The primary antigens of influenza are hemagglutinin (H) and neuraminidase (N). Recent research into the highly lethal H5N1 strain suggests that lethality of influenza may be determined by mutation in the nonstructural (NS) gene independent of antigenic shift. Fortunately, a second mutation allowing airborne transmission was not acquired by this virus and a pandemic did not occur.

Prevention

To the extent that transmission is due to hand contact and self-inoculation, hand washing and reduced finger-to-nose and eye contact should reduce infection rates. However, few studies have actually tested the utility of hand washing. Among studies using hand washing with or without an alcohol-based hand disinfectant, the data are mixed. (The better-designed studies, except for studies of day care centers with children under 2 years of age, are generally negative.) Alternatively, the evidence for a preventive effect of hand washing on the incidence of gastrointestinal infections is convincing.

To the extent that acute URIs are due to airborne infectious agents, air sanitation (through adequate supply of outdoor air, high efficiency filtration, or use of germicidal ultraviolet light) would be expected to reduce disease transmission. The application of upper-room germicidal UV light (or irradiation) was tested in a small number of military and school studies in the

1940s and 1950s. These studies focused on transmission of measles in schools and febrile respiratory illness (probably dominated by adenovirus infections) in recruits, with mixed results. In those studies where exposure to other potentially infected persons was limited outside of the treated environment, germicidal UV light (or irradiation) appears to have been effective. Formal studies of UV light (or irradiation) to prevent less explosively contagious upper respiratory infections have not been reported. Application of UV light (irradiation) to control influenza has been suggested because of the high probability that most cases are acquired via exposure to respirable infectious particles. Reduction of aerosol generation would be expected to reduce transmission, and active use of disposable tissues to cover coughs and sneezes may be helpful, but hard to enforce.

Vaccination has long been the mainstay of prevention for infection with influenza A and adenoviruses. Unfortunately, adenovirus vaccine, extensively used in the past by the military, has not been available in recent years. In Japan during the 1980s, influenza vaccination was focused on school-age children resulting in community-wide reductions in infection and reduced mortality among elderly. Unfortunately, vaccination of children against influenza became unpopular and this approach was abandoned. Recent U.S. studies are mixed on the cost effectiveness of influenza vaccination in healthy working populations. The primary drawback of vaccination is the time required to make vaccine against new strains. The current methods of surveillance and manufacture of new vaccines each year has been successful in times of antigenic drift. However, the rapidity of dissemination of pandemic influenza A seems likely to outpace the ability to produce vaccine. Thus, prevention of pandemic influenza will require widespread application of air sanitation.

Further Reading

Dick EC, Jennings LC, Mink KA, et al. Aerosol transmission of rhinovirus colds. *Journal of Infectious Diseases* 1987; 156: 442-448.

First M, Nardell E, Chaisson W, et al. Guidelines for the application of upper-room ultraviolet germicidal irradiation for preventing transmission of airborne contagion—Part I: Basic principles. *ASHRAE Transactions* 1999; 105.

First M, Nardell E, Chaisson W, et al. Guidelines for the application of upper-room ultraviolet germicidal irradiation for preventing transmission of airborne contagion-Part II: Design and operational guidance. *ASHRAE Transactions* 1999; 105.

Gern JE, Galagan DM, Jarjour NN, et al. Detection of rhinovirus RNA in lower airway cells during experimentally induced infection. American *Journal of Respiratory Critical Care Medicine* 1997; 155: 1159-1161.

Johnston SL, Pattemore PK, Sanderson G, et al. The relationship between upper respiratory infections and hospital admissions for asthma: A time-trend analysis. *American Journal of Respiratory Critical Care Medicine* 1996; 154: 654-660.

Webster RG. Influenza: An emerging disease. *Emerging Infectious Diseases* 1998; 4: 436-441. http://www.cdc.gov/ncidod/eid/vol4no3/webster.htm

Anemia

ICD-10 D50–D53, D55–D65, D69, P61.3; SHE/O

Howard Kipen

Anemia refers to a reduction in the concentration of red blood cells (RBCs) and hemoglobin in the blood. Normal ranges for the concentrations are a function of age, gender, and altitude of residence. Anemias may be classified into those due to impaired production (including aplastic anemia) and those due to increased destruction of circulating red cells (hemolysis). Both types may be caused by occupational and environmental agents.

Aplastic anemia, a special category of anemia, is a bone marrow stem cell disorder characterized by reduced hematopoietic cells and consequent deficiencies in circulating RBCs, white blood cells, and platelets. Deficiency may involve only one or two of these three blood elements, and varying degrees of pancytopenia may precede frank bone marrow failure. Some cases evolve into leukemia. The case-fatality rate is approximately 50% within 1 year of onset, primarily due to infection and hemorrhage.

Occurrence

Anemia is very common, due to many medical conditions, including blood loss, acute and chronic infections, and malignancies. There are no epidemiologic data on the incidence of anemia related to occupational or environmental exposures, but it is not likely to be high in developed economies.

Annual incidence of aplastic anemia is 5-10 per million persons in Western nations, with a rising age-specific incidence from 4 per million in children to 60 per million in people over age 65. This low rate of occurrence makes it unusually difficult to conduct epidemiological studies with adequate power to detect increased risks for specific exposures, and thus much etiologic knowledge for aplastic anemia is limited to case reports. The overwhelming majority of cases of aplastic anemia are acquired; rarely, it is caused by congenital conditions, such as Fanconi anemia. In the United States, approximately 50% of cases are clinically attributed to a defined agent, most often a therapeutic or infectious agent. Reliable rates of occurrence have not been calculated for occupational groups known to be at risk.

Causes of Aplastic Anemia

Occupational Causes

Benzene: Benzene is the classic occupational cause of aplastic anemia. There is an apparent relationship between exposure concentration of benzene and development of aplastic anemia. Prior to development of the disorder, lesser degrees of reversible depressions in one or more cell lines have sometimes been observed, such that repeated inhalations of benzene in excess of 100 ppm may be expected to cause peripheral cytopenia of at least one blood element in most individuals. In highly exposed groups, the incidence of aplastic anemia has been estimated at about 3-4%. In China, significant decreases in all types of blood cells have been reported with average exposures in the range of 30 ppm. There is probably a threshold below which aplastic anemia does not occur (less than 10 ppm TWA), although subclinical marrow toxicity (based on animal studies) and carcinogenic effects do not show such a threshold. A recent French, population-based, case-control study did not detect a role for any particular occupation, nor for exposure to solvents, although increases were seen for exposure to glues and paints.

Ionizing radiation: Aplastic anemia from ionizing radiation is also characterized by a dose–response relationship. There is a threshold for marrow toxicity of approximately 1.25 grays (1 Gy = 100 rads of energy absorbed), with a dose of 5.0 Gy causing 50% mortality from aplasia. More intense exposure also increases toxicity.

Pesticides: There are multiple case reports of aplastic anemia following relatively high exposures to Lindane (hexachlorobenzene). Animal data have not been confirmatory. Chlordane/heptachlor termiticides also have been implicated in case reports, and a recent case-control study showed a small, nonsignificant excess for exposure to pesticides.

Ethylene glycol ether: Ethylene glycol monomethyl ether and other glycol ethers have been reported to cause cytopenia and hypoplastic bone marrow in groups of workers, sometimes with relatively low exposure. More recently, less severe anemias have been reported.

Arsenic: Therapeutically used organic arsenicals are documented causes of bone marrow depression. Inorganic arsenic exposure may also cause pancytopenia, with recovery expected on withdrawal from exposure. Pancytopenia is unusual without other signs of significant inorganic arsenic poisoning.

Trinitrotoluene: This nitro-substituted benzene derivative is highly toxic to bone marrow. Large outbreaks of aplastic anemia have been reported in the munitions industry, often accompanied by hemolytic anemia, methemoglobinemia, dermatitis, and hepatitis.

Nonoccupational Causes

Chloramphenicol: For this classic example and most other pharmaceutical agents, the mechanism of disease is idiosyncratic rather than dose-dependent, occurring about once in 50 000 patients. This idiosyncratic reaction should not be confused with the predictable and mild suppression of erythropoiesis (production of red blood cells), which commonly accompanies treatment with this agent. There is no threshold dose for occurrence of aplastic anemia.

Approximately 100 other drugs have also been reported to cause aplastic anemia, including phenylbutazone, sulfa drugs, and gold compounds.

Systemic Illness: Viral hepatitis, other viral infections, miliary tuberculosis, and systemic lupus erythematosus have all been demonstrated to result in aplastic anemia.

Chemotherapeutic Agents: Alkylating agents, anti-metabolites, mitiotic inhibitors, and anthracyclines (alone or with ionizing radiation) predictably cause marrow aplasia in a dose-dependent fashion. Risk is potentially present for pharmaceutical workers, although the apparent dose dependence of this toxicity probably offers protection.

Causes of Hemolytic Anemia

There are a number of occupational and environmental causes of hemolysis. Among them are exposures that promote the formation of oxidized methomoglobin in the ferric state, rather than in the ferrous state that helps to transport oxygen. Clinically significant methemoglobinemia has been not uncommon with exposure to nitrate-contaminated well water, nitrous gases (in welding and in silos), aniline dyes, moth balls (naphthalene), and potassium chlorate, nitrobenzenes, phenylenediamine, and toluenediamine. At high levels, these same oxidizing compounds cause nonspecific oxidative denaturation of hemoglobin leading to Heinz-body hemolytic anemia.

Arsine gas is a very potent cause of intravascular hemolysis. Inhalation of relatively small amounts of arsine leads to swelling and eventual bursting of circulating RBCs. Liberation of hemoglobin readily leads to renal failure. While arsine gas is used commercially, particularly in the microelectronics industry, most reports of acute hemolytic episodes are from the unexpected liberation of the gas as a byproduct of industrial processes, particularly when acid is added to a material which contains arsenic, such as many metals and even coal. The NIOSH IDLH level is 3 ppm. Prevention of arsine poisoning depends most importantly on avoiding unintended introduction of acid onto metals or other arsenic-containing materials.

Lead poisoning also affects RBC formation and survival among workers in lead smelters and foundries, firing range personnel, automobile radiator workers, iron workers, painters, and welders. The interference by lead with heme biosynthesis leads to both decreased RBC production and, because of membrane abnormalities, decreased survival time. Either normochronic microcytic or normochromic nomocytic anemia may occur, usually limited to individuals chronically exposed to lead with a blood lead level more than 50 µg/dL.

Pathophysiology of Aplastic Anemia

Pathogenesis of aplastic anemia is complex and not completely understood. The disorder is most commonly thought to be caused by injury or destruction of a common pluripotential stem cell, which thus affects all RBCs, white blood cells, and platelets. In recognized occupational cases, aplastic anemia has been diagnosed soon after the responsible exposure. In contrast, in recognized cases of occupational leukemia that are associated with some of the same exposures, there is a longer latency between exposure and diagnosis of disease. Only for benzene, ionizing radiation, and cytostatic drugs is the pancytopenia established to be dose-dependent, although, even for benzene exposure, only a fraction of those people who have been highly exposed progress to aplastic anemia. Intermittent high exposures on a background of lower exposures have been implicated in animals and humans as potentially having etiologic importance. Recovery from aplastic anemia associated with benzene and other chemicals is substantially more frequent than when the disorder is due to idiopathic and idiosyncratic causes.

Pathophysiology of Hemolytic Anemia

When RBC survival time decreases below normal (120 days), anemia may result if production does not adequately compensate. Such hemolysis can be intravascular, leading to immediate release of hemoglobin into in the circulation, or extravascular, with destruction of RBCs in the liver or spleen.

Prevention

For the idiosyncratic causes of aplastic anemia, prevention is problematic. The two major occupational causes, benzene and ionizing radiation, have well-developed primary and secondary preventive approaches. Because both agents are also carcinogenic, exposure is regulated more strictly than would be required for prevention of aplasia alone.

For both ionizing radiation and benzene, strict monitoring of dose, with environmental and personal sampling, is indicated. Biological monitoring of blood for benzene or of urine for the benzene metabolite phenol are not practical or effective, especially at the current, relatively low exposures in the United States. However, measurement of urinary phenol may allow the retrospective verification of acute benzene overexposures.

Serial monitoring with complete blood counts is of limited sensitivity at current levels of exposure in developed countries. If aggregate blood counts in a population or individual were depressed, this would certainly indicate a myelotoxic, or potentially myelotoxic, problem. However, because of the apparent threshold discussed above, this effect is not likely to occur until exposures reach levels between 10 and 100 times the current exposure limit (OSHA PEL: 1 ppm, TWA). Thus, normal counts are reassuring in terms of aplastic anemia, but they are not adequate to ensure compliance with the standard and to reduce carcinogenic risk to a minimum.

Other Issues

Prognosis of aplastic anemia is poor, with a median survival of 3 months and a 50% case-fatality rate at 1 year. Quick removal from exposure of a patient with developing aplasia may enhance the chance of recovery.

Progress is being made with immunologic therapies and bone marrow transplantation. Both during recovery and later, there is a substantially increased risk (1-5%) for development of acute leukemia.

Further Reading

Cullen MR, Tado T, Waldron JA, et al. Bone marrow injury in lithographers exposed to glycol ethers and organic solvents used in multicolor offset and ultraviolet curing printing processes. *Archives of Environmental Health* 1983; 38: 347-354.

Fowler BA, Weissberg JB. Arsine poisoning. *New England Journal of Medicine* 1974; 291: 1171-1174.

Guiguet M, Baumelou E, Mary JY. A case-control study of aplastic anaemia: occupational exposures. The French Cooperative Group for Epidemiological Study of Aplastic Anaemia. *International Journal of Epidemiology* 1995; 24: 993-999.

Jandl JH. Aplastic anemias. In Jandl JH (Ed.) *Blood: Textbook of Hematology.* Boston: Little Brown, 1987, pp. 115-152.

Kyle RA, Pease GL. Hematologic aspects of arsenic intoxication. *New England Journal of Medicine* 1965; 273: 18.

Rothman N, Li GL, Dosemeci M, et al. Hematotoxicity among Chinese workers heavily exposed to benzene. *American Journal of Industrial Medicine* 1996; 29: 236-246.

Welch LS, Cullen MR. Effect of exposure to ethylene glycol ethers on shipyard painters: III. Hematologic effects. *American Journal of Industrial Medicine* 1988; 14: 527-536.

Arrhythmias and Arrhythmia-Induced Sudden Death

ICD-10 44.0-44.9

Kenneth D. Rosenman

Arrhythmias are irregular heart rhythms that may be identified by physical examination or electrocardiogram (ECG). Symptoms include thumping in the chest, sensation of an irregular heartbeat or pause between heartbeats, anxiety, and fatigue. If cardiac function is sufficiently compromised, individuals may present with syncope (fainting), dyspnea (shortness of breath), or angina (chest pain). Arrhythmias may form blood clots that travel from the heart and occlude blood vessels in the brain (causing a stroke) or elsewhere, with resultant symptoms.

History, physical examination, chest x-ray, and echocardiogram may indicate the underlying disease. The diagnosis may be confirmed with a resting ECG or with further testing, such as an exercise tolerance (stress) test or long-term ECG (Holter) monitoring. Invasive electrophysiological studies, with actual induction of the arrhythmia, may be required.

There are many different types of arrhythmias, with characteristic pre-sentations, and different causes and prognoses. Major categories include deviations of sinus rhythm, supraventricular arrhythmias, sick sinus syndrome, ventricular arrhythmias, heart block, atrioventricular junctional variants, and escape rhythms.

"Sudden death," often caused by an arrhythmia, may be defined as a rapid, unexpected death occurring in an apparently healthy individual without a history of pertinent disease or in an individual with a stable, underlying, chronic disease.

Occurrence

In the United States each year, arrhythmias are (a) the underlying cause of death for 39,000 people; (b) listed on another 500,000 death certificates; (c) among the hospital-discharge diagnoses for an estimated 750,000 people (for 25% of these discharges, an arrhythmia is listed as the primary cause of admission); and (d) responsible for two million physician office visits. The prevalence of arrhythmias, including benign rhythm changes, in the general population is high: as high as 28% in normal children, 60% in college and graduate students, and approximately 90% in individuals over age 60. Significantly lower percentages are reported for arrhythmias with poor prognoses. For example, people have a 2% chance of developing atrial fibrillation in their lifetimes after age 30. The incidence of arrhythmias is greater in men than women.

"Sudden death" is the single largest cause of death among middle-aged men in the U.S. It is responsible for approximately 50% of all deaths due to ischemic heart disease, and its incidence is similar to that of ischemic heart disease. Of all deaths that occur at work, "sudden death" is the leading cause, outnumbering traumatic fatalities by two to one in one descriptive epidemiological study. Most sudden death is caused by an arrythmia.

Causes

The most common underlying cause of arrhythmias and sudden death is artherosclerosis. Other less-common causes include congenital malformations, connective tissue diseases with arteritis, infectious diseases, heart tumors, alcoholism, iatrogenic disorders, and electrolyte disorders.

Exposure to carbon monoxide (CO) and to many different types of solvents has been associated with both arrhythmias and sudden death. No particular type of arrhythmia is specific to chemical exposures. Solvents that have been reported in case reports to cause both arrhythmias and sudden death include fluorocarbons, benzene, chloroform, trichloroethylene, 1,1,1-trichloroethane, perchloroethylene, toluene, phenol, and gasoline. Methylene chloride as a solvent is a weak sensitizing agent but it causes arrhythmias because it is converted into CO in the body. There are no data on the incidence of arrhythmogenic effects from these widely used substances. There is one study associating a prolongation in the QT interval on the ECG, which would be a risk factor for sudden death among shift workers.

With regard to sudden death, artherosclerotic change contributes significantly to the cause of the fatal arrhythmia. Most autopsies of individuals after sudden death do not reveal a myocardial infarction.

Pathophysiology

Arrhythmias involve a disturbance in the normal sequence of cardiac activation and/or a change in rate or regularity beyond conventionally defined limits of normal. Abnormalities in impulse formation or conduction underlie the different types of arrhythmias. The sinus node, from which impulses originate in the normal heart, is under the influence of the nervous system. Special ionic mechanisms and differentials are present in cells that initiate and conduct electrical impulses. Damage occurs to these cells from ischemic changes and, less frequently, from congenital, infectious, or immune abnormalities. The type of arrhythmia depends on the particular nodal or conduction cells that are affected.

Solvents have been shown to (a) sensitize the heart to epinephrine, which lowers the initiation threshold; and (b) have a direct negative inotropic effect. CO has both a direct and indirect ischemic effect and a direct effect on mitochondrial enzymes.

Studies to determine the mechanism for the association between air pollution, particularly fine particulates, and increased cardiovascular mortality have identified an effect on the autonomic control of the heart. This change in autonomic control is reflected in decreased heart-rate variability, as measured by Holter monitoring of heart rhythm.

Prevention

Primary prevention measures include both (a) reducing exposures to substances that cause ischemic heart disease, hypertension, or direct arrhythmogenic effects; and (b) modifying personal lifestyle habits, such as cigarette smoking, that are risk factors for ischemic heart disease. A wellness program that combines reduction of workplace exposures and stress with lifestyle modifications can focus on several factors that cause or contribute to arrhythmias.

Whether individuals with underlying heart disease should be allowed to work in jobs where significant exposure to solvents or CO occurs, even within legal limits, needs to be assessed on an individual basis. The adequacy of current standards to protect against arrhythmias needs further investigation. Because only about half of all people who die from cardiac deaths have a clinical history of heart disease, medical screening cannot identify all potential victims.

Other Issues

Individuals with underlying ischemic heart disease are theoretically more at risk than others to manifest arrhythmias or die suddenly with exposure to CO and/or solvents. CO levels from all sources, including cigarette smoking, are additive to CO inhaled at the workplace.

Exposure to levels of CO and/or solvents that are high enough to cause sudden death is most likely to occur in enclosed spaces, such as in a reactor vessel.

The following principles from animal studies need to be considered: (a) The threshold for initiation of an arrhythmia is independent of duration of exposure but dependent on dose. (b) The heart remains sensitized until the solvent level in the blood is reduced below the threshold of initiation—not until exposure in the air ceases. (Therefore, arrhythmias after work may be due to elevated levels of workplace chemicals in the blood.) (c) Halogenated solvents are more arrhythmogenic than aliphatic solvents. (d) Other stresses, such as noise, lower the initiation threshold for arrhythmias. (See also chapters on Stress and Work Organization.)

Further Reading

Abedin Z, Cook RC, Milberg RM. Cardiac toxicity of perchlorethylene (a dry cleaning agent). *Southern Medical Journal* 1980; 73: 1081-1083.

Kaufman JD, Silverstein MA, Moure-Eraso R. Atrial fibrillation and sudden death related to occupational solvent exposure. *American Journal of Industrial Medicine* 1992; 25: 731-735.

Murata K, Yano E, Shinozaki T. Cardiovascular dysfunction due to shift work. *Journal of Occupational and Environmental Medicine* 1999; 41: 748-753.

Robinson CC, Juller LH, Perper J. An epidemiologic study of sudden death at work in an industrial county, 1979-1982. *American Journal of Epidemiology* 1988; 128: 806-820.

Sheps DS, Herbst MC, Hinderliter Al, et al. Production of arrhythmias by elevated carboxyhemoglobin in patients with coronary artery disease. *Annals of Internal Medicine* 1990; 113: 343-351.

Steffee CH, Davis GJ, Nicol KK. A whiff of death: Fatal volatile solvent inhalation abuse. *Southern Medical Journal* 1996; 89: 879-884.

Zakhari S. Cardiovascular toxicology of halogenated hydrocarbons and other solvents. In Acosta D (Ed.), *Cardiovascular Toxicology*. New York: Raven Press, 1992, pp. 409-454.

Arsenic, Adverse Effects

ICD-10 J57.0

Rose H. Goldman

Acute arsenic poisoning can occur after the ingestion of arsenic trioxide or lead arsenate, or after the inhalation of the toxic arsine gas (AsH_3). Acute symptoms can begin minutes to hours following ingestion of contaminated food or drink, and include vomiting, abdominal pain and cramping, diarrhea, sometimes gastric hemorrhage. These initial symptoms can then lead to dehydration and leg cramps, followed by irregular pulse, cardiac toxicity, shock, and, in extreme cases, death. If those poisoned survive the initial illness, they usually develop hepatitis and pancytopenia within 1 week, and may also experience peripheral neuropathy 1-3 weeks after the exposure. The blood changes are usually reversible once exposure ceases.

Arsine gas is highly toxic. It has been reported that one half-hour of exposure to 25-50 ppm can be lethal. A patient may present initially with dizziness, headache, abdominal pain, hemolysis, hyperbilirubinemia, jaundice, and kidney failure.

The diagnosis of acute arsenic poisoning is supported with urine confirmation of arsenic exposure. In an emergency situation, a spot urine sample is sufficient. In the setting of acute symptoms due to poisoning, the spot urine samples would be expected to exceed 1000 µg/L reference values usually are less than 50 µg/L or 50 µg/gram creatinine). Recent ingestion of seafood may elevate total urinary arsenic levels markedly for the subsequent 48 hours due to the presence of a relatively nontoxic organic form of arsenic, so a diet history is critical. If diet information is not available, then the total urine arsenic can be speciated into total and elemental arsenic. Whole blood arsenic may be elevated early in an acute poisoning event, but will decline faster than the urine levels.

Most human problems with arsenic are due to chronic ingestion or inhalation, rather than to acute poisoning. One of the characteristic developments is that of a sensorimotor peripheral neuropathy. In mild cases, the patient may present with paresthesias. In more serious cases, there may be stocking-and-glove sensory deficits, painful dysesthesias (painful burning sensations in the soles of the feet), loss of vibration, and positional sense, gait disorders, motor weakness, and loss of deep tendon reflexes. Some central nervous system effects have been reported, including confusion, delirium, encephalopathy, and seizures. More severe cases of chronic intoxication can cause hematological effects, kidney damage, and liver disease. Skin manifestations of chronic exposure, particularly after long-term ingestion of arsenic in drinking water or for medicinal purposes, include chronic eczema, patchy hyperpigmentation, keratoses (raised warty lesions especially on palms and soles), and skin cancers (squamous and basal cell carcinomas). Nails may be marked by horizontal white bands known as Mees lines. Recent studies and analyses have suggested that chronic ingestion of arsenic is associated with an increased incidence of bladder and lung cancer at arsenic concentrations in drinking water that are below 50 µg/L (50 ppb), which was the previous EPA maximum contaminant level (MCL) for arsenic in drinking water. In early 2002, the EPA adopted a new standard for arsenic (10 ppb) to replace the previous 50 ppb standard; water systems must comply with the new 10 ppb standard by January 23, 2006.

Chronic inhalation of arsenic compounds can cause conjunctivitis, inflammation of the mucous membranes, and sometimes perforation of the nasal septum at very high exposure levels. Human studies have reported that inhalation of inorganic arsenic is strongly associated with lung cancer.

Urine arsenic levels are useful for documenting recent exposures within the last few days after exposure, but are less useful for documenting more distant exposures. Screening urine tests are typically measured for total arsenic, so persons should be advised to eat no fish or shellfish for at least 48 hours in order to eliminate measurement of organic (nontoxic) forms of arsenic found in fish. In that way, the urine arsenic is more reflective of the more toxic inorganic forms of arsenic. Inorganic arsenic is also bound in hair and nails and may persist for months after urine arsenic values have returned to baseline. However, hair and nail assessments are problematic as a reflection of internal dose because of the difficulty in removing the exogenous arsenic and problems with reliability of testing in commercial laboratories.

Occurrence

The number of cases of adverse effects of arsenic and the proportion that is occupationally related are not known.

Causes

Arsenic is a naturally occurring element found in the earth's crust and is found in numerous ferrous and nonferrous ores. It is classed as a metalloid, or transition element (Group V in the periodic table) because it complexes with metals but also reacts with other elements such as oxygen, hydrogen, chlorine, carbon, and sulfur. Arsenical compounds can be grouped as inorganic, organic, and arsine gas. The most common valence states are the metalloid (elemental, 0), arsenite (trivalent, +3), and the arsenate (pentavalent, +5). Toxicity of these arsenical compounds vary, with the most toxic being the inorganic trivalent arsenic compounds followed by organic trivalent arsenic compounds, inorganic pentavalent arsenic compounds, organic pentavalent compounds, and elemental arsenic. Some fish and crustaceans contain large amounts of forms of organic arsenic ("fish arsenic")—arsenobetaine (a trimethylated arsenic compound) and arsenocholine, both of which are thought to be of negligible toxicity.

Exposures come from naturally occurring and human-made sources. Natural sources include volcanic eruptions, and leaching of arsenic from rocks and soil into drinking water. Arsenic compounds are used in a number of commercial and industrial endeavors. Metallic arsenic is used as an alloy, such as for hardening lead in bearings and to improve the toughness and corrosion resistance of copper. Arsenic trioxide and arsenic pentoxide are used in the manufacture of calcium, copper, and lead arsenate pesticides. Arsenic compounds are also used in herbicides. Sodium arsenate is used in ant killers and in animal dips acting as insecticides. Arsenic and arsenic trioxide are used in the manufacture of low-melting glasses. Fowler's solution (sodium arsenite) was used in the past to treat leukemia, psoriasis, and other diseases; arsanilic acid is still used in some veterinary medicines. Some arsenic compounds are found in Chinese and Indian traditional medicines. Copper acetoarsenite is used as a wood preservative, and exposure can occur from burning plywood treated with an arsenic wood preservative or from dermal contact with treated wood. Arsenic has been used in some pigments ("Paris green") and also in preservatives in tanning and taxidermy. Recent new uses for arsenic compounds have been found in the semiconductor industry. Crystals of gallium arsenide (GaAs) have been found to be better superconductors than silicon, and so are being used in semiconductors, integrated circuits, and scientific instruments. The highly toxic arsine gas (AsH_3) is used to make gallium arsenide.

Occupational exposure to arsenic can occur during metal smelting, where arsenic occurs as a contaminant or when heated ores give off arsenic trioxide (As_2O_3, or "white arsenic"). Dust can gather in flue dust, and therefore furnace and flue maintenance operations have a high risk of exposure. Fly ash in coal boilers may also contain arsenic, and thus present a hazard during maintenance. Workers involved in the manufacture or use of arsenic-containing pesticides may also be exposed. Persons who handle wood treated with chromated copper arsenate, such as in making house decks or playgrounds, may be exposed. Inhalation of fumes from burning arsenic-treated wood has caused arsenic poisoning. Epidemiolgic studies of pesticide production workers, sprayers, smelter workers, residents near polluting industries, and patients treated with arsenicals have demonstrated an association between respiratory exposure to arsenic and an increased risk of lung cancer. A current episode of mass arsenic poisoning recently began in West Bengal, India, in which high levels of arsenic have been leaching from natural underground sources to contaminate newly drilled wells, leading to more than 1 million people drinking arsenic-contaminated water (above 50 µg/L). Thousands have been found to have arsenic-related skin lesions and liver problems.

Pathophysiology

Arsenic and its compounds enter humans through ingestion or inhalation. The rate of absorption is highly dependent on the solubility of the compound, and also on the valence state of the arsenic. After absorption, arsenic is initially bound to proteins in blood, but then rapidly cleared from blood as it is redistributed to the liver, spleen, kidneys, lungs, and gastrointestinal tract within about 24 hours. A major mechanism of toxicity is through trivalent arsenic binding to SH-groups (such as proteins, glutathione, and cysteine) and interfering with numerous enzyme systems, including those involved in cellular respiration and DNA repair. Another mechanism of toxicity is through "arsenolysis," in which pentavalent arsenate substitutes for phosphate in biochemical reactions, leading to an uncoupling of oxidative phosphorylation, disruption of cellular oxidative processes, and consequently endothelial cellular damage. The clinical manifestations of these cellular changes are loss of capillary integrity, increased permeability of blood vessels, generalized vasodilation, transudation of plasma, hypovolemia, and shock. Pentavalent arsenic (+5) and arsine are converted to trivalent arsenic (+3) in vivo.

Most trivalent arsenic is metabolized to dimethylarsinic acid (DMA) and monomethylarsonic acid (MMA). DMA, MMA, and unchanged inorganic arsenic are then excreted in the urine, with an overall half-life of about 10 hours. Organic arsenic compounds are excreted unchanged in the urine.

Ingested inorganic arsenic crosses the placenta in humans. Studies of women who work in, or live near, metal smelters have shown an association between arsenic and higher than normal spontaneous abortion rates, but these studies may be limited because of methodological questions.

Prevention

Engineering controls can be used to control fumes in smelting and manufacturing. Personal protective equipment and clothing should be worn when performing maintenance work. Environmental exposures can be controlled by regulations/and or guidelines that limit the amount of arsenic in air and water. Since some forms of arsenic are carcinogens, there is no totally safe level. In 1993, WHO adopted a provisional guideline value of 10 µg/L as a realistic limit, given measurement capabilities. In 2001, EPA lowered the maximum level of arsenic permitted in drinking water to 10 µg/L. The current acceptable levels for reducing adverse health effects due to air exposures are the ACGIH TLV of 0.01 mg/m^3, the OSHA PEL of 0.01 mg/m^3 TWA, and the NIOSH REL of 2 µg/m^3 (ceiling limit for 15 minutes).

Further Reading

Agency for Toxic Substances and Disease Registry. *Toxicological Profile for Arsenic.* Atlanta, GA: ATSDR, 2000.

Environmental Protection Agency Air Toxics Website. Arsenic Compounds: http://www.epa.gov/cgi-bin/epaprintonly.cgi

Grandjean P. Health significance of metal exposures. Wallace RB (Ed.). *Maxcy-Rosenau-Last: Public Health & Preventive Medicine* (14th Ed.), Stamford, CT: Appleton & Lange, 1998, pp. 493-508.

Chowdhury UK, Biswas BK, Chowdhury TR, et al. Groundwater arsenic contamination in Bangladesh and West Bengal, India. *Environmental Health Perspectives* 2000; 108: 393-397.

Ford, MD. Arsenic. In: Goldfrankl, Flomenbaum N, Lewin N, et al (Eds.). *Goldfrank's Toxicological Emergencies.* New York; McGraw-Hill, 2002.

LaDou J (Ed.). *Occupational and Environmental Medicine* (3rd Ed.). New York: Lange Medical Books/McGraw-Hill, 2004, pp. 432-435.

Lauwerys RR, Hoet P. *Industrial Chemical Exposure: Guidelines for Biological Monitoring.* Boca Raton, FL: Lewis Publishers, 2000, pp. 36-49.

Rossman TG. Arsenic. In Rom WM (Ed.), *Environmental and Occupational Medicine* (3rd Ed.). Philadelphia: Lippincott-Raven, 1998, pp. 1011-1019.

Subcommittee to Update the 1999 Arsenic in Drinking Water Report, National Research Council. *Arsenic in Drinking Water.* Washington, DC: National Academy Press, 2001, p. 225.

WHO Fact Sheet No 210: Arsenic in Drinking Water, Revised May 2001. (accessible at: http://www.who.int/inf-fs/en/fact210.html)

Yip L, Dart RC. Arsenic (Chapter 71). In Sullivan JB, Krieger GR (Eds.), *Clinical Environmental Health and Toxic Exposures.* Philadelphia: Lippincott Williams & Wilkins, 2001, pp. 858-866.

Asbestos-Related Diseases

ICD-10 J61, J92.0

James L. Weeks and David C. Christiani

Exposure to asbestos causes asbestosis, nonmalignant pleural disease, lung cancer, and mesothelioma. Exposure has also been associated with increased risk of cancer of the larynx, and pharynx and certain cancers of the gastrointestinal tract. Some studies have also associated cancer of the kidney with asbestos exposure, but other studies have not. Adverse effects of asbestos have been known since about the late 19th century. The purpose of this chapter is to describe disease control measure common to all asbestos-related diseases. (See Asbestosis, Laryngeal Cancer, Lung Cancer, Mesothelioma, and Pleural Diseases, Asbestos-Related.)

Asbestos is a naturally occurring class of silicate fibers mined primarily by open-cast mining. More than half of the world's production comes from mines in the Soviet Union and Canada. Other producing countries include South Africa, Zimbabwe, the United States, Italy, China, and Australia. Worldwide production peaked at about 6 million tons in 1973. Currently, U.S. production and consumption has declined dramatically because of regulation of the use of asbestos and litigation because of the health effects of asbestos exposure. The U.S. no longer mines asbestos, and U.S. consumption of imported (mostly Canadian) asbestos was down to 9000 tons in 2002. Despite what is known about the devastating effects of asbestos exposure, use has continued and even expanded in many developing countries.

Properties and Users

Asbestos has very high tensile strength and durability. It resists both physical and chemical corrosion. It has been sprayed, woven, and molded. Because of its properties, it has been and is being used in many ways: for fire and electrical insulation, as filler in cement and other products, such as pipes, floor tiles, wall coverings, and roofing, to add durability, and in friction products such as brake shoes and clutch plates. It also has been used in filters and gaskets.

Exposure

The route of exposure is primarily by inhalation of airborne fibers and secondarily by ingestion. Following inhalation, fibers are transported by the lung's clearance mechanisms. They may then enter the gastrointestional tract by being coughed up and swallowed. They may also be ingested when contaminated items are consumed. Consequently, any process or activity that generates airborne asbestos dust—spraying, grinding, demolition, milling, drilling, sawing, packaging—creates a risk of exposure.

Occupational exposure has occurred among miners and millers of asbestos, makers of asbestos products, construction workers (such as insulation workers, plumbers, and pipe fitters), electricians, and sheet metal workers), and shipbuilders. In the past, major clusters of exposure and of asbestos-related disease have occurred in building construction workers, makers of asbestos products, and shipbuilders. There are lower rates of asbestos-related disease in miners than there are in people who work later in the cycle of mining, milling, manufacture, use, renovation, and disposal. Asbestos insulation was sprayed onto buildings during construction for fire protection until the EPA banned spray-on asbestos insulation in 1973. Asbestos has been commonly used as insulation on steam pipes and boilers. In shipbuilding, asbestos was used as thermal electrical insulation on pipes and conduits and has been sprayed onto bulkheads. Current exposure can occur during building demolition and renovation, asbestos removal, ship breaking and repair, and boiler/furnace repairs, and motor vehicle maintenance.

Use of asbestos in the U.S. was very common as a result of shipbuilding during World War II; however, it has declined since reaching a peak in 1973. Thus, although occupational exposure to asbestos in the U.S. was widespread, it is now decreasing. In the U.S., most exposures occur in contact with or removal of asbestos in place in the construction trades. As of 2002, approximately 1.3 million workers were potentially exposed. Exposure in workers in the developing world and Eastern and Southern Europe continues, and may be increasing in some areas.

Exposure also has occurred secondary to occupational exposure when asbestos workers have come home contaminated with fibers in their clothing or hair. Children, spouses, and others have been exposed in this way, as well as by living near establishments that use asbestos. Exposure has been reported near asbestos mines in Canada and California and near surface deposits in several locations, including Turkey and Corsica. Community and occupational exposures to asbestiform fibers also occur in the area of Libby, Montana.

Because of their long latency, asbestos-related cases of mesothelioma and lung cancer are expected to persist far into this century. Past exposure may account for 2000 mesothelioma deaths and 4000 to 6000 lung cancer deaths per year in the U.S.

Fiber Types

There are two main types of asbestos: chrysotile and amphibole. These differ primarily in their shape and molecular structure. Chrysotile, also called white asbestos, has curled, snake-like fibers typically occurring in bundles and fragments about 10 micrometers or more in length. Chrysotile asbestos accounts for more that 90% of consumption of commercial asbestos, so exposure to chrysotile is widespread. Amphiboles include the subtypes of amosite (brown asbestos), crocidolite (blue asbestos), actinolite, tremolite, and anthophyllite (the latter three of which occur in fibrous and other forms as a result of different patterns of crystalline growth). Amphiboles, which are straight and needle-like, are found in some deposits of chrysotile, talc, vermiculite, and other minerals. Amosite and actinolite are common in North America, and crocidolite, anthophyllite, and tremolite are more common in Britain. Amosite is also found in soil in Turkey. Tremolite and anthophyllite fibers are not used commercially but frequently contaminate other minerals.

Although both chrysotile and amphiboles can apparently cause cancer, there is some evidence that amphiboles are more likely to produce mesothelioma and lung cancer. In some epidemiologic studies, the occurrence of mesothelioma is more frequent among workers exposed primarily to amphibole fibers than to chrysotile. However, in other investigations, mesothelioma also occurs among workers exposed only to chrysotile. Both fiber types produce mesothelioma in exposed laboratory animals. Some laboratory studies show that chrysotile fibers may be more toxic than amphiboles.

Other evidence for this hypothesis comes from the distribution of fiber types in lung tissue. Amphiboles are more common in lung tissue of people autopsied for mesothelioma. Asbestos bodies (iron-rich deposits around an asbestos core, also called ferruginous or coated bodies) are also found in lung tissue and are more likely to contain a core of amphiboles than of chrysotile. Uncoated fibers are more numerous than asbestos bodies in lung tissue. The reason for these differences in the occurrence of coated and uncoated fibers may be that chrysotile fibers are physically cleared more efficiently or are dissolved in lung tissues.

Either chrysotile or amphibole asbestos can also produce asbestosis and nonmalignant pleural disease. The mechanisms whereby these fibers produce fibrosis, nonmalignant pleural disease, or malignancy are unknown. Either shape or surface characteristics of the fibers may play an important role. Based on investigations in animals, which involved both inhalation and intrapleural implantation, carcinogenesis is influenced by fiber length and width. Thus, long fibers (8 μm or longer) with a high aspect ratio (length to width) tend to be more carcinogenic than shorter (5 μm or less) and thicker fibers. An alternative mechanism attributes carcinogenicity to fiber surface characteristics and immunologic mechanisms. If shape is the most important factor for producing disease, then substitute products such as fibrous glass or mineral wool, which have similar shape, could pose a similar risk.

Prevention

Primary prevention of asbestos-related disease can be achieved by eliminating the use of this substance. Substitutes exist for most applications. Given the risk of disease and death even with apparently trivial exposures, reliance on even the most fastidious industrial hygiene techniques does not entirely eliminate risk. Additional use of asbestos has been banned or significantly restricted in Sweden, the U.S. (by the EPA), and other countries.

When elimination is not possible—as, for example, during asbestos removal from buildings, demolition of buildings containing asbestos, or ship breaking or repair—careful attention must be paid to hygienic practices. These should include total enclosure of an area under negative pressure, use of specially designed disposable clothing, wetting and other dust control techniques, and vacuum removal. Asbestos-removal protocols have been developed by the EPA and the National Institute of Building Sciences, and worker protection regulations are enforced by OSHA.

An important part of the successful control of exposure in the U.S. has been achieved by litigation. Many former asbestos workers with asbestos-related disease have sued asbestos manufacturers as third parties. Manufacturers have been found negligent for not warning asbestos workers of health risks about which the manufacturers knew or should have known. When these suits have been successful, awards have been significantly larger than any amounts that would have been paid under workers' compensation claims. Adverse publicity, the magnitude of awards, and projected future liabilities have prompted some asbestos manufacturers to claim bankruptcy because projected liabilities exceeded their net worth. Although litigation has done nothing to prevent disease for those people exposed in the past, it has served as an important object lesson and incentive for makers and users of asbestos both to inform and educate workers and to control exposure now in order to avoid similar liabilities in the future.

Opportunities for secondary prevention based on medical screening for early signs of asbestos-related disease are limited because treatment is often ineffective. Medical screening of workers known to have been or suspected of having been exposed to asbestos should focus on early signs of nonmalignant disease (pleural thickening, pleural plaques, and parenchymal fibrosis).

Workers with pleural thickening, pleural plaques, pleural calcification, or early signs of parenchymal fibrosis are at increased risk of mesothelioma, lung cancer and progressive fibrosis. Bilateral pleural thickening on chest x-ray is a reasonably sensitive and specific indicator of asbestos exposure. In one investigation, the predictive value of bilateral pleural thickening for past exposure to asbestos was 80% when nonasbestos causes of pleural thickening (chronic renal failure, chest surgery or trauma, severe rheumatoid arthritis, bilateral empyema) were eliminated. Recovery of asbestos fibers or asbestos bodies in bronchoalveolar lavage fluids, though not a screening test, is also an indicator of past exposure. Successful prevention for workers with positive evidence of exposure is limited to restricting additional exposure to asbestos and other pulmonary hazards and advising workers of the significantly added risk of lung cancer if they smoke. Only smoking cessation has been shown to be helpful.

Because knowledge about how health effects differ with fiber type is incomplete, prevention strategies should be uniform for exposure to all types of asbestos. Exposure to a homogeneous type of fiber is rare in any case; most often, one encounters a mixture. Both OSHA and NIOSH recommend an exposure limit of 0.1 fibers per cubic centimeter (cc) (100 000 fiber per m^3), regardless of fiber type. The ACGIH, however, recommends a TLV of 0.5 fibers per cc for amosite, 0.2 for crocidolite, and 2.0 for all other forms. The OSHA standard (29 CFR 1910.1001), in addition to setting an exposure limit, also requires exposure monitoring, medical surveillance, worker training and education, shower and change facilities, and use of personal protective equipment.

In the past, limits on exposure were much more lenient. Prior to 1972, occupational exposure to 12 fibers per cc of air was permitted. OSHA reduced the limit to 2 fibers per cc in 1972 to prevent asbestosis. Then in 1985, the limit was reduced to 0.1 fibers per cc. Some malignancies will occur even at this exposure level.

Industrial Hygiene Measurement and Identification of Exposure

Airborne dust concentrations are measured not gravimetrically, as most other dusts, but by counting the number of fibers for a given volume of air. Specific mineralogical characterization of asbestos fibers in samples usually requires both transmission or scanning electron microscopy to identify the shape, and x-ray dispersive analysis to identify the chemical composition. Fibers are difficult to identify positively with the ordinary light microscope, but phase-contrast light microscopes are often used for screening purposes.

Further Reading

Levin SM, Kunn PE, Lax MB. Medical examination for asbestos-related disease. *American Journal of Industrial Medicine* 2000; 37: 6-22.

Osinubi OY, Gochfeld M, Kipen HM. Health effects of asbestos and non-asbestos fibers. *Environmental Health Perspectives* 2000; 108 (Suppl. 4): 665-674.

Wagner GR. Asbestosis and silicosis. *Lancet* 1997; 3: 1311-1315.

Asbestosis

ICD-10 J61

David C. Christiani

Asbestosis is a pneumoconiosis produced by inhalation of asbestos fibers. It is a chronic disease with slow onset that usually requires several years of exposure, depending on the intensity of exposure. Clinically, it is character-ized by diffuse interstitial pulmonary fibrosis, often accompanied by thicken-ing and sometimes calcification of the pleura. Shortness of breath on exertion is the most common presenting symptom. A chronic dry cough is common, but the cough may be productive, especially among smokers. Finger clubbing may appear in advanced cases.

In most cases, the first and often the only physical sign are crackles, also known as rales, usually detected near the end of a full inspiration. Chest x-rays reveal small, irregular opacities commonly distributed in the middle and lower lung fields. With disease progression, all lung zones may be affected, and honeycombing, especially in the lower zones, is not unusual. Late mani-festations can include an irregular diaphragm and cardiac border, which is associated with pleural plaques and diffuse pleural thickening.

Chest radiographs are often classified according to a standardized method developed by the International Labor Organization. Chest x-ray find-ings may be interpreted as normal in close to 20% of asbestos workers with fibrosis on microscopic examination of lung tissue.

In diffuse interstitial fibrosis, particularly cases of low levels of profusion, other etiologies should be considered. These include usual interstitial pneu-monitis leading to idiopathic pulmonary fibrosis; interstitial lung disease associated with connective tissue disorders; and lipoid pneumonia, all of which can present with irregular opacities in the lower lung fields. If nodular opacities are prevalent, the possibility of exposure to silica or coal dust should be considered.

Lung function tests can help to quantify the level of pulmonary dysfunc-tion and extent of impairment. In advanced asbestosis, vital capacity, func-tional residual capacity, and hence, total lung capacity are reduced. Diffusing capacity is also decreased, and, in more advanced cases, resting arterial oxy-gen tension is reduced. Earlier in the course of disease, arterial oxygen ten-sion may be reduced with exercise. Carbon dioxide exchange is usually not affected. Obstructive defects may occur, even in the absence of smoking. Currently, the most common abnormality seen in patients with asbestosis is a mixed restrictive and obstructive pattern in lung function. This is consistent with pathological observations that show peribronchiolar fibrosis early in the course of asbestosis, with distortion of the airways in advanced cases.

Diagnosis usually requires a history of sufficient exposure to asbestos established by work history, by the presence of pleural thickening or plaques, or by asbestos bodies in sputum or bronchoalveolar lavage, and by sufficient latency (usually 15 or more years), combined with some or all of the following: (a) rales or crackles; (b) positive chest x-ray or high-resolution CT findings for fibrosis (which may only appear later in the course of disease); (c) reduced lung function (vital capacity, total lung capacity, diffusing capacity, or arterial oxygen tension reduction with exercise); and (d) shortness of breath on effort.

Occurrence

The incidence of asbestosis depends on both duration and intensity of exposure. Prevalence increases among stable groups of asbestos workers with increasing length of exposure. Higher exposure produces asbestosis earlier. Estimates of prevalence have been reported as 2.5% to 4.0% with exposure between 50 and 99 fiber-years/cc (fibers/cc × years exposed) and as 6.0% to 8.5% between 100 and 149 fiber-years/cc. Even brief intense exposures such as occurred during temporary employment in shipbuilding or repair can result in disease years later.

Prevalence studies have reported variable rates of pulmonary parenchymal fibrosis among asbestos-exposed workers. A 1965 cross-sectional survey of asbestos insulation workers found a 50% prevalence of radiographically determinable pulmonary fibrosis overall, with increasing rates of fibrosis associated with longer employment. Surveys of asbestos textile workers have found that between 6% and 40% of workers reveal fibrosis on x-ray, with higher rates associated with longer periods between first exposure and examination. Surveys of asbestos miners, millers, cement workers, railroad repair workers, plumbers, pipe fitters, and maintenance workers have also demonstrated significant levels of asbestosis. Studies in the 1970s and 1980s showed that among insulation workers in New York and New Jersey, 7.7% of all deaths were due to asbestosis, with an average latency of 45 years.

New cases are continuing to be recognized. A network of clinics known as the Association of Occupational and Environmental Clinics reported about 2000 new cases of asbestosis from 1991 to 1996.

Causes

Asbestosis is caused by exposure to airborne asbestos dust, including both chrysotile and amphibole fibers. There are no other causes, although the diagnosis can be mistaken, as indicated above.

Pathophysiology

Although the pathogenesis of asbestosis is unknown, a variety of pathological and physiological changes have been identified as a result of asbestos fiber inhalation in laboratory animals and in humans. Asbestos fibers deposited in the air exchange units are ingested by macrophages, resulting in release of inflammatory mediators, inflammation, and eventually, scar formation. The pattern of scar formation is indistinguishable from that of other fibrotic lung diseases except for the presence of asbestos bodies, which are seen as markers of exposure. With progressive fibrosis, there is destruction of the normal lung architecture. Fibrosis is generally concentrated at the lung bases but may spread to include the entire lung.

Prevention
See Asbestos-Related Diseases.

Other Issues
Asbestos exposure has also been associated with respiratory tract cancer, mesothelioma, cancer of the gastrointestinal tract, and, in some reports, kidney cancer. These conditions are discussed elsewhere.

There has been considerable litigation and contention concerning all aspects of asbestos-related disease. Controversy has spilled over into the scientific literature. The ambiguity that results from disagreement among qualified experts should not be used to discourage prudent public health actions in favor of health protection.

Further Reading
International Labour Office. *Guidelines for the Use of the ILO International Classification of Radiographs of Pneumonconioses (Occupational Safety and Health Series No. 22 [Rev. 2000])*. Geneva: International Labor Office, 2002.
Levin SM, Kunn PE, Lax MB. Medical examination for asbestos-related disease. *American Journal of Industrial Medicine* 2000; 37: 6-22.
Osinubi OY, Gochfeld M, Kipen HM. Health effects of asbestos and non-asbestos fibers. *Environmental Health Perspectives* 2000; 108 (Suppl. 4): 665-674.
Wagner GR. Asbestosis and silicosis. *Lancet* 1997; 3: 1311-1315.

Asphyxiants

ICD-10 J70, T17, T20, T27, T52-T60, T70.2

Michael R. Grey and Marc Croteau

There are two broad categories of asphyxiants: simple and chemical. Simple asphyxiants are physiologically inert gases. These gases cause inadequate tissue oxygenation by displacing oxygen from respirable air: the higher the concentration of a simple asphyxiant, the lower the concentration of oxygen. With simple asphyxiants, the concentration of oxygen in air is the primary determinant of its physiological effect. In contrast, chemical or toxic asphyxiants interfere with cellular metabolism causing cells to become oxygen starved.

Occurrence

Occupational deaths from asphyxiation are not common; however, accurate incidence data for morbidity and mortality from asphyxiation are lacking. The most common cause of asphyxiation from work appears to be mechanical, such as in a trench cave-in or immersion in a suffocating material such as grain.

In one analysis of poisoning fatalities in construction workers over a 10-year period, 87 deaths (1.3% of all construction fatalities) were attributed to inhalation of hazardous substances or an oxygen-deficient atmosphere. All victims of these events were men; there was no association between age and risk. On-site OSHA investigators determined that failure to appreciate the hazardous nature of a situation constituted the most common human factor to have contributed to fatalities.

Manufacturing, oil and gas industries, construction, and utilities appear to account for most fatalities due to asphyxiation. Death due to simple asphyxiants most commonly occurs in relation to entry into confined spaces, not intended for continuous worker occupancy, having inadequate natural ventilation and containing or producing dangerous air contaminants. In confined spaces, oxygen can be replaced by other gases, leading to an oxygen-deficient atmosphere or to an atmosphere with high concentrations of toxic gases. Workers can be overcome, and death can occur upon entry, unless adequate fresh air or a self-contained breathing apparatus is available. Would-be rescuers are often overcome as well, leading to multiple fatalities.

Causes and Pathophysiology

Simple Asphyxiants

As a class, simple asphyxiants (such as argon, nitrogen, hydrogen, helium, methane, ethane, and carbon dioxide) are biologically inert. They affect health by decreasing the percentage of oxygen in inspired air. This displacement causes the partial pressure of oxygen in the alveoli to decrease, and, ultimately, less oxygen is delivered to tissues. Many of the clinical effects of an oxygen-poor atmosphere relate to the central nervous system (see Table 1). Environmental conditions and individual personal factors contribute to the effects of simple asphyxiants. For example, the effects of asphyxiants can be accelerated or exacerbated by personal factors such as increased work pace (which increases minute ventilation), underlying medical conditions such as chronic lung disease, or the absence or improper use of personal protective equipment. Environmental factors, such as inadequate dilution ventilation, high ambient temperatures, and elevated altitude, may also impact suscepti- bility. Ultimately, all these factors operate either to decrease the amount of oxygen available to tissues or to augment the tissues' oxygen requirements, which cannot be matched by the available oxygen in the inspired atmosphere.

Table 1 implies that levels of inspired oxygen of 16% to 21% can be toler- ated without adverse effects. This may not hold true in workers with under- lying medical disorders or in circumstances with increased tissue oxygen requirements, such as a heavy workload. In recognition of increased oxygen requirements with exertion compounded by certain physical environmental conditions, such as high-altitude exertion and heat, OSHA requires a mini- mum of 19% oxygen. Sources of simple asphyxiant exposures include a vari- ety of liquefied gases that are widely used throughout industry. For example, noble gases (argon, helium, neon, and xenon) form cryogenic liquids when compressed. Uncontrolled release of these agents in enclosed spaces may cause freezing injuries as well as asphyxiation.

Chemical Asphyxiants

Chemical asphyxiants produce symptoms by interfering with cellular oxygen transport or utilization. The mechanisms by which each toxic asphyxiant interrupts cellular metabolism vary and will be addressed indi- vidually. As with simple asphyxiants, other variables influence the severity of clinical effects for any given toxic asphyxiant. Concentration of the gas, length of exposure, and adequacy of ventilation are important environmental

Table 1.   Physiological Effects of Oxygen Deficiency

Oxygen Concentration (% inspired)	Clinical effects
16-21	Asymptomatic
12-16	Tachypnea, tachycardia, motor incoordination
10-14	Emotional lability, fatigue, exertional dyspnea
6-10	Nausea/vomiting, lethargy, possible unconsciousness
<6	Seizure, apnea, asystole

conditions influencing toxicity. Individual factors, such as baseline health status, such as cardiac or pulmonary disease, or the improper use of personal protective equipment affect toxicity as well. Concurrent exposure to other asphyxiants or gases may augment the potential toxicity of any given asphyxiant. Such is the case for firefighters exposed to high levels of carbon monoxide and hydrogen cyanide in fires in which pyrolysis of plastics occurs.

CARBON MONOXIDE
OSHA PEL 35 ppm (8-hour TWA), STEL 200 ppm
NIOSH REL 35 ppm (10-hour TWA)

Carbon monoxide (CO) is regarded as a leading cause of poisoning worldwide. It is a colorless, odorless, tasteless, and nonirritating product of incomplete combustion of hydrocarbons. Carbon monoxide is ubiquitous in our environment. However, the atmosphere's concentration of CO is usually less than 0.001%. Occupations with well-recognized CO exposure include fire fighters, tollbooth operators, traffic police, coal miners, coke oven workers, and smelter workers. Propane-operated forklifts within warehouses have been implicated as a cause of CO-related headache among workers. Propane-fueled ice rink resurfacers have been known to generate toxic levels of CO. Ambient CO levels as high as 117 ppm have been detected during hockey games in six arenas resurfaced by such machines. In 1996, an event involving a poorly maintained ice resurfacer powered by a propane engine resulted in the evacuation of 300 persons and the transport of 67 persons to local emergency departments. This incident was associated with ambient CO levels measured at 354 ppm inside the ice arena. Motor vehicle exhaust accounts for 57% of the 11 547 unintentional carbon monoxide deaths reported between 1979 and 1988. Methylene chloride found in some solvents and paint strippers is another source of carbon monoxide. It is absorbed through skin and/or inhaled and is then metabolized to CO in the liver. The health effects of carbon monoxide poisoning range from subtle nonspecific neurobehaviorial affects to death. Estimates of annual mortality and morbidity due to CO exposure, along with those due to other asphyxiants, are notoriously inadequate and probably grossly underreported. The majority of deaths attributed to smoke inhalation are probably caused by CO exposure.

Pathophysiology

Inhaled CO affects a number of important molecular biological systems. It is readily absorbed into blood where it forms a tight complex with hemoglobin known as carboxyhemoglobin (COHb). The amount of CO absorption depends upon minute respiratory ventilation, duration of exposure, and relative concentrations of CO and O_2 in the environment. Hemoglobin has up to 250 times the affinity for carbon monoxide than for oxygen. Consequently, very small concentrations of CO in inspired air can have significant clinical effects (Table 2). The altered properties of COHb prevent oxygen loading onto hemoglobin. Furthermore, as blood concentration of COHb increases, the oxygen dissociation curve shifts to the left, resulting in impaired oxygen release and cellular hypoxia. CO also binds to myoglobin thereby reducing myocardial oxygen extraction and further increasing the risk of ischemia. Carbon monoxide is thought to interfere with the mitochondrial cytochrome oxidase system, precipitating a complex cascade of events leading to ischemic re-perfusion injury to the brain. A number of other mechanisms for toxicity are probable.

While COHb levels correlate well with ambient CO levels and can be predicted using the Coburn-Forster-Kane differential equation, the signs and symptoms of CO exposure do not correlate well with COHb levels. As a practical matter, COHb should not be relied upon either to indicate clinical severity or to determine prognosis. Duration of exposure appears to be a significant determinant of toxicity. At 200 ppm (or 0.02% inspired CO by volume), headache, tinnitus, and a sense of discomfort are frequent after several hours of exposure. At 800 ppm (0.08% inspired CO), frontal headache, nausea, and dizziness occur within an hour. Concentrations above 1600 ppm (0.16% inspired CO) can lead to narcosis, coma, and, eventually, death. Table 2 indicates at what COHb level various clinical effects are expected. Severity of poisoning is best judged clinically rather than relying solely on COHb level or environmental CO concentrations, both of which may over- or underestimate risk. Cigarette smokers often have high baseline levels of COHb (up to 14%) and may be more susceptible than nonsmokers to adverse health effects of ambient CO exposure at lower concentrations. Exposure to CO at levels of activity that lead to an increase in minute ventilation, as occurs with heavy labor or high-altitude work, also increases COHb levels and can produce

Table 2. Relationship of Carbon Monoxide Levels and Symptoms

Severity	%COHb	Symptoms
Mild	< 20	nausea, tinnitus, dyspnea on exertion, headache
Moderate	20 - 40	fatigue, poor judgment, mental obtundation
Severe	> 40	arrhythmias, death

(Adapted from Rosenstock L (Ed.). Occupational pulmonary disease. *State of the Art Reviews in Occupational Medicine* 1987; 2 : 297.)

symptoms and signs more rapidly. Despite some debate, experimental animal data support the position that coronary artery disease can be accelerated by CO exposure. Many case reports and some human and animal experimental data have shown that individuals with known coronary disease may have symptoms provoked by low levels of inspired CO. Pregnant women and their fetuses are also more vulnerable at any given CO level. CO crosses the placenta readily and is teratogenic. Further, COHb levels are higher in the fetus compared to the mother because fetal hemoglobin retains CO up to five times longer than adult hemoglobin. Exposure to CO in utero can cause permanent central nervous system impairment including mental retardation.

Initial management of CO poisoning consists of prompt removal from exposure and the administration of 100% supplemental O_2 delivered via a non-rebreather face mask. Obtunded patients may require endotracheal intubation. Standard ACLS protocols apply to cardiovascular complications. An insidious consequence of acute CO poisoning is the development of delayed neuropsychiatric impairment. Within 2-28 days after poisoning, up to 40% of victims develop persistent cognitive deficits including impaired judgment, poor concentration, and memory loss. The risks of cognitive deficits are thought to increase in patients with a history of loss of consciousness. Timely recognition and treatment of acute CO poisoning is critical in preventing this complication. Although precise indications are not established, the use of hyperbaric oxygen treatment within a 24-hour period of acute carbon monoxide poisoning is thought to reduce the development of cognitive deficits. Hyperbaric oxygen treatment has demonstrated efficacy in patients with loss of consciousness, COHb >25%, age >50, or base excess <-2 mEq/L. Hyperbaric oxygen is considered to be most effective if given within 6 hours following exposure. The half-time elimination of CO is 320 minutes while breathing air compared to 80 minutes with 100% O2 supplementation at normobaric pressure, and 23 minutes with hyperbaric oxygen at 3 atmospheres absolute. However, it is important to realize that in certain cases, even prehospital normalization of carboxyhemoglobin may be an insufficient endpoint for treatment. Therapeutic benefits have been demonstrated from a three-treatment course of hyperbaric oxygen even after near normalization of COHb. Recent research suggests that delayed cognitive deficits may arise from lipid peroxidation and ischemic re-perfusion injury within the brain. In support of this mechanism, multiple animal studies demonstrate benefit with hyperbaric oxygen and re-perfusion injury. New treatment techniques involving the induction of normocapnic hyperpnea for the treatment of acute CO poisoning are under active investigation.

The effects of chronic low-level exposure to CO are less well studied. Highway toll workers have difficulty performing on parallel processing tasks. Health care providers must remain alert to nonspecific recurrent symptoms of headache and nausea which could suggest ongoing CO exposure.

The risk of occult poisoning can be reduced by the use of CO detection devices. More importantly, prevention strategies should include the proper use, maintenance, and ventilation of fuel burning heating systems, vehicles, and machinery.

HYDROGEN CYANIDE (HCN)

OSHA PEL 10 ppm (8-hour TWA)

NIOSH REL 4.7 ppm (skin); (10-minute ceiling limit)

Cyanide is a systemic poison with a long and notorious history. It has been used in chemical warfare and as a genocidal agent. Industrial uses for cyanide are found in electroplating, metal extraction (such as gold from ore), fumigation, metal (steel) hardening, photography, blast furnaces, nitrile preparation, chemical manufacture, and in pharmaceuticals. Thermal combustion of polyurethanes and plastics can release HCN, representing a risk to firefighters as well as to aircraft crew and passengers during on-board fires. Acetonitrile, a chemical used commonly in a variety of industrial processes, undergoes biotransformation via the cytochrome P450 to form cyanide and formaldehyde. The pits of certain fruit, including apricots, peaches, and cherries, contain substances that can also become biotransformed to cyanide.

The toxic dose of cyanide depends upon its physical state—for example, salt versus gas—as well as the route and duration of exposure. It is a broad inhibitor of enzymes, most importantly mitochondrial cytochrome oxidase. As such, cyanide interrupts aerobic metabolism and prevents cells from utilizing oxygen for the production of ATP. The pathophysiologic consequences of cyanide poisoning include rapid cessation of cellular respiration and profound metabolic acidosis. Clinical effects lead to progressive malfunction of the central nervous system and the cardiovascular system. Other organ systems are also affected. As with CO poisoning, survivors of significant cyanide poisoning may develop delayed neurological sequelae. With cyanide these typically manifest as a refractory form of parkinsonism.

The effects of chronic low-level exposure to cyanide are both less well understood and more controversial. Chronic HCN exposure is thought to be the cause of the insidious neurological disorder tobacco amblyopia. The major physiological route for cyanide detoxification is its conversion to thiocyanate which is then excreted in urine. Urinary thiocyanate may serve as a biomarker of exposure. The limiting factor in cyanide detoxification revolves around the availability of sulfur donors. Treatment includes decontamination, supportive care, 100% oxygen supplementation, and the cyanide antidote kit which contains nitrites and thiosulfate. Thiosulfate donates sulfur atoms for biotransformation to thiocyanite, a more soluble and therefore more easily excreted less toxic metabolite. Nitrites induce methemoglobinemia thereby facilitating the preferential binding of cyanide to the hemoglobin molecule, thus competing with cytochromes for CN binding and restoring cytochrome function.

HYDROGEN SULFIDE (H₂S)

OSHA PEL 20 ppm (ceiling limit)

OSHA PEL 50 ppm (peak <10 minutes), NIOSH REL 10 ppm for 10 minutes

Descriptions of diseases afflicting miners and sewer workers suggest that H_2S is an industrial hazard with a long history. However, only with the advent of oil drilling and refining in the 20th century has widespread industrial exposure to H_2S occurred. Hydrogen sulfide is a colorless, dense, malodorous gas produced through the anaerobic decomposition of organic matter, such as might naturally be found in geothermal stream fields, in oil drilling, mining (where it is known as "stink damp"), landfills, and wastewater treatment facilities. It is also used in or given off as a by-product of a variety of industrial processes. Industries in which H_2S is commonly found include the oil and gas industry, heavy water production, sewage treatment, rayon manufacturing, leather tanning, mining (coal and metal), and rubber vulcanization.

Accurate current figures on the annual incidence of intoxication, asphyxiation, and fatalities are not available. However, numerous industrial case series document H_2S intoxication and fatalities. From 1984 to 1994, there were 80 reported fatalities in 57 incidents; 86% of these involved confined spaces; 22/80 (28%) of the fatalities involved the petroleum industry; and 25% of the fatalities involved rescue workers. OSHA cited the workplace for violations in respiratory protection and confined space standards in 60% of the fatalities. A number of environmental disasters have occurred secondary to industrial release of hydrogen sulfide. In 1950, 320 people were hospitalized and 22 died in Poza Rica, Mexico, when a local oil refinery released a large cloud of H_2S. In 1975, a cloud of hydrogen sulfide gas resulted in nine fatalities in a Texas community when a gas well failed.

At very low concentrations (< 1 ppm) concentrations, H_2S is a malodorous air pollutant without significant irritancy or systemic toxicity. Its odor is characterized invariably as that of rotten eggs. Above 10 ppm H_2S is a potent respiratory and mucous membrane irritant. At higher concentrations, H_2S can be a rapid and deadly asphyxiant (Table 3). Despite its potent odor, smell is an unreliable means of warning. As the concentration of H_2S increases, the brain's olfactory center becomes desensitized to the characteristic odor. Survivors of near-fatal asphyxiation often do not recall any "warning" odor. Subacute intoxication due to acute low-level (50-100 ppm) or chronic low-level exposure has resulted in delayed pulmonary edema and death. Controversial claims about the chronic health effects of extremely low (ppb) levels of H_2S exposure have been made, but beyond nuisance odors, convincing injury at these extremes is lacking.

Hydrogen sulfide causes tissue asphyxiation by binding to cytochrome oxidase and uncoupling oxidative phosphorylation. This results in an interruption of aerobic metabolism and in profound metabolic acidosis. Hydrogen sulfide affects directly the brain's respiratory center and the carotid body, causing initial respiratory stimulation, followed by depression, apnea, and death. Hydrogen sulfide is denser than air and tends to accumulate at the bottom of pits, tanks, or other enclosed spaces. This fact has obvious ramifications for the occurrence and prevention of H_2S asphyxiation.

Prevention

Clinically significant exposure to toxic asphyxiants is preventable by appropriate and readily implementable environmental controls. Management and worker education about these hazards and how to prevent asphyxiation is critically important. Dilution ventilation and effective maintenance of machinery, pipes, ventilation, and exhaust systems serve to maintain ambient levels of asphyxiants below hazardous levels. Because most asphyxiant deaths occur without warning, engineering and safety programs must anticipate potential hazards if they are to be effective in preventing accidents. For example, most suffocations due to trench cave-ins are preventable by using correct trenching and sloping techniques. Potentially confined, poorly ventilated spaces should have available means of egress in the event of unanticipated asphyxiant exposures. Confined, non-ventilated spaces should never be entered without prior air monitoring using a number of readily available and accurate field devices to determine oxygen content and content of toxic or explosive gases in the air. If the levels of oxygen or toxic gases are dangerous, trained workers with air-supplied (not cartridge) respirators should be assigned. Finally, the "buddy system" with appropriately trained workers should always be in effect when confined spaces are entered or rescue efforts are attempted. Fatalities still occur even when meticulous attention to worker education, hazard evaluation, and safety precautions are present. If asphyxiants are generated within a workplace, a nonoccupational environment, or some other enclosed space, adequate preparation for entry into these zones must include the above precautions.

Table 3. Relationship of Hydrogen Sulfide Levels and Clinical Effects

Hydrogen Sulfide Concentration (ppm)	Clinical Effect
0.1-0.2	Odor threshold
10-100	Eye and upper respiratory tract irritant
>200	Anosmia, delayed pulmonary edema
>500	Hyperpnea, apnea
>1000	Respiratory paralysis, death

(Adapted from National Research Council. *Hydrogen Sulfide*. Baltimore, MD: University Park Press, 1979; and Deng JF, Chang SC. Hydrogen sulfide poisonings in hot-spring reservoir cleaning: Two case reports. *Am. J. Indus. Med. 1987; 11: 447-451.*)

For the most part, biological surveillance for asphyxiants is neither technically possible nor reasonable, because most deaths and overexposures occur through inadvertent exposure. However, for workers with known potential CO exposure, surveillance mechanisms can be instituted to monitor for CO intoxication. For example, post-shift COHb levels or expired alveolar air for CO are reliable means of surveillance. If surveillance is instituted, adjustment for the potential confounding by tobacco smoke is necessary. Once an individual is overcome by an asphyxiant, individualized treatment of the condition depends on the suspected exposure. In all instances, the individual should be removed from the asphyxiant atmosphere before resuscitation is attempted. Would-be rescuers must use proper protective equipment, such as air-supplied respirators. Oxygen (100%) should always be administered to victims in cases of suspected asphyxiation.

High concentrations of delivered oxygen, including hyperbaric oxygen, have been shown to decrease the half-life of CO elimination. Some authors have questioned the value of oxygen therapy following asphyxiation with HCN or H_2S; they argue that H_2S and HCN prevent cellular oxygen use, so unless this action is counteracted, oxygen therapy is not useful in acute asphyxiation. Nevertheless, laboratory animal data have shown an independent protective effect of oxygen when it is given as part of an overall treatment protocol for toxic asphyxiation. Medical consensus is that oxygen should be provided in all instances. Induction of methemoglobinemia through the administration of nitrites (sodium nitrite) has been shown to be effective in reversing the binding of cyanide anions to cytochrome oxidase. The use of nitrites for H_2S poisoning has unproven efficacy. Some authors have cautioned that nitrites can cause unpredictable levels of methemoglobinemia, which in itself can complicate resuscitation. However, intravenous hydroxocobalamin has been shown to decrease the cyanide toxicity associated with nitroprusside administration by forming less toxic cyanocobalamin. It has recently found a use in the treatment of cyanide intoxication, but it remains available in the U.S. on an investigational basis only. A field kit exists for the treatment of cyanide poisoning (Lilly, Cyanide Antidote Kit). Finally, industries in which HCN is a known hazard should provide appropriate first aid training, including the administration of amyl nitrate, to employees potentially exposed to HCN.

Further Reading

Dorevitch S, Forst L, Conroy L, et al. Toxic inhalation fatalities of US construction workers, 1990 to 1999. *Journal of Occupational and Environmental Medicine* 2002; 44: 657-662.

Ernst A, Zibrak JD. Carbon monoxide poisoning. *New England Journal of Medicine* 1998; 339: 1603-1608.

Goldfrank LR (Ed.). *Goldfrank's Toxicologic Emergencies* (7th ed.), New York: McGraw-Hill, 2002.

Milby TH, Baselt RC. Hydrogen sulfide poisoning: clarification of some controversial issues. *American Journal of Industrial Medicine* 1999; 35: 192-95.

National Institute for Occupational Safety and Health. Criteria for a recommended standard: working in confined spaces. Publication No. 80-106. Washington, DC: NIOSH, 1979.

Penney, DG (Ed.). *Carbon Monoxide Toxicity.* Boca Raton: CRC Press, 2000.

Raub JA, Mathieu-Nolf M, Hampson NB, et al. Carbon monoxide poisoning: a public health perspective. *Toxicology* 2000; 145: 1-14.

Asthma

ICD-10 J45, J46

David C. Christiani

Occupational asthma is defined as a reversible, generalized airway narrowing as a result of exposure to airborne dust, gases, vapors, or fumes in the work environment. There are different forms of occupational asthma: one type requires a period of sensitization, hence latency period; another occurs without latency and includes irritant-induced (nonsensitizing) asthma. These two major types of occupational asthma reflect differing exposures and different pathogenic mechanisms.

Diagnosis is made by confirming the presence of asthma and establishing a relationship with work exposures. Patients may present with typical symptoms of episodic dyspnea, chest tightness, and wheezing associated with airflow limitation, which is reversible after administration of a bronchodilator. Patients may also present with recurrent attacks of bronchitis. A detailed work history is important, with particular attention to the periodicity of symptoms. Suggestive patterns are symptoms that occur only at work or in the evening of workdays, worsen at the start or in the course of the workweek, improve on weekends or vacations, or resolve with changes in the working environment.

Pulmonary function tests may be normal or may show obstruction with improvement after an inhaled bronchodilator is administered. In patients with normal pulmonary function, standardized tests of airway reactivity can be done using methacholine, a very sensitive indicator of airway hyperresponsiveness. Spirometry can also be done before and after a work shift to determine the relationship of airflow obstruction to work; a 10% or greater decrease in forced expiratory volume (FEV) is considered significant. However, a smaller decrease, or no decrease at all, does not exclude work-related asthma. Although general skin or serologic testing for IgE to common inhalants and food allergens can be used to define the atopic status, this usually has little relevance to most cases of occupational asthma.

With some exposures, such as high-molecular-weight compounds, specific skin tests may help to identify the responsible agent in asthma with a latency period. They have been used in such exposures as flour, dust, coffee, and some animal products. Serological tests for IgE antibodies, such as the radioallergosorbent test (RAST) or enzyme-linked immunosorbent assay (ELISA), to various occupational allergens can be done, but these tests can yield positive results after exposure alone, with no evidence of disease. However, they may also be useful in associating symptoms with specific workplace exposures.

Generally, it is unnecessary to resort to specific inhalation challenge testing of workers using a suspected agent in order to diagnose occupational asthma. Challenge testing can be a valuable research tool that should only be performed in specialized centers with personnel experienced in recognizing and treating complications that can arise from such procedures, such as severe or delayed reactions. As with all testing, an informed decision must be made that the information which can be derived from the test is worth the risk to the test subject.

A case definition for surveillance purposes has been published by the CDC. It consists of "a physician diagnosis of asthma [as above], an association (patterns of which may vary) between symptoms of asthma and work and either (1) workplace exposure to an agent or process previously associated with occupational asthma or (2) significant work-related changes in forced expiratory volume in 1 second (FEV_1) or peak expiratory flow rate or (3) significant work-related changes in airways responsiveness as measured by nonspecific inhalation challenge or (4) positive response to inhalation provocation testing with an agent to which the patient is exposed at work." In addition, the number of co-workers with exposures similar to those of the reported case-patient, and the number of co-workers with respiratory symptoms should also be ascertained.

Occurrence

About 10 million American adults have asthma. Public health surveillance in four states (California, Massachusetts, Michigan, and New Jersey) have identified over 2 500 cases of work-related asthma over a recent seven-year period (1993-1999). The overall prevalence of occupational asthma is unknown, but recent studies have estimated that occupational asthma accounts for between 5% and 37% of all asthma. An expert panel that reviewed the relevant literature concluded that occupational exposures contribute to 15% of asthma among adults. Prevalence of the disease depends on the nature of the agent and the level of exposure.

Causes

More than 250 agents are reported to cause occupational asthma. Examples of hazards and occupations associated with occupational asthma are presented in Table 1.

Pathophysiology

Pathological appearance of occupational asthma is not different from that of asthma of any cause. The bronchial lumen is narrowed and filled with mucus ("mucus plugging"). Both the mucus and the bronchial wall are infiltrated with inflammatory cells. There is epithelial denudation, thickening of the basement membrane, and bronchial muscle hypertrophy. The mechanisms of occupational asthma include the following:

1) Reflex bronchoconstriction, such as by cold air, noxious gases, and irritants.
2) Acute inflammation, such as by irritant gases and vapors, like acid mists, or by pyrolysis products of polyvinyl chloride.
3) Pharmacological reaction such as by isocyanates.
4) Allergic reaction by grain, baker's flour, detergent enzymes, isocyanates, wood, and metals, such as nickel, platinum, and chromium.

Asthmatic reactions can be immediate, late (delayed), or both (dual). The immediate reaction can be produced by nonallergic or allergic mechanisms. Nonallergic reactions, via reflex bronchial constriction, can occur in persons with preexisting bronchial hyperreactivity. Allergic asthma is usually mediated via IgE reaginic antibodies, which have a high affinity for membrane receptors of circulating basophils and tissue mast cells.

The late type of reaction can occur alone or as a sequel to the immediate reaction. When occurring as a sequel, it is part of a dual reaction. The late or dual reaction may be induced by allergens or by a number of low-molecular-weight compounds.

The late reaction is also associated with an ongoing inflammatory process that probably causes the nonspecific bronchial reactivity that is seen in asthma induced by any of the above mechanisms.

Prevention

There is sufficient understanding of the etiology of occupational asthma to allow for aggressive measures in primary and secondary prevention Preemployment screening and prophylactic desensitization have not been shown to be effective for occupational asthma. Instead, primary prevention means elimination or reduction of exposure to the causative agent, where substitution with less-sensitizing materials is often the ideal intervention. For example, a decline in latex-induced occupational asthma in several developed countries has been attributed largely to the substitution of low-protein or powder-free latex gloves, or nonlatex gloves, in place of powdered latex gloves. Where substitution is not possible, primary preventive strategies should concentrate on (a) exposure reduction through efficient environmental control of exposures (ventilation, enclosure, allergen monitoring); (b) safer work practices (spill containment, good housekeeping, personal protective equipment);

and (c) ongoing worker training and education. The detergent and laboratory animal industries provide useful examples in preventing occupational asthma and allergy by instituting improved dust control measures accompanied by ongoing educational programs. Routine medical surveillance of workers was also used to assess the impact of preventive measures.

Routine medical surveillance that includes questionnaires, chest examinations, and spirometry, to detect early signs and symptoms of asthma, is a useful means of secondary prevention. Early intervention, such as controlling or eliminating the exposure to the causative agent, can take place to effectively stop the disease from progressing. Failure to control exposure may result in the development of irreversible airways obstruction. When officials in the province of Ontario in Canada mandated exposure monitoring and medical surveillance for workers exposed, the results were earlier diagnosis and favorable prognosis. Once sensitized to an agent, a worker may respond to minute quantities of it. For these workers, job change with wage rate retention may be the only reasonable preventive strategy, but occupational options and economic costs for the worker must be considered.

Other Issues

Predisposing host factors may be important in the development of occupational asthma. Atopy is the capacity to develop immediate sensitivity after exposure to common environmental allergens. Atopic workers become sensitized more readily than nonatopic workers when high-molecular-weight compounds are used, such as in the enzyme detergent industry, in bakeries, and in industries in which animals are handled. However, because nonatopic workers also become sensitized at significant rates, it is inappropriate to exclude all atopic individuals from employment in these industries.

The role of cigarette smoking in the development of occupational sensitization and asthma is unknown. Research findings have been contradictory, and there is little evidence to suggest that smokers are more predisposed to asthma in most exposure settings. Nearly all workers with symptomatic occupational asthma have nonspecific bronchial hyperreactivity, which is most likely a consequence of workplace exposures rather than a predisposing factor.

Work-relatedness of asthma may be obscured if a worker becomes sensitized and develops symptoms after a long asymptomatic period in the same workplace, or if a worker chronically exposed to a causative agent develops relatively fixed airways obstruction that does not lead to recognition of an association with work.

Airway obstruction may resolve over months to years following cessation of exposure, or it may remain fixed.

Further Reading

Balmes J, Becklake M, Blanc P, et al. American Thoracic Society Statement: Occupational contribution to the burden of airway disease. *American Journal of Respiratory and Critical Care Medicine* 2003; 167: 787-797.

Bernstein IL, Chan-Yeung M, Malo JL, et al. *Asthma in the Workplace* (2nd Ed.). New York: Marcel Dekker, 1999.

Crapo RO, Casaburi R, Coates AL, et al. American Thoracic Society Statement: Guidelines for methacholine and exercise challenge testing, 1999. American Journal of Respiratory and Critical Care Medicine 2000; 161: 309-329.

Cullinan P, Tarlo S, Nemery B. The prevention of occupational asthma. *European Respiratory Journal* 2003; 22: 853-860.

Jimenez GF, Beckett WS, Szeinuk J, et al. Clinical evaluation, management, and prevention of work-related asthma. *American Journal of Industrial Medicine* 2000; 37: 121-141.

Malo JL, Chan-Yeung M. Occupational asthma. *Journal of Allergy and Clinical Immunology* 2001; 108: 317-328.

National Institute for Occupational Safety and Health. *Work-Related Lung Disease Surveillance Report (2002)*, (DHHS NOISH Publication No. 2003-111). Washington, DC: NIOSH, 2003.

Wagner GR, Wegman DH. Occupational asthma: Prevention by definition. *American Journal of Industrial Medicine* 1998; 33: 427-429.

Table 1.

Hazard	Occupation	Prevalence
Substances of Animal Origin		
Hair, epidermal squamae, animal products, urine protein, insects, birds, laboratory animals, danders, egg protein	Hair dressers, research lab workers, animal handlers, grain and poultry workers, veterinarians, laboratory workers, egg producers	3% to 30%
Bee toxin and squamae, hairs, chitin	Entomologists, fish bait breeders, apiarists	?
Marine organisms	Oyster farmers, crab and prawn processors, food processors, fishermen	4% to 36%
Silk hair and larva, butterfly squamae	Sericulture workers, silkwork cutters	0.2% to 34%
Locusts, river flies, screwworm flies, sewage flies, fruit flies, sheep blowfly	Research lab workers, flight crews, outdoor workers, technicians	3% to 70%
Tetranychus macdanieli, *Tetranychus urticae*	Vine growers, outdoor workers	11% to 35%
Grain mite	Farmers, grain-store workers	12% to 33%
Chironimid midges	Aquarists, fish food	45%
Arthropods	Technicians	23%
Substances of Plant Origin		
Grain dust	Grain handlers, millers, grain elevator workers	40% to 47%
Wheat/rye/soya flour, buckwheat	Bakers, millers	20% to 100%
Coffee bean, caster bean	Coffee workers, oil extractors, food processors	9% to 34%
Tea, herbal tea	Tea workers, herbal tea workers	?
Tobacco manufacturers		69%
Hops *(Humulus lupulus)*	Brewery chemists, farmers	?
Wood dust	Carpenters, construction workers, sawmill workers, cabinetmakers, furniture making, wood workers, bowmakers	3% to 33%
Cotton, flax, and hemp dust	Textile workers, weavers	?

Hazard	Occupation	Prevalence
Gum acacia	Printers	19%
Gum tragacanth	Gum manufacturers	?
Gum karaya	Hair dressers	4%

Substances of Chemical Origin

Hazard	Occupation	Prevalence
Toluene diisocyanate	Polyurethane industry, plastics, varnish workers	5% to 38%
Diphenylmethane diisocyanate, furfuryl alcohol	Foundry workers	5% to 27%
Hexamethylene diisocyanate, dimethylethanolamine	Spray painters	45%
Phthalic trimetalic and tetrachlorophthalic anhydrides	Epoxy resin, plastic, paint workers	14% to 36%
Methyl tetrahydrophthalic anhydride, hexahydrophtalic anhydride, himic anhydrides	Electrical workers, flame retardant workers	5% to 35%
Hexachlorophene, formalin	Hospital workers	29%
Paraphenylene diamine, dioazonium salt	Dye workers, manufacturer of fluorine polymer precursor	37% to 56%
Pyrethrins, organophosphates	Insecticide manufacturers, farmers	?
Azoformamide	Plastics, rubber workers	19%

Metals

Hazard	Occupation	Prevalence
Platinum	Platinum refinery workers, jewelers, electroplating workers	29% to 57%
Nickel	Metal-plating workers, stainless steel workers	?
Chromium	Manufacturers of pigments, tannery workers, precision casters, stainless steel welders	?
Cobalt	Metalworkers, cobalt refinery or alloy workers	?
Vanadium	Metalworkers, mineral ore processors	33%
Tungsten carbide	Metalworkers	?
Nickel sulfate	Nickel workers, metal-plating workers	?
Aminoethyl ethanolamine and colophony soldering flux	Electricians and electronics manufacturing workers	4% to 21%

Hazard	Occupation	Prevalence
Biological Enzymes		
B. subtilis	Detergent industry or laundry workers	3%
Trypsin	Plastic and pharmaceutical workers	29%
Pancreatin, pepsin, bromelin, flaviastase	Pharmaceutical workers	11%
Fungal amylase	Manufacturing workers, bakers	?
Papain	Laboratory and food technicians, packagers, pharmaceutical workers	45%
Penicillium casei	Cheese production workers	?
Drugs		
General pharmaceutical compounds	Nurses, health personnel	4%
Pencillinis, methyldopa, cephalosporins, spiramycin, sabutamol intermediate, phenylglycine acide chloride, tetracycline	Pharmacists, nurses, physicians, factory workers	8% to 29%
Piperazine hydrochloride	Chemists	?
Psyllium	Laxative manufacturers	?
Amprolium HCl	Poultry feed mixtures	?
Sulphone choramides	Brewery workers	?
Ipecacuanha	Pharmaceutical workers	48%
Opiate compounds	Pharmaceutical workers	14% to 26%
Non-Sensitizing Agents		
Household cleaning products	Janitors, cleaners, housekeepers, nurses and nurses' aides, clerical workers	3% to 14%
Chlorine, ozone	Bleachery workers	?
Sulfur Dioxide (SO_2)	Pulp mill workers	?
Welding fumes	Welders	?

Source: Prevalence range of work-related asthma (NHANES III) or current asthma (ECRHS) among cleaners taken from Arif [*American Journal of Industrial Medicine* 2003; 44: 368-376] and Zock [*European Respiratory Journal* 2002; 20: 679-685].

Atopic Dermatitis

ICD-10 L20

Michael E. Bigby

Atopic dermatitis is a chronic, pruritic eruption often occurring in association with a personal or family history of hay fever, asthma, and multiple allergies. The primary lesion in atopic dermatitis is an ill-defined plaque, with scaling, erythema, vesicles, exudate, and honey-colored crusting. This constellation of features is an eczematous dermatitis (which can also occur in allergic and irritant contact dermatitis). In chronic cases, thick scale, lichenification (exaggeration of skin folds), and hyperpigmentation are characteristic features. Atopic dermatitis usually begins in infancy where it affects the scalp, face, and torso. The neck, as well as the antecubital and popliteal fossae, are common sites of involvement in older children and adults. The disease lessens in severity or remits in over 50% of patients after adolescence.

Atopic dermatitis is commonly accompanied by or followed by post-inflammatory hyperpigmentation or hypopigmentation, a problem that is most severe and bothersome to patients with dark skin. One not uncommon variant of atopic dermatitis is seen in children and young adults who have annular (round), hypopigmented eczematous plaques on their face and arms (pityriasis alba).

Occurrence

The prevalence of atopic dermatitis is estimated between 10% and 20% in industrialized countries. The onset of atopic dermatitis is usually during infancy. A family history of atopy is the most important risk factor for the development of atopic dermatitis. Atopic dermatitis is more prevalent in higher socioeconomic classes and is inversely related to family size in several studies.

Causes

Atopic dermatitis is a genetic disorder commonly occurring in association with asthma or seasonal allergic rhinitis. Patients with atopic dermatitis frequently have elevated IgE levels, colonization of the skin with *Staphylococcus aureus,* and are scratch-test-positive to a wide variety of antigens.

Pathophysiology

Atopic dermatitis is a genetic disorder. The T cell response to allergens in atopic dermatitis is skewed toward TH-2 responses. (T cells predominantly secrete IL-4, IL-5, and IL-10.) This cytokine profile augments humoral responses (particularly IgE) and inhibits some cell-mediated responses. Patients with atopic dermatitis have increased susceptibility to infections with some bacteria, such as *Staphylococcus aureus,* and some viruses, such as herpes simplex infections.

Prevention

Prevention of flares of atopic dermatitis involves hydrating the skin and avoiding irritants. Most patients with atopic dermatitis have dry skin. Skin dryness is exacerbated during the winter months because of the low humidity. Water softens the skin and makes it more flexible. A major problem in atopic dermatitis is that the normal barrier formed by an intact stratum corneum is damaged. Increased transepidermal water loss is therefore a common and significant problem.

Rehydration of the skin with frequent baths or showers, followed by the application of emollients, is beneficial in the management of patients with atopic dermatitis. This advice has not been proven by direct clinical evidence; there has been no randomized clinical trial of emollients for treating atopic dermatitis. However, emollients plus corticosteroids have been shown to be more effective than topical corticosteroids alone.

Exposure to irritants should be avoided. Workers with a history of atopic dermatitis are 13 times more likely to develop occupational irritant contact dermatitis. Exposure of atopic workers to irritants should be prevented or severely limited, but workers with atopic dermatitis should be excluded from a job only if avoidance of irritants is impossible.

Use of clothing with cotton or synthetic smooth fibers is important for some patients. Coarse fibers and wool should be avoided. The role of food allergies is uncertain, but they appear to play more of a role in infancy and early childhood than in late childhood and adulthood. Food allergens, if identified by a strong clinical history or blind food challenges, should be avoided.

A practice guideline for the management of atopic dermatitis has been published. Patients with mild to moderate atopic dermatitis can usually be managed with mid- to high-potency topical corticosteroids. Stronger topical corticosteroids produce more rapid response, but should only be used for short periods of time.

Corticosteroids in an ointment base are absorbed better than those in a cream base. Ointments should be used for chronic atopic dermatitis, especially in exposed areas of the skin. Creams are preferable in skin folds and in acute atopic dermatitis, especially if there is transudation within affected areas.

Complications of topical corticosteroid use are the development of striae and localized atrophy. Chronic application of topical corticosteroids to the face can cause rosacea. Topical corticosteroids should be used for as short a time as possible to avoid these adverse effects.

Topical tacrolimus ointment is an alternative treatment for atopic dermatitis in patients over the age of 2 years. It is also an alternative for patients with chronic disease because it can be used long term without the fear of atrophy. Approximately 20% of patients experience skin burning with the initiation of topical tacrolimus treatment.

Patients with severe atopic dermatitis should be referred to a dermatologist. Moderate to severe cases can be managed with short courses of systemic corticosteroids, phototherapy, photochemotherapy, cyclosporin, and/or chemotherapy.

Further Reading

Adams RM. *Occupational Skin Disease* (3rd ed.). Philadelphia: W.B. Saunders, 1999.
Beltrani VS, Boguneiwicz M. Atopic dermatitis. *Dermatology Online Journal* 2003; 9:1-28.
Best Treatments. Clinical Evidence for Patients and Doctors. https://www.besttreatments.org/Unified/CDA/CN/tmpl/CNSection/index/both/0,1025,cid%253D-1%2526aid%253D2,00.html.
Hanifin JM, Cooper KD, Ho VC, et al. Guidelines of care for atopic dermatitis. *Journal of the American Academy of Dermatology* 2004; 50: 391-404.
Hoare C, Li Wan Po A, Williams H. Systematic review of treatments for atopic eczema. *Health Technology Assessment* 2000; 4: 1-191.
Larsen FS, Hanifin JM. Epidemiology of atopic dermatitis. *Immunology and Allergy Clinics of North America* 2002; 22: 1-25.
Leung DY, Hanifin JM, Charlesworth EN, et al. Disease management of atopic dermatitis: A practice parameter. Joint Task Force on Practice Parameters, representing the American Academy of Allergy, Asthma and Immunology; the American College of Allergy, Asthma and Immunology; and the Joint Council of Allergy, Asthma and Immunology. Work Group on Atopic Dermatitis. *Annals of Allergy, Asthma, and Immunology* 1997; 79: 197-211.

Beryllium Disease

ICD-10 J63.2

Kathleen Kreiss

Inhalation of beryllium-containing dusts can cause a chronic interstitial lung disease (chronic beryllium disease, or berylliosis), characterized by non-caseating granulomas associated with cell-mediated immunity to beryllium. In addition to chronic beryllium disease, dermatitis can occur without sensitization in the beryllium extraction industry, and skin granulomas can occur when insoluble beryllium salts enter the skin.

As a result of workplace medical surveillance with a blood test for beryllium sensitization, chronic beryllium disease is frequently diagnosed before symptoms appear. Beryllium sensitization is a cell-mediated immune reaction to beryllium identified through the beryllium-specific lymphocyte proliferation test, an in vitro assay on blood or bronchoalveolar lavage mononuclear cells that is currently available at six laboratories in the United States. Workers identified as sensitized undergo bronchoscopy with lavage and transbronchial biopsy to establish whether they have chronic beryllium disease. Chronic beryllium disease is diagnosed in sensitized workers when granulomas or mononuclear cell infiltrates are identified in biopsy samples.

The clinical disease usually begins with slow onset of dyspnea on exertion, which may be accompanied by dry cough, weight loss, chest pain, and fatigue. Basilar rales may be present on physical exam. In more advanced disease, finger clubbing and signs of cor pulmonale can occur. Pulmonary function tests may show a decrease in diffusing capacity, an increase in alveolar-arterial oxygen gradient, airflow limitation, restrictive lung volumes, or a decrease in compliance. The chest x-ray may show irregular or rounded interstitial opacities and, in some cases, hilar adenopathy. These abnormalities are not diagnostic for chronic beryllium disease, which is differentiated from similar diseases, such as sarcoidosis, by the demonstration of a beryllium-specific cell-mediated immune response.

Historically, a second lung disease, acute beryllium disease, was associated with beryllium exposure in parts of the industry dealing with beryllium extraction and soluble beryllium salts. Acute beryllium disease was thought to be a form of chemical pneumonitis caused by inhalation of high concentrations of beryllium (>100 µg/m^3). The disease had rapid onset and was characterized by dyspnea, cough and sputum, chest pain, tachycardia, crackles, and cyanosis. Chest x-rays showed diffuse or localized infiltrates, which evolved over a few weeks. Pulmonary function tests showed decreased lung volumes and hypoxemia at rest. The hallmark of acute beryllium disease was its reversibility over a period of months, in contrast to chronic beryllium disease, although about one-third of cases were reported to eventually progress to chronic beryllium disease. It is possible that acute beryllium disease was a manifestation of reversible granulomatous disease which resolved when the soluble salts were excreted. Acute beryllium disease has not been diagnosed in the United States for decades. Present-day exposures in excess of 100 µg/m^3 have not been recognized to cause acute beryllium disease, and the nature of acute beryllium disease remains unclear. A number of other entities, such as tracheobronchitis, were described historically, but these are not recognized today in the beryllium industry.

Occurrence

With the advent of worker surveillance programs in the primary beryllium industry in the late 1980s, understanding of the epidemiology of beryllium sensitization and chronic beryllium disease has changed dramatically. A substantial subset, as many as 20%, of new employees in the primary industry became sensitized within months of employment. This early incidence of sensitization did not appear to decrease after stringent control of air concentrations of beryllium. Skin protection to prevent another potential route of beryllium sensitization is being evaluated.

Cross-sectional studies of beryllium sensitization and disease in the primary beryllium industry have documented sensitization rates of about 10%, whether exposures were to beryllium metal, copper beryllium alloy, or beryllium oxide. Rates of chronic beryllium disease among the sensitized have ranged from about 30% to 100% in various facilities. Cumulative incidence rates of sensitization in defined cohorts, however, are much higher than cross-sectional rates. For example, a 10-year follow-up of a beryllium ceramics cohort documented that 18% had demonstrated sensitization and 13% had developed chronic beryllium disease. The rate of beryllium disease among sensitized workers is related to the duration of beryllium exposure; sensitized workers with longer tenure are much more likely to also have chronic beryllium disease compared to sensitized workers with shorter tenure.

Although estimates of cumulative beryllium exposure correlate poorly with beryllium sensitization and disease risk in most plants and across the primary industry, almost every plant studied has high-risk processes. These process-related risks can only be explained by unique environmental characteristics of the beryllium exposures. Gravimetric (mass-based) measures of beryllium exposure measured for compliance purposes have not been predictive of risk. Other measures, perhaps more pertinent to bioavailability and deposition, are being evaluated as alternative metrics of hazard. However, in the absence of a predictive metric of exposure, medical surveillance using the blood lymphocyte proliferation test can indicate where protective measures have been inadequate and which subgroups of workers in a particular plant are at higher risk. Processes associated with higher risk of sensitization have included machining of beryllium ceramics and metal, laboratory work, metal production, annealing, and drawing of copper beryllium alloy wire. Medical surveillance results can allow priority setting for interventions to prevent further sensitization and disease.

In addition to worker subgroups with greater process-related risks, chronic beryllium disease can occur among those who have had only unrecognized or brief exposure to beryllium. Examples include secretaries and other office workers in beryllium plants, security guards, building-trades workers, and end-product inspectors. In the 1940s, cases of beryllium disease were recognized in residents living near beryllium refineries and fluorescent light factories. Family members of beryllium workers have developed beryllium disease, perhaps by having contact with contaminated clothing. Individual susceptibility to immune sensitization likely plays a role in these unexpected cases.

Sensitized workers without current beryllium disease are at high risk for eventual development of chronic beryllium disease, but whether all sensitized workers will develop disease is unclear. The early recognition of sensitization in new employees with subsequent exposure reduction may result in smaller lung burdens of beryllium, but only longitudinal follow-up will establish the merits of early recognition. Similarly, sensitized workers with subclinical chronic beryllium disease are at risk for developing clinical disease, but longitudinal studies to determine the natural history of the disease have just begun. The latency for chronic beryllium disease ranges from months from first exposure to decades after last exposure. Beryllium workers carry a risk of beryllium disease for the remainder of their lives, and many clinical cases have been identified long after workers left beryllium industry employment.

Workers at risk for developing beryllium disease include those engaged in all operations that produce or use beryllium and its compounds, excluding beryl ore mining. Beryllium production workers have been thought to have the highest prevalence of disease. However, prevalence of disease in workers downstream of primary production has not been well studied. NIOSH estimates that as many as 134 000 workers in the United States are currently exposed to beryllium in diverse industries (Table 1). Exposure occurs in operations that involve melting, casting, grinding, machining, and drilling of beryllium-containing products, and it is in these industries that most U.S. workers exposed to beryllium are employed. Because beryllium is a neutron moderator, nuclear workers are frequently exposed to beryllium. Exposures to beryllium also occur in the aerospace, scrap metal reclaiming, specialty ceramics, dental technology, and electronics industries.

Causes

One theory currently under investigation is that beryllium sensitization may result from skin exposure to submicron particles of beryllium compounds, since control of air concentrations does not appear to prevent sensitization. Once sensitized, a worker is at risk for chronic beryllium disease if exposed to respirable beryllium dust or its compounds. No cases of chronic beryllium disease have been recognized in persons exposed only to beryllium ores, but the number of such workers is small.

Pathophysiology

Chronic beryllium disease is unusual among toxic metal diseases in that its pathophysiological mechanism is a cell-mediated immune response. People who are exposed to beryllium and who do not develop a delayed-type hypersensitivity response to it do not develop this interstitial lung disease. Workers with a glutamic acid in the 69th position of the *HLA-DPβ1* molecule have increased risk of beryllium sensitization and disease. However, approximately one-fourth of patients with chronic beryllium disease lack this genetic characteristic. Since the frequency of this characteristic is between 30% and 40% in the population, genetic testing has poor predictive value. Research

Table 1. Beryllium Industries and Products

Industry	Products
Aerospace	Altimeters, braking systems, bushings and bearings for landing gear, electronic and electrical connectors, engines, gyroscopes, mirrors (such as the Hubble Telescope), precision tools, rockets, satellites, and structural components.
Automotive	Air-bag triggers, anti-lock brake system terminals, electronic and electrical connectors, steering wheel connecting springs, and valve seats for drag racing engines.
Biomedical	Dental crowns, bridges, partials, and other prostheses, medical laser and scanning electron microscope components, and x-ray tube windows.
Defense	Heat shields, mast-mounted sights, missile guidance systems, nuclear reactor components and nuclear triggers, submarine hatch springs, and tank mirrors.
Energy and electrical	Heat exchanger tubes, microelectronics, microwave devices, nuclear reactor components, oil field drilling and exploring devices, and relays and switches.
Fire prevention	Non-sparking tools and sprinkler system springs.
Instruments, equipment, and objects	Bellows, camera shutters, clock and watch gears and springs, commercial speaker domes, computer disk drives, musical instrument valve springs, pen clips, and commercial phonograph styluses.
Manufacturing	Injection molds for plastics.
Sporting goods and jewelry items	Golf clubs, fishing rods, naturally-occurring beryl and chrysoberyl gemstones such as aquamarine, emerald, and alexandrite, and synthetic gemstones, such as emeralds with distinctive colors.
Scrap recovery and recycling	Various beryllium-containing products.
Telecommunications	Cellular telephone components, electromagnetic shields, electronic and electrical connectors, personal computer components, rotary telephone springs and connectors, and undersea repeater housings.

continues on specific alleles which may confer higher risk, other genes pertinent to antigen recognition and granulomatous response, and the interaction between genes and beryllium exposure.

Prevention

The current OSHA PEL of 2 µg/m^3 is not protective. The DOE has established an action level of 0.2 µg/m^3. Control of beryllium air concentrations has not prevented sensitization among newly employed workers, who have not had higher historical exposures. Scrupulous housekeeping may be necessary to avoid skin contamination from surfaces, and a comprehensive program of keeping beryllium off the skin, out of the lungs, at its source, and in the workplace is advisable as the effectiveness of primary preventive measures is being evaluated. Whenever possible, less hazardous substances should be substituted for beryllium.

Medical surveillance with the beryllium lymphocyte proliferation test can provide employers with data regarding which work process subgroups are at higher risk and where preventive interventions should be directed. At present, environmental measures of beryllium exposure are difficult to interpret since mass-based exposure correlates poorly with risk of sensitization or disease. Only health screening for beryllium sensitization can demonstrate whether preventive measures are needed. However, the blood beryllium lymphocyte proliferation test has limitations as a screening test. Intra- and interlaboratory reproducibility is poor, and sensitization may not be accompanied by concurrent chronic beryllium disease. No controlled clinical trials exist to show whether intervention among the sensitized, such as removal from beryllium exposure, changes the natural history of sensitization or of subclinical chronic beryllium disease. Nevertheless, sensitization is an adverse outcome of beryllium exposure since it confers an increased risk of developing beryllium disease. In the absence of data to the contrary, medical prudence suggests that sensitized individuals not be further exposed to beryllium.

At present, no role for workplace genetic screening is justifiable in a prevention program, since the *HLA-DPβ1*^{Glu69} marker is not sufficiently predictive of disease, and absence of the marker does not confer absolute protection from disease. As genetic research proceeds, more specific markers may be identified that might hold promise for predicting which workers are likely to develop beryllium disease with beryllium exposure. Employment and insurance discrimination may arise from genetic testing conducted by employers or without stringent safeguards for test results. Prospective beryllium workers might benefit from being able to obtain confidential genetic testing and counseling before accepting employment in the industry.

Other Issues

Beryllium is an animal carcinogen and is associated with elevated risk of lung cancer in humans.

Further Reading

Henneberger PK, Cumro D, Deubner D, et al. Beryllium sensitization and disease among long-term and short-term workers in a beryllium ceramics plant. *International Archives of Occupational and Environmental Health* 2001; 74: 167-176.

Kreiss K, Mroz MM, Zhen B, et al. Risk of beryllium disease related to work processes at a metal, alloy, and oxide production plant. *Occupational and Environmental Medicine* 1997; 54: 605-612.

Wang Z, Farris GM, Newman LS, et al. Beryllium sensitivity is linked to HLA-DP genotype. *Toxicology* 2001; 165: 27-38.

Weston A, Ensey J, Kreiss K, et al. Racial differences in prevalence of a supratypic HLA-genetic marker immaterial to pre-employment testing for chronic beryllium disease. *American Journal of Industrial Medicine* 2002; 42: 457-465.

Bladder Cancer

ICD-10 C67

Elizabeth Ward

Bladder neoplasms appear clinically with microscopic or gross hematuria, with or without pain. Other symptoms may include a change in the time and frequency of urination, and pain on urination. Diagnosis is confirmed with cystoscopy and biopsy. Because of a long preclinical period, early detection by urine cytology and identification of red blood cells in urine (hematuria) may be considered in high-risk groups.

Occurrence

An estimated 57 400 new cases of bladder cancer and 12,500 bladder cancer deaths occurred in the United States in 2003. The male-to-female incidence ratio was 4:1, and the white-to-black incidence ratio was 2:1. Over 90% of bladder cancers in the United States are transitional cell carcinomas; in areas of the world where *Schistosoma hematobium* is endemic, squamous cell carcinomas predominate. In the United States, the population attributable risk for bladder cancer related to occupational exposure has been estimated to be 21% to 25% for white males, 27% for nonwhite males, and 11% for white females.

Causes

The major nonoccupational risk factor is cigarette smoking, which accounts for approximately 47% of bladder cancers in men and 37% in women. The high risk of bladder cancer associated with exposure to certain aromatic amines (including β-naphthylamine, benzidine, and 4-aminobiphenyl) is well documented. Although these compounds are no longer produced or used industrially, other aromatic amines with known or suspected association with bladder cancer in humans include benzidine-based dyes, *o*-toluidine, 4-chloro-*o*-toluidine (4-COT), and 4,4'-methylenebis (2-chloroaniline). Known or suspected bladder carcinogens present in cigarette smoke include 4-aminobiphenyl and *o*-toluidine.

Occupations with substantial evidence for excess bladder cancer include painters, workers in the rubber industry, Soderberg process workers in the aluminum industry, foundry workers, machinists, drill press operators, dry cleaners, hairdressers, truck drivers, motor vehicle operators, mechanics and repairers, dyestuff production and textile workers, and leather industry workers. Associations have also been reported for pesticide manufacture and application, chemical processing, plumbing, heating and air conditioning, tobacco workers, field crop and vegetable farm workers, tailors and dressmakers, carpenters, construction workers, miners, gas workers, coke-plant workers, petroleum workers, railroad workers, engineers, butchers, clerical workers, cooks and kitchen workers, food processing workers, electricians, gas station attendants, medical workers, pharmacists, glass processors, nurserymen, photographic workers, security guards and watchmen, welders, sailors, firefighters, stationary firemen or furnace operators, stationary engineers, paper and pulp workers, roofers, gardeners, bootblacks, and asbestos workers.

Pathophysiology

Transitional cell bladder carcinomas consist of two pathologically distinguishable, but overlapping, types: papillary and nonpapillary. Individual low-grade papillary tumors have a relatively good prognosis, with only a small percentage of cases developing invasive disease. Increased risk of invasive disease is associated with multiple and higher-grade papillary tumors. Most invasive carcinomas of the bladder are of the nonpapillary type and develop in a background of diffuse epithelial abnormalities, including carcinoma in situ and markedly atypical urothelium. Both papillary and nonpapillary bladder cancers have a long preclinical phase, during which papillary tumors may be detected by urinalysis (presence of red blood cells in the urine) and cystoscopy, and carcinoma in situ may be detected by urine cytology.

Prevention

Primary prevention entails eliminating or reducing exposure to known or suspected bladder carcinogens. In occupational settings, this is achieved optimally by substitution and engineering controls. Many aromatic amines can be absorbed through the skin, so it is important to protect workers from dermal exposure. Measuring urinary aromatic amine levels is useful to monitor exposure by all routes. Secondary prevention includes worker education about early symptoms and medical screening of workers with increased risk of bladder cancer. The U.S. Preventive Services Task Force concluded that there was insufficient evidence that screening for hematuria and cytologic abnormalities improved the prognosis for those found to have cancer, even in high-risk groups. However, data on the efficacy of screening in occupational high-risk groups are limited and do not demonstrate absence of benefit. A consensus panel recommended in 1990 that workers with high exposure to known bladder or suspect carcinogens should be offered cytology and urinalysis (to look for microscopic hematuria) every 6 months. In addition to offering potential benefit to the individuals, occupational screening programs may confirm a risk in populations exposed to suspect carcinogens, thus triggering interventions to reduce exposure. Treatment depends on the pathologic type, grade, and stage of cancer. Treatment modalities include surgery, intravesical chemotherapy, immunotherapy, intravenous chemotherapy, and radiation therapy. The 5-year survival rate for bladder cancer cases in the United States for the 1992-1998 time period was 81.5%; among the 74% of cases diagnosed at a localized stage the survival rate was 94%. Due to the high survival rate for bladder cancer, epidemiologic mortality studies of cohorts exposed to bladder carcinogens may yield false-negative results. Measuring cancer incidence through linkage with population registries or other sources is therefore recommended.

Other Issues

The site specificity of aromatic amines for the bladder is related to their metabolism. In the liver, aromatic amines may be N-acetylated by NAT2, rendering them less active, or may be N-hydroxylated by CYP1A2. The hydroylated form can be subsequently metabolized by multiple pathways, including O-acetylation by NAT1 in the bladder epithelium, to a highly reactive species that can form DNA adducts. Genetic polymorphisms in these metabolic enzymes may play a role in susceptibility to aromatic amine-induced bladder cancer.

Further Reading

American Cancer Society. All About Bladder Cancer: Detailed Guide. Learn about cancer [serial online]. ACS, Atlanta [updated 01/04/2001]. Available from URL: http://www.cancer.org/docroot/CRI/CRI_2_3x.asp?dt=44

National Institutes of Health. What you need to know about bladder cancer. What you need to know about cancer publications [serial online]. National Cancer Institute, National Institutes of Health, Bethesda, MD. [updated 9/16/2002]. Available from URL: http://www.nci.nih.gov/cancerinfo/wyntk/bladder

Ruder AM, Carreon T, Ward EM, et al. Bladder cancer. In Rosenstock, L, Cullen M, Brodkin C, Redlich C (Eds.). *Textbook of Occupational and Environmental Medicine* (2nd ed.). London: Elsevier, 2004.

Schulte P, Halperin W, Ward EM. Final discussion: Where do we go from here? *Journal of Occupational Medicine* 1990; 32: 936-945.

Silverman DT, Morrison AS, Devesa SS. Bladder cancer. In Schottenfeld D, Fraumeni JF Jr. (Eds.). *Cancer Epidemiology and Prevention* (2nd ed.). New York: Oxford University Press, 1996, pp. 1156-1179.

Silverman DT, Levin LI, Hoover RN, et al. Occupational risks of bladder cancer in the United States: I. White men. *Journal of the National Cancer Institute* 1989; 81: 1472-1480.

Silverman DT, Levin LI, Hoover RN. Occupational risks of bladder cancer in the United States: II Nonwhite men. *Journal of the National Cancer Institute* 1989; 81: 1480-1483.

Silverman DT, Levin LI, Hoover RN. Occupational risks of bladder cancer among white women in the United States. *American Journal of Epidemiology* 1990; 132: 453-461.

Taylor JA, Umbach DM, Stephens E, et al. The role of N-acetylation polymorphisms in smoking-associated bladder cancer: evidence of a gene-gene-exposure three-way interaction. *Cancer Research* 1998; 58: 3603-3610.

Brain Cancer

ICD-10 C71

Barry S. Levy

Brain cancers are identified clinically on the basis of symptoms, signs, and specific diagnostic tests. They are also identified based on cell pathology. Symptoms vary considerably, depending on the type and location of tumor, but often include headache, confusion, fatigue, weight loss, and a variety of neurological abnormalities. Some symptoms are due to a direct effect of the tumor, such as destruction or irritation of surrounding nerve tissue in an acoustic neuroma; some are due to indirect effects, such as increased intracranial and intraspinal pressure in cerebral or spinal glioblastomas. Suggestive signs are primarily focal neurological deficits, as in limb weakness or numbness or language difficulties. Definitive diagnosis is based on some combination of computed tomography (CT) scanning, magnetic resonance imaging (MRI), arteriography of the cerebral blood vessels, open biopsy, and other diagnostic tests. The brain is a frequent site of the metastatic spread of cancers from other sites, most notably lung cancer in men and breast cancer in women; therefore, primary brain tumors need to be differentiated from the secondary spread of tumors from elsewhere in the body.

Occurrence

The annual incidence of brain cancer in the United States is approximately 5 per 100 000 males and approximately 4 per 100 000 females. The overall 5-year survival rate is 32%. The two main histological groups of brain cancer are gliomas and meningiomas. The occurrence of occupationally-related brain cancer in the general population is not known.

Causes

For most brain cancers, a cause is never established. Occupational exposure to ionizing radiation is a risk factor for brain cancer. Studies have also found increased occurrence of brain cancer among electrical, electronics, and petrochemical workers, farmers, synthetic rubber manufacturing and fabrication workers, polyvinyl chloride (PVC) production workers, and health professionals; and among workers with exposure to solvents, chlorinated aliphatic hydrocarbons, vinyl chloride, formaldehyde, cutting oils, polycyclic aromatic hydrocarbons (PAHs), lead and other heavy metals, and pesticides. A study of workers in a nuclear-fuels fabrication plant found excess brain cancer, despite the fact that cancer and death rates overall were less than expected. Excesses of glioblastoma multiforme have been demonstrated among workers exposed to vinyl chloride, solvents, and other chemicals. Some studies have shown relationships between brain cancer and head trauma, and brain cancer and x-ray radiation to the head. In these studies, authors have concluded that brain gliomas appear to be more likely to be occupationally related than other types of brain cancer.

Several studies suggest an association between childhood brain tumors and parental exposure to solvents, metals, and ionizing and nonionizing radiation. In utero exposure to N-nitroso compounds, several of which are potent neurocarcinogens, may increase the risk of childhood tumors of the central nervous system. In addition, genetic factors seem to play an important role in the etiology of some rare brain cancers, such as bilateral retinoblastoma.

Brain cancer has been induced in experiments on animals by bischloromethyl ether (BCME), vinyl chloride, and acrylonitrile.

Pathophysiology

Because of the various types of brain cancer, it is beyond the scope of this book to describe the pathophysiology of this condition.

Prevention

Because brain tumors are often incurable, emphasis is on primary prevention, prohibiting or, if this is not feasible, limiting exposure to factors known to cause, or suspected of causing, brain tumors.

Further Reading

Cocco P, Dosemeci M, Heineman EF. Brain cancer and occupational exposure to lead. *Journal of Occupational and Environmental Medicine* 1998; 40: 937-942.

Cocco P, Dosemeci M, Heineman EF. Occupational risk factors for cancer of the central nervous system: A case-control study on death certificates from 24 U.S. states. *American Journal of Industrial Medicine* 1998; 33: 247-255.

Cocco P, Heineman EF, Dosemeci M. Occupational risk factors for cancer of the central nervous systems (CNS) among U.S. women. *American Journal of Industrial Medicine* 1999; 36: 70-74.

Kheifets LI, Afifi AA, Buffler PA, et al. Occupational electric and magnetic field exposure and brain cancer: a meta-analysis. *Journal of Occupational and Environmental Medicine* 1995; 37: 1327-1341.

Navas-Acien A, Pollan M, Gustavsson P, et al. Occupation, exposure to chemicals and risk of gliomas and meningiomas in Sweden. *American Journal of Industrial Medicine* 2002; 42: 214-227.

Schlehofer B, Kunze S, Sachsenheimer W, et al. Occupational risk factors for brain tumors: Results from a population-based case-control study in Germany. *Cancer Causes and Control* 1990; 1: 209-215.

Viel J-F, Challier B, Pitard A, et al. Brain cancer mortality among French farmers: the vineyard pesticide hypothesis. *Archives of Environmental Health* 1998; 53: 65-70.

Zheng T, Cantor KP, Zhang Y, et al. Occupational risk factors for brain cancer: A population-based case-control study in Iowa. *Journal of Occupational and Environmental Medicine* 2001; 43: 317-324.

Bronchitis, Chronic

ICD-10 J41, J42, J44.8

David C. Christiani

Chronic bronchitis is defined as the presence of chronic cough with sputum production most days (at least 4 per week) for at least 3 months per year for at least 2 consecutive years. Its hallmark is chronic production of sputum (either clear or colored). In the clinical setting, chronic bronchitis is often accompanied by emphysema and may include shortness of breath and airways obstruction.

Diagnosis of chronic bronchitis depends on the presence of symptoms, often identified by questionnaire. The British Medical Research Council and American Thoracic Society questionnaires on respiratory symptoms are internationally recognized as a valid and reliable epidemiologic methods for eliciting symptoms of chronic bronchitis.

Chronic bronchitis can occur with or without dyspnea (shortness of breath) on exertion. Although the presence of excess mucus and cough is a response to airways insult, it does not, by itself, define the seriousness of the disorder or its prognosis. Dyspnea implies more serious chronic airways disease, and affected people develop respiratory insufficiency more rapidly than people with chronic bronchitis alone. Chronic bronchitis in the absence of airflow obstruction or emphysema is not associated with premature mortality.

The chest sounds of chronic bronchitis include rhonchi or coarse rales on inspiration and, in some people, wheezing on expiration. Expiratory time is increased.

Pulmonary function tests are usually normal when dyspnea is absent. In chronic bronchitis with airflow obstruction, mid-maximum expiratory flow rates (MMEF or FEF_{25-75}) are decreased early in the course of disease. Later, forced expiratory function in 1 second (FEV_1) is also decreased, and total lung capacity (TLC) is generally normal or increased. Single-breath diffusion capacity is generally normal, but if emphysema is also present, the diffusion capacity will also be decreased.

Occurrence

By the late 1990s, chronic bronchitis affected about 9 million people in the United States. This is probably an underestimate, because a common practice is to include only by implication a diagnosis of chronic bronchitis in chronic obstructive pulmonary disease (COPD), which typically includes both emphysema and bronchitis. Chronic bronchitis is more common among smokers, men, people over 40, and urban residents. It is usually preceded by 10 to 20 years of exposure to causative agents, with this time period declining with age. In some occupationally-exposed populations, such as miners and textile workers prevalence is 10% to 40%. Prevalence has increased steadily in the past 25 years, and currently chronic lung disease ranks third in number of deaths in the United States. National surveys revealed self-reports of chronic bronchitis among 14.2 million persons (53.2 per 1000) in 1996, an 84% increase over 1982.

There were over 112,000 COPD deaths in the United States in 1998. Because chronic airways obstruction often includes chronic bronchitis and emphysema, chronic bronchitis was likely a contributing factor to these deaths. Among those with chronic bronchitis, premature mortality is 2.4 times greater in men than in women, although in those aged 45 to 54, mortality for women approaches that for men. The risk of dying from chronic bronchitis is 10 times greater among smokers than among nonsmokers, and is even greater for male cigarette smokers.

Causes

Chronic bronchitis is multifactorial in origin. Enviromental causation is prominent, with summation of all harmful environmental factors. Family studies indicate the presence of genetic predisposition, with gene–environment interaction. Acute and chronic viral infections of the airways may play an important role in the origin and persistence of chronic bronchitis. Nonetheless, most affected persons have clearly identifiable inhalation exposures as major causative factors. In the population as a whole, cigarette smoking is by far the most common cause. Among certain groups of workers, however, occupational exposures are also important causes. Occupational causes have been identified primarily on the basis of clinical and epidemiologic data. Many gases, fumes, and aerosols are acutely irritating to the respiratory tract, causing inflammation of the airways lining and other parts of the lung. (See Respiratory Tract Irritation.) Repetitive or continuous exposure to such irritants can result in chronic disease. Occupational exposures leading to chronic bronchitis can be grouped into the following broad categories: chronic bronchitis caused by (a) specific chemicals, such as ammonia, arsenic, chlorine, cadmium oxide, chromium, oxides of nitrogen, osmium tetroxide, phosgene, tungsten carbide, vanadium pentoxide, sulfur dioxide, isocyanates, and aerosols, such as smoke; (b) those dusts and aerosols found in brickworks, cement plants, coke ovens, foundries, coal and other mines, rubber plants, quarries, rock-crushing operations, smelters, and welding operations; and (c) dusts of cotton, flax, soft hemp, mixed potash, and phosphates.

Pathophysiology

The pathophysiology of chronic bronchitis due to occupational exposures is the same as that due to inhalation of tobacco smoke. It is an inflammation of the large airways in response to repeated physical and chemical irritation. It is characterized by enlargement of goblet cells and mucus glands and by mucus obstruction of the small airways.

The pathogenesis of chronic bronchitis is not well established. In addition to increases in sputum production and mucus obstruction of small airways, there may also be airway wall scarring produced by a variety of agents.

Prevention

Chronic bronchitis can be controlled by reducing exposure to respiratory irritants using standard means of industrial hygiene, such as positive engineering to eliminate hazards, ventilation to remove hazards from the workplace, changes in work processes to reduce exposures, and use of respirators in some circumstances.

Medical surveillance by periodic questionnaires is recommended for workers at risk of developing occupational chronic bronchitis. If exposed individuals develop symptoms, similarly exposed workers should be examined and exposure should be monitored. Pulmonary function testing may also be helpful.

Other Issues

Efforts are needed to reduce pollution of the personal environment by tobacco smoke. Cigarette smoke usually is additive to other risks for chronic bronchitis, but it may also be synergistic, as it is with cotton dust exposure.

Further Reading

Balmes J, Becklake M, Blanc P, et al. American Thoracic Society Statement: Occupational Contribution to Airway Disease. *American Journal of Respiratory and Critical Care Medicine* 2003; 167: 787-797.

Hnizdo E, Vallyathan V. Chronic obstructive pulmonary disease due to occupational exposures. *Occupational and Environmental Medicine* 2003; 60: 237-243.

Building-Related Illness

ICD-10 T75.8

Kathleen Kreiss

Workers with building-related symptoms generally have either nonspecific symptoms of unclear etiology or specific clinical diagnoses with objective findings. Sick building syndrome consists of nonspecific symptoms of headache, mucous membrane irritation, and difficulty concentrating that occur in close temporal association with being in an implicated building. In contrast, building-related illnesses, such as asthma and hypersensitivity pneumonitis, have symptoms that may be exacerbated by building occupancy, but generally persist away from the building. (See Table 1.)

In addition, building-related illnesses exist where symptoms have no temporal relation to building occupancy, but are due to building-related exposures. Examples are infectious diseases, such as Legionnaires' disease and tuberculosis, when epidemiologically linked with a building. There is some evidence that transmission of common communicable respiratory infections may be increased in relation to relatively low ventilation rates. The spector of bioterrorism from bacterial diseases, such as anthrax, and from viral diseases, such as smallpox, has increased interest in the role of ventilation and resuspension of particles from indoor surfaces in transmission of infections.

Rarely, specific toxic exposures lead to serious illness that is associated with a building environment. Examples include carbon monoxide poisoning from unvented combustion products and lung cancer from radon gas infiltrating from soil. However, public health authorities usually solve acute toxic problems rapidly, and unsuspected exposures rarely present to clinicians as building-related complaints, so these will not be considered further here.

Occurrence

The EPA has supported cross-sectional studies of occupants from 115 buildings without a history of publicized indoor air quality complaints, and found that more than 40% of workers experienced frequent work-related symptoms consistent with sick building syndrome. On the basis on this estimate, 35 to 60 million indoor workers in the United States are suspected to have at least one weekly building-related symptom. Prevalence rates in publicized "sick buildings" are sometimes higher. The proportion of the building stock in which occupants commonly have sick building syndrome (however defined) is unknown. No national surveillance system exists for nonspecific building-related symptoms. However, public health authorities at local, state, and federal levels commonly receive building-related complaints from office

Table 1. Building-Related Conditions

Condition	Etiologic Agent(s)	Basis of Diagnosis
Sick building syndrome	Unknown	Clinical history in setting of similar complaints by co-workers
Asthma	Bioaerosols associated with water-damaged building materials; latex antigen in health care settings	Clinical history, work-related airflow limitation or airways hyperresponsiveness
Hypersensitivity pneumonitis	Fungi contaminating HVAC systems and humidifiers; bioaerosols associated with water-damaged building materials	Clinical history, clinical responses to removal from exposure, epidemiologic pattern surrounding sentinel cases, objective tests for restrictive and granulomatous lung disease
Respiratory infection	Legionella species, *Mycobacterium tuberculosis*, respiratory viruses, anthrax	Culture, appropriate stains, serology
Carbon monoxide intoxication	Carbon monoxide from combustion sources, vehicle exhaust	Elevated carboxyhemoglobin level

workers and school personnel. Indoor air quality complaints constituted 37% of the requests for health hazard evaluations received by NIOSH in the 1999-2002 period.

Although no national data are available with which to estimate incidence or prevalence of building-related asthma, increasing proportions of occupational asthma are attributed to the nonindustrial built environment or indoor air quality. Building-related asthma accounted for 28% of all work-related asthma identified in the 1993-1999 period by four state health departments engaged in work-related asthma surveillance. AOEC reports an increase from about 6% of occupational asthma cases being building-related in 1991-1992 to 30% in 1995-1996. Finally, NIOSH has observed an increasing proportion of asthma health hazard evaluation (HHE) requests from the public in which the exposure listed in the request is an indoor air quality concern. None were reported in the 1970-1978 period, but the proportion increased relentlessly by subsequent 4-year intervals to 75% in 1999-2002. Among requests for HHEs for indoor air quality, however, only a minority mention asthma or chest symptoms. In the second half of the 1990s, only about 20% of indoor air quality requests mentioned asthma, which is consistent with the observation that respiratory symptoms are not typical of sick building syndrome alone. Collectively, these physician-diagnosis and public-request databases suggest that asthma in relation to the indoor environment is increasingly recognized. (See Asthma.)

In comparison to asthma, hypersensitivity pneumonitis is uncommon in the general population. Many occupational causes are known in diverse industries with bioaerosol exposure. A case of hypersensitivity pneumonitis linked to a nonindustrial work environment represents a sentinel event for risk among co-workers. Most well-documented outbreaks of hypersensitivity pneumonitis have been traced to microbially contaminated ventilation system components. However, current cases often occur in the setting of water-damaged buildings from roof leaks, other building envelope water intrusions, or below-grade moisture. As with building-related asthma, no surveillance data exist with which to estimate its prevalence or incidence. (See Hypersensitivity Pneumonitis.)

National incidence data are available for some infections that may be building-related. In 2002, a total of 1 163 cases of Legionella pneumonia in the United States were reported to the CDC, which estimated an incidence of 8 000 to 18 000 cases per year, most being sporadic. National surveillance data for tuberculosis exist, but nonoccupational transmission among family members and close contacts accounts for a large portion of tuberculosis. The proportion of all tuberculosis that could be considered to be occupational, for example, to health care workers and prison staff, is unknown. (See Tuberculosis.)

Recent work on sick leave among office workers, largely due to common communicable respiratory infections (colds and influenza), suggests that office ventilation with outdoor air has a substantial effect. Several studies have shown that some characteristic of a building or indoor environment has been associated with a 50% or greater difference in a metric of illness, such as rates of febrile respiratory disease requiring infirmary care. Since most workers suffer from a cold or influenza every year (0.58 cases per person per year in the working age population), even small effects of building characteristics on infection rates provide opportunities for prevention of sickness and for increasing productivity. Of the approximately 52 million cases of influenza or the common cold occurring annually among indoor workers in the United States, it has been estimated that 10% to 14% could probably be prevented with changes in indoor environments.

Causes

The cause of sick building syndrome remains unknown. Historically, many complaints were attributed to mass psychogenic illness. However, hysteria does not explain the tight temporal association with building occupancy, nature of the symptoms, lack of visual chain of transmission, and endemic nature of building-related complaints. European investigators showed two decades ago that the prevalence of building-related complaints was associated with building characteristics, such as air conditioning. Sick building syndrome came to public health attention in the late 1970s, following an energy crisis and change in consensus building ventilation standards that limited recommended outdoor air intake into mechanically ventilated buildings.

In the 1960s, building construction practices changed significantly. New construction emphasized the use of prefabricated exterior sections mounted on a steel frame, creating a much tighter building envelope. Windows were made inoperable, and mechanical ventilation, often under centralized control, replaced natural ventilation under individual control. Often, changing use of spaces after the original design leads to inadequate ventilation. Inadequate maintenance of building heating, ventilation, and air-conditioning (HVAC) systems may also threaten indoor air quality. At the same time that ventilation has decreased, the number and variety of potentially toxic agents in the office environment have increased. The sources are diverse (Table 2). Outdoor pollutants may enter through improperly placed ventilation intakes. Bacteria, fungi, and other biological agents may grow on moist surfaces in the building and in air conditioners, ducts, filters, and humidifiers. Building materials and furnishings may release formaldehyde and other volatile organic compounds; sources include building insulation, adhesives, tiles, vinyl wall coverings, rugs, carpets, copying machines, and furniture. Cleaning products and pesticides may also contaminate the air. Ozone from office machines may transform volatile organic chemicals indoors into short-lived irritants, such as aldehydes. Cigarette smoke is a significant source of noxious gases and respirable particles. Human bioeffluents, such as butyric acid and body odor, may also contribute to an unsatisfactory work environment.

Decades of research have pursued hypothetical causes of sick building syndrome, including volatile organic compounds, bioaerosols, and inadequate outdoor air ventilation, which limits dilution of indoor concentrations of suspected agents. Some evidence exists that outdoor air ventilation rates are related to sick building syndrome prevalence, particularly for ventilation rates below 30 cubic feet per person per minute. Since ventilation is difficult to measure, indoor air quality consultants commonly measure carbon dioxide levels in buildings with complaints. Carbon dioxide concentrations reflect ventilation effectiveness in relation to human occupancy (which is not likely the cause), and they are not predictive of sick building syndrome.

Although no measurement can assure occupants that they are not at risk of sick building syndrome, epidemiologic studies of occupants of buildings selected without regard to known indoor air quality complaints have contributed information about causes. The variation of complaint prevalence among buildings suggests remediable causes. Air conditioning is associated with higher building-related complaint prevalences than natural ventilation

Table 2. Common Indoor Pollutants and Their Sources in Commercial Buildings

Source	Pollutants
Tobacco smoke	Carbon monoxide, particles, organics
Gas boilers, furnaces, cookers	Carbon monoxide, nitrogen dioxide, particles, organics
People	Carbon dioxide, organics (bioeffluents, perfumes, fragances), bacteria, viruses
Standing water, water damage	Biological agents
Furnishings, building materials	Formaldehyde and other organics, fibers
Computers, copiers, correction fluids, typesetting equipment, cleaning agents	Organics, particles, ozone
Air from garages and loading docks	Carbon monoxide, particles, organics
Outdoor air	Carbon monoxide, nitrogen dioxide, ozone, sulfur dioxide, particles, organics, pollen, fungi
Soil gas	Radon, biocides, organics

or mechanical ventilation that does not alter air temperature or humidity. Studies of air quality acceptability have shown that the ventilation system itself may be the origin of poor indoor air quality. Other environmental correlates of sick building syndrome include carpeting, high occupancy, and video display terminal use. Personal factors associated with symptoms include female gender, job stress or dissatisfaction, and allergies. The perception of comfort may depend also on the movement of air, temperature, lighting, and humidity. Workers' responses to indoor air quality problems may also be affected by the quality of labor-management relations.

Building-related respiratory diseases of asthma and hypersensitivity pneumonitis are commonly associated with building dampness. Many epidemiologic studies have demonstrated an association of respiratory symptoms and asthma to residential dampness, and this work is now being extended to the nonindustrial work environment. The presumed causes are bioaerosols. The specific microbial cause is seldom evident in specific case patients or outbreaks. Rather, indices of bioaerosols, such as evident moisture, visible mold, moldy odor, and water stains, correlate better with respiratory disease risk than counts of viable colony forming units, for example. Most available bioaerosol measurements are limited in repeatability by short sampling times and wide fluctuations in concentration, making exposure classification suspect. In some outbreaks of hypersensitivity pneumonitis, the illness cannot be linked to a specific biological agent, but the affected people may have precipitating antibodies to extracts of the biological mixture contaminating an implicated humidifier or HVAC system. Research continues on more promising bioaerosol measurement techniques addressing biomass indices, such as endotoxin and fungal wall constituents, rather than culturable microorganisms.

Bacteria and viruses that are disseminated in the indoor environment can cause infections. In the case of bacteria, the source of amplification may be an environmental reservoir, as in Legionella growth in cooling towers or in hot water systems. Alternatively, the source of communicable infections, such as tuberculosis, is infected human building occupants. The marker for risk to some communicable respiratory infections may well be carbon dioxide concentration, which reflects the potential to rebreathe other occupants' exhaled breath. Viral and bacterial agents of infection vary in transmissibility by the airborne route.

Pathophysiology

Sick building syndrome mucous membrane symptoms reflect irritation of short duration. Efforts to establish pathophysiologic correlates of irritation have been meager. Objective evidence has been documented in tear film breakup time, which correlates with eye symptoms that are building-related. Some work has been initiated on nasal resistance measurements. The pathophysiologies of headache and difficulty concentrating in implicated buildings remain unexplored. However, no evidence exists that sick building syndrome leads to a serious or chronic condition that persists after employees leave an implicated building.

Asthma symptoms in relation to building occupancy reflect both exacerbation of pre-existing asthma and new-onset asthma. Exacerbation of asthma might reflect inflammation from irritant exposure or allergen-mediated inflammation, but the phenomenon has not been studied. Similarly, the mechanism of new-onset asthma with work-related patterns of symptoms and medication use has not been studied. Asthma is diagnosed by establishing the existence of bronchial hyperresponsiveness in the presence of asthma symptoms. Bronchial hyperresponsiveness is reflected by reversible airflow limitation, response to bronchodilator, or an abnormal methacholine bronchial provocation test.

Hypersensitivity pneumonitis is a lymphocyte-mediated immune system response to an inhaled antigen to which a patient has become sensitized. In acute hypersensitivity pneumonitis, symptom onset is delayed 4 to 12 hours after onset of exposure, characteristic of a cell-mediated reaction. Macrophages presenting antigen to sensitized lymphocytes participate in formation of giant cells characteristic of granulomatous inflammation in the lung. Lymphocytic alveolitis is characteristic of hypersensitivity pneumonitis, and can be demonstrated in bronchoalveolar lavage cell proportions. Unrecognized hypersensitivity pneumonitis can progress to permanent lung fibrosis and accompanying permanent impairment.

The pathophysiology of building-related infections is as diverse as the number of microbial agents and immune system responses to infection.

Prevention

Primary prevention of building-related complaints and illness requires implementing health protective practices that are known but often not implemented. These include attention to indoor air quality in architectural design and inspection of completed construction; choice of building materials and furnishings; control of pollution generated by occupant activities, cleaning products, and office machines; and design and maintenance of the HVAC system. Redesign of interior spaces needs to consider ventilation requirements for new occupants, the impact of modular office design, and the load of lights, computers, and equipment.

Building-related complaints and illnesses result when primary prevention has failed. The class of complaint should guide the investigation of building-related problems. Multidisciplinary "building diagnostics" teams, including physicians or epidemiologists, industrial hygienists, and ventilation engineers, can be an advantage in resolving the problems prompting health complaints.

Sick building syndrome does not include lower respiratory symptoms. Physicians can establish whether respiratory symptoms exist that may reflect asthma or hypersensitivity pneumonitis. These more serious medical conditions require attempts to identify the symptomatic workers for medical evaluation and consideration of work restriction from the implicated environment. Early recognition and diagnosis can provide a more favorable prognosis and even cure by removing the affected persons from continuing exposure to an environment in which there are antigens to which they react. The corollary environmental investigation should focus on identifying and remediating sources of bioaerosol amplification and dissemination. For at least hypersensitivity pneumonitis cases, remediation sometimes fails to prevent disease recurrence when affected workers return to the implicated environment.

Some infections, such as Legionella pneumonia, prompt examination of water reservoirs which disseminate aerosols. Cooling towers, whose mists sometimes are entrained in fresh air ventilation systems, are a classic source of Legionella organisms. Other infections, such as fungal infections in hospitalized immune-deficient patients, may arise in transplant units which have water-damaged building components that become sources of microbial amplification. Risk of transmissible infections, such as tuberculosis, may be influenced by negative pressure rooms, upper air ultraviolet irradiation, and room ventilation rate. Research is beginning on ways to control risk of common airborne respiratory infections that are not usually considered building-related, but may be influenced by the built environment.

Remediation of buildings with occupants having sick building syndrome is challenging because there is no environmental measure to establish its likelihood or to assure its resolution. The complaints are commonly contested or trivialized by management, resulting in workplace conflict and distrust. A multidisciplinary team can assist by (1) establishing the existence of complaints in a substantial portion of the workforce; (2) evaluating the heating, ventilation, and air-conditioning system design, performance, and maintenance; (3) checking the building envelope for integrity against water damage; and (4) reviewing housekeeping measures.

Further Reading

Bornehag CG, Blomquist G, Gyntelberg F, et al. Dampness in buildings and health. Nordic interdisciplinary review of the scientific evidence on associations between exposure to "dampness" in buildings and health effects (NORDDAMP). *Indoor Air.* 2001; 11: 72-86.

Mendell MJ, Fisk WJ, Kreiss K, et al. Improving the health of workers in indoor environments: Priority research needs for a national occupational research agenda. *American Journal of Public Health* 2002; 92: 1430-1440.

Institute of Medicine: Committee on the Assessment of Asthma and Indoor Air, Richard B. Johnston, Jr. (Chair). *Clearing the Air: Asthma and Indoor Air Exposures.* Washington, DC: National Academy Press, 2000.

Mendell MJ. Nonspecific symptoms in office workers: A review and summary of the epidemiologic literature. *Indoor Air* 1993; 3: 227-236.

Milton DK, Glencross PM, Walters MD. Risk of sick leave associated with outdoor air supply rate, humidification, and occupant complaints. *Indoor Air* 2000; 10: 212-221.

Burn Injury

ICD-10 T20-T32

Annette MacKay Rossignol and Reza Alaghehbandan

Thermal burns are classified according to the depth of the wounds and the percent of the body surface area affected. First-degree burns, the least severe category, affect only the epidermis and are characterized by redness and swelling of the affected areas. Sunburns exemplify this depth of burning. Second-degree burns affect not only the epidermis but also the dermis and are characterized by blistering of the affected areas. If uninfected, second-degree burns usually heal by themselves in 7 to 10 days. Third-degree burns ("full thickness" burns) affect both the epidermis and the dermis, but, unlike first- or second-degree burns, third-degree burns larger than approximately 1 inch in diameter will not heal by themselves. These burns require skin grafting because all of the dermis in the affected area has been destroyed. Fourth-degree burns, affecting the muscle, constitute the most severe category and, like third-degree burns, require skin grafting to close the wounds.

Occurrence

According to the American Burn Association (ABA), approximately 4500 fire and burn deaths (occupational plus nonoccupational) occur each year in the United States. This total includes an estimated 3750 deaths from fires and 750 deaths from motor vehicle and aircraft crashes, contact with electricity, chemicals or hot liquids and substances, and other sources of burn injury. Because the respective role of flame and smoke in fire deaths often is not determined by autopsy, "burn" death totals cannot be distinguished from deaths that result from smoke inhalation. The number of fire and burn deaths in the United States declined by about 50% between 1971 and 1998. The decline in the death rate was over 60% because the U.S. population grew by 25% over this period.

Between 1992 and 1999 in the United States, there were 1189 occupational thermal burns that resulted in death, for a mortality rate of 0.11 deaths per 100 000 worker-years. This number represents 2.4% of the total number of 49 944 fatal occupational injuries during this time interval. More recent data indicate that worker deaths resulting from electrocutions and from fires and explosions increased to levels of the late 1990s after falling to a near 10-year low in 2000 (Table 1).

At least an additional 45 000 persons are hospitalized annually for the treatment of nonfatal but serious burns, of which an estimated 20% to 30% are occupationally related.

Table 2 shows the number of occupational injuries and illnesses (in 1 000s) involving time away from work for selected injuries and illnesses, 1995-2001.

The total number of work-related burn injuries treated in hospital emergency departments each year in the United States is estimated to be 150 000, based on data derived from the NIOSH-CPSC National Electronic Injury Surveillance System (NEISS). Most of these injuries are treated on an outpatient basis.

Deep burns can cause severe, permanent disability and disfigurement; for this reason, the social and economic costs of burns exceed expected costs based on incidence alone. Thus, while large occupational burns are uncommon, they are costly when they occur. While no national data on the costs of occupational burn injuries are available, one epidemiologic study conducted in Washington State provides data on the direct costs of occupational burns for

Table 1. Numbers of Fatal Occupational Burns, 1996-2001

Type of Burn	Average Number of Fatalities, 1996-2000	Number of Fatalities in 2000	Number of Fatalities in 2001 (Percentage)*
Contact with an electric current	290	256	285 (5%)
Contact with temperature extremes	40	29	35 (1%)
Fires and explosions	196	177	188 (3%)

* Percentage of all fatal occupational injuries (excluding fatalities from the terrorist attacks on September 11, 2001).
Source: Bureau of Labor Statistics. *Census of Fatal Occupational Injuries Summary. Report No. USDL 02-541.* Washington, DC: U.S. Department of Labor, 2002.

Table 2. Number of occupational injuries and illnesses (in 1000s) involving time away from work for selected injuries and illnesses, 1995-2001

	1995	1996	1997	1998	1999	2000	2001
Total cases	2,040.9	1,880.5	1,833.4	1,730.	1,702.5	1,664.0	1,537.6
Sprains, strains	876.8	819.7	799.0	760.0	739.7	728.2	669.9
Bruises, contusions	192.1	174.9	165.8	153.1	156.0	151.7	136.4
Cuts, lacerations	153.2	133.2	133.6	137.6	132.4	121.3	114.8
Fractures	124.6	120.5	119.5	115.4	113.7	116.7	108.1
Back pain	59.0	52.0	48.7	42.4	43.2	46.1	42.7
Carpal tunnel syndrome	31.5	29.9	29.2	26.3	27.9	27.7	26.8
Heat burns	36.1	29.0	30.0	28.4	27.1	24.3	25.1
Tendonitis	22.1	17.4	18.0	16.9	16.6	14.4	14.1
Chemical burns	13.9	11.6	12.2	11.7	11.6	9.4	9.5
Amputations	11.3	10.2	10.9	10.2	10.0	9.7	8.6

Source: Bureau of Labor Statistics. *Lost-Worktime Injuries and Illnesses: Characteristics and Resulting Time Away from Work, 2001. Report No. USDL 03-138.* Washington, DC: U.S. Department of Labor, 2003.

the period 1994-1998. In this study, the average, annual costs to the Washington State fund insurance system for occupational burns was approximately $5,080,000, and the average, annual number of lost workdays was 7600.

The peer-reviewed literature on burn epidemiology and control focuses primarily on domestic burns, such as scalds to young children and flame burns involving clothing ignition, flammable liquids, or structural fires, even though 20% to 30% of serious burns are work-related. While there have been several large, population-based studies of occupational burns, most of the literature on occupational burns is limited to the evaluation of burn victims treated at one or several hospitals or to an unusual sequence of events leading to a burn injury.

For occupational burns, fatal thermal burns are more common among males, accounting for 95% of cases and a mortality rate of 0.20 deaths per 100 000 worker-years. Only 5% of burn deaths occur to females (0.01 deaths per 100 000 worker-years). The burn mortality rate among workers increases with age, with the highest rate among workers aged 65 years and over (0.20 deaths per 100,000 worker-years). Although white workers account for the majority of burn fatalities (82.7%), both white and black workers experience nearly the same burn mortality rate (0.11 deaths and 0.12 deaths per 100 000 worker-years, respectively) (Table 3).

Fatal occupational burns, given their often-large size, usually affect multiple body sites. Hospitalized, nonfatal burns tend to affect the hands, fingers, and wrists. In addition, about half of the hospitalized chemical burns involve the eyes.

Substantial variation in burn rates occurs by occupation. The highest rates of thermal burn fatality occur among the extractive occupations that include oil well drillers, explosives workers, mining machine operators, other mining occupations, and supervisory personnel (1.15 deaths per 100 000 worker-years) (Table 4). Employees working in "transportation and materials moving" account for 30.7% of all fatal occupational burns and have a high rate of mortality (0.83 deaths per 100 000 worker-years). Farming, forestry, and fishing occupations also have high burn mortality rates (0.27 deaths per 100 000 worker-years).

By industry type, workers in mining experience the highest mortality rate (0.77 per 100 000 worker-years). The industries with the highest proportions of fatal thermal burns are construction and "transportation and public utilities" (13% and 24%, respectively). Workers in these industries also have high burn death rates (0.22 and 0.38 deaths per 100 000 worker-years, respectively). Workers in the "agriculture, forestry, and fishing" industry also have a high burn mortality rate (0.24 deaths per 100 000 worker-years) (Table 5).

Occupational burn mortality rates are similar among all regions in the United States except the Northeast. This region had a mortality rate less than half the rates for the other regions (Table 6). States included in the Northeast Region are Connecticut, Maine, Massachusetts, New Hampshire, New Jersey, New York, Pennsylvania, Rhode Island, and Vermont. The lower burn fatality rate for the Northeast is attributable to lower rates than in other regions for workers in agriculture, forestry, and fishing; mining; and transportation and public utilities.

Table 3. Fatal occupational thermal burns by sex, age, and race, United States, 1992-1999

	Number	Percent	Mortality Rate*
GENDER			
Male	1132	95.2	0.20
Female	57	4.8	0.01
AGE GROUP (years)			
< 20	23	2.0	0.03
20-24	58	7.2	0.08
25-34	283	23.8	0.10
35-44	321	27.0	0.11
45-54	251	21.1	0.13
55-64	155	13.0	0.15
> or = 65	67	5.6	0.20
Unknown	4	0.3	-
RACE			
White	983	82.7	0.11
Black	142	12.0	0.12
Asian	19	1.6	0.05
Other	45	3.8	-

* Deaths/100,000 worker-years.
Source: Quinney B, McGwin G Jr., Cross JM, et al. Thermal burn fatalities in the workplace, United States, 1992 to 1999. *Journal of Burn Case and Rehabilitation* 2002: 23:305-310.

Causes

It is important to determine the causative agent in the burn injury because, in many cases, the causative agent determines the depth and extent of the burn. Scalding liquids or steam cause most of the superficial burns at work although both scalding liquids and steam can cause large burns. Flame

Table 4. Distribution of patients with fatal thermal burns by occupation, United States, 1992-1999*

Occupational Description	Number	Percent	Rate*
Extractive occupations	15	1.3	1.15
Transportation and material moving	365	30.7	0.83
Farming, forestry, fishing	75	6.3	0.27
Handlers, equipment cleaners, laborers	110	9.3	0.24
Precision craft and repair	90	7.6	0.23
Construction trades	106	9.0	0.23
Machine operators, assemblers, inspectors	138	11.6	0.20
Technical sales, administrative support	48	4.0	0.14
Precision production	48	4.0	0.14
Service occupations	74	6.2	0.05
Managerial and professional	71	6.0	0.02
Sales occupations and administrative support	38	3.2	0.01
Other, not reported	11	1.0	-
Total	1189	100	-

* Deaths per 100,000 worker-years.
Source: Quinney B, McGwin G Jr., Cross JM, et al. Thermal burn fatalities in the workplace, United States, 1992 to 1999. *Journal of Burn Case and Rehabilitation* 2002: 23:305-310.

injuries will more often result in full-thickness burns especially if clothing ignition is involved. Flashes, explosions, electricity, or chemicals also can cause significant burns.

Table 5. Distribution of patients with fatal thermal burns by industry, United States, 1992-1999*

Industry Description	Number	Percent	Rate*
Mining	44	3.7	0.77
Transportation and public utilities	282	23.7	0.38
Agriculture, forestry, and fishing	80	6.7	0.24
Construction	160	13.5	0.22
Manufacturing, non-durable goods	113	9.5	0.16
Public administration	65	5.5	0.14
Manufacturing, durable goods	142	12.0	0.13
Wholesale and retail trade	143	12.0	0.06
Services	135	11.4	0.04
Finance, insurance, real estate	11	1.0	0.02
Other, not reported	14	1.2	-
Total	1189	100.2	-

* Death per 100,000 worker-years.
Source: Quinney B, McGwin G Jr., Cross JM, et al. Thermal burn fatalities in the workplace, United States, 1992 to 1999. *Journal of Burn Case and Rehabilitation* 2002: 23:305-310.

There are six broad types of burn injuries. These types and their approximate proportions among people who are hospitalized for work-related burn injuries are scalds (40%); flame/flash/smoke inhalation (30%); chemical (10% to 15%); contact (10%); electrical (5% to 10%); and, only rarely, radiation burns, most of which are caused by non-ionizing radiation. Radiation burns, such as flash burns to the cornea and conjunctiva associated with arc welding, are more highly represented among work-related burns that do not require in-patient care at a hospital. (See Eye Injuries.) Table 7 shows the fre-

Table 6. Numbers and Rates of Occupational Burn Death by Region in the United States, 1992-1999.

Region	Number of deaths	Percentage	Rate*
Northeast	135	11.4	0.06
Midwest	299	25.2	0.11
South	469	39.4	0.13
West	286	24.1	0.12

*Deaths per 100,000 worker-years
Source: Quinney B, McGwin G Jr., Cross JM, et al. Thermal burn fatalities in the workplace, United States, 1992 to 1999. *Journal of Burn Case and Rehabilitation* 2002: 23:305-310.

Table 7. Event/exposure involved in fatal, occupational thermal burns, United States, 1992-1999.

Event/Exposure	Number	Percent
Fire	457	38.4
Transportation accidents	442	37.2
Explosion	151	12.7
Harmful substances/environment	113	9.5
Assaults/violent	15	1.3
Fires/explosion, unspecified	11	0.9

Source: Quinney B, McGwin G Jr., Cross JM, et al. Thermal burn fatalities in the workplace, United States, 1992 to 1999. *Journal of Burn Case and Rehabilitation* 2002: 23:305-310.

quency of fatal occupational burns by event/exposure. Thirty-five percent of these burns occurred on "industrial places and premises" and 29% occurred on a "street or highway."

The epidemiology of people hospitalized for the treatment of occupational burns reveals the following:

1. The large majority of scald and contact burns involve either food preparation/consumption or motor vehicles. Products such as stoves, ovens, grills, cups, and dishwashers feature prominently among the environmental hazards contributing to food preparation/consumption-related burns. More females than males suffer from such burns, consistent with their more frequent work activities using these products. Usual types of motor vehicle-related burns include scalds from radiator fluids, flame burns from gasoline or from post-crash fires, contact burns from tailpipes or exhaust manifolds, and chemical burns from battery acid (see #3 below). More males than females suffer from motor vehicle-related burns, again consistent with their occupational preferences.
2. Approximately 75% of all flame/flash burns or cases of smoke inhalation involve flammable liquids, particularly gasoline. (In other countries, the flammable liquid most often related to large burns is kerosene.) Only 10% to 15% of flame burns occur in structural fires.
3. Many of the chemical burns (in contrast to occupational chemical toxicity) are caused by contact with motor vehicle battery acid. Another common cause of chemical burns is contact with household cleaning products.
4. Most of the electrical burns occur while people are working with outdoor electrical supply lines.
5. Few work-related burns are known to be associated with alcohol use.

Pathophysiology

Despite continuing, intense research, the immediate changes in tissues after thermal injuries are still poorly understood. The skin, the largest organ in the body, has three main functions: (1) to prevent water loss by maintaining a constant temperature, (2) to act as a barrier against microbial invasion, and (3) to house nerve endings. The devastating effects of a burn injury are the result of changes in or loss of these functions.

The human body can tolerate temperatures up to 40°C; at higher temperatures, protein denaturation occurs and exceeds the capacity of cellular repair. After a burn, there is a transient phase of vasoconstriction followed by opening of precapillary sphincters. This change results in increased transcapillary filtration, vasodilation, and edema formation, that reaches a maximum level 18 to 24 hours post-injury.

Human skin is damaged by an immediate heat injury and by a delayed injury that occur secondary to progressive dermal ischemia. Many other physiologic changes also occur. The degrees of change in the normal functioning of all physiologic systems are directly proportional to the size (percent and depth of the body surface affected) of the burn injury, making a serious burn a potentially devastating illness. Large burns affect the functioning of every body system and place enormous demands on the body's ability to regulate normal body temperature, protect against infection, and repair damaged tissue.

Prevention

Primary prevention of occupational burn injury entails adherence to various engineering and administrative control strategies, including the following:

1. In-depth investigations of the burn injury occurrence should be conducted to identify the subtle factors that alter the balance between safety and tragedy in a given type of situation. Promptness in conducting an investigation is imperative if accurate information about the events leading to injury is to be obtained. In addition, the tendency to "blame the victim" for his or her injury must be resisted if environmental, behavioral, and ergonomic hazards for burn injury are to be identified and either controlled or eliminated.

2. Unambiguous warning labels should be placed on containers of flammable liquids and industrial equipment and processes that use flammable liquids. Adequate pictorial warnings to convey the degree of flammability of the material should be used. Flammable liquids and chemicals that burn, such as strong acids and bases, should be stored in safety containers and clearly labeled (in languages understood by the work force) with their contents. Substances that react with one another (for example, acids and bases) should have separately vented storage areas. (OSHA regulations that pertain to fire prevention and fire suppression are in Chapter 30 of the Code of Federal Regulations [30 CFR] § 1910, Standards for General Industry: §106 - 110 covers storage and handling of hazardous liquids; Subpart L covers fire protection in general, §1200 covers hazardous communications, and §450 covers hazardous chemicals in laboratories.)

3. Ignition sources, such as lighted cigarettes and pilot lights, should be eliminated from areas where flammable liquids are being used. "Intrinsically safe" or "permissible" equipment, such as sparkless electrical devices, should be substituted for less safe equipment whenever the possibility of exposure to explosive substances exists.

4. Products and work areas associated with food preparation and consumption should be modified to incorporate sound ergonomic design criteria. For example, cups should be designed to prevent easy spillage; stoves, ovens, and grills should be designed to discourage users from reaching across hot surfaces and burners to reach control knobs. Control knobs should be positioned either to the side or in front of hot surfaces and burners, not behind them. Barriers (guards and enclosures) can be used to prevent contact with hot products and surfaces. In addition, work areas should be designed to avoid a

       potentially hazardous "functional triangle" resulting from the place-
ment of refrigeration, cooking, and dishwashing equipment.
5. Supervisors and other individuals who work in occupations associated with an increased risk of burn injury should receive focused,
evaluated training programs that cover the fundamental principles
and practices of burn injury control.

Secondary prevention of burn injuries requires prompt treatment in the
form of cooling (not icing) the burned skin and, depending on the size of the
burn injury, fluid replacement therapy in the early post-burn period to prevent complications arising from the burn shock phase of the injury. Fluid
replacement therapy should be initiated immediately (in the ambulance) for
any burn affecting more than 20% to 30% of the body surface area. For all
burns, infection prevention and control is of paramount importance to minimize the physiological load from this source and to promote healing.
Prevention of infection by applying dry sterile dressings to the burn wounds
should be accomplished immediately after the burned areas have been
cooled. In addition, burns to the face and head, or knowledge that inhalation
of smoke and toxic gases has occurred, may suggest the need to provide
prompt respiratory support to the victim.

In addition, because even a 5 percent full-thickness burn to an older
person can be fatal, people aged 50 years or older who have sustained
burns should be evaluated in a hospital. People of any age who have sustained second-degree or full-thickness burns to the hands, face, or genitalia
should receive prompt in-hospital care to minimize disfigurement and disability.

As mentioned previously, third- and fourth-degree burns require skin
grafting to close the burn wounds. The benefits of prompt skin grafting are
many and include control of infection, prevention of the increased metabolic
and pathophysiologic responses to open wounds, and reduced lengths of
hospital stays.

Counseling for people disfigured by burns is indicated during the initial
hospital stay, especially as the date of discharge approaches. Additional counseling may be necessary over the next several years as part of the patient's follow-up care.

Treatment of large burns and treatment of smoke inhalation are among
the most difficult tasks in medicine. Specialized burn treatment units and centers have been established to treat large or difficult burn injuries. The cost for
care in these specialized facilities often is high, because long hospital stays
may be required to treat the acute burn injury and because the staff-to-patient
ratio is high. In addition, reconstructive procedures, if necessary, may require
additional hospital admissions over an extended period of time. The
American Burn Association (ABA) has established guidelines that address
the need to transfer a burn victim to a specialized burn treatment unit or cen-

ter. According to the ABA's Burn Unit Referral Criteria, burn injuries that should be referred to a specialized burn unit or center include the following:

1. Partial thickness burns greater than 10% total body surface area (TBSA);
2. Burns that involve the face, hands, feet, genitalia, perineum, or major joints;
3. Third-degree burns in any age group;
4. Electrical burns, including lightning injury;
5. Chemical burns;
6. Inhalation injury;
7. Burn injury in patients with pre-existing medical disorders that could complicate management, prolong recovery, or affect mortality;
8. Any patients with burns and concomitant trauma (such as fractures) in which the burn injury poses the greatest risk of morbidity or mortality. In such cases, if the trauma poses the greater immediate risk, the patient may be initially stabilized in a trauma center before being transferred to a burn unit. Physician judgment will be necessary in such situations and should be in concert with the regional medical control plan and triage protocols;
9. Burned children in hospitals without qualified personnel or equipment for the care of children; and/or
10. Burn injury in patients who will require special social, emotional, or long-term rehabilitative intervention.

In addition, the decision of whether to transfer a patient to a specialized burn unit or center should take into account the range of treatment options available to the patient at the initial treatment facility.

Further Reading

Baggs, J., Curwick, C., Silverstein, B. Work-related burns in Washington State, 1994 to 1998. *J Occup Environ Med.* 2002, 44: 692-699.

Gurdak C. Thermal injuries in the workplace. *AAOHN J,* 1990 38: 492-496.

Inancsi W, Guidotti TL. Occupation-related burns: five-year experience of an urban burn center. *J Occup Med.* 1987, 29: 730-733.

McCullough JE, Henderson AK, Kaufman JD. Occupational burns in Washington State, 1989-1993. *J Occup Environ Med.* 1998: 40: 1083-1089.

Munnoch DA, Darcy CM, Whallett EJ, Dickson WA. Work-related burns in South Wales 1995-96. *Burns.* 2000: 26: 565-570.

Rossignol AM, Locke JA, Burke JF. Employment status and the frequency and causes of burn injuries in *New England. J Occup Med* 1989; 31: 751-757.

Rossignol AM, Locke JA, Boyle CM et al. Epidemiology of hospitalized, work-related burn injuries in Massachusetts. *J Trauma* 1986; 265: 1097-1101.

Taylor AJ, McGwin G Jr, Cross JM, Smith DR, Birmingham BR, Rue LW 3rd. Serious occupational burn injuries treated at a regional burn center. *J Burn Care Rehabil.* 2002: 23: 244-248.

Byssinosis

ICD-10 J66.0

David C. Christiani

Byssinosis is an acute and chronic airways disease caused by exposure to the dust of cotton, flax, hemp, or sisal. The acute response is reversible and is characterized by a sensation of chest tightness and/or shortness of breath upon return to exposure following a weekend or holiday break. In societies in which Monday is the first workday of the week, it is sometimes referred to as "Monday morning syndrome." Cough may be present and may become productive. Ventilatory function (specifically forced expiratory volume in 1 second, or FEV_1) often decreases from beginning to end of the work shift on the first workday of the week. For most affected people, these effects will decrease or disappear on the second and subsequent days of work. However, if workers are exposed virtually every day of the workweek, as is the case in many developing countries, or if they work on rotating shifts, the periodicity may be absent or altered.

Onset of chest tightness usually occurs 2-3 hours after start of exposure, distinguishing byssinosis from asthma, which usually has either immediate or delayed (6 or more hours) onset following exposure to allergens.

With prolonged exposure, both chronic cough (productive or nonproductive) and decline in ventilatory function become more severe. Dyspnea becomes a prominent complaint and acute declines in expiratory flow rates or FEV_1 over the work shift can be marked. In advanced stages, clear clinical and physiological evidence of irreversible chronic obstructive lung disease emerges, and cross-shift declines are often more pronounced.

A standard definition for use in epidemiology was developed by Schilling and adopted by the British Medical Research Council:

Grade 0: No symptoms of chest tightness or breathlessness on Monday (or on return to exposure).

Grade 1/2: Occasional chest tightness or mild symptoms, such as respiratory tract irritation, on return to exposure.

Grade 1: Chest tightness and/or breathlessness on return to exposure on all first workdays of the week.

Grade 2: Chest tightness and/or breathlessness on return to exposure on first and subsequent workdays of the week.

There are no characteristic signs on physical examination unless severe bronchospasm is present. In this case, wheezing on expiration may be heard.

Among those with advanced disease, decreased breath sounds and a prolonged expiratory phase will be noted. These findings are characteristic of chronic obstructive lung disease and are not specific to byssinosis. Therefore, an accurate diagnosis depends on well-characterized exposure, a compatible history, and documentation of cross-shift declines in FEV_1 and expiratory flow rates.

Although the chest x-ray can produce no specific information to associate impairment with occupational exposure, it is important to eliminate other pulmonary pathologies, such as tuberculosis, lung cancer, and pulmonary fibrosis. In cases in which symptoms and functional changes are inconsistent, a complete set of pulmonary function tests, including static lung volumes and diffusion capacity, may be useful.

Occurrence

Byssinosis occurs worldwide, with several million workers occupationally exposed. Prevalence averages about 20% among exposed workers and has been reported to be as high as 83% among active workers with the highest exposure. It is most often associated with occupational exposure to dust in cotton textile mills, but it also occurs with exposure to the dust of flax, hemp, and sisal and among workers in the cotton waste-processing industries. In the United States, approximately 500 000 workers are at potential risk, with perhaps half this number in high-exposure work, such as yarn manufacture and agricultural processing of cotton.

In the cotton textile industry, the prevalence of byssinosis has typically been found to be highest among workers employed in the initial stages of cotton processing (opening, picking, carding, and stripping and grinding of the cards) in mills that process a coarse grade of cotton and produce higher concentrations of dust. Prevalence is lower among workers in yarn processing and lowest among workers in slashing and weaving; it is virtually nonexistent in handling of finished cloth.

Byssinosis also occurs among cotton gin workers and among workers in the garnetting or bedding industry. It has been reported among workers in the delinting process of cottonseed oil extraction plants. It has not been reported among workers who process washed and scoured cotton in the medical cotton industry or among workers who process cotton fibers reclaimed from waste fabric.

Prevalence among active cotton textile workers increases with the duration of exposure up to about 20 years, and then it decreases. This decline is attributed to workers who retire early because of impairment, leaving a survivor population of active workers. Prevalence has significantly declined among cotton textile workers in the United States, and is now very low among those first exposed after the OSHA PEL was phased in from the late 1970s and early 1980s.

There is no evidence that race or ethnic background affects the risk of byssinosis. After controlling for smoking and dust exposure, there is no difference in risk between males and females.

Causes

The actual etiologic agent in cotton dust that causes the syndrome has not been identified, but current research has focused much attention on the role of gram-negative bacterial endotoxins. In experimental models, cotton dust extracts combined with purified *Escherichia coli* endotoxins in rabbits over a 20-week period have resulted in severe bronchitis and bronchiolitis. When humans selected for their sensitivity to cotton dust were exposed to cotton dust in a model card room, acute declines in lung function correlated well with the gram-negative endotoxin content of the air. Chronic airflow obstruction is associated with long-term exposure to cotton-dust-related endotoxin, possibly in association with other constituents of cotton dust.

Pathophysiology

Byssinosis appears to be pathologically similar to chronic bronchitis. It remains unclear whether emphysema is a significant part of the pathological picture in late stages of byssinosis. Pathological studies that include good occupational and smoking histories are needed. Potential pathophysiological mediators and mechanisms include histamine, histamine-like substances, and endotoxin-associated inflammation.

Prevention

Reduction of risk depends on reduction of cotton dust concentrations with the standard industrial hygiene practice of local exhaust ventilation and process enclosure. A second control technology is cotton washing, which, although effective, may not be technically feasible on a large scale.

The OSHA cotton dust regulation (29 CFR 1910.1043) is complex. It requires exposure monitoring, control of exposure, and medical surveillance, and it prescribes conditions under which respirators, as opposed to engineering controls, are permitted. The current PELs for cotton dust in cotton textile mills range from 200 to 750 $\mu g/m^3$ and are dependent on work location. In early stages of processing (from opening, picking, and carding to yarn preparation), the PEL is a time-weighted average (TWA) of 200 $\mu g/m^3$. In slashing and weaving, the PEL is 750 $\mu g/m^3$. In the nontextile cotton-processing industry—involving cottonseed oil, batting, and bedding—and in waste processing in textile mills, the PEL is 500 $\mu g/m^3$. Employers are required to conduct medical surveillance and to monitor exposure regularly. If exposure reaches an "action level" of half the PEL, more frequent surveillance and monitoring are required.

Medical surveillance should include use of questionnaires for symptoms of byssinosis following return to work, physical examination, and spirometry before and after the work shift on Monday or on return to work. These examinations will identify many of those people who are acutely affected and probably all of those with significant impairment. Because smoking contributes significantly to the risk of chronic bronchitis and airways obstruction, exposed workers should be encouraged to stop smoking.

Acute responsiveness should be treated as a sentinel health event. Workers in the same jobs or with the same exposure as those with stage 1/2 or higher byssinosis should be evaluated, and exposure on these jobs should be assessed and, if indicated, reduced. If these initial preventive measures are ineffective, affected workers should be allowed to transfer, with wage retention, to jobs with less exposure. They should then be reevaluated. Respirators are useful as temporary protective measures or when there is no alternative, but they are not practical for permanent protection. If workers are symptomatic, employers in the United States are required by OSHA to monitor both exposure and workers' health more frequently.

The OSHA cotton textile regulation requires that a worker in a job with exposure that exceeds the PEL must be offered the opportunity to transfer with wage protection to a job with exposure at or below the PEL, if the worker is found by a physician to be unable to wear a respirator [see 29 CFR 1910.1043(f)].

Other Issues

Smokers are at increased risk for chronic obstructive lung disease, which adds to the risk of byssinosis. There is no evidence that asthma is an additional risk factor for development of byssinosis, though acute bronchospasm can be triggered in asthmatic individuals by exposure to cotton dust.

Until recently, data have largely come from cross-sectional studies of actively employed workers. Out-migration of affected workers leaves a survivor population and such cross-sectional studies result in an underestimated risk. Several prospective studies of textile workers are now under way in various parts of the world. Prospective studies have shown that chronic airflow obstruction in textile workers is not always preceded by the acute response, although the acute (cross-shift) response predicts chronic lung function loss, even in the absence of symptoms. It remains unclear whether the natural history of byssinosis is a steady progression from grade 1/2 to higher grades.

Further Reading

Christiani DC, Wang XR. Respiratory effects of long-term exposure to cotton dust. *Current Opinions in Pulmonary Medicine* 2003; 9: 151-155.

Christiani DC, Wang XR, Pan LD, et al. Longitudinal changes in pulmonary function and respiratory symptoms in cotton textile workers: A 15-year follow-up study. *American Journal of Respiratory and Critical Care Medicine* 2001; 163: 847-853.

Held HD, Uhlig S. Mechanisms of endotoxin-induced airway and pulmonary vascular hyperreactivity in mice. *American Journal of Respiratory and Critical Care Medicine* 2000; 162: 1547-1552.

Wang XR, Eisen EA, Zhang HX, et al. Respiratory symptoms and cotton dust exposure: results of 15-year follow-up observation. *Occupational and Environmental Medicine* 2003; 60: 935-941.

Wang XR, Pan LD, Zhang HX, et al. A longitudinal observation of early pulmonary responses to cotton dust. *Occupational and Environmental Medicine* 2003; 60: 115-121.

Carpal Tunnel Syndrome

ICD-10 G56.0
SHE/O

Laura Punnett and Judith E. Gold

Carpal tunnel syndrome (CTS) is a nerve compression disorder affecting the median nerve, which passes through the carpal tunnel from the forearm into the hand. The nerve may become compressed by internal or external pressure within the wrist.

CTS usually begins with gradual onset of symptoms, specifically tingling and numbness in the fingers, initially at night. The area of the median nerve distribution includes part of the thumb, the index, third, and part of the fourth fingers, although many variations occur. The symptoms may progress to burning pain, severe and painful numbness, a sensation of swelling without objective signs, or a sensation of grip weakness (due to loss of sensory nerve function). Loss of manual dexterity and motor control often creates difficulty in performing hand movements such as opening bottles, turning doorknobs or keys, or grasping small items.

Symptoms may be provoked or aggravated by Phalen's test (sustained wrist flexion for 1 minute) or Tinel's test (percussion over the median nerve at the carpal crease). There may be objective loss of strength in the thumb or in any of the affected fingers, or loss of sensation noted on palpation or two-point discrimination in the affected fingertips. Wasting of the muscles at the base of the thumb is observed in severe cases.

The NIOSH surveillance case definition for CTS requires both typical symptoms and findings on physical examination or electrodiagnostic tests, plus evidence of relevant occupational exposure. However, the clinical tests have relatively low sensitivity, even when performed in a carefully standardized manner. Hand pain diagrams, on which affected individuals mark the location and quality of symptoms for comparison with classic patterns, have better predictive value than most other clinical findings, although their agreement with electrodiagnostic test results is modest.

Nerve conduction testing can determine if a nerve impulse is conducted more slowly (decreased nerve conduction velocity) or less efficiently (decreased nerve conduction amplitude) through the wrist in comparison with other nerve segments or in the corresponding area of the other hand, if only one hand is symptomatic. Although considered the clinical "gold standard," some have questioned its sensitivity, especially in early stages. It is particularly difficult to interpret in workplace screenings, as there is a high percentage of false-positive tests.

Other test modalities include magnetic resonance imaging and ultrasound, both of which are unlikely to be used routinely in most workplaces. The utility of these tests remains to be demonstrated; however, they are used increasingly to diagnose patients with ambiguous electrodiagnostic and physical examination findings in the clinical setting. Improvements in resolution may increase the diagnostic utility of these imaging modalities in the future, but may not improve their utility for screening.

Although CTS is sometimes used by the general population as a generic term for all work-related upper extremity musculoskeletal disorders, CTS comprises only a small percentage of these disorders.

The differential diagnosis includes compression of the median nerve at other locations, such as at the forearm near the elbow (pronator teres syndrome) or the shoulder (see Peripheral Nerve Entrapment Syndromes), cervical radiculopathy, tendonitis, osteoarthritis, and rheumatoid arthritis. However, it is not uncommon for more than one of these conditions to exist simultaneously.

Occurrence

Data are sparse on the incidence of CTS in the general population. Newly diagnosed CTS cases in Wisconsin have occurred at an incidence rate of 4.6 cases per 1000 person-years. The prevalence in Sweden of symptoms typical of CTS has been estimated at 14.4%, but only 3.8% for "clinically certain" CTS (symptoms plus findings). Surprisingly, 18% of asymptomatic persons had median nerve damage identified on electrodiagnosis. The prevalence was somewhat higher in women than men and increased with age, although middle-aged men had a higher prevalence than their female counterparts. This last finding was unexpected and requires further exploration, as it may be difficult to distinguish age from the duration of cumulative exposure to ergonomic stressors. In any case, estimates for the general populations in different places may be greatly affected by differences in case definition, distribution of age, gender, occupation, and industry.

Healthcare providers in Santa Clara County, California, judged that 47% of the cases that they diagnosed in 1987 may have been work-related. Similarly, the U.S. National Health Interview Survey (NHIS) has suggested that as many as 50% of all CTS cases are work-related.

The NHIS found a CTS prevalence of 2.8 per 1000 among the U.S. working population. Estimates from workplace administrative data vary widely among data sources and are likely to suffer from underreporting. CTS accounts for 13% of all work-related repetitive trauma disorders reported to the BLS by private sector employers. The U.S. incidence rate of reported work-related CTS in 2000 was 3.0 cases per 1000 person-years, but may be as high as 100 cases per 1000 person-years in high-risk occupations. Industries with the highest number of CTS cases have included manufacturing, services, retail trade, and finance, insurance, and real estate.

Across a variety of workplaces, workers in ergonomically stressful jobs have been found to have a CTS prevalence ranging from 2% to 53% and an annual CTS incidence rate ranging from less than 1% to 26%. These rates depend on both the criteria used to define a case of CTS (self-reported symptoms, signs, and/or test results) and the intensity of exposure to repetitive or forceful exertions. Among men and women performing the same or nearly identical work, little difference in risk has been observed; the effect of ergonomic factors, especially repetitive work, appears to be more important than gender.

NIOSH has found evidence for separate associations between CTS and repetition, force, and vibration. Strong evidence was noted for the effect of combined risk factors: force and repetition, and force plus non-neutral wrist posture. In the best studies of repetitive manual work, relative risks of CTS have ranged from 1.1 to 6.7 compared with workers jobs in non-repetitive jobs. High quality studies of work involving segmental vibration have shown relative risks of 5.3 to 10.9. Work that was both forceful and repetitive increased the risk of CTS 15-fold. CTS risk due to repetitive motion, forceful loading, and segmental vibration was tripled again when occupational exposure lasted for more than 20 years.

CTS is associated with extended lost work time and job turnover in 20% to 30% of affected people. Surgical treatment has been reported to be successful in a large proportion of patients, but well-designed studies are few, and outcomes are usually less favorable among work-related cases. The direct costs of workers' compensation claims for CTS (medical treatment and lost wages) in Washington State have been estimated to average almost $13,000 per case, with a median more than $4,000. In addition, employers and workers bear additional expenses that are not compensated.

Causes

Occupational activities linked to CTS include typing and other data-input operations; many manufacturing, assembly, and packing jobs; work with vibrating hand tools; and a wide variety of other activities that involve highly repetitive or forceful manual tasks. (See Upper Extremity Musculoskeletal Disorders.) Specific risk factors include frequent, prolonged, or forceful use of a pinch grip or other finger activity, and bending of the wrist in the direction of either the palm or the back of the hand (palmar or dorsal flexion). The hand has more biomechanical advantage in a power grasp when the object is in contact with the full hand, not just the fingertips, than in a pinch grip, when greater forces are generated internal to the carpal tunnel. Repetitive or forceful pinching or other finger activities while the wrist is flexed are particularly likely to cause strain within the carpal tunnel. Exposure to local or environmental cold may intensify the effects of posture, force, and repetitiveness by decreasing finger sensitivity and requiring a worker to increase the grip force to be sure that an object is held securely. Vibration has a similar effect through loss of sensation as well as direct damage to the nerves. In addition, prolonged contact of the wrist with hard or sharp edges or other sources of mechanical compression may directly damage the median nerve, along with its circulatory supply. Many hand-intensive jobs involve simultaneous exposure to more than one of these factors. For example, vibration or cold, by interfering with sensory nerve function and neuromuscular feedback, may result in increased grip forces compared with objective requirements.

Common nonoccupational associations with CTS include diabetes mellitus, rheumatoid arthritis, thyroid disease, menopause, pregnancy, obesity, gout, alcoholism, possibly osteoarthritis, estrogenic agents, and gynecologic surgery, such as bilateral oophorectomy. Although these conditions are prevalent in the general population, they do not explain many cases in workforces with important ergonomic exposures and should not deflect attention from preventable workplace causes. Especially in high-exposure occupations, the contribution of nonoccupational factors, such as age, gender, and medical conditions, to overall CTS risk appears to be relatively small. Uncommon nonoccupational associations include Paget's disease of bone, gout, acromegaly, possibly severe vitamin B6 (pyridoxine) deficiency or excess, renal disease, toxic shock syndrome, sarcoidosis, tuberculosis, systemic lupus erythematosus, and multiple myeloma and other neoplasms.

Pathophysiology

The mechanism of damage is compression or irritation of the median nerve as it passes into the hand between the carpal (wrist) bones and the transverse carpal ligament across the front of the wrist. This causes slowed or incomplete transmission of sensory and motor nerve signals through the wrist to the part of the hand supplied by the median nerve, resulting in discomfort and impaired neuromuscular function. Compression may result directly from using the hand with the wrist flexed, so that the space within the carpal tunnel is narrowed and the nerve is squeezed against the other tissues. Intracarpal tunnel pressure is also increased with extreme wrist extension and, to a lesser extent, with ulnar and radial deviation. In a pinch grip, high forces are generated internally to the carpal tunnel, increasing the pressure on the median nerve. It is also possible that tendons within the tunnel become irritated and swollen as a result of repetitive or forceful finger activity, so that they crowd the carpal tunnel and compress the nerve. There is evidence that the mechanism involves ischemia (loss of blood supply) to the nerve itself and/or thickening of tendons, synovia, or other connective tissue. Segmental vibration and low temperatures may also interfere with the microcirculation of the nerve and exacerbate the effects of repetitive motion and postural stressors. (see Hand-Arm Vibration Syndrome.) When vibration is the predominant occupational exposure, mechanical compression of the nerve in the carpal tunnel may not also be necessary for development of the syndrome.

Prevention

Primary prevention involves the use of engineering controls to reduce exposure to occupational ergonomic stressors, such as high work pace and repetitiveness, hand force requirements, and local or environmental exposure to cold temperatures. (See Upper Extremity Musculoskeletal Disorders.) Particular attention should be paid to eliminating the need for pinch grips or wrist flexion by redesigning tools, equipment, and work methods. A keyboard, workbench, or other work surface may need to be raised or lowered so that the wrist can be kept straight. A tool with a bent handle may eliminate the need to bend the wrist, but there is no such thing as an "ergonomic tool" per se; each tool should be selected and installed appropriately for the particular application. Sources of external mechanical compression of the wrist should be eliminated by such measures as padding the edges of a desk or bench where the wrists rest. Where vibrating hand tools are used or where there is other exposure to segmental vibration, mechanical isolation and damping should be used to reduce the amplitude of the vibration transmitted to the hand and arm. (See Hand-Arm Vibration Syndrome.)

As a secondary preventive measure, health care providers should be trained in the appropriate interview and clinical examination procedures to identify work-related CTS. Workers who report symptoms typical of CTS, such as numbness and tingling in the relevant part of the hand especially at night, should receive immediate attention, even if the physical examination and laboratory tests are negative or inconclusive. Once reported, cases should be treated conservatively, and jobs should be analyzed for ergonomic features that can be modified. Treatments frequently include nonsteroidal anti-inflammatory drugs (NSAIDs), splint use at night, and physical or occupational therapy. Splint use on the job should be considered only if it does not interfere with work, or require the worker to exert more force or strain another joint in order to perform the task. Removal from sources of exposure is essential to prevent the nerve damage from becoming irreversible. Follow-up is important to ensure that job modifications have been effective, that "light-duty" jobs have been correctly selected to avoid continuing stress to the wrist, and that symptoms and signs do not progress. There is no evidence that administration of pyridoxine is effective in treating the occupationally induced syndrome. Although yoga, transcutaneous electrical nerve stimulation (TENS), and oral steroids are advocated by some, their efficacy for CTS is unknown.

Concerning tertiary measures, surgery may be only temporarily effective if the worker is returned too soon to an ergonomically stressful job that has not been modified. Possible loss of grip strength, build-up of scar tissue, and increased vulnerability of the carpal tunnel to mechanical insult following surgery make it imperative that job assignments be selected carefully to avoid recurrence. Surgical outcomes have been reported as promising, although there are few long-term follow-up studies addressing outcomes in people who continue to work, especially in ergonomically stressful jobs. Surgery is highly questionable when there has been occupational exposure to segmental vibration, because mechanical compression at the carpal tunnel may not be the primary etiologic mechanism.

Other Issues

No data are available concerning the effects of combined exposures to ergonomic stressors and to other peripheral neurotoxins, such as heavy metals or organic solvents.

Further Reading

Bernard BP (Ed.). *Musculoskeletal Disorders and Workplace Factors: Critical Review of Epidemiologic Evidence for Work-Related Musculoskeletal Disorders of the Neck, Upper Extremity, and Low Back.* Cincinnati, OH: NIOSH, 1997.

Cailliet R. Hand *Pain and Impairment* (3rd ed.). Philadelphia: F.A. Davis, 1982.

Harrington JM, Carter JT, Birrell L, et al. Surveillance case definitions for work-related upper limb pain syndromes. *Occupational and Environmental Medicine* 1998; 55: 264-271.

Herbert R, Gerr F, Dropkin J. Clinical evaluation and management of work-related carpal tunnel syndrome. *American Journal of Industrial Medicine* 2000; 37: 62-74.

Jarvik JG, Yuen E. Diagnosis of carpal tunnel syndrome: Electrodiagnostic and magnetic resonance imaging evaluation. *Neurosurgery Clinics of North America* 2001; 12: 241-253.

Kannan N, Sawaya R. Carpal tunnel syndrome: Modern diagnostic and management techniques. *British Journal of General Practice* 2001; 41: 311-314.

Lincoln AE, Vernick JS, Ogaitis S, et al. Interventions for the primary prevention of work-related carpal tunnel syndrome. *American Journal of Preventive Medicine* 2000; 18: 37-50.

Rempel DM, Evanoff B, Amadio PC, et al. Consensus criteria for the classification of carpal tunnel syndrome in epidemiologic studies. *American Journal of Public Health* 1998; 88: 1447-1451.

Sluiter JK, Rest KM, Frings-Dresen M Criteria document for evaluation of the work-relatedness of upper extremity musculoskeletal disorders. *Scandinavian Journal of Work, Envirnoment, and Health 2001; 27 (Suppl 1): 1-102.*

Viikari-Juntura E, Silverstein BA. Role of physical load factors in carpal tunnel syndrome. *Scandinavian Journal of Work Environment and Health* 1999; 25: 163-185.

Cataracts

ICD-10 H25-H28

Satya B. Verma and Andrew S. Gurwood

A cataract is the opacification of the crystalline lens of the eye. There are many forms of cataracts, each traditionally defined by shape, location, age of onset (congenital or acquired), or causative factor.

In the circumstance that the cataract is produced by a slowly progressive process, such as age-related cataract or medicinally induced cataracts, as the opacity first develops visual acuity may remain relatively unaffected, allowing the patient to be asymptomatic. As the opacities mature in their density or move into areas that impinge on the visual axis, appropriate and consistent symptoms develop. Symptoms depend on the density, shape, size, and location of the obstruction. For example, patients with posterior subcapsular cataract often complain of glare (defined as disability glare) from bright lights at night or during the day. Cortical cataracts are characterized by translucent grayish spokes, flakes, and dots arranged radially in the cortex that surrounds the nucleus. Subcapsular opacities usually appear as granules, vacuoles, and crystals of various colors. Nuclear cataracts present as a central lenticular yellowing in the lens nucleus. The lens swells and causes an increase in myopic shift.

In general, as the opacities progress, vision may fluctuate with the time of the day (as the visual demand changes) and become progressively poor. The amount of vision left or lost is dependent not only upon the extent and location of lens opacities, but also on the rate of change. In general, the hallmark of all cataracts is painless progressive loss of vision. Classically, the rate of loss is variable and nonlinear. Exceptions include diabetic cataract; drug-induced cataract, such as from some antipsychotic medications, some antiseizure medications, and steroids; and congenital cataract, which may produce severe and profound amblyopia. Typically, both distance and near acuities are affected.

Cataracts are most often first identified and diagnosed by eye care professionals who evaluate the crystalline lens using an instrument known as a slit lamp or biomicroscope. Lens opacities may also be seen using a retinoscope or direct or indirect ophthalmoscope. When the opacities are larger in size, dilation of the pupil is essential for obtaining an appropriate view of the posterior segment of the eye, since the opacities tend to obscure the view of the retina and optic nerve.

Occurrence

Acquired cataract is considered a malady of the old age, although a leading cause of blindness in developing countries. All people would eventually acquire a cataract if they lived long enough. According to the Center for Medicare and Medicaid, the total cost paid for cataract surgeries is second only to heart surgeries. The prevalence of cataracts among 65- to 74-year-olds is 127 per 1000; it increases to 243 per 1000 among those over age 75. The proportion of cataracts due to occupational causes is not known.

Causes

Occupational causes of cataracts include ionizing and microwave radiation and a form of ultraviolet radiation (UV-B). Sun exposure is a risk factor for cataracts. Other causes include congenital, toxic, traumatic, and metabolic factors. In addition, senescence is associated with cataracts. The biochemistry of the lens is complex and the actual causes of opacities in some instances remain unclear. In most cases, cataract formation is thought to be the result of oxidative damage to lens proteins that reduces the solubility of substances that ordinarily leave the tissue, allowing it to remain transparent. Risk factors that increase the chances of lens opacification or the speed at which it occurs include smoking, poor nutrition, corticosteroid therapy, and exposure to ultraviolet light radiation (more common in sunny climates).

Pathophysiology

While a cataract has various clinical appearances, its histopathology is fairly uniform: degeneration and atrophy of lens epithelium, water clefts in the cortex, lens fiber fragmentation, and deposits of calcium crystals throughout the tissue. The lens has a high potassium and low sodium concentration, which is maintained by active epithelial pumps. This too may be compromised during the process leading to loss of clarity.

Prevention

Primary prevention include avoidance of the occupational causes of cataract, such as ionizing radiation.

Since all people who live long enough eventually develop this anomaly of life, it is really not a question of "if," but instead a question of "when," to treat an individual's cataracts. Patients with cataracts can continue to enjoy good vision for a long time. Patients may need reassurance that they are not going blind from this naturally occurring phenomenon. Those who complain of glare can obtain a prescription for antireflective coating on their lenses or sunglasses. Nonoptical suggestions include the use of sun visors.

Patients with cataracts need to be monitored on a regular basis. When eyeglass prescription changes no longer improve vision to suitable levels for daily living or the tasks at hand, a surgical solution should be discussed. Surgery includes removing the opacified crystalline lens from its capsule and inserting an intraocular lens into the eye in its place. This has become an outpatient procedure with a very high success rate. Patients often find that their vision returns to levels of their youth, provided they have no other ocular pathology causing vision loss. Patients may need reading glasses after the surgery because implants do not have the capacity to change shape and focus.

Other Issues

One of the common complications of cataract surgery is cystoid macular edema, or "Irvine Gass syndrome." The incidence of this complication is on the decline as cataract surgery procedures have been improved with new "small incision technologies" and new anti-inflammatory medications. Sometimes, infection or delayed-onset inflammation ensues; this complication is called endophthalmitis and is serious, but rare. Occasionally following the procedure, an individual's eye pressure may rise or fall. Either complication requires the attention of a trained eye care professional. Finally, in some instances, from 3 months to years after surgery, a patient's visual acuity may decrease several lines due to the growth of fibers left behind inadvertently over the capsule (Elschnig's pearls, or after-cataract). This can easily be corrected by a procedure known as a YAG capsulotomy, in which a laser beam, fired from outside the eye, is aimed at the cloudy structure and used to open a clearing, allowing the light to again pass through uninterrupted.

The decision for surgery should be made based on age, occupation, avocation, and lifestyle. Most patients today get a lens implant (pseudo-phakia). If for any reason an intraocular lens implant is not indicated (aphakia), the patient will need a contact lens for that eye so that one thick eyeglass lens is not required. If for any reason a contact lens is not advisable, the patient will have difficulty using both eyes together as the spectacle correction for the aphakic eye would produce an image that is 70% larger than the normal eye, resulting in diplopia.

Further Reading

Belkacemi Y, Ozsahin M, Pene F, et al. Cataractogenesis after total body irradiation. *International Journal of Oncology, Biology, and Physics* 1996; 35: 53-60.

DeBlack SS. Cigarette smoking as a risk factor for cataract and age-related macular degeneration: A review of the literature. *Optometry* 2003; 74: 99-110.

Lipman RM, Tripathi BJ, Tripathi RC. Cataracts induced by microwave and ionizing radiation. *Survey of Ophthalmology* 1988; 33: 200-210.

Lodi V, Fregonara C, Prati F, et al. Ocular hypertonia and crystalline lens opacities in healthcare workers exposed to ionising radiation. *Arh Hig Rada Toksikol* 1999; 50: 183-187.

Neale RE, Purdie JL, Hirst LW, et al. Sun exposure as a risk factor for nuclear cataract. *Epidemiology* 2003; 14: 707-712.

Negahban K, Chern K. Cataracts associated with systemic disorders and syndromes. *Current Opinions in Ophthalmology* 2002; 13: 419-422.

Nelson ML, Martidis A. Managing cystoid macular edema after cataract surgery. *Current Opinions in Ophthalmology* 2003; 14: 39-43.

Nordlund ML, Marques DM, Marques FF, et al. Techniques for managing common complications of cataract surgery. *Current Opinions in Ophthalmology* 2003; 14: 7-19.

Richer SP, Yonatan E, Harper CK, et al. A clinical review of non-age-related cataracts. *Optometry* 2001; 72: 767-778.

Schull WJ. Late radiation responses in man: Current evaluation from results from Hiroshima and Nagasaki. *Advances in Space Research* 1983 3: 231-239.

Taylor HR. Ocular effects of UV-B exposure. *Documenta Ophthalmologica* 1994-95; 88: 285-293.

Wilson ME, Pandey SK, Thakur J. Paediatric cataract blindness in the developing world: Surgical techniques and intraocular lenses in the new millennium. *British Journal of Ophthalmology* 2003; 87: 14-19.

Coal Workers' Pneumoconiosis

ICD-10 J60

Michael Attfield and Gregory R. Wagner

Coal workers' pneumoconiosis (CWP) is one of the lung diseases arising from inhalation and deposition of respirable coal mine dust in the lungs, and from the reaction of the lungs to the dust. It is a chronic, irreversible disease of insidious onset, usually—but not always—requiring 10 or more years of dust exposure before appearing on chest x-ray. It is characterized by abnormalities visible as small or large opacities (spots) on chest x-ray. When only small opacities are present, the condition is called chronic or simple CWP. Complicated CWP or progressive massive fibrosis (PMF) are the terms used when opacities greater than 1 centimeter attributable to coal dust exposure are present on x-ray. Obsolete terms applied to the same conditions include anthracosis, anthracosilicosis, miners' phthisis, and miners' asthma.

For purposes of compensation, various jurisdictions define pneumoconiosis differently. For example, in the U.S. federal compensation system for miners, pneumoconiosis is defined as "a chronic dust disease of the lung arising out of employment in an underground coal mine" [Federal Mine Safety and Health Act, Section 402(b)]. Thus, eligibility for benefits under the federal act is not limited to dust effects visible on chest x-ray. Disease definitions delineating eligibility for state workers' compensation benefits also vary from state to state.

The ILO disseminates a conventional method for x-ray classification that is employed in the recognition and categorization of the pneumoconioses, including CWP. This method classifies opacities according to their shape, size, location, and profusion. Profusion is determined by comparing the miner's film with "standard" ILO films. There are four major categories of increasing profusion of opacities: 0, 1, 2, and 3. Whichever standard film most closely matches that of the miner determines the "major category" of profusion. If the film is in a border area between two major categories, both categories are noted, with the category most like the film noted first. For example, a film that shows a higher profusion of opacities than the category 1 standard film, but that is more like the 1/1 standard than like the 2/2 standard, is classified 1/2. PMF is classified as category A, B, or C, depending on the size of the large opacities.

The ILO system was originally established to achieve consistency in film interpretation during the conduct of health surveillance or epidemiological investigations. In the United States, readers trained in this method of interpretation who pass a competency test administered by NIOSH are designated as B readers. However, despite efforts to achieve standardized interpretations of chest x-rays through use of the ILO system, significant variability is often found between and within readers regarding the presence and severity of CWP.

Coal workers' pneumoconiosis has characteristic pathological features, which can be seen on autopsy or biopsy and are described in standard reference works. Tissue examination is not necessary for the diagnosis of the disease.

Occurrence

The prevalence of CWP increases with increasing dust exposure and also varies with coal rank (essentially, the age of the coal). Prevalence data from a federal underground coal miner medical monitoring program for 1996-2002 show an overall prevalence of category 1/0 or greater CWP of 3.2%, and a prevalence of PMF of 0.2%. The prevalence of CWP varies around the US, with rates in Virginia and West Virginia being about eight times higher than in Illinois and Alabama. CWP is more prevalent in workers in small mines. The data also indicate that about 6% of underground miners with 25 years or more mining experience had category 1/0 or greater CWP. These statistics are very consistent with some projections based on British data suggesting that 6.2% of miners will have simple CWP and 0.4% will develop PMF after 35 years of exposure at the current US limit of 2 mg/m^3. In the U.S., about 1000 coal miners died with CWP noted on their death certificates in 1999. Surface coal workers are also at risk of developing lung diseases from dust exposure. This risk is increased for certain workers such as drillers, as well as for surface miners with prior underground experience.

Causes

Respirable coal mine dust causes CWP. Respirable dust is any dust that is small enough to be deposited in the terminal bronchioles or alveolar airspaces (generally less than 5 μm in diameter).

Coal mine dust is a mixed dust consisting mostly of coal particulate but also including other minerals found in mines, particularly silica. Certain coal mining jobs, such as surface drilling and underground roof bolting, may involve high concentrations of silica dust, and give rise to silicosis (see Silicosis) in addition to CWP. On autopsy or biopsy examination of lungs, findings of CWP and silicosis may coexist. However, the x-ray appearances of CWP and silicosis are virtually identical, and the separate diseases typically cannot be distinguished radiographically.

Pathophysiology

Inhaled fine particles of coal dust are scavenged by specialized lung cells (macrophages). These cells and dust particles accumulate deep in the respiratory tree near or in the air exchange units (alveoli). Abnormal fibrotic material (reticulin and collagen) may develop. Localized areas of tissue destruction (focal emphysema) may also occur. The areas demonstrating a combination of these abnormalities are called coal macules. In some cases, nodules consisting of macrophages and a greater quantity of abnormal fibrotic tissue are present.

These can clump together (coalesce) to form the large lesions of PMF. Generally, there is significant destruction of lung tissue as the nodules coalesce.

The definition of the exact physiological abnormalities resulting from the development of CWP has been complicated, in part, by the diversity of pulmonary responses to coal mine dust. Increasing dust exposure has been associated with progressive loss of lung function, resulting in the development of obstructive lung disease. This loss appears similar in magnitude to that caused by regular cigarette smoking. Miners have increased rates of emphysema and chronic bronchitis.

Coal workers' pneumoconiosis is itself an effect of exposure to dust. When PMF destroys and distorts lung tissue, it is typically associated with loss of lung function that can have predominantly obstructive, but also restrictive, or combined patterns of abnormality. Simple CWP alone may or may not be associated with diminished function in any particular individual. Nevertheless, individuals with simple CWP may experience significant loss of lung function as a result of the same exposure (coal mine dust) that caused the development of CWP (see chapters on Emphysema and Chronic Bronchitis).

Prevention

Primary prevention of CWP is achieved by reducing exposure to coal dust through improved ventilation and dust suppression supported by enforcement of strict dust control standards. Feasible and effective engineering dust controls exist for surface and underground coal mining operations. Current preventive efforts focus on suppressing the "respirable" fraction of coal mine dust (less than 10 μm in diameter) that appears to cause CWP. In the US, the permissible exposure level is a time-weighted average of 2.0 mg/m^3 measured as a personal sample. NIOSH has recommended improved prevention through an exposure limit of 1 mg/m^3 for respirable coal mine dust and a 0.05 mg/m^3 limit for respirable crystalline silica. Control of other lung diseases associated with dust exposure in mining which may be caused by larger dust particles may or may not be achieved by suppressing the respirable dust fraction.

Prevention of pulmonary impairment associated with PMF has been based on an assumption that PMF almost invariably develops on a background of advanced simple CWP; therefore, secondary prevention efforts have been directed at identifying miners with early simple CWP (category 1) and at further reducing the dust exposure for these miners. Later reports have brought that assumption into question, with the finding that significant numbers of PMF cases were developing among miners with early simple CWP.

Coal workers' pneumoconiosis does not resolve or improve with the elimination of exposure to coal dust, and may, in some cases, progress from simple CWP to PMF in the absence of additional exposure. Rehabilitation efforts (tertiary prevention) are the same as those employed for anyone with disabling lung disease: elimination of adverse environmental exposures; immunization against influenza and pneumococcal infection; early recognition and treatment of infection; education directed at improved levels of self-care; graded exercise; and consideration of medications such as bronchodilators.

Other Issues

Some miners develop disabling lung disease in the absence of PMF. Periodic pulmonary function testing might be useful in early disease recognition and secondary prevention.

Efforts to control CWP through the monitoring and reduction of respirable dust exposure may not be adequate to control the development of pulmonary impairment in miners. The absence of a positive finding for CWP on x-ray does not ensure the absence of significant disease from coal mine dust exposure.

Preventive strategies directed toward eliminating PMF through early identification of miners with simple CWP may be inadequate.

Coal mining communities are often isolated, with few local alternative sources of employment. Miners developing lung disease are reluctant to eliminate exposure to dust by leaving the industry when unemployment is the only alternative.

Cigarette smoking has no apparent effect on the formation of the coal macule, but adds to the miner's risk of developing emphysema, chronic bronchitis, and airway obstruction.

Modern underground mining techniques such as longwall mining result in high levels of coal mine dust being generated and thereby requiring careful identification of sources of dust and attention to control methods.

Dust regulations in the U.S. assume a 5-day, 8-hour-per-shift workweek; however, other considerations have resulted in increasing numbers of shifts per worker in some areas. The impact of these changes on health has not been evaluated.

Further Reading

Attfield MD, Wagner GR. Respiratory disease in coal miners. In *Environmental and Occupational Medicine* (3rd ed.). Philadelphia: Lippincott Raven, 1998, pp. 413-434.

ILO. Guidelines for the Use of the ILO *International Classification of Radiographs of Pneumoconioses*, 2000 edition, Geneva: International Labor Office 2002 (Occupational Safety and Health Series, No. 22 [rev. 2000]).

Kleinerman J, et al. Standards for the pathology of coal workers' pneumoconiosis. *Archives of Pathology* 1979; 103: 375-432.

Kissell FN (Ed.). *Handbook for Dust Control in Mining.* IC 9465. (DHHS [NIOSH] Publication No. 2003-147). Washington, DC: NIOSH, 2003.

National Institute for Occupational Safety and Health. *Occupational Exposure to Respirable Coal Mine Dust: Criteria for a Recommended Standard.* Cincinnati, OH: NIOSH, 1995.

National Institute for Occupational Safety and Health. *Work-related Lung Disease Surveillance Report*, 2002. Cincinnati, OH: NIOSH, 2003.

Weeks JL, Wagner GR. Compensation for occupational disease with multiple causes: the case of coal miners' respiratory diseases. *American Journal of Public Health* 1986; 76: 58-61.

Collective Stress Disorder

ICD-10 F43

James L. Weeks

Collective stress disorder (also referred to as mass psychogenic illness, mass sociogenic illness, or epidemic hysteria [see Other Issues, below]) is the occurrence of (a) usually similar symptoms among a highly variable number of individuals, (b) in a shared environment, (c) in the absence of an identifiable organic or physical cause, and (d) with little or no clinical or laboratory evidence of disease. Symptoms may include headache, fatigue, dizziness, hyperventilation, nausea, vomiting, and occasionally fainting. Although symptoms may appear to be spectacular, no serious or life-threatening signs are observed. Individuals typically have shared beliefs about the causes. This phenomenon typically occurs in groups under stress and follows an event that acts as a trigger, such as an odor, sound, visual event, or manifest illness. Symptoms may arise very rapidly or over a matter of days and many decline quickly, although in some episodes symptoms persist for 2 to 3 months. Relapses are common. (See also entries on Stress and on Multiple Chemical Sensitivities.)

Occurrence

Knowledge about occurrence is limited to published reports of episodes. Episodes are reported to occur most commonly among schoolchildren and workers, but also among prisoners and other groups that are in situations of stress and dependency. From 1973 to 1993, 70 episodes were reported in the scientific literature: 50% occurred in schools, 29% at work, and the remainder in a wide variety of organizational settings.

It is more common among women than any other segment of the workforce. This is likely a consequence of the kinds of jobs and work environments in which women work rather than any innate hypersusceptibility. There are some episodes reported among all-male military recruits.

Causes

Almost without exception, published incidents of collective stress disorder in occupational settings occur in a workforce undergoing stress in the form of, for example, job insecurity, rigidly paced work, pressure to work faster, or poor labor-management relations. Nonoccupational episodes, such as among schoolchildren, share similar characteristics: underlying stress, lack of control, shared beliefs, in highly structured (even rigid) social settings. Such conditions, combined with lack of control over them, a sense of isolation, and lack of social support, seem to predispose workers to react to adverse precipitating events, such as frank illness, foul odor, noise, or social conflict.

When investigating an episode, the possibility of organic or physical causes should carefully be considered, based on existing medical knowledge and experience. They are more likely to be contributing factors if the occurrence of symptoms matches exposure in time and place. A mismatch suggests the absence of a common response to a toxic exposure. In other words, if groups of people sharing the same environment remain totally unaffected, this should raise the suspicion of mass psychogenic illness.

Pathophysiology

The pathophysiology is unknown. It seems likely, however, that if a group of workers is acutely under stress, in a rigid social environment, over which they have little control, then some (perhaps many) individuals may have heightened vigilance and be hyperresponsive. (See chapter on Stress.) If so, then exposure even to low concentrations of acutely hazardous agents, particularly if they are odorous, may be sufficient to precipitate symptoms. If beliefs about hazards are shared, this may lead to similar symptoms among others.

Prevention

Given the unpredictability of events that might precipitate an episode of collective stress, a strategy specifically to prevent these kinds of events is virtually impossible. Rather it makes more sense to address underlying stress-creating work conditions that predispose to collective stress and give rise not only to this kind of event but are also associated with many other adverse outcomes: musculoskeletal disorders, back pain, injuries, stress, sleep loss, depression, suicide, Raynaud's phenomenon, and others. Readers are urged to consider chapters on the disorders noted above in Part II plus the chapter on Work Organization in Part III.

Response of health professionals can affect secondary prevention efforts. The inability to identify organic or physical causes may delay investigations and delay itself may add to stress in the affected workers. Rapidity of the response, a high degree of transparency, and a recognized independence on the part of the investigators are important conditions of success in solving an acute episode of collective stress disorder. Concluding that there is no organic cause, however, can obscure instances where there is or was exposure to toxic or physical agents. One should not conclude that an organic cause is absent without a thorough investigation. Even then, it is a conclusion reached by process of elimination. Whether the cause was stress or a toxic or physical agent are, however, not mutually exclusive. It is entirely possible that a population under stress—and therefore hyperresponsive—that is exposed to a manifest hazard could experience adverse effects that were magnified by existing conditions.

Other Issues

Labeling events as psychogenic, hysterical, or delusional can exacerbate conditions by implying that the victims themselves are malingering and, in some sense, responsible for their own illnesses. Investigations of some episodes have shown that personality traits such as neuroticism, extroversion, and introversion do not distinguish the affected from others. Thus, we suggest the term "collective stress disorder," aiming to focus attention on the stressful work environment and precipitating events rather than the victims' state of mind or personality.

Further Reading

Boss LP. Epidemic hysteria: A review of the published literature. *Epidemiology Reviews* 1997; 19: 233-243.

Bowler RM, Mergler D, Rauch SS, et al. Stability of psychological impairment: two year follow-up of former microelectronic workers' affective and personality disturbance. *Women's Health* 1992; 18: 27-48.

Colligan MJ, Pennebaker JW, Murphy LR (eds.). *Mass Psychogenic Illness: A Social Psychological Analysis*. Hillsdale, NJ: Lawrence Erlbaum Associates, 1982.

House RA, Holness DL. Investigation of factors affecting mass psychogenic illness in employees in a fish-packing plant. *Amerian Journal of Industrial Medicine* 1997; 32: 90-96.

Johnson JV. Control, Collectivity and the Psychosocial Work Environment. In Sauter SL (ed.). *Job Control and Worker Health*. New York: John Wiley & Sons, 1989, pp. 55-74.

Magnavita N. Individual mass psychogenic illness: the unfashionable diagnosis. *British Journal of Medical Psychology* 2000; 73: 371-375.

Nemery B, Fishler B, Boogaerts M, et al. The Coca-Cola incident in Belgium, June, 1999. *Food and Chemical Toxicology* 2002; 40:1657-1667.

Struewing JP, Gray GC. An epidemic of respiratory complaints exacerbated by mass psychogenic illness in a military recruit population. *American Journal of Epidemiology* 1990; 132: 1120-1129.

Colorectal Cancer

ICD-10 C18-C20

Elizabeth Ward

Abdominal pain, weight loss, symptoms related to anemia, the presence of blood in the stool and bowel, and stool changes are the leading symptoms of colorectal cancer, depending on tumor location and stage of disease. Symptoms are uncommon in the early stages of disease. Early detection is possible through fecal occult blood testing (FOBT), sigmoidoscopy, colonoscopy, and barium enema.

Occurrence

An estimated 147 500 new cases of colorectal cancer and 57 100 colorectal cancer deaths occurred in 2003 in the United States. Rectal cancers comprise an estimated 42 000 (28.5%) of the new cases. Colorectal cancer is the third most common cancer in men and women (excluding skin cancer), and accounts for about 10% of cancer deaths. Because of the large variations in incidence around the world, colon cancer is thought to have major environmental determinants, particularly from diet and physical inactivity. Risk factors may differ for colon and rectal cancers. There are no formal estimates of the proportion of colorectal cancers that are related to occupation.

Causes

Physical inactivity is consistently associated with increased risk of colon, but not rectal, cancer. Many studies that have examined occupational physical activity report 1.5 to 2.5-fold higher incidence or death rates among the most sedentary, compared to the most active, occupations. The evidence regarding other occupational determinants of colorectal cancer is more limited and/or less consistent. Although the relationship between asbestos exposure and colorectal cancer has been extensively investigated, results are inconsistent, with some strongly positive and a number of negative studies. A meta-analysis published in 1994 found a relationship for amphibole, but not serpentine, asbestos, and another meta-analysis reported in 1995 found no increased risk. Several studies have found that wood dust exposure may be protective for colorectal cancer, although consistent increases in colon cancer risk have been observed among pattern and model workers in the automobile industry who have exposure to wood dusts but also adhesives, resins, paints, and solvents. Increased colon cancer has been reported among workers involved in the manufacture of cellulose acetate fiber and polypropylene, and workers exposed to mineral oils used in printing press operations, typesetting, and textiles. A recent death certificate study reported that occupational exposure to sunlight was protective against colon cancer, independent of physical activity. Two studies found excess colon cancer in drycleaning workers, but other studies have not.

Increased risks of colon cancer have also been observed in one or more studies for farmers (although other studies report decreased risks), workers in the petroleum and coal products industry, coke-gas furnace workers, machinists and job setters, leather and tannery workers, shoemakers, manufacturers of meat, poultry, and fish products, meatpackers, chemical processors, textile workers, electrical workers and electricians, stainless steel grinding, conservationists, iron and steel workers, metalsmiths and foundry workers, and workers in lens production and metal frame manufacturing in the optical industry, printing and publishing, and manufacturing, business and services.

An increased risk of rectal cancer associated with exposure to acrylonitrile has been reported in a case-control study, but not in cohort studies. Rectal cancer has been associated with machining in general and specifically with exposure to straight (mineral) oils, employment in dusty jobs, and farming. A case-control study found associations of rectal cancer with rubber dust, rubber pyrolysis products, cotton dust, wool fibers, rayon fibers, a group of solvents, polychloroprene, glass fibers, formaldehyde, extenders, and ionizing radiation; however, many of the exposures were correlated with each other, so independent effects could not be discerned.

A variety of nonoccupational exposures, such as obesity, diabetes mellitus, inflammatory bowel disease, meat consumption, and cigarette smoking, are associated with increased risk, while regular use of nonsteroidal anti-inflammatory drugs (NSAIDs), including aspirin, postmenopausal hormones, and folic acid supplementation are associated with reduced risk. Genetic factors predispose to the development of colorectal cancer. There is an increased incidence in first-degree blood relatives of patients. The hereditary conditions familial adenomatous polyposis (FAP) and hereditary non-polyposis colon cancer (HNPCC) account for approximately 5% of cases. The association between ulcerative colitis and higher frequency of colorectal cancer is well known. There is conflicting evidence regarding the relation of dietary fiber and fat consumption to colorectal cancer risk.

Pathophysiology

Most colon cancers develop from adenomatous polyps. It is thought that in persons without an inherited genetic predisposition, early mutations that disrupt function of both alleles of the adenomatous polyposis coli (APC) gene cause abnormalities in cell proliferation, migration, and adhesion, resulting in the formation of an adenomatous polyp. This initial event may involve a bloodborne carcinogen, as the stem cells are not in contact with the luminal surface. Formation of a polyp or microadenoma involves clonal proliferation of cells that have lost APC function, increasing the probability of additional mutations and progression towards invasive cancer.

Prevention

Primary prevention includes reducing or eliminating exposure to known or suspected carcinogens. Workplace policies that promote a healthful diet and regular physical activity can have broad health benefits. Worksite-based health promotion and opportunities for recreational physical activity are especially important for workers whose occupations require them to be sedentary.

Screening is effective in reducing colon cancer incidence through removal of precancerous lesions, as well as in detecting cancers at an early stage. The American Cancer Society recommends several options for colorectal cancer screening in average-risk individuals age 50 and older. These include fecal occult blood testing (FOBT) annually and flexible sigmoidoscopy every 5 years (which is preferred to FOBT or flexible sigmoidoscopy alone); or colonoscopy every 10 years; or a barium enema examination every 5 years. Colonoscopy at a younger age and at more frequent intervals is recommended for individuals with a previous adenomatous polyp or family history of colorectal cancer. There are no specific recommendations for high-risk occupational groups, but a prudent course would be to begin screening at age 40. Employer-sponsored insurance plans should cover a full range of colon cancer screening examinations. Surgery is the main treatment for colon cancer. Both radiation and chemotherapy may also be recommended. Use of chemotherapy after surgery has been shown to increase the survival rate for some stages of colon and rectal cancer. The five-year survival rate in the U.S. for the 1992-1998 period was 62%; for the 37% of cases diagnosed at a localized stage the five-year survival rate was 90%.

Further Reading

All about colon and rectum cancer: Detailed guide. Learn about cancer [serial online]. American Cancer Society, Atlanta, GA [updated 01/04/2001]. Available from URL: http://www.cancer.org/docroot/CRI/CRI_2_3x.asp?dt=10

Goldberg MS, Parent ME, Siemiatycki J, et al. A case-control study of the relationship between the risk of colon cancer in men and exposures to occupational agents. *American Journal of Industrial Medicine* 2001; 39: 531-546.

Homa DM, Garabrant DH, Gillespie BW. A meta-analysis of colorectal cancer and asbestos exposure. *American Journal of Epidemiology* 1994; 139: 1210-1222.

Hsing AW, McLaughlin JK, Chow WH, et al. Risk factors for colorectal cancer in a prospective study among U.S. white men. *International Journal of Cancer* 1998; 77: 549-553.

Longnecker MP, Gerhardsson le Verdier M, Frumkin H, et al. A case-control study of physical activity in relation to risk of cancer of the right colon and rectum in men. *International Journal of Epidemiology* 1995; 24: 42-50.

Potter JD, Slattery ML, Bostick RM, et al. Colon cancer: A review of the epidemiology. *Epidemiologic Reviews* 1993; 15: 499-545.

Potter JD. Colorectal cancer: molecules and populations. *Journal of the National Cancer Institute* 1999; 91: 916-932.

Schottenfeld D, Winawer SJ. Cancers of the large intestine. In Schottenfeld D, Fraumeni JF, Jr. (Eds.) *Cancer Epidemiology and Prevention* (2nd ed.). New York: Oxford University Press, 1996, pp. 813-840.

National Institutes of Health. What you need to know about cancer of the colon and rectum. What you need to know about cancer publications [serial online]. National Cancer Institute, National Institutes of Health, Bethesda, MD [updated 9/16/2002]. Available from URL: http://www.nci.nih.gov/cancerinfo/wyntk/colon-and-rectum

Congestive Heart Failure

ICD-10 50.0-50.9

Kenneth D. Rosenman

Congestive heart failure (CHF) is a clinical syndrome due to heart disease and is characterized by breathlessness and abnormal sodium and water retention, resulting in edema. The congestion may occur in the lungs and/or the peripheral circulation, depending on whether the heart failure is left-sided, right-sided, or general.

Symptoms include dyspnea (shortness of breath), fatigue, weakness, nocturia (urination at night), and, with advanced impairment, confusion, memory loss, and other symptoms that reflect decreased brain perfusion. The most common physical findings are crackles, a gallop heart sound (S_3), and edema. The diagnosis is confirmed by enlargement of the heart and a change in bronchovascular markings on chest x-ray.

Electrocardiographic (ECG) findings may reflect the underlying cause of the congestive heart failure, but there are no ECG findings specific to heart failure. Abnormalities in kidney and liver function may be detected in severe late-stage heart failure. Tests for the cause of heart failure may include echocardiography, cardiac catheterization, and endomyocardial biopsy.

Occurrence

In the United States, the reported prevalence of heart failure is 5 million and the reported incidence is 550 000 annually, with approximately 1 million hospitalizations. Incidence increases with age: five times higher for men in the 60s than for men in their 40s. The percentage of cases that is occupationally related is not known.

Causes

Coronary artery disease and hypertension are, by far, the two most common causes of CHF, accounting for at least 75% of all cases. There are many less common causes, including infections, poor nutrition (both vitamin deficiency and excess alcohol), connective tissue problems, and genetic diseases. CHF has been associated with ingestion of beer contaminated with arsenic (in 1900) and cobalt (in the 1960s) and, very rarely, with occupational inhalation of cobalt. CHF may be a secondary occupational disease if occupational exposures have caused ischemic heart disease or hypertension, or if occupational exposures have caused either chronic obstructive or restrictive lung disease that caused cor pulmonale (with hypertrophy of the right ventricle). (See chapters on Ischemic Heart Disease and on Hypertension for more discussion on causation.)

Pathophysiology

CHF is the failure of the heart to pump blood at a sufficient rate to meet the physiological requirements of the organs. Although it is usually caused by failure of the heart muscles (most commonly due to ischemic heart disease and hypertension), similar symptoms can be engendered by conditions that cause increased blood volume (such as renal failure) or inadequate filling of the ventricular chambers of the heart (such as constrictive pericarditis). Direct damage to myocardial muscle can occur from infectious agents, certain medications, nutritional excesses or deficiencies, and toxins (such as arsenic and cobalt). Cobalt myocardiopathy causes unique pathological findings. The synergistic effect of alcohol, cobalt, and a protein-poor diet on enzyme metabolism has been suggested as the mechanism responsible for this condition.

Prevention

Prevention of the disease underlying CHF is necessary for primary prevention. This means reduction in exposures to substances that cause direct myocardial toxicity and reductions in exposures and physical and psychological stress that cause ischemic heart disease or hypertension. Although alcohol ingestion and a protein-poor diet are thought to play a role in cobalt cardiomyopathy, not enough is known about actual amounts to recommend specific dietary guidelines other than alcohol in moderation and a well-balanced diet for industrial workers exposed to cobalt. A wellness program that combines reduction of workplace exposures and stress with lifestyle modifications can focus on several factors that cause or contribute to CHF.

Other Issues

Arsenic cardiomyopathy has been described only with ingestion of contaminated beer. Ischemic heart and peripheral vascular disease have been described with industrial and environmental exposure to arsenic. Carbon monoxide and solvent exposure may precipitate an acute exacerbation of CHF. Individuals with underlying ischemic heart disease will be more susceptible to the effects of carbon monoxide and solvents.

Further Reading

Hertz-Picciotto I, Arrighi HM, Hu SW. Does arsenic increase the risk for circulatory disease? *American Journal of Epidemiology* 2000; 151: 174-181.

Jarvis JO, Hammond E, Meier R, et al. Cobalt cardiomyopathy. A report of two cases from mineral assay laboratories and a review of the literature. *Journal of Occupational Medicine* 1992; 34: 620-626.

Seghizzi P, D'adda F, Borleri D, et al. Cobalt myocardiopathy. A critical review of literature. *The Science of the Total Environment 1994*; 150: 105-109.

Conjunctivitis

ICD-10 H10, H11, H13

Satya B. Verma and Andrew S. Gurwood

Toxic/allergic conjunctivitis results from acute exposure of the conjunctiva to a dusty or vapor-laden atmosphere. Atopic conjunctivitis is an allergic reaction to one (or more) of many allergens. Symptoms of both include tearing and itching eyes, and possibly edema and nasal discharge.

Occurrence

Workers in agriculture and related industries may suffer allergic reactions while exposed to various plant or animal material and other allergens, including chemicals and pesticides. For example, workers exposed to pollens during artificial pollination and workers handling green plants, such as the weeping fig (*Ficus benjamina*), have been reported to suffer from conjunctivitis and/or rhinitis and asthma. There are no reliable data on the occurrence of occupationally induced conjunctivitis.

Causes

Toxic/allergic conjunctivitis can be caused by acids, alkalis, aerosols, solvent vapors, and airborne dusts. Atopic conjunctivitis may be caused by exposure of susceptible individuals to one (or more) of many allergens, most of which are airborne. Pollens are most common; other allergens include animal epidermal products, nonpathogenic fungi, vegetable and animal proteins, hair, wool, feathers, industrial chemicals, and pesticides. Workers with occupational exposure to polychlorinated biphenyls (PCBs) have been reported to suffer from hypersecretion of the meibomian glands, swelling of the upper eyelids, and hyperpigmentation of the conjunctivae.

Pathophysiology

The presentation of toxic/allergic conjunctivitis varies in intensity from hyperemia to necrosis, depending on the irritant and exposure. It is particularly marked in the palpebral aperture when due to vapors or dusts, and in the lower fornix area when due to liquids. It is accompanied by watery discharge in mild cases and by mucopurulent discharge in severe cases.

Atopic conjunctivitis is mediated by histamine release. In its most severe state, it can result in swelling of the eyelids, superficial necrosis of the eyelids, involvement of regional lymph nodes, and formation of a psuedomembrane of the conjunctivae.

Prevention

Toxic/allergic conjunctivitis can be prevented by applying standard prevention strategies, including industrial hygiene evaluation and control methods, medical surveillance, and medical monitoring. Prevention of atopic conjunctivitis is more difficult because reactions vary significantly among individuals and may be provoked by very low concentrations of airborne allergens.

Treatment in both instances requires substance identification, preferably without challenging atopic individuals. Ideally, the toxin should be removed from the workplace. Alternatively, workers may have to be transferred to areas that are clear from contamination. Use of protective eyewear that forms an airtight seal around the eyes or face (if the reaction is not limited to conjunctivitis) may be useful, but the best method is avoidance.

Other Issues

Although conjunctivitis is usually not considered a serious medical problem, it can have a significant adverse effect on workers. It can be debilitating and can reduce attentiveness, which can, in turn, result in injuries and impaired performance.

People with a history of atopy, allergic rhinitis, or asthma are at increased risk. However, excluding such people from certain jobs or workplaces a priori, based only on a history of atopy, is not warranted.

Further Reading

Duane DD, Jaeger EA (Eds.). *Clinical Ophthalmology. Vol. 4: External Disease.* Philadelphia: Harper & Row, 1987.
Friedlander MH, Okumoto M, Kelly J. Diagnosis of allergic conjunctivitis. *Archives of Ophthalmology* 1982; 100: 1275.

Contact Dermatitis

ICD-10 L23, L24
SHE/O

Michael E. Bigby

Irritant contact dermatitis is defined as cutaneous inflammation that develops as a result of the direct effect of chemicals on the skin. Prior sensitization is not required, and antigen-specific immune responses cannot be detected. Irritant contact dermatitis will develop in all workers exposed to adequate concentrations for an adequate length of time. The clinical spectrum of irritant contact dermatitis is very broad, ranging from slight erythema to large bullae, necrosis, and ulceration. Mild irritation is characterized by localized erythema, edema, papule, and vesicle formation, crusting, and scaling. Thickening of the skin may occur with chronic exposure or by scratching (lichenification). Exposure to strong irritants causes acute injury to the epidermis and sometimes to the dermis and may cause the formation of bullae and ulceration. Irritant contact dermatitis usually affects exposed surfaces, mainly the hands and forearms. If an irritant soaks into clothing, the skin underlying the soaked clothing will become affected.

Allergic contact dermatitis is defined as the development of a T-cell-mediated, antigen-specific response of a patient to a hapten or antigen applied to the skin. The development of dermatitis requires an antecedent period of exposure (induction phase), during which the patient develops an immune response. For most antigens, only a small percentage of the population is capable of developing the immune response.

The clinical spectrum is broad. Pruritus (itching) is the primary symptom. Typically, there are grouped or linear papules, vesicles, or bullae. Erythema (redness), edema (swelling), exudation, crusting, and scaling are typically present. As with irritant contact dermatitis, lesions occur in exposed or contact areas.

Occurrence

According to Bureau of Labor Statistics (BLS) data, the rate of reported skin diseases has fallen steadily over the past decade. The rate of cases per 10,000 full-time workers fell steadily from 8.2 in 1992 to a nadir of 4.3 in 2001. The overall rate was 5.1 in 2002; the highest rate (18.8) was in agriculture, forestry, and fishing (Table 1). Within this category, rates were highest for production of crops (26.4 cases per 10,000 full-time workers), forestry (25.7), and agricultural services (17.1).

Cases of occupational skin disease, approximately 80% of which are irritant contact dermatitis, represent the second most common occupational disease reported to the BLS. Available national data sources do not contain information on the occurrence of specific dermatologic conditions. Incidence rates based on workers' compensation claims differ, sometimes more than several fold, from those measured from nationally reported sources, but the rank order of major industrial groupings remains the same. The annual cost of occupational skin diseases, including lost production, medical costs, and lost wages, is estimated to range from $222 million to $1 billion (in 1984 dollars).

Most contact allergens produce sensitization in only a small percentage of exposed individuals. Exceptions to this rule are poison ivy, poison oak, and poison sumac, from which more than 70% of those who are exposed develop allergic contact dermatitis.

According to workers' compensation claims reported to a register of occupational skin disease in Northern Bavaria, Germany, from 1990 to 1999, the annual incidence rate of irritant contact dermatitis was 4.1 cases per

Table 1. Reported incidence rates for occupational skin disease, United States, 2002.

Industry	Rate of reported skin diseases per 10,000 full-time workers
Agriculture, forestry, fishing	18.8
Manufacturing	8.4
Construction	3.9
Transportation and public utilities	3.8
Wholesale trade - durable goods	3.1
Wholesale and retail trade	2.5
Mining	1.5

10,000 workers. The highest annual irritant contact dermatitis rates (per 10,000 workers) were in hairdressers (47), bakers (24), and pastry cooks (17); results of a questionnaire indicated that workers were frequently exposed to detergents (52%), disinfectants (24%), and acidic and alkaline chemicals (24%).

Causes

Among the substances causing irritant contact dermatitis are strong alkalis and acids, soaps and detergents, and many organic compounds. Aggravating factors include reduced humidity at work, friction, occlusion, and excess environmental heat. Sometimes airborne factors, such as dusts, fibers, acids, alkalis, gases, or vapors, cause irritant contact dermatitis.

Contact allergens commonly found in the workplace are nickel salts, epoxy resins, chromium salts, paraphenylenediamine, and formaldehyde. Sometimes airborne substances cause allergic contact dermatitis.

Pathophysiology

Contact irritants cause direct injury or necrosis of epidermal cells and cells within the dermis. Injury leads to the release of inflammatory mediators, including histamines, prostaglandins, leukotrienes, and cytokines. Inflammatory mediators attract a mixed cellular infiltrate into the skin, which may cause further damage and is responsible for the typical histological picture seen in acute irritant dermatitis.

Allergic contact dermatitis is a form of delayed-type hypersensitivity. It results from the development of T-cell-mediated immunity in individuals to contact allergens. The period from initial contact to the mounting of an immune response (the induction phase) ranges from 5 to 21 days. The time period to develop an allergic contact dermatitis after reexposure (the elicitation phase) is 1 to 3 days. The typical reaction of a sensitized person after reexposure to a contact sensitizer is the appearance of an eczematous dermatitis in 1 to 3 days and its disappearance within 2 to 4 weeks. With heavy exposure, exposure to potent sensitizers, or continual exposure, lesions appear more quickly (within 6 to 12 hours) and heal more slowly.

Prevention

Prevention is of primary importance in contact dermatitis. Skin contact with the irritant or allergen should be eliminated or significantly reduced. This goal can be accomplished by substituting a less irritating substance or nonallergen; implementing engineering measures, such as enclosing the process in which the substance is used; changing work practices to reduce the likelihood of worker exposure to the substance; and using gloves and protective clothing. Cleanliness of the worker and good housekeeping of the workplace are important, as are good ventilation, optimal temperature (17°C to 22°C) and humidity (about 50%), education of workers, and labeling of irritant substances. Availability of showers and wash facilities and time to use them at the workplace are important.

Dermatitis from metalworking fluids, especially from water-soluble oils, is common. Exposure can be controlled by changing fabrication methods, keeping the fluids as clean as possible, and judiciously selecting and using germicides (which are known contact allergens). Personal protective devices, such as gloves and aprons, must be selected in accordance with specific job requirements as well as with the properties of the irritants from which protection is needed. The effect of barrier creams is controversial.

In agriculture, including forestry, substitution and other engineering controls are difficult to implement for vegetables, or in the environment, such as poison ivy and poison oak. Exposure can be reduced by substituting mechanical for manual handling of these products or by using personal proactive equipment (PPE).

Other Issues

Fibrous glass causes a pruritic rash on exposed areas, usually the hands and forearms. The rash is often nonspecific, with erythema, lichenification, and small papules. Diagnosis requires a high index of suspicion. A history of occupational exposure to fibers must be sought, such as in industries where fibrous glass or asbestos is found. The small fibers are not normally visible to the naked eye, but may be visualized with a microscopic examination of tape-stripping samples from the affected area or by an examination of a biopsy specimen with polarized light.

Atopy is a constellation of clinical findings, including asthma, hay fever, and atopic eczematous dermatitis. There are no pathognomonic features of atopic disease; atopy is the major genetic condition predisposing to irritant contact dermatitis. Because workers with a history of atopic dermatitis are 13 times more likely to develop irritant contact dermatitis in the workplace, preplacement screening should always include eliciting a history of atopy. Exposure of atopic workers to irritants should be prevented or severely limited, but workers with atopic dermatitis should be excluded from a job only if avoidance of irritants is impossible.

Patch testing is an in vivo bioassay used to determine the cause of contact allergy. Patch tests are performed by placing patches with nonirritating concentrations of potential contact sensitizers onto the skin surface (usually on the upper back). The patches remain in contact with the skin for 48 hours. The patches are then removed and discarded at 48 hours after placement and evidence for a delayed-type hypersensitivity reaction (scored on erythema, edema, vesiculation, bullae formation, or frank necrosis) is sought at that time. Additional readings, usually at 96 hours and 7 days after placement, improve test sensitivity. Properly performed, patch testing is a valuable tool in evaluating workers with contact dermatitis. Considerable experience with patch testing is required to perform and interpret the test adequately.

Further Reading

Adams RM. *Occupational Skin Disease* (3rd ed.). Philadelphia: W.B. Saunders, 1999.

Rietschel RL, Fowler, JF (Eds.). *Fisher's Contact Dermatitis* (5th ed). Philadelphia: Lippincott Williams & Wilkins, 2001.

O'Malley MA, Mathias CG. Distribution of lost-work-time claims for skin disease in California agriculture: 1978-1983. *American Journal of Industrial Medicine* 1988; 14: 715-720.

O'Malley M, Thun M, Morrison J, et al. Surveillance of occupational skin disease using the supplementary data system. *American Journal of Industrial Medicine* 1988; 13: 291-299.

Taylor JS. Occupational dermatoses. *Dermatologic Clinics* 1988; 6: 1-156.

US Department of Labor, Bureau of Labor Statistics. Workplace injuries and illnesses in 2001. Washington, DC: U.S. Government Printing Office; 2002. http://www.bls.gov/iif/home.htm

Lushniak BD. The importance of occupational skin diseases in the United States. *International Archives of Occupational and Environmental Health*, 2003; 76: 325-330.

Dickel H, Kuss O, Schmidt A, et al. Importance of irritant contact dermatitis in occupational skin disease. *American Journal of Clinical Dermatology* 2002; 3:283-289.

Depression

ICD-10 F03, F20, F21, F32-F34,
F41-F44, F53

Joe Hurrell

Depressive episodes are serious medical conditions that affect thoughts, feelings and the ability to function in everyday life. In typical mild, moderate, or severe depressive episodes, an individual suffers from a lowering of mood, reduction of energy, and decreases in activity level. Capacity for enjoyment, interest, and concentration is reduced, and marked fatigue, after even minimal effort, is common. Sleep is frequently disturbed and appetite diminished. Self-esteem and self-confidence are almost always reduced and feelings of guilt or worthlessness are often present. The lowered mood is generally unresponsive to circumstances and varies little from day to day. Episodes may be accompanied by various somatic symptoms including psychomotor retardation, agitation, loss of appetite, weight loss, and loss of libido. Depending on the number, variety, and severity of the symptoms, depressive episodes can be labeled as mild, moderate, or severe.

Occurrence

Depression has become one of the most prevalent, debilitating, and costly disorders of modern times, unsurpassed by any chronic health disorder in terms of the number of people afflicted at any point in time. Nationally representative studies show a 12-month prevalence of 8% for men and 13% for women, and a lifetime prevalence of 13% for men and 21% for women. Chronic mild depression affects at least an additional 2-3% of the U.S. population annually. Depression rates are much higher for women, decline with increasing socioeconomic status, and are highest in the 15-to-54 year age group. Prevalence in the workforce (4%) is similar to that in the community (5%). Morbidity rates appear to vary considerably by occupation, even after adjustment for such potentially confounding factors as age, race, gender, and job tenure. Blue-collar workers, as compared with professional and managerial workers, appear to be at excess risk.

Judged in terms of years of life lost, premature death, and years lived with disability, depression will, by 2020, surpass all health disorders worldwide, except for ischemic heart disease. The total cost of depression to the U.S. economy alone is more than $43 billion, including direct medical costs, suicide-related mortality costs, and productivity losses—exceeding the cost of all cases of cancer, respiratory disease, AIDS, coronary heart disease, and arthritis combined.

Causes

Depression results from abnormal functioning of the brain. The specific causal mechanisms are not known, but interaction between genetic predisposition and life history appears to determine an individual's level of risk. Episodes of depression can be triggered by stress, illnesses, difficult life events, various medications, or other environmental exposures. Job stress, in particular, has been found, in both cross-sectional and prospective studies, to be linked to symptoms of depression and physician-diagnosed psychiatric morbidity.

Consistent with the hypothesis that depression is secondary to the experience of stressful working conditions, numerous studies examining diverse occupational groups have identified various job stressors that appear to be associated with depression. Convergent evidence from many epidemiologic studies implicates high level of job demand, low levels of job control, and limited work-related social support as risk factors for depression. The potency of these stressors in causing depression may be gender-dependent. The presence of high levels of job control and/or social support may buffer the deleterious effects of high levels of job demand. Whether the effects on depression of job demand, job control, and social support are interactive or additive is not clear.

Specific job stressors, such as time pressures, physical demands, conflict with co-workers, long work hours, poor supervision, and ambiguous and conflicting work expectations, are thought to contribute to depression susceptibility.

Pathophysiology

The precise pathophysiologic mechanisms of depression are not fully understood. The consistent observation that central norepinephrine, dopamine, and serotonin are functionally reduced during episodes of depression has led to the belief that depression is due to reduced functional activity of any of the endogenous monoamines in the brain. In this view, antidepressants relieve depression by inhibiting the reuptake of central monoamines, thereby correcting this deficiency. The mechanisms through which external stressors, such as those encountered on the job, affect brain function have been studied extensively. Disorganization of the sympathetic system and the pituitary-adrenal axis appears to be involved. The "stress hormone" corticotropin-releasing factor appears to play an important role in the regulation of central norepinephirne. Thus, there seems to be a close link between changes in the pituitary-adrenal axis, activation of the central norepinephrine system, stress responses, and depression.

Prevention

Despite the accumulating evidence of a causal relationship between job stress and depression, very few studies have been able to provide strong evidence for the efficacy of primary prevention efforts. The body of literature on interventions to change aspects of job design or organizational practices to reduce exposure to job stressors is small and tends to be beset by various methodological problems, especially the absence of strong study designs involving randomized trials, making evaluations and attributions of outcomes difficult. Conducting workplace intervention research using strong designs is difficult because of practical, ethical, legal, and other constraints. Much critical thinking is needed to close the gaps between basic and applied research. Several well-designed studies suggest that primary prevention can be effective, such as by engineering, job design, training, and work process features designed to reduce exposures to job stressors.

Stress management interventions have been shown to be effective in reducing symptoms of depression among groups of workers, but the effects of these interventions are generally thought to be short-lived. In designing overall stress reduction programs aimed at reducing the health consequences of job stress in general, most experts recommend combining organizational change with stress management.

Major depression, known to the public as clinical depression, can come on slowly, but, once recognized, requires immediate referral and treatment. Psychotherapy and antidepressant medications are the most common treatments of major depression and are believed to work well in combination. Common job-related signs of depression include tardiness, absenteeism, decreased productivity, complaints of fatigue, safety problems and accidents, unexplained aches and pains, and alcohol or drug abuse. Fostering an awareness of the signs and symptoms of depression among managers and employees alike may help in the early identification of affected workers. While there are few studies in the literature regarding the efficacy of worksite-based depression screening, voluntary test screening seems to offer a good opportunity to identify previously unidentified and untreated people with depression.

Other Issues

Heart disease affects an estimated 12.2 million Americans and is the leading cause of death in the United States. Research over the past two decades has shown that people with heart disease are much more likely to suffer from depression than healthy people. While the notion that depression increases one's risk for developing heart disease remains controversial, nearly all recent studies of the heart disease-depression relationship document increased cardiovascular morbidity and mortality in patients with depressive symptoms or major depression, thereby implicating depression as a potential independent risk factor in the pathophysiologic progression of heart disease, rather than merely a secondary affective response to the illness. Understanding the relationships between these two illnesses represents a major challenge.

Depression is an important, if not the most important, risk factor for suicide. It is estimated that over 20% of all people with recurrent depressive disorders will attempt suicide and many will die. In the United States, it is believed that more than 20 000 people commit suicide each year. In addition to increasing the risk for self-directed violence, depression seems to increase the risk of violence directed at family members, co-workers, and others.

Further Reading

Klein J, Sussman L. An executive guide to workplace depression. *Academy of Management Executive* 2000; 14: 103-114.

Leonard BE. Stress norepinepherine and depression. *Journal of Psychiatry and Neuroscience* 2001, 26: S11-S16.

National Institute for Occupational Safety and Health. *Stress at Work (DHHS [NIOSH] Publication No. 99-101.)* Washington, DC: NIOSH, 1999.

Tennant C. Work-related stress and depressive disorders. *Journal of Psychosomatic Research* 2001; 51: 697-704.

Emphysema

ICD-10 J43, J68.4, J98.2, J98.3
P25.0, P25.2, T79.7, T81.8

David C. Christiani

Pulmonary emphysema is a disorder of lung anatomy defined as a permanent, abnormal increase in the size of airspaces distal to the terminal bronchiole accompanied by destruction of the lung tissue. Because the tissue destruction is nonuniform, the orderly appearance of the airspaces is disrupted and may be lost altogether.

The clinical presentation of emphysema can vary from mild shortness of breath to severe breathlessness and respiratory failure. The level of dyspnea depends on the demands placed on the respiratory system. When the disease has progressed to the point where the forced expiratory volume in 1 second (FEV_1) is about 40% of what is predicted, carbon dioxide retention and cor pulmonale (right heart failure) may complicate the picture. Because emphysema is a process destructive to lung tissue, the diffusing capacity is typically reduced in moderate to severe disease. Both emphysema and bronchitis can result from the same exposures; therefore, many people with one condition will also have signs or symptoms of the other.

X-ray changes are absent in early emphysema. Later, the characteristic changes are of two general types: (1) an increase in the volume of the thorax occupied by the lung, and (1) a decrease of the overall pulmonary vascular pattern.

Occurrence

Emphysema, with or without associated chronic bronchitis, affects at least 3 million Americans. It is the most costly disease that the U.S. Veterans

Administration treats currently. The reported prevalence is about 15.0 per 1000 persons.

The precise number of workers with emphysema as a result of occupational exposure, in whole or part, is not known. Groups known to be at increased risk include workers exposed to cadmium oxide or coal mine dust. Other exposures contributing to emphysema may be difficult to isolate because of concurrent tobacco smoking.

Causes

The most common agent indicated in causing emphysema is tobacco smoke. However, there are several important occupational causes that, alone or in combination with cigarette smoking, will produce emphysema. Since the 1950s, reports have described emphysema due to cadmium fumes, particularly in workers exposed to cadmium oxide for long periods of time. In addition, emphysema occurs in coal workers. If these workers also smoke, there is more emphysema and it is more advanced when compared with matched controls. Emphysema also occurs occasionally in relation to other pneumoconiosis, such as asbestosis and silicosis. Nitrogen oxides can cause emphysema in laboratory animals, but these findings have not been confirmed in humans.

Pathophysiology

Emphysema is defined and classified in anatomical terms, depending on how the acinus (the portion of the lung distal to the terminal bronchioles) is involved. This classification is available in textbooks of pathology or medicine.

However, the severity rather than the type of emphysema is the important variable in relation to clinical dysfunction. Decreases in expiratory flow rates and increases in residual volume are generally well correlated with the severity of emphysema. An increase in the total lung capacity with decrease in the diffusing capacity shows the highest correlation with anatomic changes.

Prevention

Aggressive control of exposures associated with increased risk of emphysema is the only reasonable preventive strategy. No current screening tool is sufficiently sensitive to identify the development of emphysema at an early stage. The FEV_1 can be used to identify early irreversible disease. Surveillance of exposed workers for associated symptoms and lung function changes may, in some instances, be appropriate. In some special situations, personal protective devices, such as respirators, may also be appropriate.

Other Issues

There is a need to determine the prevalence of emphysema in various occupational groups. This determination should include study of groups having a high prevalence of obstructive airway disease and studies in which the type of agent, severity of exposure, or evidence from experimental animals suggests that obstructive airway disease may occur. Interactions between cigarette smoking and various workplace agents need to be examined more closely, and interactions between non-tobacco particulate exposures in the development of emphysema should be studied.

Homozygous persons with homozygous alpha-1 antitrypsin deficiency are at significant risk for development of emphysema and should not work in areas where there is exposure to dust, respiratory irritants, or causes of emphysema. Heterozygotes do not appear to be at abnormally increased risk, however.

Further Reading

Balmes J, Becklake M, Blanc P, et al. American Thoracic Society Statement: Occupational Contribution to Airway Disease. *American Journal of Respiratory and Critical Care Medicine* 2003; 167: 787-797.

Hnizdo E, Vallyathan V. Chronic obstructive pulmonary disease due to occupational exposures. *Occupational and Environmental Medicine* 2003; 60: 237-243.

Encephalopathy, Toxic

ICD-10 G92

Robert G. Feldman, Patricia A. Janulewicz, and Marcia H. Ratner

Toxic encephalopathy occurs after exposure to toxic chemicals and is characterized by fatigue and CNS symptoms of affect lability, irritability, depression, attention deficits and memory disturbances. A WHO working group has proposed a scheme for classifying this syndrome (Table 1). Clinical manifestations usually reflect degree and duration of exposure to chemicals that pass through the blood-brain barrier. Acute intoxication follows a single toxic exposure to organic solvents or inhalant anesthetics, and presents with dizziness, stupor, lightheadedness, incoordination, and poor balance; the signs of CNS depression may last from a few minutes to several hours. A more severe and potentially fatal acute toxic encephalopathy may follow acute exposure of several minutes to hours to hydrogen sulfide, carbon monoxide, organic solvents and certain metals, such as trimethyltin, and result in acute symptoms of seizures, coma, abnormal reflexes, and EEG slowing and deficits in cognitive function that persist after the acute symptoms have resolved. Acute toxic encephalopathy is often associated with exposures in confined spaces. Chronic toxic encephalopathy is insidious in onset, occurring with continued or repeated acute exposures to toxic chemi-

Table 1. Toxic Central Nervous System Disorders

Acute Organic Mental Disorders

Acute Intoxication
> Pharmacological effect: duration of minutes to hours
> No sequelae
> Clinical: acute CNS depression, psychomotor impairment
> Agents: solvents

Acute Toxic Encephalopathy
> Rare
> Pathophysiology: cerebral edema, CNS capillary damage
> May cause permanent deficits
> Clinical: coma, seizures
> Agent: lead

Chronic Organic Mental Disorders

Organic Affective Syndrome
> Mood disturbance: depression, irritability, loss of interest in daily activities
> Pathophysiology: unclear
> Course: days to weeks
> No sequelae
> Agents: solvents, metals, pesticides

Mild Chronic Toxic Encephalopathy
> Clinical symptoms: fatigue, mood disturbances, memory complaints, attention complaints
> Course: insidious onset; duration of weeks to months
> Pathophysiology: unclear; reversibility uncertain
> Reduced CNS function
> Psychomotor function (speed, attention, dexterity)
> Short-term memory impairment
> Other abnormalities common
> Agents: solvents, metals, pesticides

Severe Chronic Toxic Encephalopathy
> Clinical manifestations
> Loss of intellectual abilities of sufficient severity to interfere with social
>   or occupational functioning
> Memory impairment
> Impairment of abstract thinking
> Impairment of judgment
> Other disturbances of cortical function
> Personality change
> Course: insidious onset, irreversible
> Pathophysiology: unclear
> Reduced CNS function
> Types of abnormalities similar to mild chronic toxic encephalopathy but more pronounced
> Some neurophysiological and neuroradiological tests abnormal
> Agents: controversial—solvents, lead, carbon disulfide

cals, as neurological damage progresses from mild to severe. Organic affective syndrome, the mildest form of chronic toxic encephalopathy in which neurobehavioral disturbances predominate, is manifested by depression and mood changes, sleep disturbances, apathy, and fatigue. Mild chronic toxic encephalopathy presents with similar mood disorders plus impairment of cognitive functions, measurable by performance deficits on neuropsychological tests of visuospatial ability, abstract concept formation, attention, and short-term memory that may impair social and work activities. Severe chronic toxic encephalopathy is a permanent impairment of CNS function, with severe cognitive impairment, short-term memory difficulties, diminished attention span, inability to adapt to situations requiring the learning of new information, and, often, personality changes, impairment of reasoning judgment and abstract thinking, and depression.

Diagnosis of toxic encephalopathy depends on medical histories and on diagnostic procedures that rule out other causes of organic mental syndromes. Mild and severe forms of chronic toxic encephalopathies are generally irreversible; symptoms of acute intoxication and organic affective syndrome are reversible and subside following termination of exposure.

Occurrence

Knowledge about occurrence of toxic encephalopathy is limited to case series and individual case reports. Population-based occurrence data are not available. Out of 220 studied chemicals, 149 cause neurotoxicity in humans. Among workers' compensation awards for occupational conditions of the nervous system, the vast majority are for "diseases of the nerves and peripheral ganglia."

Causes

Metals (such as lead), organic solvents (such as carbon disulfide), gases (such as carbon monoxide), and pesticides (such as organophosphates) have been implicated in causing acute and chronic encephalopathic syndromes (Table 2). More intense and longer exposure increases the risk for persistent impairment. Viscose rayon workers exposed to carbon disulfide have persistent irritability, fatigue, memory loss, and problems with intellectual processing. Workers exposed to mixed solvents have demonstrated impairments in reasoning, visuoconstructive abilities, short-term memory, motor coordination, speed, and attention, as well as EEG abnormalities.

Nonoccupationally related causes of toxic encephalopathy include drug abuse, alcoholism, and domestic hobby-related exposures to lead-based paint, household solvents, and pesticides. Exposures to solvents and metals may also occur from well water that has been contaminated by industrial wastes. It has been postulated that these exposures may lead to alterations in neuropsychological functions. Other clinical syndromes to be considered in the differential diagnosis of cognitive impairment include neurodegenerative

---

Table 2. Exposures Associated with Encephalopathy

Neurotoxin	Major Uses or Sources of Exposure
Metals	
Arsenic	Pesticides
	Pigments
	Antifouling paint
	Electroplating industry
	Semiconductors
	Seafood
	Smelters
Lead	Solder
	Illicit whiskey
	Storage battery manufacturing plants
	Foundries, smelters
	Lead-based paint
	Lead shot
	Insecticides
	Auto body shops
	Lead-stained glass
	Lead pipes
Manganese	Iron, steel industry
	Metal-finishing operations of high- manganese steel
	Manufacturers using oxidation catalysts
	Manufacturers of fireworks, matches
	Manufacturers of dry cell batteries
	Fertilizers
	Welding operations
Mercury	Scientific instruments
	Amalgams
	Photography
	Taxidermy
	Pigments
	Electrical equipment
	Electroplating industry
	Felt making
	Textiles

(continued)

nontoxic causes (such as Alzheimer's disease), hydrocephalus, multiple cerebrovascular infarctions, depression, and previous closed head injury.

Pathophysiology

Acute delirium and persistent dementia result from cerebral disturbances, induced by chemical exposure. The acute symptoms reflect disturbances of neurotransmission that resolve with cessation of exposure. Persistent symptoms reflect neuronal loss and demyelination, which may be

Table 2 *(cont.)*. Exposures Associated with Encephalopathy

Neurotoxin	Major Uses or Sources of Exposure
Solvents	
Carbon disulfide	Manufacturing of viscose rayon
	Preservatives
	Electroplating industry
	Paints, varnishes
	Rubber cement
Perchloroethylene	Paint removers
	Degreasers
	Extraction agents
	Dry-cleaning and textile industries
Toluene	Rubber solvents, glues
	Paints, lacquers
	Manufacturers of benzene
	Cleaning agents
	Automobile, aviation fuels
	Paint thinners
Trichloroethylene	Degreasers
	Painting industry
	Paints, lacquers
	Varnishes
	Process of extracting caffeine from coffee
	Adhesive in shoe and boot industry
	Rubber solvents
	Dry-cleaning industry
Gases	
Carbon monoxide	Exhaust fumes of internal combustion engines, incomplete combustion
	Acetylene welding
Insecticides	
Organophosphates	Agricultural industry

secondary to disturbances of oxygen transport and utilization, free-radical damage, disruption of axonal transport, and/or edema. Researchers have observed edema involving astrocytes, axons, and myelin sheaths in organotin intoxication. Cerebral swelling in the absence of myelin degeneration has been observed in lead-induced encephalopathy in animals. Encephalopathy symptoms are frequent in chronic carbon disulfide poisoning as a result of diffuse vascular damage. Fat-soluble agents are proposed to cross the blood-brain barrier more commonly and disrupt the phospholipids bilayer of cells in the CNS. Carbon monoxide, hydrogen sulfide, and cyanide disrupt oxidative metabolism and cause neuronal death. Free radicals associated with exposures to heavy metals such as manganese may lead to alterations in membrane permeability due to lipid peroxidation.

Prevention

Exposure to neurotoxins should be minimized by effective industrial hygiene monitoring. Control of respiratory, gastrointestinal, and dermal routes of chemical entry is discussed in Part I. Biological monitoring of workers through personal and air sampling for chemicals may be used to check exposure levels.

Other Issues

Drug abuse and alcohol abuse confound the diagnosis of encephalopathy. Furthermore, workers with multiple and different exposures may increase metabolism such that more of a toxic metabolite may result in toxicity. Lastly, health problems such as liver disease may predispose a person to health effects from toxicity such as in manganese exposure, since an agent may be processed slower or less effectively. Also, workers who have previously experienced somatic symptoms following a toxic exposure are at increased risk for developing symptoms of encephalopathy as a post-traumatic stress disorder. Severe chronic toxic encephalopathy should be considered an entity distinct from other progressive dementias, such as Alzheimer's disease, and should also be evaluated with caution in older persons because of the age-related deterioration in intellectual functioning.

Further Reading

Anthony DC, Montine TJ, Graham DG. Toxic responses of the nervous system. In Klaasen CD, Amdur MD, Doull J (Eds.). Casarett and Doull's Toxicology: *The Basic Sciences of Poison*, (5th ed.). New York: McGraw-Hill, 1996, pp. 463-486.

Baker EL, White RF, Murawski BJ. Clinical evaluation of neurophysiological effects of occupational exposure to organic solvents and lead. *International Journal of Mental Health* 1985; 14: 135-158.

Feldman RG. *Occupational and Environmental Neurotoxicology*, Philadelphia: Lippincott-Raven, 1999, pp. 20-25.

Feldman, RG. Treatment of the Neurotoxic Effects of Organic Solvents. In Noseworthy J (Ed.). *Neurological Therapeutics: Principles and Practice*. London: Martin Dunitz, 2003.

Johnson BL. *Prevention of Neurotoxic Illness in Working Populations*. New York: John Wiley & Sons, 1987.

Ratner MH, Feldman RG, and White RF: Neurobehavioral Toxicology. In: Ramachandran V.S. (Ed); *Encyclopedia of the Human Brain*. New York, Elsevier Science, Vol. 3, pp 423-439, 2002.

Spencer PS, Schaumburg HH, Ludolph AC (Eds.). *Experimental and Clinical Neurotoxicology*, (2nd ed.). New York: Oxford University Press, 2000.

White RF, Feldman RG, Travers PH. Neurobehavioral effects of toxicology due to metals, solvents, and insecticides. *Clinical Neuropharmacology* 1990; 13:392-412.

Esophageal Cancer

ICD-10 C15

Elizabeth Ward

The two main histological types of esophageal cancer are squamous cell carcinoma, generally in the upper two-thirds of the esophagus, and adenocarcinoma in the distal esophagus. Early esophageal cancer usually does not cause any symptoms. The first symptom of esophageal cancer is often progressive difficulty in swallowing, first for solids and then for liquids. Other symptoms are painful swallowing; significant weight loss without dieting; pain in the throat or back, behind the sternum (breastbone) or between the shoulder blades; hoarseness or chronic cough; vomiting; and coughing up blood.

Occurrence

Approximately 13 900 new cases of esophageal cancer and 13 000 deaths occurred in the United States in 2003. The male-to-female ratio is 3:1, and the black-to-white ratio is 1.8:1. Age-adjusted mortality rates are comparable to incidence rates. The incidence of adenocarcinoma of the esophagus has increased in the United States from 1974 to 1998, while the incidence of squamous cell cancer has decreased. Esophageal cancer rates in Iran, Northern China, India, and Southern Africa are 10 to 100 times higher than in the United States. There are no formal estimates of the proportion of esophageal cancers that are related to occupation.

Causes

There is limited information on occupational causes of esophageal cancer, due, in part, to its low incidence in the U.S. and other industrialized countries. Occupational exposures found in one or more studies to increase the risk of esophageal cancer include perchloroethylene (tetrachloroethylene), mustard gas, silica dust, metal dust, asbestos, combustion products, sulfuric acid, carbon black, and ionizing radiation. Occupational groups reported to be at increased risk include workers in vulcanization and reclaim processes in the rubber industry, workers in metal-grinding operations, metal polishers and platers, drycleaners, chimney sweeps, brewery workers, butchers, food-service workers, bookbinders, and printers. Some studies have found associations between esophageal cancer and wood dust exposure, but results have been inconsistent. No estimates are available for the proportion of all esophageal cancers that may be due to occupational exposures. Alcohol and cigarette use are the primary nonoccupational risk factors. Hot fluids, such as tea, may be a risk factor worldwide. The rising incidence of adenocarcinoma in U.S. males may be related to increasing obesity, a risk factor for this type of cancer.

Pathophysiology

Chronic irritation from reflux of acid and bile is thought to be the main predisposing factor for esophageal adenocarcinoma. People with Barrett's esophagus, a condition of the lower esophagus present in less than 1% of the population, are between 40 and 125 times more likely to develop esophageal adenocarcinoma than people without this condition. Chronic acid reflux increases risk, but to a lesser extent. Key components of the mechanism through which exogenous agents cause esophageal cancer include intimate contact between the esophageal wall and a carcinogenic agent, transport through the mucosa to the basal layer, and increased cell turnover caused by irritants.

Prevention

Primary prevention includes eliminating or reducing exposure to agents with known or suspected carcinogenicity. Avoidance of tobacco smoking, excessive alcohol consumption, and weight gain will reduce risk due to nonoccupational factors; consumption of fresh fruits and vegetables may also be of benefit. Secondary prevention involves treating reflux, whether through smoking cessation, dietary modification, or other means, but these methods have not been shown to lead to lower rates of esophageal cancer. Observational studies have shown decreased risk of esophageal carcinoma among individuals who use aspirin and nonsteroidal anti-inflammatory drugs (NSAIDs). Clinical trials are testing these agents to determine if they may be effective for chemoprevention of esophageal cancer in high-risk patients. There are no specific recommendations for screening of individuals who may be at increased risk due to their occupation; screening with endoscopy is recommended for patients with Barrett's esophagus and has been used for population screening in high-risk areas in China. Surgery is the most common treatment for esophageal cancer. Radiation and chemotherapy sometimes may be used to shrink the tumor before surgery or for palliation in inoperable cases. The five-year survival rate was 13.3% in the U.S. for the 1992-1999 period; for the 25% of esophageal cancers diagnosed at a localized stage, the five-year survival rate was 27.1%.

Further Reading

All about esophagus cancer: detailed guide. Learn about cancer [serial online]. American Cancer Society, Atlanta [updated 01/04/2001]. Available from URL: http://www.cancer.org/docroot/CRI/CRI_2_3x.asp?dt=12

Brown LM, Devesa SS. Epidemiologic trends in esophageal and gastric cancer in the United States. *Surgical Oncology Clinics of North America* 2002; 11: 235-256.

Chow WH, McLaughlin JK, Malker HS, et al. Esophageal cancer and occupation in a cohort of Swedish men. *American Journal of Industrial Medicine* 1995; 27: 749-757.

Lynge E, Anttila A, Hemminki K. Organic solvents and cancer. *Cancer Causes and Control* 1997; 8: 406-419.

Munoz N, Day NE. Esophageal cancer. In: Schottenfeld D, Fraumeni JF Jr. (eds.). *Cancer Epidemiology and Prevention* (2nd ed.). New York: Oxford University Press, 1996, pp. 681-706.

National Cancer Institute. What you need to know about cancer of the esophagus. What you need to know about cancer publications [serial online]. National Cancer Institute, National Institutes of Health, Bethesda, MD [updated 9/16/2002]. Available from URL:
http://www.nci.nih.gov/cancerinfo/wyntk/esophagus

Neugut AI, Wylie P. Occupational cancers of the gastrointestinal tract. I. Colon, stomach, and esophagus. *Occupational Medicine* 1987; 2: 109-135.

Pan G, Takahashi K, Feng Y, et al. Nested case-control study of esophageal cancer in relation to occupational exposure to silica and other dusts. *American Journal of Industrial Medicine* 1999; 35: 272-280.

Parent ME, Siemiatycki J, Fritschi L. Workplace exposures and oesophageal cancer. *Occupational and Environmental Medicine* 2000; 57: 325-334.

Shaheen N, Ransohoff DF. Gastroesophageal reflux, Barrett esophagus, and esophageal cancer: scientific review. *JAMA* 2002; 287: 1972-1981.

Eye Injuries

ICD-10 S05.9

Satya B. Verma and Andrew S. Gurwood

An occupational eye injury is a work-related compromise to the eye and supportive tissue, mechanical or otherwise, that results in physical damage or in temporary or permanent vision impairment.

Injuries categorized as mechanical include (a) superficial injuries (non-penetrating damage to the adnexa, conjunctiva, or cornea), which are usually painful and accompanied by tearing and photophobia (sensitivity to light); (b) penetrating injuries (complete thickness damage usually caused by sharp, small objects capable of producing permanent visual disability); and (c) blunt object trauma (injuries caused by direct or indirect transmission of force to the globe or its adnexa, affecting all structures surrounding the eye and the eye itself).

Injuries categorized as burns include thermal, chemical, and radiation burns. Thermal burns represent damage caused by direct contact with heat. Tissue destruction may range from edema of the skin, conjunctiva, and cornea to complete tissue necrosis (tissue death), depending on the heat source, duration of exposure, and area of contact. Thermal injuries include the damage that occurs to the crystalline lens, forming heat or electric-shock cataract and retinal-related damage, such as solar maculopathy. Chemical burns represent tissue that is compromised by solid, liquid, or gaseous compounds. Common offenders include alkalis, acids, and organic solvents. Caustic agents can produce permanent disability in seconds, making these events true emergencies requiring immediate actions. Radiation burns represent injury transmitted most commonly via ultraviolet (UV) radiation (short-wavelength light energy). Snow blindness (from prolonged, unprotected exposure to the bright reflection of sunlight off of white snow) and "welder's flash" (from arc welding) are both on the continuum known as phototoxic or photochemical keratopathy. Both of these problems result from cumulative exposure to UV radiation. Both are significant risk factors for cataract formation.

Occurrence

About one-fourth of all eye injuries are work-related. Based on eye injury registry data, eye injuries account for about one-eighth of all cases of compensated occupational impairment. Compared with nonoccupational eye injuries, work-related eye injuries have the lowest rate of enucleation and loss of light perception. Most reported work-related eye injuries occur in men; nearly 60% occur among people in their 20s.

About 80% of occupational eye injuries occur in manufacturing and construction, with fewer in agriculture, mining, and transportation. Metalwork, excavating and foundation work, and logging are also high-risk occupations for eye injury.

Chemical burns of the eye occur most often in manufacturing, followed by services, retail trade, and construction. About five-sixths of all chemical burns affect men, and two-thirds occur to individuals in their 20s.

Causes

About three-fourths of occupational eye injuries are caused by flying objects associated with machines or hand tools. About 80% of all eye injuries are due to superficial mechanical trauma. About 10% are caused by chemicals. Other causes are UV burns of the cornea, thermal and chemical eye burns, blunt object trauma, penetrating wounds, and complications secondary to post-traumatic infection. About half of reported chemical burns to the head and neck also affect the eyes. Foreign bodies in the eye, contusions, and open wounds of the eye and orbit are also common. Approximately 40% of these injuries result in permanent visual impairment.

Pathophysiology

Mechanical injuries produce physical trauma to the eye and surrounding structures. Injuries caused by chemical burns vary with the type of chemical. Many alkalis can penetrate through and between the cells of the cornea to enter the anterior chamber and other intraocular tissues before being neutralized. Alkaline substances with a pH above 11 are very dangerous to the eye. Organic solvents cause considerable eye irritation, but rarely cause permanent damage. Injuries caused by UV radiation can be acute or chronic. They arise as the conjunctival, corneal, lenticular, and retinal structures absorb short-wavelength light energy, producing (a) damage to mechanisms of metabolism and (b) unstable free radicals.

Prevention

Primary prevention of mechanical trauma from flying physical objects can be achieved by engineering controls designed to prevent objects from becoming airborne. This can be accomplished by identifying possible sources and changing or shielding the process. For chemical burns, chemicals should be clearly labeled and handled in such a way as to minimize splashing and spills. Radiation burns can be prevented by marking the work area and modifying the welding process, such as by using submerged arc welding, as well as providing welding booths to protect nearby people. Other sources of photochemical injury can be minimized by providing shade from the sun for outdoor workers via canopies or personal devices.

Workplaces with potential for eye injury should have a written, well-communicated eye safety policy for employees. All at-risk workers should wear protective eyewear, which, according to OSHA, must be made available at no cost to workers. All ophthalmic equipment/devices should meet ANSI standard Z87.1-1989 (Standard Practice for Occupational and Educational Eye and Face Protection). Eyewear that meets this standard is easily identifiable by a "watermark" etched onto one of the lenses, the clearly labeled zyle frame and the lip on the eyewires of the front that prevent lenses from ejecting posteriorly. A supervisor, health and safety committee member, union representative, or other person should constantly be reassessing the work environment to ensure that at-risk workers are in compliance by wearing their protective eyewear. Inspections should also include ensuring that engineering controls designed to mitigate injury sources, such as flying objects, chemical spills or splashes, and UV radiation, are in place and properly functioning.

Prophylactic vision screening can be conducted on-site periodically. Screening helps detect uncorrected visual acuity and other eye problems, and it provides an opportunity for education and installation of eye/face and drench showers at places where chemical burns may occur.

Other Issues

Although ANSI standard protective eyewear is essential, it is not "failsafe." It only works if it is worn. It is more likely to be worn if it is required. In a survey of workers who had sustained occupational eye injuries, 41% reported that they were wearing some form of eye protection at the time of injury. In these cases, while damage occurs, experts and treating clinicians agree, had the protective shields been absent, it is almost certain that the damage sustained would have been greater. Interestingly, of those not wearing eye protection, 60% were not required to do so by their employers.

Some professional athletes, such as racquet sports athletes, boxers, and basketball players, are at increased risk for eye injury, especially from blunt-object trauma. Boxing injuries to the eye can be prevented by the use of thumbless boxing gloves and mandatory protective headgear. Recreational athletes, such as softball and touch-football players, are also at risk.

Some healthcare professionals, including dentists, dental hygienists, and laser users, are also at risk. Injuries to these individuals are often the result of sloppiness or noncompliance. Efforts must be made to improve worker education and workplace rule enforcement.

Finally, individuals with pseudophakia (implanted intraocular lenses) are at increased risk for both UV and blue-light injury, as the old lens that blocked these wavelengths of the spectrum is no longer present. Advances in technology have led to UV protective coatings being added to the surface and chemistry of glass lenses, plastic lenses, contact lenses, and intraocular lens implants.

Further Reading

Bureau of Labor Statistics. *Accidents Involving Eye Injuries* (Report 597). Washington, DC: U.S. Government Printing Office, 1984.

Taylor HR, West SK, Rosenthal FS, et al. Effects of ultraviolet radiation on cataract formation. *New England Journal of Medicine* 1988; 319: 1429-1433.

White MF Jr, Morris R, Feist RM, et al. Eye injury: Prevalence and prognosis by setting. *Southern Medical Journal* 1989; 82: 151-158.

Eye Strain (Asthenopia)

ICD-10 H53.1

Satya B. Verma and Andrew S. Gurwood

Eye strain, or asthenopia, is among the most common maladies and presents with symptoms that include, but are not limited to, photophobia, headaches, blurred vision, and periorbital pain. It typically manifests after ordinary work under dim light, after or during excessive close work, or after or during persistently vigilant work.

Occurrence

Reliable data on the occurrence of occupational eye strain are not available. Eye strain commonly occurs in workers who need to perform prolonged visual tasks or visual tasks in an environment that lacks appropriate lighting, contrast, or suitable magnification. It appears to be more common among older workers, and among women than men. Although most prevalence estimates are based on symptoms, symptoms are well correlated with objective measurements of impairment, such as reduced productivity.

Causes

Eye strain may be caused by tasks requiring prolonged attention to visual detail, the visual demand of observing small objects, poor lighting, poor contrast, undetected or uncorrected refractive error, untreated or undiagnosed binocular vision abnormalities, poor posture, poor workstation set-up, and presbyopia. It is prevalent among video display terminal (VDT) operators, microscopists, workers who monitor radar screens, accountants, students, drivers, and jewelers. When illumination level, luminance, and contrast are adequate, comfort can be improved.

Pathophysiology

The mechanism of asthenopia depends on the cause. Uncorrected or undiagnosed refractive error, an uncorrected or undiagnosed binocular vision anomaly, early or beginning presbyopia, poor lighting, poor contrast, poor posture, and increased work intensity or difficulty are all potential triggers. Whether acute symptoms develop into permanent impairment is unknown. In most cases, resolution of the underlying cause results in symptom reduction or complete dissolution.

Prevention

Work requiring prolonged visual vigilance should be designed with flexible requirements and deadlines, where possible. This could be accomplished in part by frequent work breaks (at least once per hour), glare-reduction visors, larger computer fonts, larger computer monitors, and louvered lights. Workstations should be designed to reduce glare and reflection to provide adequate illumination, luminance, and contrast and to allow for proper posture (ergonomics). Keyboards should be movable so that they can be placed in a comfortable position and the visual distance to both screens and work should be customizable and adjustable. Regular on-site consultations and eye exams are helpful, not only to detect eye strain, but also to ensure proper visual acuity.

Other Issues

Everyone is predisposed to eye strain. Patients with hyperopia and astigmatism, intolerance to a new correction, or even a first correction with astigmatism or presbyopia, anisometropia, or aniseikonia are especially susceptible. Extraocular muscle imbalance, improperly set bifocals, and improperly adjusted frames are also contributors. Middle-aged workers are more likely to have problems adjusting to varying focal distances and may require special glasses (occupational spectacles) while working on VDTs.

Further Reading

DeGroot JP, Kamphius A. Eyestrain in VDT users: Physical correlated and long-term effects. *Human Factors* 1983; 25: 409-413.

Goussard Y, Martin B, Stark L. A new quantitative indicator of visual discomfort. *Human Factors* 1986; 28: 347-351.

Iwaski T, Kurimoto A. Objective evaluation of eye strain using measurements of accommodative oscillation. *Ergonomics* 1987; 30: 581-587.

Rose L. Workplace video display terminals and visual fatigue. *Journal of Occupational Medicine* 1987; 29: 321-324.

World Health Organization. Visual display terminals and workers' health. *WHO Offset Publ.* 1987; 99: 1-206.

Falls (to a lower level)

ICD-10 W17-W19

James L. Weeks

Falls are usually classified as falls to a lower level or falls to the same level. In this entry, we discuss falls to a lower level.

Occurrence

Falls are the third most common cause of fatal occupational injuries for the U.S. labor force as a whole, the leading cause of fatal injuries in construction, second most frequent in agriculture, and third most frequent in manufacturing and wholesale trade. Falls occur from roofs, ladders, scaffolds, structural metal, vehicles, trees, aerial work platforms, ships' superstructure, towers, tanks, and myriad other elevated places where people work. Workers in agriculture fall most frequently from vehicles and trees. The highest rate of fatal falls among the construction trades occurs among structural metalworkers. The rate of fatal falls among roofers, painters, and carpenters is also higher than that for other workers in construction.

From 1992 to 2000, there were an average of 678 fatal falls each year in the United States. About half of these occurred in the construction industry. Among these, 33% were from roofs, 19% from scaffolds, 14% from ladders, and 10% from building girders. There is an overall decline in the incidence rate of fatal falls over this 8-year time period.

In 1997, there were 313 000 nonfatal falls resulting in many days away from work in the U.S., with the average incidence rate equal to 36 per 10 000 FTE workers. The highest rate was in the construction industry (72 per 10 000

FTE workers), followed by transportation (65), public utilities (65), agriculture, forestry, and fishing (55), and mining oil and gas extraction (49).

Older workers (above age 55) have a significantly higher rate of fatal falls than do younger workers. The reasons are unclear, but could be due to a combination of reduced sensory acuity, reduced physical agility, higher prevalence of other impairment, and reduced capacity to recover from injuries that would not otherwise be fatal. The rate of fatal falls for workers up to age 44 ranges from 0.23 to 0.40 deaths per 100 000 workers. The annual rates among U.S. workers aged 45-54 are 0.54 deaths per 100 000 workers; those age 55-64, 0.86; and those over 65, 1.57.

The duration of disability from nonfatal falls is typically longer than the average duration of disability from other injuries. The frequency of permanent disability, with related loss of earnings and degraded quality of life, is also higher for nonfatal falls than it is for other lost-time injuries.

Causes

The causes of falls vary by the place from which workers fall. When falling from roofs, about half of fatal falls occur from the edge of an unguarded roof and the remainder typically fall through unguarded or poorly guarded holes. Workers may lose balance, become disoriented, or trip over obstacles. Falls from scaffolds occur because of lack of guarding or inadequate planking of the scaffold and many occur when the worker is ascending, descending, erecting, or disassembling the scaffold. Falls from or with ladders occur when a straight or extension ladder slips out from under the user or falls over backwards or when a step ladder tips over. Falls from structural steel occur when workers lose their footing on narrow elevated walkways. Fatal falls from vehicles are usually associated with failure to use seat belts and sometimes workers are tossed from vehicles. The cause of falls from trees is not known with certainty but is likely due to loss of balance. Falls from ships' superstructure occur because of slippery surfaces or rough seas.

The severity of injury and the risk of fatal injury is proportional to the height of the fall. From Newton's laws of motion, the potential energy at height "h" is equal to the product of the worker's mass (or weight), the gravitational constant "g," and h ($E = mgh$). When a worker falls, this potential energy is converted to kinetic energy which must be absorbed by the worker and what he or she lands on. Although fatal falls have occurred at minimal height and some workers have survived falls from over 100 feet, the risk of injuries resulting in death increases significantly at heights above about 10 feet.

The severity of injury is also dependent on the features of the surface on which the person lands. If the energy accumulated with a fall is distributed over a large area and absorbed at a lower rate, such as with an automobile air bag or its equivalent, the severity of injury will be significantly less. Persons who have survived falls from over 100 feet, for example, managed to fall on surfaces that distributed enough energy in time or space to allow them to survive.

Environmental conditions may also contribute to the risk of falls. Thus, exposure to electrical hazard, heat stress, organic solvents or other inhalation hazards, noise, or adverse weather increases the risk of falling. Adverse weather resulting in wet, icy, or snow-covered surfaces also increases the risk of falls.

Failure to use fall prevention technology is a significant factor in many fatal falls. In a series of 91 fatal falls investigated by NIOSH, failure to use fall protection or failure to use it correctly was associated with nearly half (41% or 45%) of all fatalities; in only 18% of fatal falls was fall protection equipment known not to be available.

Pathophysiology

Acute trauma to the head and neck accounts for more than half of all fatal injuries from falls. Improved headgear may reduce the risk of fatal injuries from falls. Traumatic injury to multiple body systems is the second most common cause of fatalities. Nonfatal injuries include fractures, dislocations, contusions, strains, and sprains.

Prevention

Primary prevention of injuries from falls is conceivable by reducing the need to work at elevated places. This is possible in some instances. For example, if a worker must climb a ladder to read a gauge that indicates pressure at the top of a tank, the gauge can be relocated at eye level rather than the top of the tank.

Other preventive measures are considered standard practice at construction job sites. Holes in roofs should be securely covered or guarded, the edge of a roof or floor, (the so-called "leading edge") should be clearly marked and guarded, secure guardrails should be erected on scaffolds, scaffolds should be assembled and disassembled in a systematic and orderly fashion, ladders should be in good repair, and set up properly, aerial work platforms should operate on solid uncluttered surfaces, workers in aerial work platforms should not leave the work platform without fall protection, and personal fall protection should be used wherever possible. Some technological innovations, such as system-scaffolding, are inherently safer.

Current OSHA standards requiring fall protection are an improvement but are not consistently followed or enforced. Fall prevention regulations for the construction industry were promulgated by OSHA in 1994 (29 CFR 1926.500-503) and, in general, require fall protection in the form of guardrails, personal harness and lanyard, and other means for workers at elevations of 6 feet or more. Similar rules for general industry are pending.

Recent evaluation of OSHA's rules in construction demonstrate a significant reduction in the rate of fatal injuries for workers that come under the standard's jurisdiction. The State of Washington promulgated rules similar to OSHA's in 1991; from 1989 to 1998, the annual incidence rate of all nonfatal injuries from falls there declined from 4.4 to 1.2 injuries per 100 FTE workers. The decline in the rate of injuries remained, but was reduced after adjusting for age and simultaneous reduction in overall injury rates.

Other Issues

Falls occur from places that are simultaneously the product of work and the place of work. Roofs, scaffolds, and ladders, for example, are often erected as temporary work platforms, often by the same workers who use them. Workers also construct guardrails at leading edges and on scaffolds, cover or guard holes, place or find secure anchorage places, erect scaffolds, and set up ladders. Since much of the hazards of work at elevated places derive from their structure, workers therefore need to be trained not only about using fall protection but also about constructing it on safe work platforms.

Further Reading

Agnew J, Suruda AJ. Age and fatal work-related falls. *Human Factors* 1993; 35: 731-736.

Derr J, Forst L, Chen HY, et al. Fatal falls in the US construction industry, 1990 to 1999. *Journal of Occupational and Environmental Medicine* 2001; 43: 853-860.

Gillen M. Injuries from construction falls. Functional limitations and return to work. AAOHN Journal 1999; 47: 65-73.

Hsiao H, Simeonov P. Preventing falls from roofs: A critical review. *Ergonomics* 2001; 44: 537-561.

Janicak CA. Fall-related deaths in the construction industry. *Journal of Safety Research* 1998; 29: 35-42.

Lipscomb H, Leiming L, Dement JM. Work-related falls among union carpenters in Washington State before and after the vertical fall arrest standard. *American Journal of Industrial Medicine* 2003; 44: 157-165.

Loomis D, Dufort V, Kleckner RC, et al. Fatal occupational injuries among electric power company workers. American Journal of Industrial Medicine 1999; 35: 302-309.

National Institute for Occupational Safety and Health. *Worker Deaths by Falls: A Summary of Surveillance Findings and Investigative Case Reports* (DHHS [NIOSH] Publication No. 2000-116). Washington, DC: NIOSH, 2000.

Occupational Safety and Health Administration. Safety Standards for Fall Protection in the Construction Industry (59 FR 40672-753), August 9, 1994.

Occupational Safety and Health Administration. Fall Protection in Construction. (OSHA Publication No. 3146) (revised). Washington, DC: OSHA, 1998.

Occupational Safety and Health Administration. Safety and Health Standards for the Construction Industry (29 CFR 1926.500-503).

Occupational Safety and Health Administration. Walking and Working Surfaces; Personal Protective Equipment (Fall Protection) (68 FR 23527-68), May 2, 2003.

Rivara FP, Thompson DC. Prevention of falls in the construction industry: evidence for program effectiveness. American Journal of Preventive Medicine 2000; 18 (Suppl. 4): 23-26.

Webster T. Workplace Falls. Compensation & Working Conditions, Spring 2000, pg. 28-38. (Reprinted in Fatal Occupational Injuries in 1995: A Collection of Data and Analysis (Report 954). Washington, DC: U.S. Bureau of Labor Statistics, 1998.)

Fatty Liver Disease and Cirrhosis

ICD-10 A52.7, K70.3, K74, K76.0, K76.1, P78.8

Susan Buchanan and Michael Hodgson

Fatty liver disease or nonalcoholic steatohepatitis (NASH) is characterized by findings on liver biopsy indistinguishable from those of patients with alcoholic hepatitis. NASH usually occurs without symptoms or merely with right upper quadrant fullness, vague abdominal discomfort, or fatigue and malaise. Although up to one-sixth of cases will progress to cirrhosis, most follow a benign course. Diagnosis may be made on the basis of (a) elevations of alanine aminotransferase (ALT) or aspartate aminotransferase (AST); or (b) ultrasonography, CT, or MRI revealing fatty infiltration. Unlike alcohol-related fatty liver disease, the AST/ALT ratio in patients with NASH is usually less than 1. Alkaline phosphatase and gamma glutamine transferase (GGT) may be elevated, but bilirubin is normal. Unequivocal diagnosis is made by liver biopsy. Biopsy findings include macrovesicular fatty changes, parenchymal inflammation, and fibrosis. Physical examination of the liver is difficult, but hepatomegaly with a rounded liver edge may be found. The development of cirrhosis is associated with a firm liver with a rounded or nodular rim, spider nevi, and ascites.

Occurrence

Fatty liver disease has been found in up to 9% of patients undergoing liver biopsy. The prevalence in the general population may be as high as 24%, possibly due to high rates of obesity and insulin resistance. The proportion of these cases in which hepatotoxin exposure plays a contributing role is unknown but is generally assumed to be low. Cirrhosis is the 12th leading cause of death and the 10th leading cause of years of potential life lost before the age of 65 in the United States. It is estimated that alcohol causes or contributes to the vast majority of cirrhosis deaths. An overall death rate from cirrhosis was estimated at 9.6 per 100 000 per year in 2000. In developing countries, the proportion of disease related to occupational exposures is substantially higher. Although some experts consider disease more likely to occur after additional risk factors, such as obesity, or through combinations of alcohol and occupational exposures, others document that occupational exposures alone are adequate to explain disease. Occupational disease will often improve if the individual is removed from exposure before cirrhosis is established.

Causes

There are generally six classes of hepatotoxins that have been found via either animal or human studies to cause steatosis and cirrhosis. By far, the most common substances are the halogenated hydrocarbons. Some examples from each class include (a) halogenated aliphatic and aromatic hydrocarbons: carbon tetrachloride, chloroform, trichloroethylene, tetrachloroethane, and pentachlorophenol; (b) polychlorinated biphenyls: chlorodiphenyl; and chloronaphthalene; (c) nitro compounds: dinitrotoluene; (d) metals: arsenic, cadmium, and thallium; (e) inorganics: phosphorus; (f) cyanides: diisocyanate; (g) others: dioxane and aflatoxin.

Pathophysiology

Hepatocellular steatosis (fatty deposition), necrosis, and cirrhosis result from oxidative stress and lipid peroxidation. Chronic insult leading to cirrhosis results from the combined effects of hepatocellular necrosis, collapse, fibrosis, and regeneration.

Highly hepatotoxic chemicals produce direct injury to key subcellular structures like the mitochondria and plasma membrane. Most hepatotoxins, however, require metabolic activation to exert damaging effects. Induction of the cytochrome P450 (CYP) enzyme system is the most common Phase I biotransformation step. This chemical oxidation process results in free-radical formation which can cause hepatocellular injury. Phase II conjugation reactions are less likely to result in toxic reactive metabolites.

Prevention

Prevention of exposure prevents disease. The hierarchy of controls described in Chapter 1 provide guidance of exposure control methods, with engineering controls considered first. Inhalation is probably the most common route of exposure. For some agents with low vapor pressures, skin absorption is the primary mode of absorption. Appropriate contact prevention must then be considered. General preventive measures include maintenance of ideal body weight, control of diabetes, and abstention or moderation of alcohol consumption. Although genetic polymorphisms of drug metabolism have been identified, inadequate information is available to determine whether some individuals are at greater risk for hepatotoxicity than others. Antioxidants, such as Vitamin E or N-acetyl cysteine, have not been shown to reduce the incidence of disease in humans but are recommended by some health care practitioners.

Further Reading

Cotrim HP, Andrade ZA, Parana R, et al. Nonalcoholic steatohepatitis: A toxic liver disease in industrial workers. *Liver* 1999; 19: 299-304.

Feldman M, Friedman LS, Sleisenger MH (Eds.). *Sleisenger & Fordtran's Gastrointestinal and Liver Disease* (7th ed.). St. Louis, MO: Saunders, 2002.

Hodgson MJ, Van Thiel DH, Goodman-Klein B. Hepatotoxin exposure and obesity as risk factors for fatty liver disease. *British Journal of Industrial Medicine* 1991; 48: 690-695.

Lundqvist G, Flodin U, Axelson O. A case-control study of fatty liver disease and organic solvent exposure. *American Journal of Industrial Medicine* 1999; 35: 132-136.

Reid AE. Nonalcoholic steatohepatitis. *Gastroenterology* 2001; 121: 710-723.

Sheth SG, Gordon FD, Chopar S. Nonalcoholic steatohepatitis. Annals of Internal Medicine 1997; 126: 137-145.

Zimmerman HJ. *Hepatotoxicity*. Philadelphia: Lippincott Williams & Wilkins, 1999.

Glomerulonephritis

ICD-10 N00-N08

Carl-G. Elinder and C. Michael Fored

Glomerular kidney disease from occupational exposures should be suspected from the history of relevant exposures along with one or more of the following urinary findings: hematuria, red or granulated cell casts, and proteinuria, which may be in the nephrotic range (greater than 3 g/day). Patients typically display few subjective symptoms, but often have hypertension. Measurement of plasma creatinine gives a rapid clue to the degree of functional loss of renal function. The glomerular filtration rate (GFR) can be estimated from several simple formulas based on age, weight, gender, and plasma creatinine. Different types of glomerulonephritis develop very differently. Some types of glomerulonephritis are self-limited, some respond to therapy, and others progress relentlessly to renal failure. Occasionally glomerulonephritis is diagnosed in patients with signs of systemic disease and acute, influenza-like, symptoms, including high fever, cough, and general discomfort. A special type of aggressive glomerulonephritis (rapidly progressive glomerulonephritis) progresses fast to end-stage renal disease (ESRD), if not treated properly. Although history as well as blood and urine examinations can narrow down the differential diagnosis, a renal biopsy is necessary in order to specifically identify the type of glomerulonephritis. Clinical progression varies greatly according to the pathological picture and has important implications for treatment. Glomerular disorders underlie 20% to 40% of end-stage renal failure.

Occurrence

In general, glomerulonephritis is a rare disease with an annual recognized incidence rate of 20 to 100 biopsy-diagnosed cases per million. The actual rate, however, is possibly about twice as high since biopsies are not performed on all patients. Incidence and type vary with age and gender. IgA-nephropathy, focal glomerulosclerosis nephropathy, and membranous nephropathy are most common in middle-aged individuals. Rapidly progressive glomerulonephritis, with typical crescentic changes in the glomerulus, is most often diagnosed in people over age 50. The proportion of glomerulonephritis caused by occupational exposures is probably small, at least in countries where occupational exposures are relatively well controlled.

Occupations at risk for glomerulonephritis include chlor-alkali workers; manufacturers of thermometers, barometers, and other instruments; fur preservers; and gold extractors—all of whom may be exposed to mercury. Some people may be susceptible to initiation of the immune response that causes glomerular disease; organic solvents may possibly play a precipitating role. However, tests are not currently available to identify such people.

Silica exposure has been associated with glomerulonephritis and ESRD. Approximately 2 million people are occupationally exposed to silica in the United States, approximately 100 000 of them at more than twice the NIOSH REL of 0.05 mg/m³.

Causes

Glomerulonephritis is caused by one of several immune processes triggered by factors that are usually unidentified, but that may include infections, neoplasms, and toxic exposures. Although occupational exposures are the suspected precipitants of both immune complex disease and anti-glomerular basement membrane (anti-GBM) antibody, the evidence is circumstantial.

Occupational associations include the following:

Hydrocarbon Solvents

Exposure to organic solvents has long been thought to cause and exacerbate glomerulonephritis. Concern initially involved Goodpasture's syndrome (a rare immunologic lung and kidney disease), but has been extended to more common glomerular disorders. More than a dozen case-control studies, performed over the last 25 years, have examined the relationship between organic solvents and glomerulonephritis. Most of them indicate a significantly increased relative risk, often between 2 and 5, among persons with occupational exposure to organic solvents. However, almost all of the case-control studies suggesting an increased risk suffer serious methodological flaws and limitations, and the associations found between exposure to organic solvents and renal disease have to be interpreted carefully. In two recent case-control studies, where exposure to organic solvents has been assessed in great detail, no significant risk for either renal disease in general or glomerulonephritis was found.

Organic solvent exposure often entails exposure to several different compounds and mixtures. Thus, associations between possibly nephrotoxic subclasses of solvents and chronic renal failure may have been undetected in case-control studies. Therefore, an association between high exposure of certain subclasses of solvents and specific types of glomerulonephritis cannot be ruled out.

Silica

Several case reports have described glomerulonephritis in workers who are heavily exposed to silica, in particular rapidly progressive glomerulonephritis in association with elevated antineutrophil cytoplasmatic antibodies (ANCA). In support of these case reports are several cohort studies of silica-exposed workers that have shown an excess of ESRD and glomerulonephritis. In a large cohort study of 4626 silica-exposed workers in the United States, the SMR for acute renal disease was 2.6 and for chronic renal disease, 1.6. In addition, there was an elevated SIR for ESRD, which was highest for glomerulonephritis (3.8). In strong support for a causal association, an increasing incidence of ESRD was noted with increasing cumulative exposure to silica.

Mercury

The toxicity of mercury depends on its chemical form. However, all forms of mercury may cause transient proteinuria not associated with renal disease development. In genetically selected rodents, mercury produces immunological effects, with glomerular lesions. This finding may provide a model for the glomerular damage, manifested as membranous nephropathy, rarely seen in humans following exposure to mercury. Inorganic mercury compounds accumulate in the kidneys and are recognized to cause nephrotic syndrome and/or tubular injury with tubular dysfunction. The tubular lesions are probably dose-related, with large doses causing acute tubular necrosis. Albuminuria appears in some workers exposed to elemental mercury vapor at levels associated with tremor. Nephrotic syndrome may occur in association with exposure to organic mercury, mostly with spontaneous recovery after removal of exposure. Membranous glomerulonephritis has been described in people who have taken organomercurial diuretics, used mercury-containing skin creams, or have had accidental exposure to mercury vapor. Nephrotic syndrome may also result from taking medication containing other metals, such as rheumatoid arthritis patients taking gold salts, and as an idiosyncratic response to lithium and bismuth medication.

Pathophysiology

Goodpasture's anti-GBM antibody is believed to result from deposition of circulating IgG on the basement membrane in a linear pattern, initiating immunologically mediated damage of the basement membrane. Immune-complex disorders are believed to involve deposition of antigen-antibody complexes in the glomerulus, initiating damage. It has been speculated that silica initiates an inflammatory response, which, in turn, may initiate production of autoantibodies, such as ANCA, which attack the glomerulus.

Prevention

Preventive measures are aimed at reducing exposure to agents that may cause this disease.

Workers with mercury-induced nephrotic syndrome improve when they are removed from exposure. It is unclear whether removal from exposure improves the prognosis for people with disease caused by other agents, such as silica, but controlling and reducing exposure should reduce disease risk.

Further Reading

Asal NR, Cleveland HL, Kaufman C, et al. Hydrocarbon exposure and chronic renal disease. *International Archives of Occupational and Environmental Health* 1996; 68: 229-235.

Calvert GM, Steenland K, Palu S. End-stage renal disease among silica-exposed gold miners: a new method for assessing incidence among epidemiologic cohorts. *JAMA* 1997; 277: 1219-1223.

Elinder CG, Fowler BA. Mercury. In de Broe ME, Bennett WM, Porter GA, Verpooten GA (Eds.). *Clinical Nephrotoxins: Renal Injury from Drugs and Chemicals.* Dordrecht, The Netherlands: Kluwer Academic Publishers, 1998, pp. 363-370.

Fored CM, Nise G, Ejerblad E, et al. Absence of association between organic solvent exposure and risk of chronic renal failure: A nationwide population-based case-control study. *Journal of the American Society of Nephrology* 2004; 15: 180-186.

Osorio AM, Thun MJ, Novak RF, et al. Silica and glomerulonephritis: case report and review of the literature. *American Journal of Kidney Disease* 1987; 9: 224-229.

Ravnskov U. Hydrocarbons may worsen renal function in glomerulonephritis: A meta-analysis of the case-control studies. *American Journal of Industrial Medicine* 2000; 37: 599-606.

Steenland K, Sanderson W, Calvert GM. Kidney disease and arthritis in a cohort study of workers exposed to silica. *Epidemiology* 2001; 12: 405-412.

Wedeen RP. Occupational and environmental renal disease. *Seminars in Nephrology* 1997; 17: 46-53.

Hand-Arm Vibration Syndrome

ICD-10 T75.2
SHE/O

Laura Punnett and Judith E. Gold

Hand-arm vibration syndrome (HAVS) is also known as vibration-induced white finger, traumatic vasospastic disease, or secondary Raynaud's phenomenon of occupational origin. It is a disorder of the blood vessels and nerves in the fingers that is caused by vibration transmitted directly to the hands ("segmental vibration") by tools, parts, or work surfaces. Reduced handgrip strength and hearing loss have also been noted in some groups of exposed workers.

The condition is primarily characterized by numbness, tingling, and blanching (loss of normal color) of the fingers. Initially, there is intermittent numbness and tingling; blanching is a later sign, first in the fingertip and eventually over the entire finger. Symptoms usually appear suddenly and are often precipitated by exposure to cold. Attacks usually last 15 to 60 minutes, but in advanced cases, they may last up to 2 hours. Recovery, often accompanied by pain, begins with a red flush ("reactive hyperemia"), usually starting at the wrist and palm and moving down to the fingers.

The grading system for severity has been revised to account for injuries to nerves and blood vessels that appear to develop independently. The grading system defines up to four stages, based on (a) the frequency of tingling, numbness, and blanching; and (b) the extent of the loss of function in each hand. With continuing exposure to vibration, signs and symptoms become more severe and are eventually irreversible. In later stages, there is reduced sensitivity to heat and cold, with accompanying pain. Tactile sensitivity and neuromuscular control are impaired and finger joints become increasingly stiff, so precise manual tasks, such as picking up small objects and fastening buttons and zippers, become difficult. In the most advanced and severe cases, the fingers have a dusky, cyanotic (bluish) appearance, and there is complete obliteration of the arteries.

Diagnosis is based on symptoms and history of hand-arm vibration exposure. There are no reliable, objective diagnostic tests; there is a need for standardization of present methods. The available tests examine peripheral vascular function (plethysmography, arteriography, and cold-air or cold-water provocation), and neurological function (two-point discrimination, aesthesiometry, pinprick, touch, and temperature tests). They may differentiate affected from unaffected workers on a group basis, but have poor validity on an individual basis unless symptoms are moderate or severe. Thermal and vibratory perception thresholds are increased in vibration-exposed subjects. An ISO standard is being developed for vibrotactile threshold measurements, and discussions are under way to standardize cold-stress tests.

The differential diagnosis includes peripheral nerve compression at the wrist, forearm, or shoulder (see Carpal Tunnel Syndrome and Peripheral Nerve Entrapment Syndromes); non-occupational vascular or connective tissue diseases; and cancer.

Occurrence

The industries with the highest numbers of workers probably exposed to sources of segmental vibration are construction, farming, and truck and automobile manufacturing. Any worker using a powered hand tool, such as a chain saw, pneumatic drill, chipping hammer, jack hammer, grinder, buffer, or polisher, should be considered at risk.

According to NIOSH estimates, approximately 1.5 million workers in the United States were potentially exposed to vibrating hand tools or other sources of segmental vibration in 1989 and therefore at risk of developing HAVS. A study in 2000 found 14% of the working population in Great Britain affected with HAVS, 32% in men and 4% in women. Other prevalence estimates range from 20% to 100% among workers exposed to segmental vibration for at least 1 year; however, the intermittent nature of symptoms in the early stages leads to substantial underreporting. Both the intensity (acceleration level) and the duration of exposure determine (a) how short the latency is until the first episode, (b) how often the attacks occur, and (c) the rate of occurrence.

Causes

The cause of this syndrome is the direct physical transmission of vibration from a mechanical object to the hand and arm. This occurs through the use of vibrating hand tools or through other hand or arm exposure to segmental vibration, such as that transmitted through a truck or bus steering wheel or a part held to a grinding wheel. Both oscillatory (continuous) and impact sources, in the frequency range of 4000 to 5000 Hz, may be injurious. Higher frequencies are now known to be more deleterious than was previously believed. Depending on vibration intensity (acceleration), exposure for as little as 1 month is sufficient to initiate the disease process. Latency from first exposure to onset of blanching can range from 1 to 30 years, depending on the daily duration, frequency, and intensity of vibration and on work practices. Therefore, the shorter the latency period until onset, the more severe the expected syndrome if exposure continues.

Twenty studies reviewed by NIOSH have provided strong evidence of a direct relationship between the intensity and duration of segmental vibration exposure and the risk of developing HAVS. Among the highest quality studies evaluated, odds ratios ranged from 6.5 to 11.8, when comparing exposed to unexposed workers. Prevalence rates as high as 90% in logging and forestry workers decreased markedly following the introduction of anti-vibration saws, although HAVS still occurs among those using anti-vibration saws.

Chronic exposure to cold temperatures or damp conditions, especially during exposure to vibration, exacerbates its effects.

Occupational exposure to vinyl chloride monomer, organic solvents, epoxy resin, and rapeseed oil have been reported to produce a secondary Raynaud's phenomenon. However, there appear to be no data regarding the effects of combined exposure to segmental vibration and vinyl chloride or other peripheral neurotoxins, such as lead. (See Peripheral Nerve Entrapment Syndromes.)

Primary Raynaud's syndrome occurs spontaneously in 5% to 10% of the general population; the female-to-male ratio has been estimated to be as high as 5:1. Non-occupational secondary Raynaud's phenomenon may be idiopathic, hereditary, or associated with acute injury (frostbite, fracture, or laceration) or with medical conditions, such as connective tissue diseases (such as scleroderma or rheumatoid arthritis), vascular disorders (such as Buerger's disease or arteriosclerosis), neurogenic causes (such as poliomyelitis), or long-term hypertension. Nicotine is considered to aggravate or precipitate Raynaud's phenomenon because it is a vasoconstrictor that reduces the blood supply to the hands and fingers.

Pathophysiology

Both circulatory and neurological effects result from exposure of the hand to vibration, although the exact physiological mechanism is not known. Vibration appears to cause direct injury to peripheral nerves, resulting in the numbness of fingers. Decreased sensation in the hands may be secondary to constriction of the blood vessels, causing ischemia of peripheral nerves. Other physiological and chemical blood vessel changes have been documented, but their causal role is not clear.

An additional mechanism, the tonic vibration reflex (TVR), appears to contribute to soft-tissue damage by affecting tendon function. Vibration interferes with the sensitivity of nerve endings that enable one to sense the force exerted by a tendon. When the hand holds a vibrating object and this sensory feedback is disrupted, muscles are signaled to exert more force than is necessary to grip the object, increasing the strain on the tendons. (See Tendonitis.) The tighter grip also increases the amount of vibration transmitted to the hand, resulting in more nerve and blood vessel damage. Tendon nerve endings then lose even more sensitivity and the control reflex is even further disrupted.

Prevention

Primary prevention consists of measures to reduce exposure to sources of vibration. Redesign of production processes and work methods can help minimize use of vibrating hand tools or equipment. For example, improved quality of metal castings could reduce the need for later grinding and polishing. Where vibration cannot be eliminated from the workplace, engineering controls, work practices, and administrative controls should be considered to reduce the intensity and duration of exposure.

At the source of exposure, engineering controls consist of the redesign of powered hand tools to minimize vibration generated or transmitted during operation. Mechanical isolation and damping should be used to reduce the acceleration (intensity) of the vibration transmitted to the hand and arm. It is often recommended that tools should be selected that vibrate well above the natural resonant frequency of the hand and arm (30 to 300 Hz), which is where the hand and arm are most vulnerable. Recent studies suggest, however, that these high frequencies may be more hazardous than previously thought.

In the path of exposure, installation of a tool on an articulating arm or an overhead suspended balancer can help to reduce vibration transmission. The same principle can be applied to reduce vibration exposure of workers who hold parts to be ground or polished against a grinding wheel. Vibration transmission to the hand will also be reduced if workers can use less force to grip objects and operate tools; this can be accomplished by improving friction between the object or tool and the hand (or glove); having low torque with a cut-off, rather than a slip-clutch mechanism; and reducing the weight of the object or tool. Padded gloves or pads on tool handles may reduce vibration, although their effectiveness should be demonstrated for the particular tool and task. Commercially available isolation materials vary tremendously, even in a laboratory setting, in their effectiveness across the range of typical tool vibration frequencies. In any case, gloves should fit well to avoid increasing the grip force required to hold and operate a tool. Work practices and administrative controls may include running power tools at lower speeds, ensuring regular tool maintenance, engaging in frequent rest breaks or alternate work without vibration exposure, keeping the hands warm and dry, and maintaining a stable core body temperature.

"Consensus standards" for control of HAVS have been published by ACGIH, ANSI, and ISO (see Pelmear and Leong in Further Reading). Each considers the frequency spectrum of the vibration exposure to be a key determinant of the limits on intensity and duration of exposure. NIOSH chose not to incorporate such "frequency weighting" into its criteria document; instead, it recommended prospective medical and exposure monitoring, with the explicit goal of developing a better epidemiological basis for any future quantitative exposure limits. The draft updated ISO standard assesses vibration exposure, based on weighted acceleration in three directions. Despite the high prevalence of HAVS in users of impact tools, instantaneous peak accelerations and impact forces are ignored in current standards. A precautionary approach is warranted at this time.

Secondary preventive measures include encouraging workers to report symptoms to their physicians or to the workplace medical service. All workers who use vibrating hand tools should be examined for signs and symptoms of HAVS; work histories should specifically include questions about previous exposure to segmental vibration. Workers with preexisting signs or symptoms should not be assigned to work with vibrating tools. Exposed workers and their supervisors should be informed of the symptoms of HAVS. If tingling, numbness, or blanching occurs, the worker should seek medical attention promptly and should be reassigned to work with little or no exposure to vibration. The jobs of affected workers should be evaluated for implementation of engineering controls. Health care providers should be trained in interview and clinical examination procedures necessary to identify occupational HAVS.

Tertiary measures include calcium channel antagonists to produce peripheral vasodilation. Carpal tunnel syndrome and HAVS sensorineural symptoms are frequently confused. However, carpal tunnel syndrome surgery has shown poor results for cases of neuropathy induced through vibrating hand tools. Reduced grip strength following this surgery may be particularly burdensome for such workers attempting to return to employment.

Other Issues

Workers with exposure to vibration should be encouraged to reduce or stop smoking, because nicotine has a separate effect on peripheral circulation that makes blood vessels more vulnerable to the effects of vibration.

Further Reading

Gemne G, Pyykko I, Taylor W, Pelmear P. The Stockholm workshop scale for the classification of cold-induced Raynaud's phenomenon in the hand-arm vibration syndrome (revision of the Taylor-Pelmear Scale). *Scandinavian Journal of Work Environment and Health* 1987; 13: 275-278.

NIOSH. *Criteria for a Recommended Standard. Occupational Exposure to Hand-Arm Vibration.* Washington, DC: NIOSH, 1989 (DHHS [NIOSH] Publication No. 89-106).

Palmer KT, Griffin MJ, Syddall H, et al. Prevalence of Raynaud's phenomenon in Great Britain and its relation to hand transmitted vibration: A national postal survey. *Occupational and Environmental Medicine* 2000; 57: 448-452.

Pelmear PL. Vibration (hand-arm and whole body). In Baxter PJ, Adams PH, Aw T-C (Eds.). *Hunter's Diseases of Occupations* (9th ed.). New York: Oxford University Press, 2000.

Pelmear PL, Leong D. Review of occupational standards and guidelines for hand-arm (segmental) vibration syndrome (HAVS). *Applied Occupational and Environmental Hygiene* 2000; 15: 291-302.

Piligian G, Herbert R, Hearns M, et al. Evaluation and management of chronic work-related musculoskeletal disorders of the distal upper extremity. *American Journal of Industrial Medicine* 2000; 37: 75-93.

Hard Metal Interstitial Lung Disease

ICD-10 J84.8

Barry S. Levy and David H. Wegman

This disease is a rare form of diffuse, interstitial pulmonary fibrosis that results from inhalation of cobalt-containing aerosolized particles during manufacture or grinding of hard metal. It is distinguishable from similar conditions only by occupational history. Progressive dyspnea can be either insidious, developing after protracted cobalt exposure, or relatively abrupt, developing after brief and intense exposure. Common early symptoms consist of a dry cough with scant mucoid sputum. Spontaneous pneumothorax has been reported in one hard metal worker. Chest x-ray shows diffuse or patchy infiltrates. Pulmonary function testing shows restrictive changes. Asthma and other obstructive diseases are also associated with cobalt exposure, so one cannot rely on pulmonary function testing alone to determine whether a cobalt-induced disease is present. High resolution CT scans demonstrating reticulation, traction bronchiectasis, and large peripheral cystic spaces in a mid and upper lung distribution strongly suggest the diagnosis.

Occurrence

There is little information on the occurrence of cobalt-induced interstitial lung disease. Approximately 30,000 workers in the United States may be exposed to tungsten and its compounds. Between 1% and 10% of exposed workers may develop disease. Those at risk are involved in the manufacture or grinding of tungsten carbide tools and diamond polishing (where polishing disks containing microdiamonds and cobalt are used). Because the condition bears the initial features of a hypersensitivity pneumonitis (extrinsic allergic alveolitis), it is possible that cases often are diagnosed without an assignment of etiology; therefore, unrecognized cases may occur.

Causes

The cause of this condition is cobalt, a binder used in combination with tungsten carbide in the manufacture of cemented carbide metals. These metals are used as abrasives or as cutting tips for tools used in the high-speed cutting of metals, very hard woods, cement, or other hard materials. Tungsten carbide, the major component of these products, is considered to be biologically inert; however, when cobalt exposure occurs along with tungsten carbide exposure (or diamond dust exposure) hard metal interstitial lung disease may result. Studies in cobalt refineries where these other exposures are not present fail to identify cases of hard metal interstitial lung disease. Cobalt content in hard metal is usually less than 10% but may be as high as 25%. For workers to be at risk, it has been suggested that the content of cobalt in the alloy needs to be at least 2%.

Pathophysiology

Cobalt metal and metallic carbides interact to produce this disease. This combined exposure results in increased uptake of cobalt by macrophages compared to isolated exposure to cobalt. Cobalt may act by both immunologic and cytotoxic mechanisms, capable of provoking release of a fibrogenic agent from macrophages. There is little correlation between the amount of cobalt recovered from the lung and the amount of disease in cases that have undergone detailed study. Cobalt is highly soluble in biologic fluids and combines readily with protein, which suggests that cobalt may act as a hapten and promote immunologic reactions. Cobalt-induced apoptosis of alveolar macrophages may be a mechanism of cobalt-induced lung injury. A second condition often seen in workers with hard metal disease is occupational asthma associated with cobalt exposure. This is further evidence that the interstitial disease may result from an immunologic process.

Prevention

Dust suppression or elimination is essential in cemented tungsten carbide manufacturing facilities and in locations where tungsten carbide tools are ground. Use of exhaust hoods with adequate airflows has been shown to be useful in reducing exposure. Because cobalt is soluble in certain cutting fluids, such as water-soluble oils, those fluids with the lowest capacity for dissolving cobalt should be selected when machining or grinding tools or other materials made from cemented tungsten carbide. Annual chest x-rays and pulmonary function tests are indicated for workers who may be exposed.

Other Issues

Although there is a clear immunologic component to this disease, the associated risk factors are not understood. Therefore, excluding certain individuals, such as those who are atopic, from exposure is not justifiable on the basis of current scientific evidence. Certain exposures to cobalt have also been associated with increased risk of cardiomyopathy.

Further Reading

Araya J, Maruyama M, Inoue A, et al. Inhibition of proteasome activity is involved in cobalt-induced apoptosis of human alveolar macrophages. *American Journal of Physiology Lung Cellular and Molecular Physiology* 2002; 283: L849-L858.

Lison D. Human toxicity of cobalt-containing dust and experimental studies on the mechanism of interstitial lung disease (hard metal disease). *Critical Reviews in Toxicology* 1996; 26: 585-616.

Lison D, Lauwerys R, Demedts M, et al. Experimental research into the pathogenesis of cobalt/hard metal lung disease. *European Respiratory Journal* 1996; 9: 1024-1028.

Simcox NJ, Stebbins A, Guffey S, et al. Hard metal exposures. Part 2: Prospective exposure assessment. *Applied Occupational and Environmental Hygiene* 2000; 15: 342-353.

Headache

ICD-10 R51

Robert G. Feldman

Head pain can be a symptom of an acute neurological emergency or a chronically recurrent annoyance. It must be analyzed according to quality, location, temporal pattern, associated symptoms, and relationship to occupational factors.

Hypoxia, a decrease in oxygenation, or hypercapnia, an increase in carbon dioxide, induces an acute headache by causing vasodilation—the usual mechanism underlying headaches caused by acute chemical exposures. Cyclic and more chronic forms of headaches, such as migraines and cluster headaches and those resulting from transient acute cerebrovascular insufficiency, may also occur in workers exposed to chemicals. Migraine headaches are variable but may be classical, unilateral and beginning with visual or other neurological symptoms. Cluster headaches are also unilateral and are characterized by nonthrobbing, aching pressure. There are no diagnostic tests to determine the work-relatedness of headache. Tests, such as CT and MRI, can demonstrate an intracranial lesion, such as a brain tumor, abscess, subdural hematoma, or obstructive hydrocephalus, and thus aid in ruling in or out nonoccupational causes of headaches.

Occurrence

There are no reliable occurrence data for occupationally-induced headache. However, headache may occur in any environment where there is excessive exposures to chemicals. Of 220 chemicals reviewed, 149 have caused neurotoxicity in humans. The intensity of exposure to chemicals relates to risk and incidence of headache among exposed persons.

Causes

Acute and chronic occupational exposure to metals (such as arsenic, lead, nickel, tin, and tellurium), organic solvents, gases, and certain insecticides commonly produce headaches in exposed persons (Table 1). Inhalation of organic solvents may produce headache as a part of prenarcotic syndrome (in conjunction with cognitive impairment, depression, and/or anxiety), which, with prolonged exposure, often develops into toxic encephalopathy. Headaches brought on by carbon monoxide exposure progress in severity as exposure concentration, as measured by the percentage of carboxyhemoglobin present in blood, increases; these begin as a slight headache, with tightness across the forehead, and turn into a throbbing frontal headache and a

Table 1. Toxic Exposures Associated with Headache

Neurotoxin	Major Uses or Sources of Exposure
Metals	
Arsenic	Pesticides
	Antifouling paint
	Seafood
	Semiconductors
	Pigments
	Electroplating industry
	Smelter
Lead	Solder
	Illicit whiskey
	Storage battery manufacturing plants
	Auto body shops
	Lead-based paint
	Foundries
	Lead shot
	Insecticides
	Smelters
	Lead pipes
	Lead-stained glass
Nickel	Electroplating industry
	Paints
	Alloys
	Surgical and dental instruments
	Nickel-cadmium batteries
	Inks
	Coinage
Tin	Canning industry
	Solder
	Polyvinyl plastics
	Coated wire
	Silverware
	Electronic components
	Fungicides
Tellurium	Coloring agents in glazes and glass
	Rubber vulcanization
	Electronics industry
	Foundries
	Semiconductors
	Thermoelectric devices

(continued)

Table 1 (cont.). Toxic Exposures Associated with Headache

Neurotoxin	Major Uses or Sources of Exposure
Solvents	
Carbon disulfide	Manufacturers of viscose rayon Preservatives Rubber cement Electroplating industry Paints Textiles Varnishes
Methyl-n-butyl ketone	Paints Varnishes Quick-drying inks Lacquers Metal-cleaning compounds Paint removers
n-Hexane	Lacquers Printing inks Pharmaceutical industry Rubber cement Stains Glues
Perchloroethylene	Paint removers Degreasers Extraction agent for vegetables and mineral oil Dry-cleaning industry Textile industry
Toluene	Glues Paints Automobile, aviation fuels Paints, thinners Cleaning agents Manufacturers of benzene Lacquers Gasoline
Trichloroethylene	Degreasers Painting industry Paints Varnishes Process of extracting caffeine from coffee Lacquers Dry-cleaning industry Adhesive in shoe and boot industry Rubber solvents

(continued)

Table 1 (cont.). Toxic Exposures Associated with Headache

Neurotoxin	Major Uses or Sources of Exposure
Gases	
Carbon monoxide	Exhaust fumes of internal combustion engines, incomplete combustion Acetylene welding
Methane	Natural gas heating fuel Enclosed areas; mines, tunnels
Waste anesthetic gases	Operating rooms Dental offices
Insecticides	
Chlordecone (Kepone)	Agricultural industry

severe frontal and occipital headache. (See Asphyxiants.) Acute, high-dose exposure to trichloroethylene (TCE) can produce head and facial pain by affecting the trigeminal nerve, although chronic exposure to TCE more usually produces a subtle loss of sensation in the trigeminal nerve distribution. In both situations, headache may occur.

Vasodilation headaches can be due to alcoholic hangover, caffeine withdrawal, and exposure to histamines, nitrates, monosodium glutamate, and tyramine. Headache may occasionally be due to essential hypertension. Other causes of acute headache include trauma, sinusitis, glaucoma, temporal arteritis, trigeminal neuralgia, and extracranial inflammation, such as cellulitis of the scalp, periostitis, and osteomyelitis of the skull. Headache is present when a brain tumor reaches a critical size and causes intracranial pressure to increase; usually these are accompanied by other neurological complaints, such as tinnitus, which may be suggestive of an acoustic neuroma near the eighth cranial nerve, for example. Oral contraceptives containing progestin and estrogen may produce a constant dull headache. Muscle tension (muscle contraction) headache may be due to stress, anxiety, and depression and occurs in 40% of patients with head pain, 40% of whom report a family history of headache. Migraine may be precipitated by the ingestion of certain foods containing tyramine (cheese) and nitrate (sausage and bologna); it is seen most often in premenopausal women.

Pathophysiology

The mechanisms for headache following metal-fume exposure include brain swelling (tin and lead), transient hypoxia, and vasodilation (zinc, tellurium, manganese, and nickel). Carbon monoxide produces headache by vasodilation. The afferent fibers of the trigeminal nerve mediate most of the pain sensations in the head, but fibers of the glossopharyngeal, vagal, and upper cervical nerves can also mediate head pain. Because of the wide distribution and anastomoses of nerve fibers, referred pain is common in headache syndromes.

Prevention

Exposure to neurotoxic chemicals and other causes of headache should be minimized. Effective industrial hygiene monitoring and control of respiratory, gastrointestinal, and dermal routes of chemical entry are discussed in Part I. Headache symptoms may be relieved with analgesics, muscle relaxants, sedatives, antidepressants, and antianxiety drugs.

Other Issues

People with compromised cardiac function, hypertension, or other vascular problems may be at increased risk for headache following exposure to chemicals such as carbon monoxide, zinc, and tellurium because of the vasodilatory effects of these chemicals. Also, people being treated with diuretics or vasodilators may have increased frequency of headaches.

Further Reading

Feldman RG. *Occupational and Environmental Neurotoxicology.* Philadelphia: Lippincott-Raven, 1999, pp. 9-10.

Grandjean P, Sandoe SH, Kimbrough RD. Non-specificity of clinical signs and symptoms caused by environmental chemicals. *Human and Experimental Toxicology* 1991; 10: 167-173.

Langworth S, Anundi H, Friis L, et al. Acute health effects common during graffiti removal. *International Archives of Occupational and Environmental Health* 2001; 74: 213-218.

Prince TS, Spengler SE. Severe headache associated with occupational exposure to Stoddard solvent. *Occupational Medicine* 2001; 51: 136-138.

Taub A. Headache. In Feldman RG (Ed.). *Neurology: The Physician's Guide.* New York: Thieme-Stratton, 1984, pp. 129-146.

Hearing Loss, Noise-Induced

ICD-10 H83.3

Mark R. Stephenson

Noise-induced hearing loss is one of the most common occupational illnesses in the United States. It is characterized by a gradual worsening of high-frequency hearing thresholds over time following chronic and sometimes acute exposure to excessive noise levels. Figure 1 illustrates a typical progres-

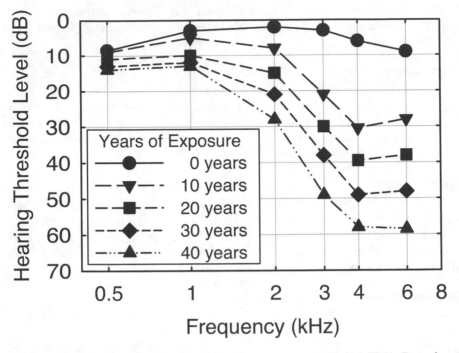

Figure 1. Progression of hearing loss following exposure to 95 dBA TWA. Data show hearing loss for white males at ages 20, 30, 40, 50 and 60 years with 0, 10, 20, 30, and 40 years of exposure, respectively (ANSI S3.44-1996).

sion over time. The pattern of noise-induced hearing loss is particularly characterized by a "notch" usually centered at 4 kHz, although the notch may be centered closer to 3 kHz or 6 kHz. In later stages, the hearing loss may spread to the middle and even low frequencies. The disease is usually bilateral; however, there may be an asymmetry between the left and right ears. It is possible to see a unilateral noise-induced hearing loss when the exposure conditions favor one side of the head. In most cases, tinnitus will also be present.

Noise-induced hearing loss is sensorineural (cochlear) in nature. The loss of hearing may be temporary at first, recovering overnight. This is referred to as a temporary threshold shift (TTS). With continued exposure, the loss becomes permanent and irreversible. Permanent noise-induced hearing loss develops gradually so that workers may lose a significant amount of hearing before becoming aware of its presence. During the early stages, sufferers often report having to turn up the volume on the television or experiencing difficulty understanding speech in groups or in the presence of modest background noise. As the hearing loss worsens, it becomes difficult to understand normal conversation, even in quiet, one-on-one situations. There is currently no clinical treatment available for permanent sensorineural hearing loss. Hearing aids may be beneficial in some cases; however, even with recent improvements in technology, hearing aids do not restore "normal" hearing.

Occupational noise-induced hearing loss is typically described in terms of shift from baseline hearing thresholds. However, there is no consensus regarding how much hearing loss constitutes a significant change from pre-existing hearing levels. The OSHA hearing conservation standard for general industry (29 CFR 1910.95) and MSHA (30 CFR 62) define a significant change (which it calls a standard threshold shift, or STS) as an average shift from baseline of 10 dB or more across the frequencies of 2000, 3000, and 4000 Hz in either ear. OSHA allows the subtraction of a correction factor for aging (presbycusis) when computing STS. NIOSH, in 1998, defined a significant threshold shift as an increase in hearing threshold of 15 dB or more at any one frequency (500, 1000, 2000, 3000, 4000, or 6000 Hz) in either ear that is confirmed for the same ear and frequency by a second test. NIOSH does not allow age correction. The Department of Defense (DOD) has adopted OSHA's definition of STS; however, it has retained its criteria for a 15-dB shift at any single frequency (1000, 2000, 3000, or 4000 Hz) as an "early warning flag" for hearing loss.

Even if a hearing loss meets the OSHA STS criteria, the hearing loss may or may not need to be reported on an OSHA Log 300. In order to be recordable, a hearing loss must (1) meet the criteria for STS, and (2) result in an absolute average threshold of 25 dB or greater (averaged across 2000, 3000, and 4000 Hz; age correction not allowed). Also, the employer may seek to have a "qualified" healthcare professional make a determination of work-relatedness. Consequently, the occurrence of occupational noise-induced hearing loss as measured by OSHA is usually less than it is by other entities.

Hearing loss may also be described in terms of absolute impairment (shift from audiometric zero). There is no universally accepted boundary demarcating the point at which hearing loss becomes a hearing impairment. NIOSH considers that a worker has sustained what it calls a "material hearing impairment" when his or her average hearing threshold level for both ears exceeds 25 dB at the audiometric frequencies of 1000, 2000, and 3000 Hz.

Occurrence

Estimates suggest that there are at least 5 million, and perhaps as many as 30 million, Americans occupationally exposed to noise levels greater than 85 dBA. Approximately 50% of all occupational noise exposures occur to manufacturing and utilities workers; 20% to transportation workers; and the remainder to workers in agriculture, construction, mining, and the military. Given at least 10 years of noise exposure at 85 dBA, 8% of workers will develop a material hearing impairment (per the NIOSH definition) by age 65. The figure rises to 22%, 38%, and 44% with exposures of 90, 95, and 100 dBA, respectively. According to the NIH, approximately one-third of all hearing losses can be attributed at least in part to noise exposure, and the most common source of excessive noise exposure is work. The impact of hearing impairment on occupational safety and health was underscored by a recent finding that sensory impairment-particularly hearing loss—is associated with a substantially increased risk of occupational injury.

Although hearing normally declines with age, the average, healthy, non-noise-exposed person will have essentially normal hearing at least up to age 60. According to the American National Standards Institute (ANSI), the median hearing level averaged across 1000, 2000, 3000, and 4000 Hz for non-exposed 60-year-old males is 17 dB HL (Hearing Level) and females, 12 dB HL. Thus, aging alone should not prevent the average person from enjoying normal hearing throughout all or most of his or her working career. The auditory effects of aging and noise are thought to be additive. Because of this and because older workers tend to have greater lifetime noise exposures, the prevalence of occupational hearing loss increases with age.

Individuals vary in their susceptibility to hearing loss. The prevalence of hearing impairment is greater among whites than blacks and higher among males than females. However, there is currently no reliable way to identify particular individuals who may be most susceptible to noise-induced hearing loss.

Causes

Occasionally, a single traumatic exposure to noise—typically having an intensity in excess of 130-140 dB SPL (Sound Pressure Level)—may cause hearing loss. More often, however, hearing loss is caused by repeated exposure to noise above 85 dBA over long periods, frequently as a mixture of impulsive and continuous-type noise. The risk of noise-induced hearing loss depends on both the intensity and duration of the exposure. As intensity increases, the length of time for which the exposure is "safe" decreases. The exchange rate is the relationship between exposure level and duration. It is defined as the change in allowable intensity level with each halving or doubling of duration. For example, a 5-dB exchange rate permits the level to be increased by 5 dBA for each halving of exposure duration, and requires the maximum level to be decreased by 5 dB for each doubling of exposure duration. OSHA and MSHA use a PEL of 90 dBA with a 5-dB exchange rate for an 8-hour, A-weighted exposure to continuous-type noise.

NIOSH, the DOD, and ACGIH use an REL of 85 dBA with a 3-dB exchange rate (also called the "equal-energy" rule). The EPA and most European countries also use the more conservative 3-dB exchange rate. The dBA scale is a logarithmic scale so that doubling the sound energy is equivalent to a 3-dBA increase; thus as noise level increases by 3 dBA, the permitted duration of exposure should decrease by half. The 3-dBA exchange rate is based on the theory that injury to the ear is proportional to the rate that sound energy is absorbed. If the sound energy doubles and the duration is decreased by half, the energy level remains constant. Hence, the rate of absorption is the same, and there is thought to be an equivalent risk of injury.

Determination of safe exposure levels is actually a value judgment based

Table 1, Comparison of models for estimating the excess risk of hearing impairment at age 60 after a 40-year working lifetime of exposure to occupational noise.

Average Exposure Level (dBA)	0.5-1-2 kHz Definition					1-2-3 kHz Definition			1-2-3-4 kHz Definition	
	1971-ISO	1972-NIOSH	1973-EPA	1990-ISO	1997-NIOSH	1972-NIOSH	1990-ISO	1997-NIOSH	1990-ISO	1997-NIOSH
90	21	29	22	3	23	29	14	32	17	25
85	10	15	12	1	10	16	4	14	6	8
80	0	3	5	0	4	3	0	5	1	1

on statistical calculations of the excess risk of hearing loss. Table 1 illustrates the percentage of an exposed population at risk of developing hearing impairment by age 60 after a 40-year working lifetime exposure to occupational noise. Risk estimates are shown as a function of exposure level for several different statistical models for risk damage and three different definitions of material hearing impairment. Although the percent risk of hearing loss varies between models, all models show little if any risk at 80-dBA exposure levels, and generally the risk of developing a material hearing impairment doubles as exposure level increases from 85 to 90 dBA.

There are many nonoccupational causes of hearing loss, including impacted cerumen, middle ear infections, and diseases such as meningitis and Ménière's syndrome. Nonoccupational noise-induced hearing loss may also result from frequent exposure to loud music and noisy hobbies, particularly those involving gunfire.

Pathophysiology

The usual mechanism of noise-induced hearing loss is gradual destruction of outer hair cells within the Organ of Corti (the sensory organ located within the cochlea). Exposure to loud noise as well as exposure to ototoxic chemicals elicits metabolic/biochemical activity within these hair cells. Metabolites known as reactive oxygen species (ROS) are generated as a byproduct of this activity. These metabolites are known to damage cell membranes, mitochondria, and DNA. With increased noise exposure, there is a corresponding increase in the production of ROS. Eventually, hair cells are unable to counter the damaging effects of the ROS. When this happens, programmed cell death occurs through a process called apoptosis. Sometimes acute traumatic exposure produces hearing loss in a single, sudden blast. In this case, the high sound pressure levels are thought to cause direct, mechanical damage to the hair cells, also leading to cell death.

Prevention

Because permanent noise-induced hearing loss is irreversible, prevention is the only way of reducing the burden of this occupational disease.

The Department of Labor has separate hearing conservation standards for general industry (29 CFR 1910.95), mining (30 CFR 62) and construction (29 CFR 1926.52). The various branches of the DOD have their own hearing conservation standards, and certain other workers fall under the jurisdiction of other federal agencies, such as the Department of Transportation. There are currently no enforceable noise standards for agricultural workers, self-employed individuals, and some public sector employees. Although each existing standard has unique provisions, most have adopted an action level of 85 dBA. In other words, workers exposed to an 8-hour TWA of 85 dBA should have available measures for protecting their hearing.

There is an increasing tendency towards using the term "hearing loss prevention" rather than "hearing conservation" in order to effect a paradigm shift towards zero occupational hearing loss. Semantics aside, most hearing loss prevention/hearing conservation programs consist of the same elements: noise measurement, engineering and/or administrative controls to limit exposures, hearing protection devices, audiometric monitoring, worker education, program evaluation, and record keeping.

Noise measurement is necessary to identify overexposed workers, and should be repeated biennially or sooner if there is a change in equipment or work practices. If hazardous exposure levels are noted, the best strategy is to use engineering controls to reduce noise at the source or shield the worker from the noise. There are many ways to apply engineering controls. Typical approaches involve reducing noise at the source, interrupting the noise path, reducing reverberation, and reducing structure-borne vibration. Engineering solutions are technologically feasible for most hazardous noise problems. However, because some solutions may be complex, the economic feasibility should be assessed on a case-by-case basis. Nevertheless, eliminating the hazard is the only way to have 100% certainty that workers will not be exposed to dangerous noise levels. The next best strategy for reducing noise exposures is to implement administrative controls, such as by scheduling activities at times or locations chosen to minimize workers' exposures, and implementing "buy quiet" policies. When engineering and administrative controls fail to reduce noise to safe levels, exposures must be reduced through hearing protective devices (HPDs), such as earplugs, earmuffs, and canal caps.

When properly selected and used, hearing protectors can be a powerful tool for preventing occupational hearing loss. An HPD must provide an appropriate amount of attenuation. Too much attenuation will interfere with a worker's ability to communicate and/or hear important sounds such as warning signals; ambient noise levels should not be attenuated below 70-75 dBA. Too little attenuation will not reduce noise exposure sufficiently. By law, manufacturers are required to label HPDs with a Noise Reduction Rating

(NRR) specifying how much attenuation their product will provide. The NRR is based on performance obtained under ideal laboratory conditions. Usually, workers obtain far less protection than the labeled rating, and there is little correlation between the labeled rating and the actual protection obtained by workers in the field (Figure 2). Because of this, OSHA has adopted a 50% de-rating of the NRR; NIOSH recommends a de-rating scheme of 25%, 50%, and 70% for earmuffs, slow-recovery formable earplugs, and all other earplugs, respectively. ANSI has approved a "subject fit" method (Method B of ANSI S12.6-1997) for estimating hearing protector attenuation which has proven an effective means for estimating the amount of protection workers can realistically expect to obtain; some manufacturers are voluntarily including this label on their products but it is not currently mandated by law. NIOSH recommends using "subject fit" (if they are available) instead of de-rated NRRs.

The best hearing protector is not the one with the highest NRR, but the one that the worker will consistently wear whenever exposed to loud noise. Hearing protectors are unlikely to be worn unless care is taken to address the "four Cs": comfort, convenience, communication, and cost. There is no single protector that will fit everyone, be universally comfortable, and be appropriate in every environment. Workers should be permitted to select from an assortment of devices that provide the proper attenuation and are appropri-

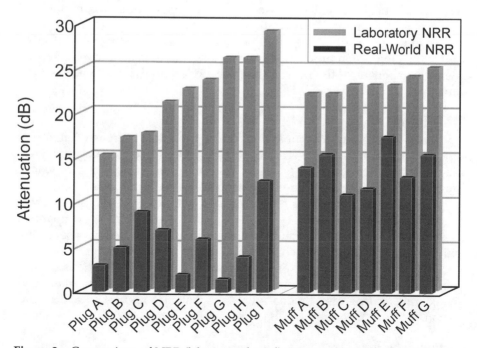

Figure 2. . Comparison of NRR (laboratory-based) attenuation values with HPD attenuation obtained in the workplace (after Berger, 2000).

ate for their work environment. Simply passing out hearing protectors is not sufficient; training must be provided in their use and care as well as on the effects of noise and hearing loss in general. Unless care is taken to instill a positive workplace safety culture, the social cost of wearing HPDs may be too great and workers may not use HPDs even when they are appropriate and readily available.

If workers are exposed to TWAs of ≥105 dBA, the use of dual hearing protection (muffs and plugs) may be advisable. In a recent review of this issue, it was concluded that there are no available methods for predicting the additional attenuation achieved by combining muffs with plugs. The gain may be as little as 0 dB at some frequencies, and as much as 15 dB at other frequencies. A good rule of thumb would be to add 5 dB to the expected attenuation of the more protective device when using dual protection. Because of bone conduction pathways to the cochlea, the maximum attenuation is about 40 to 50 dB.

Audiometric testing is necessary to monitor workers' hearing. A baseline hearing test should be obtained at the beginning of employment, and monitoring audiograms should be obtained at least annually to identify any change in hearing that might indicate underprotection from the noise. Monitoring audiograms are best obtained towards the end of the work shift in order to identify temporary threshold shifts before they become permanent. The annual hearing test is also an excellent opportunity to provide individual worker training and education.

An audiometric database should be maintained to evaluate the effectiveness of the program in preventing hearing loss among its employees. Records of each component of the hearing loss prevention program should be maintained.

Other Issues

In addition to noise, occupational hearing loss has been associated with exposure to vibration. Furthermore, some organic solvents (such as styrene and toluene), heavy metals (such as lead and mercury), and asphyxiants (such as carbon monoxide) have also been associated with hearing loss.

High noise exposures may be associated with loss of balance. Less well documented are extra-auditory effects, such as hypertension, occupational stress, and increased risk of injury. Some evidence has linked noise exposure with chronic stress hormone changes and risk of myocardial infarction.

Some noise-exposed workers may require special considerations in order to provide proper protection from hearing loss. These include workers with pre-existing hearing loss. Pregnant workers may also require special consideration.

Further Reading

ANSI. American National Standard: *Determination of Occupational Noise Exposure and Estimation of Noise-Induced Hearing Impairment* (ANSI S3.44-1996). New York:

American National Standards Institute, 1996.

Berger, EH. Noise Control and Hearing Conservation. In Berger EH, Royster, LH, Roster JD, et al. (Eds.), *The Noise Manual* (5th Edition). Fairfax, VA: American Industrial Hygiene Association, 2000, pp. 1-3.

Berger EH. Hearing Protection Devices. In Berger EH, Royster LH, Roster JD, et al. (Eds.), *The Noise Manual* (5th Edition). Fairfax, VA: American Industrial Hygiene Association, 2000, pp. 424-425.

Dobie RA. *Medical-Legal Evaluation of Hearing Loss.* New York: Van Nostrand Reinhold, Inc., 1993, pp. 1-3.

Morata TC. Effects of chemicals and noise on hearing. *Journal of Occupational and Environmental Medicine* 2003; 45: 676-682.

National Institutes of Health. *Noise and Hearing Loss: NIH Consensus Development Conference Consensus Statement 1990.* Bethesda, MD: NIH, 1990.

National Institute for Occupational Safety and Health. *Revised Criteria for a Recommended Standard: Occupational Noise Exposure* (DHHS [NIOSH] Publication No. 98-126.) Washington, DC: NIOSH, 1998.

National Institute on Deafness and Other Communication Disorders. Noise-Induced Hearing Loss (NIH Publication No. 97-4233). Bethesda, MD: NIH, 1997.

Occupational Safety and Health Administration. *Occupational Noise Exposure Standard.* 1983; 29 CFR Chapter XVII, Part 1910.95.

Prince MM, Stayner LT, Smith RJ, et al. A re-examination of risk estimates from the NIOSH Occupational Noise and Hearing Survey (ONHS). *Journal of the Acoustical Society of America* 1997; 101: 950-963.

World Health Organization. *Noise and Health.* Geneva: WHO, 2000.

Heat Stress

ICD-10 T67.8

James L. Weeks

Heat stress occurs when core body temperature is elevated. Effects can be grouped into six major types, which differ in symptoms, prognosis, and treatment: heat stroke, heat exhaustion, heat cramps, heat rash, heat syncope, and heat fatigue.

Heat Stroke

Heat stroke, or hyperthermia, is a life-threatening disorder that results from a failure of the core body temperature-regulating system which may cause core body temperature to exceed 40°C (104°F). Heat stroke is usually accompanied by hot and dry skin, mental confusion, convulsions, and unconsciousness. Death or irreversible organ damage frequently results; the fatality rate for heat stroke may be as high as 50%. A core body temperature above 42°C (108°F) for more than a few hours is usually fatal, depending on the person's health status. Early recognition and treatment of heat stroke will decrease the risk of death or damage to the brain, liver, kidneys, or other organs. Heat stroke is an emergency and medical assistance should be obtained immediately. Procedures to reduce body temperature must be initiated as early as possible. An approved first aid method for lowering body temperature is to remove the person to a cool and protected environment, remove the outer clothing, wet the skin with water, and fan vigorously. This procedure will maximize body cooling by evaporation and will prevent further body temperature increase while the patient is being transported to a hospital. Unless the person is conscious and alert, one should not administer fluids by mouth. Body temperature should be monitored to ensure that it is reduced but does not fall below normal.

Heat Exhaustion

Heat exhaustion results from the reduction of body water content or blood volume. The condition occurs when the amount of water lost by sweating exceeds the volume of water drunk. The major signs and symptoms of heat exhaustion include fatigue; extreme weakness; nausea; headache; faintness; and a cool, pale, clammy skin. Core body temperature, however, is usually normal or only slightly elevated. Treatment consists of removal to a cool area, recumbent rest, and cool fluids by mouth. Recovery usually occurs in less than 12 hours. Generally there are no permanent after-effects.

Heat Cramps

Heat cramps are painful cramps of the leg, arm, or abdominal muscles. They occur when individuals lose excessive amounts of salt while sweating during hard physical labor and high heat loads. The cramps may occur during or after work and may resolve spontaneously. If not, the person should take lightly salted fluids by mouth. Persons not acclimatized to heat may require additional salt. A normally salted diet is usually adequate to prevent heat cramps.

Heat Rash

Heat rash, commonly known as "prickly heat," is often experienced in hot, humid conditions that prevent sweat from evaporating. The extent of the condition may involve only small areas of the skin to the entire torso. When large areas of the skin are involved, sweat production is compromised, the ability to regulate body heat by sweat evaporation is reduced, and the capacity to work in the heat decreases. Even after the affected area of skin is healed, it may take

4 to 6 weeks before normal sweat production returns. If the skin can be kept clean and dry for at least 12 hours each day, severe heat rash can usually be prevented. Treatment may thus require reducing the length of time that the individual is exposed to hot, humid conditions each day. Infections of the involved skin must be prevented.

Heat Syncope

Heat syncope is alarming to the person but is the least serious of the heat-induced disorders. Heat syncope is characterized by dizziness and/or fainting while immobile, usually standing in the heat for an extended period. The condition occurs primarily in individuals who are not acclimatized to the heat, and it results from the pooling of blood in the dilated vessels of the skin and lower extremities with a resulting decrease in blood flow to the brain. Treatment consists of removal of the individual to a cooler area, if possible, and recumbent rest. Recovery is usually prompt and complete.

Heat Fatigue

Heat fatigue is a set of behavioral responses to acute or chronic heat exposure. The behavioral responses include impairment in (a) the performance of skilled sensorimotor tasks, (b) cognitive performance, and (c) alertness. These symptoms arise from the discomfort, physiological strain, psychosocial stress, and perhaps hormonal changes associated with working and living in hot climates. These aspects of heat stress are not well understood or documented.

Occurrence

Accurate measures of the occurrence of nonfatal heat stress is unknown. There is an average of 26 fatalities each year in the United States; rates are highest in agriculture and the construction industry (Table 1). In construction, those trades that require substantial manual labor and exposure to hot environments, such as roofers and construction laborers, experience the highest rate of fatal injuries. Similarly, agricultural workers who are required to perform substantial manual labor with exposure to hot environments experience an elevated rate of heat stress fatalities.

Many other workers who work in hot and humid environments are also at risk. These include miners in very deep underground mines (deeper than about 1200 meters), foundry workers, laundry workers, steel workers, rubber workers, kitchen workers and bakers, workers in hot climates, athletes, and others.

Causes

The causes of heat stress are generating or absorbing heat faster than the body can shed it, thereby increasing core temperature. In the occupational set-

Table 1. Deaths from Heat Stress, United States, 1992-2002, by Industrial Sector

Industrial Sector	Number of Deaths	% of Deaths	% of Labor Force	Ratio*
Agriculture	57	21.7	1.4	15.5
Construction	88	33.6	7.0	4.8
All other	117	44.7	91.6	0.5
Total	262			

*Ratio of percent of deaths to percent of labor force
Source: Bureau of Labor Statistics

ting, heat is most often generated by hard physical work. Heat transfer is most often impaired by hot and humid air and thus the combination of hard work in a hot and humid environment is the most common scenario resulting in heat stress. Heat transfer can also be impaired by heavy clothing and, in particular, by chemical protective clothing regardless of environmental conditions.

Environmental conditions can also be a source of heat. The most common source is the sun, but other common sources are other hot objects, such as engines, furnaces, and molten metal, transferring heat by infrared radiation. Radiant heat is transferred from the hotter to the cooler object, is attenuated by the inverse square of the distance between them, and can be blocked by any object impervious to infrared radiation. Thus, one can reduce the transfer of radiant heat by shielding the worker from the source and by increasing the distance to the source. Very hot air (>36°C) can also transfer heat to a person by conduction and convection. One can reduce heat transfer from such a source by isolating the worker from hot air and promoting bodily heat loss.

Pathophysiology

Understanding the physiology of thermoregulation is important for preventing heat stress. The body must maintain a relatively constant body temperature in the approximate range from 35° to 38°C (95° to 99°F). The principal source of body heat is metabolism and as metabolism increases, such as during physical labor, more heat is generated. The principal mechanism for heat loss, normally accounting for about 75%, is sweat evaporation. Other mechanisms include conductive and convective heat loss and radiant heat loss to the relatively cooler environment. With a rise in core temperature, cardiac output increases and peripheral blood vessels dilate, thus transporting heat from the body core to the surface. Sweat production and evaporation transfers heat to the immediate environment. Water is critical for maintaining blood volume and for the production of sweat.

Prevention

Preventing heat stress requires a threefold strategy: (1) reducing the generation of heat by reducing the work load, (2) reducing heat absorbed from the environment, and (3) promoting heat transfer to the environment.

Reduce the generation of heat. Since the principal source of heat in the occupational setting is metabolic heat, the principal means of reducing the generation of heat is by reducing the physical workload. Standards recommended by NIOSH and ACGIH both rely on this method as the principal means for preventing occupational heat stress.

Reduce heat absorbed from the environment. The most common environmental heat sources are hot objects that radiate heat and hot air that transfers heat by conduction and convection. Absorption of radiant heat can be controlled by shielding from the source and by increasing the distance from the source. Absorption of heat from hot air (>36°C) can be prevented by isolating the worker from the environment, if possible.

Promote heat transfer. Since the principal pathway for heat transfer is by sweat evaporation, it is important to promote evaporation and make sure this pathway is not obstructed. To do so, workers must have ample supplies of potable water to replace that lost by sweat. Strenuous work (>400 kCal/hr) may require a liter of water per hour to maintain body temperature. High humidity (more precisely, the partial pressure of water vapor) is a principal obstruction to sweat evaporation. Reducing humidity is an efficient means of promoting sweat evaporation, but doing so is feasible only in controlled environments, such as inside buildings, vehicles, booths, or a chemical protective suit.

Medical Evaluation and Monitoring

There is high variability between individuals in their vulnerability to heat stress. Workers not accustomed to work in the heat (unacclimatized workers) are at elevated risk. Other risk factors include heart disease including hypertension, obesity, poor physical condition, old age, prior instances of heat-related illness, thyroid disease, and use of some drugs. Persons who are poorly hydrated or have consumed diuretic beverages (with caffeine or alcohol) or have recently consumed a large meal may also be more vulnerable. Therefore, one aspect of preventing heat stress is to allow workers to become acclimatized to working in the heat (up to 2 weeks), to evaluate individuals prior to exposure, to educate workers about heat stress prevention, and to monitor them carefully during work.

One of the earliest and most subtle manifestations of heat stress is reduction in mental acuity. Therefore, it is not prudent to rely heavily on individuals monitoring themselves to take precautionary steps to reduce the risk of heat stress. Precautionary measures, such as a work-rest regimen, should be established before work and adhered to regardless of individuals' perceptions of their conditions. Similarly, workers should not work alone when there is a risk of heat stress.

Other Issues

Workers who must don chemical protective ensembles (protective clothing, respirator, gloves, and boots) may be especially vulnerable to heat stress, regardless of environmental conditions. The ensemble that prevents penetration of toxic chemicals also prevents sweat evaporation. Hard work in such an ensemble can quickly elevate the risk of heat. In addition to the usual strategies, adjusted for chemical protective clothing, some form of artificial cooling device may be needed.

Further Reading

American Conference of Governmental Industrial Hygienists. *Threshold Limit Values and Biological Exposure Indices for Substances in Workroom.* Cincinnati, OH: ACGIH, 2002.

Goldman RF. Heat stress in industrial protective encapsulating garments. In Martin WF, Levine SP (Eds.). *Protecting Personnel at Hazardous Waste Sites* (2nd ed.). Boston: Butterworth-Heineman, 1994, pp. 258-315.

Levy BS, Wegman DH (Eds.). *Occupational Health: Recognizing and Preventing Work-Related Disease and Injury* (4th ed.). Philadelphia: Lippincott Williams & Wilkins, 2000.

National Institute for Occupational Safety and Health. Criteria for a Recommended Standard: *Occupational Exposure to Hot Environments (Revised).* (DHHS/NIOSH Publication No. 86-113). Atlanta: NIOSH, 1986.

U.S. Bureau of Labor Statistics. *Census of Fatal Occupational Injuries.* Washington, DC: U.S. Department of Labor, published annually.

Hepatic Angiosarcoma

ICD-10 C22.3
SHE/O

Mohammed Saleem and Michael Hodgson

Hepatic angiosarcoma is a rare tumor. Its symptoms are nonspecific, with abdominal pain being the most common. Other symptoms include weakness, fatigue, weight loss, anorexia, and abdominal swelling. Physical examination may reveal hepatosplenomegaly, ascites, jaundice, and other nonspecific findings of hepatic dysfunction. A hepatic bruit is occasionally heard, reflecting the vascular nature of the tumor. Elevations of serum alanine aminotransferase, bilirubin, and alkaline phosphatase are common. Microangiopathic hemolytic anemia with thrombocytopenia, with or without disseminated intravascular coagulation, may occur. Serum alpha fetoprotein is normal.

Arteriography remains the standard for diagnosis. "Vascular lakes," central hypovascularity, and peripheral contrast staining on arteriography are suggestive of an angiosarcoma. A multiphasic, contrast-enhanced helical CT scan often is highly suggestive of the diagnosis. Liver biopsies are often inconclusive; the correct diagnosis is made from only 25% of percutaneous biopsies and 65% of open liver biopsies. Significant morbidity and mortality can result from a liver biopsy; the tumor's vascularity along with coagulopathy and thrombocytopenia contribute to the risk, making open biopsy preferred.

Occurrence

Approximately 10 to 20 new cases of hepatic angiosarcoma are diagnosed in the United States every year, and the prevalence varies from 0.14 to 0.25 per million. It is the most common malignant mesenchymal tumor of the liver, but accounts for only 2% of primary liver cancers. About three-fourths of cases occur in men, with frequent onset in the 40-59 age group. The relative risk ratio for hepatic angiosarcoma after occupational exposure may be 40 or higher than in the nonexposed population.

Causes

The etiology is unclear in about 70% of cases. Occupationally, it has been associated most commonly with exposure to vinyl chloride and arsenic. Nonoccupationally, exposure to Thorotrast (a contrast medium used in x-ray radiography from the 1930s to 1950s) was also a risk factor. The association with occupational exposure to vinyl chloride monomer (VCM), used in the manufacture of polyvinyl chloride, was first reported in 1974. Since then, a causal association has been established based on results of human occupational and experimental animal studies. Although angiosarcoma of the liver has occurred primarily among polymerization workers, cases have also been reported among others, including polyvinyl chloride fabrication workers exposed to low levels of VCM and among residents of neighborhoods near VCM fabrication or polymerization plants. Latency usually ranges from 11 to 37 years after start of vinyl chloride exposure.

Arsenic is also known to cause hepatic angiosarcoma in humans. Vintners who spray vines against mildew and people with psoriasis given prolonged treatment with Fowler's solution (1% potassium arsenite) are among those at increased risk.

Thorotrast (thorium dioxide), used as a radiological contrast medium (mainly for cerebral angiography) from 1928 to 1956, contains thorium-232, which emits alpha and beta radiation; alpha radiation has been implicated as a causative factor in hepatic angiosarcoma.

Pathophysiology

Human epidemiologic data have established VCM as a definite human carcinogen by the inhalation route of exposure. The EPA, IARC, and NTP have determined that VCM is a definite human carcinogen. The main route of metabolism of VCM after inhalation or oral uptake involves oxidation by cytochrome P-450 to form one or more reactive metabolites (2-chloroethylene oxide and 2-chloroacetaldehyde), which then bind to DNA, forming DNA adducts. If DNA damage is not repaired, mutations will develop, leading to tumor formation.

In VCM-exposed workers, histological lesions seen in the liver include focal hepatocytic hyperplasia, sinusoidal dilatation with hyperplasia, and focal mixed hyperplasia. There appears to be a progression from focal hepatic hyperplasia, to sinusoidal cell hyperplasia or focal mixed hyperplasia, to developing fibrosis (first parenchymal, then portal). The final stage of transformation is sinusoidal cell dysplasia with malignant transformation.

Prevention

Primary prevention involves limiting exposure to vinyl chloride using the hierarchy of controls, starting with engineering through administrative controls and appropriate respiratory protection. For secondary prevention, the OSHA vinyl chloride standard includes requirements for periodic testing (see 29 CFR 1910.1017). Older literature suggests that indocyanine green clearance may be the best indicator of underlying liver disease that predisposes to the subsequent development of this tumor, but this test is rarely done.

Thiodiglycolic acid, a major breakdown product of vinyl chloride, can be identified in the urine of exposed individuals if urine is obtained shortly after exposure. The level of this substance increases with increasing VCM exposure when the air VCM level is above 5 ppm. However, exposure to some other chemicals can produce the same breakdown product in the urine.

A specific and sensitive new method for the quantification of the vinyl chloride metabolite N-acetyl-S-cysteine by mass spectrometry may prove useful for monitoring occupational vinyl chloride exposure.

Further Reading

Baxter PJ, Anthony PP, MacSween RNM et al. Angiosarcoma of the liver in Great Britain, 1963-73. *British Medical Journal* 1977; 2: 919-921.

Baxter PJ, Anthony PP, MacSween RNM, et al. Angiosarcoma of the liver: Annual occurrence and aetiology in Great Britain. *British Journal of Industrial Medicine* 1980; 37: 213-221.

Creech JL, Johnson MN. Angiosarcoma of liver in the manufacture of polyvinyl chloride. *Journal of Occupational Medicine* 1974; 16: 150-151.

Falk H, Caldwell GG, Ishak KG, et al. Arsenic-related hepatic angiosarcoma. *American Journal of Industrial Medicine* 1981; 2: 43-50.

Maltoni C, Ciliberti A, Gianni L, et al. Vinyl chloride carcinogenesis: Current results and perspectives. *La Medicina del Lavoro* 1974; 65: 421-444.

Viola PL, Bigotti A, Caputo A. Oncogenic response of rat skin, lungs, and bones to vinyl chloride. *Cancer Research* 1971; 31: 516-519.

Hepatic Porphyrias

ICD-10 E80

Susan Buchanan and Michael Hodgson

The porphyrias are disturbances of heme synthesis, usually due to a genetic defect resulting in absence of one or more enzymes that catalyze porphyrin metabolism. They are labeled "hepatic" or "erythropoietic," reflecting the location of the enzymatic defect and the accumulation of the precursors. Chemically-induced porphyria may result from direct disruption of porphyrin metabolism by specific agents or as a result of triggered enzyme depletion in genetically susceptible individuals. The clinical manifestations of hepatic porphyries are primarily central and peripheral nervous systems, kidneys, and bone marrow. Initial neurologic symptoms include weakness, paresthesia, cogwheeling, and myotonia. Abdominal pain, constipation, nausea and vomiting, and dark urine may also be present. Psychiatric symptoms include depression, hallucinations, and psychosis. Erythropoietic porphyrias cause skin manifestations including photosensitivy, bullous lesions, scarring, and hirsutism. In general, the diagnosis requires documentation of specific metabolites and, in the inherited forms, documentation of the specific enzymatic defect.

Porphyria cutanea tarda is the most common form and the form most commonly associated with occupational or environmental exposures. Liver abnormalities include siderosis, fibrosis, necrosis, and cirrhosis. Laboratory findings include elevated levels of heme precurors in blood, urine, or stool. Liver function abnormalities are generally mild or moderate. Diagnosis is made by finding elevated levels of aminolevulenic acid (ALA) and porphobilinogin (PBG) in a 24-hour urine collection.

Occurrence

Attacks of intermittent porphyria are five times more common in women than men. Reports of chemically induced porphyria are rare. Epidemics have occurred in Turkey and Louisiana from exposure to hexachlorobenzene; in Japan, Taiwan, and Michigan from exposure to polychlorinated biphenyls (PCBs); and in New Jersey from exposure to 2,4-dichlorophenol (2,4-D) and 2,4,5-trichlorophenol (2,4,5-T). Sporadic cases have also been reported.

Causes

Lead poisoning is known to cause chemically induced porphyria. Other implicated agents include hexachlorobenzene, 2,4,5-trichlorophenoxyacetic acid, dioxin (TCDD), PCBs, and chlorophenols. Metals such as arsenic and mercury, polybrominated biphenyls (PBBs), and halogenated hydrocarbons have been found to be porphyrinogenic in lab animals.

Pathophysiology

Organochlorine compounds interfere with porphyrin metabolism by inhibiting hepatic and erythrocyte uroporphyrinogen decarboxylase. Excretion of penta- and hepta-carboxyporphyrin is increased, as is cellular concentration of toxins. Some agents produce characteristic changes in heme precursor excretion through disturbances of intracellular structure and function. The pathogenesis of clinical signs and symptoms is not well understood.

Prevention

Primary prevention can be achieved by preventing exposure to the causes. This can be accomplished by implementing the guidelines described in Chapter 1. Chelation with ethylenediaminetetraacetic acid (EDTA) has been tried with variable success on people already affected. Heme arginate may reduce symptoms if given early in the attack. Porphyrinuria may be a biologic indicator of exposure to porphyrinogenic chemicals.

Other Issues

Alcohol, certain pharmaceutical and illicit drugs, and pregnancy may all provoke attacks in genetically susceptible individuals. Other hepatotoxins and enzyme inducers, genetic factors, iron, and lead may worsen disease.

Further Reading

Daniell WE, Stockbridge HL, Labbe RF, et al. Environmental chemical exposures and disturbances of heme synthesis. *Environmental Health Perspective*s 1997; 105 (Suppl. 1): 37-53.

Downey DC. Porphyria and chemicals. *Medical Hypotheses* 1999; 53: 166-171.

Feldman M, Friedman LS, Sleisenger MH (Eds.). *Sleisenger & Fordtran's Gastrointestinal and Liver Disease* (7th ed.). St. Louis, MO: Saunders, 2002.

Zimmerman HJ. *Hepatotoxicity.* Philadelphia: Lippincott Williams & Wilkins, 1999.

Hepatitis, Toxic

ICD-10 K71.6

Sean M. Bryant and Michael Hodgson

The liver is the main organ responsible for the metabolism of xenobiotics and is considered particularly vulnerable to toxin-induced damage. It receives up to 75% of its blood supply through the portal vein; this accounts for the large first-pass effect and places it at risk after toxic exposure.

Toxic hepatitis may present in a variety of ways. Acute disease usually presents with gastrointestinal upset which may be followed by a period of well-being (1 to 3 days), then by more serious signs and symptoms, reflecting hepatic failure, renal failure, and encephalopathy, which may be fatal. The more acute the injury, the higher likelihood of elevated liver function tests, such as aspartate aminotransferase (AST) and alanine aminotransferase (ALT). Subacute injury results from persistent exposure to toxic chemicals, with more intermittent symptoms that are mild to moderate in nature. Chronic disease is characterized by a slower progression to clinically evident liver dysfunction, often taking months to years for chronic injury to become manifest.

Hepatic injury is usually classified based on the specific site of insult within the acinus, the functional unit of the liver. Toxins generally cause damage in a specific zonal distribution based on the site of their activity. Zone 1 (periportal region) is closest to the hepatic artery and therefore receives blood with the highest oxygen concentration. Specific toxins which induce damage by means of oxygen free-radical formation, in sufficient concentrations, may overwhelm this region resulting in injury. Iron, yellow phosphorus, and allyl alcohol are examples of concentration-dependent toxins that cause injury in Zone 1. Zone 3 corresponds with the centrilobular region with high cytochrome P450 activity. It is in this zone that certain parent compounds like carbon tetrachloride are metabolized by this enzyme system (P450) to reactive intermediates and thus inflict their greatest damage in the centrilobular area. Zone 2 is considered an intermediate region between Zones 1 and 3. Acetaminophen and benzene are also converted to reactive intermediates by those enzymes and cause local (centrilobular) necrosis.

In general, liver evaluation relies on tests of liver injury and imaging. Liver enzyme measurements, which represent indicators of ongoing hepatic injury, are abnormal in acute and smoldering chronic disease. Diagnostic imaging, with computerized tomography, magnetic resonance imaging, and ultrasound, may help evaluate morphological changes. Liver biopsy is a highly sensitive and specific method to identify the type of injury and shed light on possible causative factors. While this is an invasive procedure, it is often valuable, if not essential, in the patient with unexplained toxic hepatic injury.

Occurrence

Incidence, prevalence, disability, and mortality data on toxic hepatitis are not readily available. The vague and nonspecific signs and symptoms associated with liver injury, the inadequate sensitivity and specificity of routine tests, and the invasive nature of liver biopsy account for the limited number of epidemiologic studies.

Causes

More than 100 workplace chemicals have been shown to produce liver injury. Ingestion, inhalation, and dermal exposures to these toxins can result in hepatic damage. In general, inhalation is the most important portal of entry for industrial chemicals. Many studies demonstrate dose-response relationships between the degree of exposure and indices of liver impairment.

Pathophysiology

Liver injury is often classified into dose-dependent and idiosyncratic forms. Toxins may require "phase I" biotransformation, through mixed function oxidases (cytochrome P450), to cause liver injury through the formation of more toxic intermediates. These then produce cellular damage. Phase II reactions are designed to make compounds more polar, easier to excrete, and less toxic. Pathologically, classification of liver injury can be described as steatosis (fat deposition), necrosis (cellular death), cholestasis (bile accumulation), fibrosis (scar tissue development), cirrhosis (fibrosis with remodeling of liver architecture), veno-occlusive (blockage of veins), and carcinogenic (development of liver cancer). Contributing factors for hepatotoxicity include poor nutritional status, alcohol consumption, old age, genetic susceptibility, physical activity, and cigarette smoking.

Prevention

Compliance with workplace standards and implementation of engineering controls should minimize and reduce exposure to hepatotoxins. In addition to preventing airborne exposures, dermal exposures may require appropriately selected gloves since up to one-third the dose of an organic solvent may be absorbed through the skin.

Workers exposed to hepatotoxins should be screened to detect subclinical hepatic injury. Sensitive, but nonspecific, tests can result in many false-positive results. One study revealed that as many as 30% of asymptomatic workers receiving a standard battery of liver function tests had results in excess of laboratory controls. Some data suggest that such abnormalities will return to normal in a substantial proportion of individuals with injury related to exposure, suggesting both a diagnostic and a therapeutic strategy.

Because of the risk of misdiagnosis, some workers with abnormal tests may be removed unnecessarily from a workplace and be subjected to further expensive and possibly risky evaluation. Ideally, normal levels for the specific population being studied should be ascertained, followed by screening with tests of high sensitivity and subsequently with tests of higher specificity, in order to correctly identify true-positives. Some law supports screening employees to exclude those with preexisting liver disease that might place the individual at risk, but this remains controversial. Periodic surveillance to detect occupational hepatic damage may result in referring the patient with persistently abnormal studies to a specialist for consideration of biopsy and further diagnostic workup.

Occupational treatment of hepatotoxicity relies on removing the patient from the exposure. Medical treatment may include *N*-acetylcysteine, a useful antidote for acetaminophen poisoning. Studies have demonstrated benefit from using *N*-acetylcysteine in patients suffering fulminant hepatic failure caused by various etiologies other than acetaminophen overdose. Hyperbaric oxygen therapy has been explored as an option for treating liver injury due to carbon tetrachloride. Finally, and most invasively, liver transplantation is an option for severe disease. Current overall 1-year survival is greater than 85% in some centers compared to 25 years ago, when it was 30%.

Further Reading

Cotrim HP, Andrade ZA, Parana R, et al. Nonalcoholic steatohepatitis: a toxic liver disease in industrial workers. *Liver* 1999; 19: 299-304.

Dossing M, Skinhoj P. Occupational liver injury: Present state of knowledge and future perspective. *International Archives of Occupational and Environmental Health* 1985; 56: 1-21.

Goodman ZD. Drug hepatotoxicity. *Clinics in Liver Disease* 2002; 6: 381-397.

Harrison P, Wendon J, Gimson A, et al. Improvement by acetylcysteine of hemodynamics and oxygen transport in fulminant hepatic failure. *New England Journal of Medicine* 1991; 324: 1852-1857.

Tamburro CH, Liss GM. Tests For hepatotoxicity: usefulness in screening workers. *Journal of Occupational Medicine* 1986; 28: 1034-1044.

Hepatitis, Viral

ICD-10 B15-B19, B25.1, B33.8

Rosemary Sokas

A wide range of infectious agents may infect the liver and other organs, resulting in hepatocellular inflammation (hepatitis) as well as other liver tissue responses, such as granuloma, abscess, and cholestasis. However, the disease characterized as "viral hepatitis" is caused by a handful of viruses that specifically target the liver, resulting in inflammation and necrosis of hepatic cells. There has been an explosion of information about viral hepatitis over the past three decades, resulting in screening and diagnostic tests, new vaccines, and new treatments. The documented agents affecting humans include hepatitis A virus (HAV), hepatitis B virus (HBV), hepatitis C virus (HCV), hepatitis delta virus (HDV), and hepatitis E virus (HEV). HBV is a DNA virus; the other four are RNA viruses. Additional novel viruses are under investigation.

Routes of infection, incubation periods, severity, and duration of illness differ on a population basis for the various types of viral hepatitis, but there is broad overlap. Onset of symptoms may be abrupt, with fever, chills, headache, vomiting, and muscle aches, or may be gradual, with fatigue, loss of appetite, and malaise. Scleral icterus (yellowing of the whites of the eyes), jaundice (yellowing of the skin), and dark urine are characteristic of symptomatic disease, and usually occur after the initial fever has resolved. (Jaundice in the presence of fever is rarely due to viral hepatitis alone.)

HAV infection is often mild or completely asymptomatic, particularly in young children. In older children and adults, when infection results in clinically apparent hepatitis A, it is usually mild and self-limited. However, severity increases in older individuals or in those with underlying chronic liver disease, in whom it may cause hepatic failure and death. HAV does not cause chronic disease.

HBV is associated with more severe acute hepatitis and higher mortality rates. HBV may also cause persistent infection and chronic active hepatitis.

HDV can only infect patients who have HBV, either as a co-infection or a super-infection; it increases the risk of severe acute liver disease and chronic infection.

HCV infections are frequently associated with mild hepatitis. However, the infections are commonly persistent and may cause chronic disease. HCV and HBV infections may be transmitted from mother to infant during birth, and result in infections that are not apparent, but that frequently become chronic. Chronic infection with HBV and HCV are the leading causes of cirrhosis of the liver and hepatocellular cancer worldwide.

HEV infection may cause fulminant hepatic failure among young women, particularly in the last trimester of pregnancy, as well as among older patients. It does not cause chronic infection.

Occurrence and Causes

HAV (historically known as "infectious hepatitis") and HEV (historically known as "non-A, non-B, non-parenteral hepatitis") are transmitted primarily through the fecal-oral route. They are chiefly found in areas of poor sanitation. HAV is endemic worldwide; its occurrence has been steadily declining in the United States since World War II, with 13 997 cases reported in 2000 for an incidence rate of 4.9/100,000. However, the incidence equals or exceeds 20 per 100 000 in a number of southwestern states and Illinois. In those locations, the CDC recommends childhood immunization. There is no adequate information on occupationally acquired disease. Outbreaks occur sporadically with point sources identified in day care centers, in food handlers, and among institutionalized populations associated with fecal incontinence. Infection is transmitted through contaminated water or food. Shellfish, which filter large quantities of water where sewage streams empty, may be a source in the United States and in other countries.

Occupations at increased risk of HAV include those requiring travel to areas with poor sanitation; daycare workers; and healthcare workers in pediatric and chronic care settings. Sewage and wastewater-treatment workers may be exposed, although cross-sectional studies have been equivocal, in some instances because of high rates of HAV antibodies in both waste worker and control populations. Workers who handle apes and chimpanzees may be at risk for zoonotic transmission. HEV is endemic in developing countries in Asia, Africa, Latin America, and in the newly emerging states of Central and Eastern Europe. It has recently been described in Japan among individuals with no travel history. Zoonotic transmission through pigs or rats may play a role.

HBV (historically known as "serum hepatitis"), HCV (historically known as "non-A, non-B hepatitis"), and HDV are primarily transmitted parenterally. The advent of tests to screen and eliminate HBV and HCV from blood and blood products has greatly reduced their occurrence in the general population. The CDC reports a decline in HCV infections of more than 80% in the 1989-1999 period, when 35,000 new cases of HCV were estimated to have occurred. There is inadequate information to assess occupational incidence rates. Screening of the blood supply has been the major intervention responsible for this reduction. The United States is engaged in a national effort to eliminate HBV; CDC reports a decline of more than 60% during the same period, with 8036 cases of active hepatitis B reported in 2000.

The decline in HBV, one of the major public health success stories of the past century, has been enhanced by the development of effective vaccines, the implementation of universal childhood immunization recommendations, and the promulgation of the OSHA Bloodborne Pathogen Standard. Rates of new infection have declined disproportionately among health care workers, who in the years following the implementation of the standard achieved immunization rates exceeding 70%. However, the nonimmune adult population remains at risk for HBV through sexual activity and intravenous drug abuse. In Asia and Africa, where HBV and HCV are endemic, perinatal transmission is common.

There is no vaccine for HCV, and it remains problematic because of the greater likelihood of persistence in infected individuals, although transmissibility appears to be lower than for HBV. Although the prevalence of HCV in the general population is less than 2%, among patients in dialysis centers and urban emergency rooms the prevalence increases 10-fold. Thus, healthcare workers and emergency responders may encounter exposures frequently. Both viruses will occasionally escape detection because of the window in which serologic markers appear, so rare transfusion-related disease may still occur.

Because of sexual transmission and intravenous drug use, prevalence is high in prison and juvenile corrections facilities. Occupational exposures include blood-to-blood transmission through percutaneous exposures to "sharps" as well as splashes of infectious blood or other body fluids to the mucous membranes of the eyes or mouth. OSHA's Bloodborne Pathogen Standard, updated to incorporate changes based on the Needlestick Safety and Prevention Act of 2000, emphasizes primary prevention for all blood-borne pathogen exposures. Healthcare workers, first responders, and others who would be expected to encounter blood or infectious body fluids on a rou-tine basis (approximately monthly) are included; workers who experience unanticipated exposures should receive post-exposure evaluation. Police and prison guards may be at increased risk. Commercial sex workers and exploit-ed vulnerable populations, such as children or undocumented immigrants, may be at extremely high risk for these as well as other sexually transmitted diseases. Violence increases the risk.

Pathophysiology

The common pathology for all forms of viral hepatitis is viral infection of the hepatic cells, leading to inflammation and hepatocellular necrosis. The average incubation period ranges from 21 days for HAV to 70 days for HBV, but there is wide variability and overlap. Cell death releases liver enzymes; the hallmark finding of clinical hepatitis is elevation of the aminotransferas-es, often to eight or more times normal. Both aspartate aminotransferase (AST) and alanine aminotransferase (ALT) are increased, and the AST:ALT ratio is usually less than 1. Lesser increases in LDH and alkaline phosphatase are seen. Moderate increases in serum bilirubin accompany clinical findings of jaundice, scleral icterus, dark urine, and light stools. The investigation of abnormal liver function tests includes the following serologic testing: anti-HAV IgM, HBsAg (hepatitis B surface antigen), anti-HBc IgM (hepatitis B core antibody), and anti-HCV antibody. In patients with severe HBV hepati-tis, anti-HDV antibody testing should be done. Anti-HAV IgM indicates acute or recent infection with HAV; anti-HAV IgG indicates past infection with immunity or successful immunization. Anti-HBsAb (antibody to hepatitis B surface antibody) indicates (a) past infection with immunity or (b) successful vaccination. Other causes of hepatocellular injury, such as medications and toxins, may need to be considered. Abnormal coagulation tests and clinical decline in mental status indicate liver failure. Acute, fulminant hepatitis may result in coma and death. Care is supportive and may entail liver transplan-tation.

There are a variety of treatment protocols under investigation for both chronic hepatitis B and chronic hepatitis C. The FDA has approved both inter-feron and lamivudine, a reverse transcriptase inhibitor also used to treat HIV, for the treatment of chronic HBV, and has approved interferon and ribavirin for the treatment of chronic HCV. (The side effects from interferon, which must be administered parenterally, are significant and include major depres-sion.) Chronic viral hepatitis is the most common reason for liver transplan-tation in the United States.

Prevention

Primary prevention of enterically transmitted hepatitis A and hepatitis E requires good sanitation, including field sanitation for agricultural workers, as well as careful handwashing before food preparation, separation of food preparation areas from diaper-changing areas, and education for travelers concerning safe foods and beverages. Healthcare or other workers handling fecal material should wear gloves, and may need face shields if splashes are likely. For occupations where raw or partially treated sewage may be encountered, engineering and administrative controls to separate the worker from the sewage, sludge, or other exposure source should be provided, as well as appropriate PPE, such as gloves and face shields, and hand-washing facilities. Safe and effective vaccines, both alone and in combination with HBV vaccine, are available for HAV, but the first dose must be given at least 4 weeks prior to exposure. Because prior infection with HAV was common in the United States in the past, serologic testing for immunity may be considered for persons over 40 years of age in order to avoid unnecessary immunizations. For individuals with known, recent exposure to HAV, immune serum globulin should be administered.

Occupational HBV and HCV prevention has been enhanced by the Needlestick Safety and Prevention Act of 2000, which adds a focus on coordinated engineering controls to the OSHA Bloodborne Pathogen Standard. Studies suggest that integrated safety programs help both to identify safer procedures and equipment, such as needleless tubing systems and self-sheathing needles, and to facilitate their adoption. Employers are required to offer hepatitis B vaccine, which is given in three doses over a period of 6 months. It is safe and effective, although older workers and obese workers have a greater chance of vaccine failure. If exposure occurs, a series of post-exposure measures take place not only for HBV, but also for HCV and HIV. If the source patient is HBV-positive and the worker demonstrates inadequate immunity on serologic testing, both hepatitis B immune globulin (HBIG) and a repeat immunization series with hepatitis B vaccine should be administered. There is no post-exposure prophylaxis for HCV. However, because early treatment protocols are under investigation, workers exposed to blood from a source patient known to have hepatitis C should be followed closely for seroconversion. If this occurs, they should be referred to physicians with appropriate expertise for evaluation, follow-up, and potential therapy. Workers exposed to blood from an unknown source should be followed as if the source were positive, with serologic testing for immunity, HBIG and immunization for HBV, and follow-up testing for seroconversion.

Other Issues

HBV is exceptionally hardy and has been cultured from surfaces for up to 48 hours, and in intravenous tubing up to 18 inches from the insertion site.

Both HBV and HCV have been transmitted to patients through nosocomial spread, both from contaminated equipment and from infected surgeons. HAV is also transmitted to patients in healthcare settings, in restaurants, and elsewhere. Worker, patient, and community health are interrelated.

Further Reading

Berenguer M, Wright T. Viral Hepatitis. In Feldman M, Fiedman LS, Sleisenger MH (Eds.). *Sleisenger and Fordtran's Gastrointestinal and Liver Disease: Pathophysiology/Diagnosis/ Management* (7th Edition). Philadelphia: Saunders, 2002.

Centers for Disease Control. Summary of Notifiable Diseases-United States, 2000. MMWR June 14, 2002, Vol. 49. Available at
http://www.cdc.gov/mmwr/PDF/wk/mm4953.pdf

Centers for Disease Control. Appendix: Hepatitis A and B Vaccines. MMWR January 24, 2003, Vol. 52 (RR01): 34-36. Available at
http://www.cdc.gov/mmwr/preview/mmwrhtml/rr5201a2.htm

Centers for Disease Control. Notice to Readers: Recommended Adult Immunization Schedule - United States, 2002-2003. MMWR, Vol. 51: 904-908, October 11, 2002. Available at
http://www.cdc.gov/mmwr/preview/mmwrhtml/mm5140a5.htm
http://www.cdc.gov/niosh/topics/bbp/
http://www.cdc.gov/ncidod/hip/BLOOD/blood.htm
http://www.osha.gov/SLTC/bloodbornepathogens/index.html

Kawai H, Feinstone SM. Acute Viral Hepatitis. In: Mandell GL, Bennett JE, Dolin R (Eds.). *Principles and Practice of Infectious Disease*s (5th Edition). Philadelphia: Churchill Livingstone, 2000.

High-Altitude Illness

ICD-10 T70.2

James L. Weeks

High-altitude illness is a collective term that includes acute mountain sickness, high-altitude pulmonary edema, and high-altitude cerebral edema. These typically occur to unacclimatized persons who ascend rapidly to elevations above about 2000 m. The Lake Louise Acute Mountain Sickness Score is a scoring convention, based on evaluation of headache, gastrointestinal symptoms, fatigue, dizziness, sleeping problems, mental state, ataxia, and peripheral edema. The condition was first described among miners in northern Chile at altitudes above 4500 m.

Occurrence

Among unacclimatized persons, the incidence of acute mountain sickness in 1991 in Colorado was 22% at altitudes from 1850 to 2750 m, and 42% at altitudes above 3000 m. In a group of trekkers and porters in Nepal, the prevalence of high-altitude bronchitis and pharyngitis was 12% and the prevalence of acute high-altitude illness was 8%. Incidence was 28% in a group of 93 unacclimatized climbers ascending to 3900 m in Taiwan. The appearance of symptoms is usually delayed by 6 to 10 hours. Risk is higher among unacclimatized persons, at higher altitudes, with rapid rate of ascent and/or physical exertion, and for persons who ordinarily live below 900 m elevation. Occupations at risk include miners who work at high altitudes, workers in the ski industry, guides for climbers, some scientists, and some military personnel. Persons over 50 years of age appear to be at reduced risk, perhaps by selection, and children may be at increased risk. Retinopathy is common at very high altitudes (>4500 m). Physical fitness appears not to be protective.

Causes

The principal cause is rapid rate of ascent (>300-600 m per day) to high altitude (above about 2000 m) combined with individual susceptibility. Persons unacclimatized to hypoxic environments are especially vulnerable.

Pathophysiology

The pathophysiology is complex and not well understood. Hypoxia leads to right ventricular enlargement, pulmonary hypertension, increased capillary permeability, and eventually, pulmonary edema. Affected persons experience headache, shortness of breath, dry then productive cough, and reduced exercise tolerance. These effects are aggravated by cold temperature and physical exertion. Pulmonary edema can proceed rapidly to increased cerebral blood flow, increased permeability of the blood brain barrier, and eventually, cerebral edema which can be fatal unless treated promptly.

Prevention

High-altitude illness may be prevented by acclimatization at elevated altitude (1500 to 2000 m) and gradual ascent thereafter. One rule of thumb is to "climb high, sleep low." More specifically, above 2000 m, each night of sleep should average not more than 300 m above the previous with a rest day every 2 to 3 days or every 1000 m ascent. Ascending more than 300 m in one day appears not to increase risk provided that one descends to rest at a place no more than 300 m above the previous night.

Secondary prevention for persons affected with high-altitude illness is to avoid further ascent until symptoms have resolved and to rest. If there is no improvement, the person should descend. If there are any signs of cerebral edema (ataxia, altered consciousness, or other neurological deficits), immediate descent, treatment by a physician, and oxygen are essential. A hyperbaric chamber may be useful to simulate descent. (For management and treatment options, see Hackett and Roach, 2001.)

Further Reading

Bartsch P. High-altitude pulmonary edema. *Medicine and Science in Sports and Exercise* 1999; 31(1 Suppl): S23-S27.

Basnyat B, Murdoch DR. High-altitude illness. *Lancet* 2003; 361: 1967-1974.

Ge RL, Helun G. Current concept of chronic mountain sickness: Pulmonary hypertension-related high-altitude heart disease. *Wilderness and Environmental Medicine* 2001; 12: 190-194.

Hackett PH, Roach RC. High-altitude illness. *New England Journal of Medicine* 2001; 345: 107-114.

Hackett P, Rennie D. High-altitude pulmonary edema. *JAMA* 2002; 287: 2275-2278.

Purkayastha SS, Ray US, Arora BS, et al. Acclimatization at high altitude in gradual and acute induction. *Journal of Applied Physiology* 1995; 79: 487-492.

Wiedman M, Tabin GC. High-altitude retinopathy and altitude illness. *Ophthamology* 1999; 106: 1924-1926.

Histoplasmosis

ICD-10 B39
SHE/O

*Barry S. Levy and Rana A. Hajjeh**

Histoplasmosis is primarily a pulmonary disease that may at times disseminate widely throughout the body. It is caused by the dimorphic fungus *Histoplasma capsulatum*, which grows as mycelia (mold) in soil and as a yeast in living tissue. Histoplasmosis is now the most common endemic mycosis in the United States, and one of the most common opportunistic fungal infections in persons with AIDS living in areas where it is endemic.

Histoplasmosis occurs in many areas of the world. The fungus lives in soil that has been enriched by bird or bat droppings, especially in temperate climates with high humidity. In the United States, histoplasmosis is endemic in the Ohio and Mississippi River Valley areas, but outbreaks have occurred throughout much of the eastern half of the country. In addition, histoplasmosis is endemic in various parts of Mexico, Central and South America, Africa, Eastern Asia, and Australia.

---

*E.W. Chick wrote this chapter in the first edition of this book, which served as a basis for this update.

Infection occurs when soil containing Histoplasma becomes dry and is disturbed, thus raising dust-containing fungal elements. Since *H. capsulatum* is a dimorphic fungus, there is no animal-to-human transmission or human-to-human transmission. The incubation period is usually about 10 to 14 days. Initial infection provides partial immunity, but reinfection can occur following heavy exposures.

Outbreaks have occurred in families or in groups of workers with common exposure to bird or bat droppings or to recently disturbed contaminated soil. Workers at risk for developing histoplasmosis are those who come into close contact with soil, particularly soil enriched with avian or bat feces. These workers include farmers cleaning chicken coops, bat-infested lofts, and pigeon roosts; construction workers and other workers involved in earth-moving operations; workers involved in road construction, bridge scraping, tree cleaning, landscaping, or grave digging; and workers who clean or dismantle contaminated buildings. Histoplasmosis outbreaks have been also reported in cave explorers and in excavators as well. Equipment or materials containing contaminated soil may introduce the fungus into an unsuspected environment. Histoplasmosis also occurs in dogs, cats, rats, opossums, skunks, foxes, and other animals.

Occurrence

Prevalence of histoplasmosis has ranged between 20% in some surveys in the U.S. overall to 90% or even higher in some endemic areas. The percentage of a population with positive skin tests for histoplasmosis is greatest among farm dwellers, followed by rural dwellers, and, lastly, urban dwellers. Over 95% of individuals who have been infected with *H. capsulatum* can recall no clinically distinctive illness and remain free of complications of the disease.

In endemic areas, primary infection develops at an early age, increasing in incidence from childhood to 30 years of age; differences by sex are not usually observed, except for (a) the chronic pulmonary form, which is more common among middle-aged men; and (b) occupationally-exposed cases, where differences by sex are related to patterns of employment.

Causes

Infection occurs by inhalation of dust that contains spores or hyphal fragments of the fungus. Thus, people who work in dusty environments that have been inhabited by birds or bats are at increased risk of infection. *Histoplasma capsulatum* has been isolated in residential and commercial buildings, storm cellars, belfries, chicken houses, barns, and other structures in which birds or bats have been long-term occupants. Cases result from work in these areas that creates airborne dust. Epidemics have occurred in people working in, or passing through, bat caves and bird roost sites. Living or working in proximity to a bird roost site containing *H. capsulatum* increases the risk of infection.

Pathophysiology

Infection following inhalation of *H. capsulatum* spores is often asymptomatic, resulting mostly in conversion of the skin test and, usually, in focal pulmonary parenchymal and hilar calcifications.

Among the small proportion of individuals who develop a symptomatic infection, there is a wide spectrum of clinical illness. People who inhale a large dose of fungal spores may develop "epidemic histoplasmosis," an acute, usually self-limited, flu-like illness with fever, cough, and chest pain. The chest x-ray has a characteristic pattern of diffuse infiltrates resembling a snowstorm. The chronic pulmonary disease is clinically and radiologically indistinguishable from that of pulmonary tuberculosis, often following a slowly progressive pattern, with productive cough, malaise, and weight loss. Radiologically, the disease is apical in location and often bilateral. Cavitation is common. Treatment is necessary; a 10% to 13% relapse rate is not unusual.

In the very young, the aged, or immunocompromised individuals (including HIV-infected persons and transplant recipients), the fungus may spread to virtually any organ of the body. Without therapy, this disseminated form is rapidly fatal. Signs and symptoms may be quite varied and reflect the various organ systems involved. The classic pattern is fever, hepatosplenomegaly, anemia, and wasting.

Oral itraconazole has been approved for pulmonary and mild to moderate disseminated histoplasmosis. For patients with severe disseminated histoplasmosis (including patients with septic shock picture and CNS histoplasmosis), amphotericin B is the therapy of choice. Itraconazole can also be used as a chronic suppressive treatment in persons with AIDS, following initial treatment.

Prevention

In endemic areas, it is hard to prevent sporadic exposures to sources of this fungus, but one can prevent major exposures that can occur during construction or cleaning activities. CDC guidelines for preventing histoplasmosis among workers are (1) avoid sites contaminated with *H. capsulatum*, such as chicken coops, caves, and bird roosts; (2) avoid activities associated with increased risk, such as spelunking, construction, and excavation; (3) use personal protective equipment, including respirators, disposable protective clothing and shoe covering, and hoods or helmets; (4) decontaminate with disinfectants, such as formaldehyde solutions; and (5) control aerosolized dust by such methods as using water sprays.

Since initial infection provides partial immunity, a vaccine can be potentially developed. Itraconazole prophylaxis should be considered for all HIV-infected patients who live, or have lived, in endemic disease areas if their CD4+ counts are less than 100 CD4+ T cells per microliter, especially those who have high-risk exposures.

Further Reading

Cano MCV, Hajjeh RA. The epidemiology of histoplasmosis: a review. *Seminars in Respiratory Infections* 2001; 16: 109-118.

Centers for Disease Control and Prevention. *Histoplasmosis: Protecting Workers at Risk* (DHHS [NIOSH] Publication No. 97-146). Washington, DC: NIOSH, 1997.

Histoplasmosis. In: Heymann J (Ed.). *Control of Communicable Diseases Manual (18th ed.)*. Washington, DC: APHA, 2004, pp. 273-276.

Homicide and Assault

ICD-10 X85, Y09

Jane Lipscomb

A significant number of workplace fatalities and injuries occur each year in the United States. In any given week, approximately 20 workers are murdered and thousands are assaulted while working. The definition of violence in the workplace includes verbal threats, threatening behavior, physical assault, and homicide. Four basic types of workplace violence, defined by the relationship between the victim and the perpetrator, have been identified:

- Type I-Violence by strangers where the assailant has no legitimate business relationship to the workplace (for example, entering the workplace to commit a robbery).
- Type II-Violence by customers (current or former) or clients (patients, prisoners, students, passengers), usually to those who provide direct service to the public.
- Type III-Violence by current or former co-workers/employees, supervisors, or managers who often seek revenge for perceived unfair treatment.
- Type IV-Violence by an assailant who confronts an individual in the workplace with whom an outside personal relationship exists.

Occurrence

Work-related homicide was the third leading cause of fatal occupational injuries, with 639 homicides (excluding fatalities resulting from September 11) in 2001. Homicide is the leading cause of workplace death among women; however, men are at three times higher risk of becoming victims of workplace homicides than women. Most workplace homicides are robbery-related crimes (71%), with only 9% committed by co-workers or former co-workers. More than 80% of all workplace homicides are committed with a firearm. Workplace homicides have decreased from a record high of 1080 deaths in 1994, to 639 in 2001. The taxicab industry has the highest annual rate of homicides, nearly 60 times the national average rate (0.7 per 100 000 workers) followed by liquor stores, detective/protective services, and gas service. In 2001, homicides among technical sales, and administrative support workers decreased 14%, but increased sharply among workers in the service occupations, which include police and detectives, food preparation workers, and barbers and hairdressers.

The BLS Occupational Injury and Illness Survey captures the most severe nonfatal injuries, those resulting in lost-time injuries. Of all nonfatal assault and violent acts with lost workdays reported to the BLS, 35% occur within the health services sector; the rate of these injuries in health services (7.2 per 100 workers) is nearly four times the national average of 1.9 per 100 workers. According to the BLS, rates of lost-time injuries related to violence and assaults by persons are substantially greater among workers in the public sector than the private sector.

The Department of Justice National Crime Victimization Survey (NCVS) provides the most sensitive and reliable estimate of self-reported violent crimes occurring in the workplace. From 1993 though 1999, the NCVS found, on average, 1.7 million episodes of victimization per year in the United States against persons while they worked or were on duty. Of the occupations examined, police officers experienced workplace violent crime at rates higher than all other occupations (261 per 1000 workers). The victimization rate across all occupations combined was 1.26 per 100 workers.

Causes

In general, persons unknown to the victims commit most workplace homicides. Moreover, most of the victims work in retail trade, security services, or transit services occupations. These circumstances are in contrast to those that characterize nonfatal workplace assaults. Most nonfatal workplace injuries occur in settings in which the victim and the attacker are in a custodial or client-caregiver relationship, such as in health care or social services. Workplace violence and assaults are not random events. Although the occupations most at risk of workplace homicide differ dramatically from those at high risk of nonfatal assaults, both types of events share a number of common risk factors including contact with the public; exchange of money; delivery of passengers, goods, and services; having a mobile workplace, as in the case of taxi drivers and police officers; working with volatile, unstable persons; working in isolation; working late at night or in the early morning; working in high-crime areas; guarding valuables; and working in community-based settings.

Pathophysiology

Both physical and mental injuries are associated with workplace violence. Physical injuries range from death and disabling injury to mild scratches and bruises. The mental health sequelae associated with workplace violence are less well understood, but include depression and post-traumatic stress disorder.

Prevention

Workplace violence was not recognized as an occupational injury amenable to prevention via the industrial hygiene hierarchy of controls until the mid-1980s. At that time, NIOSH published surveillance data from the National Traumatic Occupational Fatality (NTOF) that defined high-risk industries and occupations and risk factors for violence. Despite the magnitude of the problem of workplace violence as demonstrated by both BLS and BJS data, OSHA has not begun rulemaking for a standard addressing workplace violence. By contrast, four states—California, Washington, Virginia, and Florida—have legislation mandating violence prevention measures. The first two states address workplace violence in the health care sector; laws of the latter two states are aimed at preventing robbery-related homicides in late-night retail establishments.

OSHA's current approach to preventing workplace violence is through voluntary guidelines. Beginning in the mid-1990s, it began issuing sector-specific guidelines for preventing workplace violence. To date, it has issued guidelines that address health-care and social service workers, workers in late-night retail jobs, and taxicab drivers. The guidelines provide a framework for addressing the problem and include the basic elements of a proactive health and safety program: management commitment and employee involvement, worksite analysis, hazard prevention and control, training, and evaluation. Hazard control strategies consist of numerous engineering or security controls, such as security alarms and lighting; administrative policies, such as hours of operation and staffing; and worker and supervisor training.

The effectiveness of individual hazard control measures in reducing workplace violence has been studied only rarely. In those few cases in which rigorous methods have been applied to their study, the measures have been found to be effective. For example, in one study, outside lighting and adequate staffing significantly reduced the risk of fatal assaults due to robberies, as did any combination of five or more administrative controls. A review of the violence prevention intervention literature yielded only nine studies (all in health care) that evaluated an intervention among the 137 papers mentioning violence prevention intervention. Five studies evaluated violence prevention training interventions, three examined post-incident psychological debriefing programs, and two evaluated administrative controls to prevent violence. All were quasi-experimental and without a formal control group.

The study findings were equivocal. Evaluation of the effectiveness of a comprehensive violence prevention program modeled after the OSHA healthcare and social service guidelines is currently under way in New York State.

Other Issues

Among industries at high risk of workplace violence, most rely on security personnel and some form of violence awareness training, often in the form of a "canned" video presentation. The need for both supervisor and worker violence prevention training is underrecognized. The range and quality of existing training materials has yet to be evaluated. OSHA recently developed training materials that could be a valuable resource for many industries. Evaluation of the content and delivery of such programs is needed if training is to be widespread and impact safety. Intervention studies of violence prevention efforts, in addition to training, are sorely needed.

Further Reading

Lipscomb J, Silverstein B, Slavin TJ, et al.: Perspectives on legal strategies to prevent workplace violence. *J Law Med Ethics* 30:166-72, 2002.

Duhart D: *Violence in the Workplace*, 1993-1999, U.S. Department of Justice, Office of Justice Programs, 2001.

U. S. Department of Labor, OSHA: Guidelines for preventing workplace violence for healthcare and social service workers. Washington, D.C., U. S. Department of Labor, Occupational Safety and Health Administration, 1996

Loomis D, Marshall S, Wolf S, Runyan C, Butts J. Effectiveness of safety measures recommended for prevention of workplace homicide. *JAMA*. 2002;287(8):1011-1017.

Runyan CW, Zakocs RC, Zwerling C: Administrative and behavioral interventions for workplace violence prevention. *Am J Prev Med* 18:116-27, 2000

Human Immunodeficiency Virus (HIV) Infection/Acquired Immunodeficiency Syndrome (AIDS)

ICD-10 B20-B24

Janice Huy and Bill Eschenbacher

The human immunodeficiency virus (HIV) is a lentivirus included within a subgroup of retroviruses. HIV can cause human infection of varying severity, with the most advanced form of infection referred to as acquired immunodeficiency syndrome (AIDS). The diagnosis of HIV infection is made by sequential laboratory tests for detection of HIV antibody; a positive result on an enzyme immunoassay followed by a confirmatory test such as a Western blot or immunofluorescent antibody titer. The CDC recommends HIV testing be expanded to identify more individuals in earlier stages of infection, at which time behavioral changes could be made that would limit the transmission. A newer and more rapid screening test for HIV has been approved by the FDA and could be used in expanded testing programs.

The surveillance case definition for AIDS used by the CDC is as follows: Positive evidence for HIV infection and one of the following: (1) CD4+ T-lymphocyte count of <200 cells/L; or (2) one of the identified clinical conditions listed as part of the AIDS surveillance case definition, which include infections such as extrapulmonary coccidioidomycosis or cryptococcosis, tuberculosis, *Pneumocystis carinii* pneumonia, and histoplasmosis, as well as some noninfectious conditions, such as Kaposi's sarcoma, some forms of lymphoma, and invasive cervical cancer.

Because of its potential for transmission through a workplace exposure, HIV infection is considered, in part, an occupational illness. This entry focuses on HIV infections that occur in the workplace and the prevention of these infections.

Occurrence

As of the end of 2001, there were approximately 360,000 reported prevalent cases of AIDS in the United States, and another 160,000 reported prevalent cases of HIV infection. The actual prevalence of HIV infection would be much higher if expanded testing was performed. Through 2001, there were 462,653 people in the United States who had died from AIDS. According to the WHO, more than 40 million people have HIV infection worldwide, and more than 20 million people have died from AIDS.

Worker populations in the United States who are at increased risk for HIV infection and AIDS include health care workers, emergency response workers, public safety workers, and community workers.

At the end of 2001, there were approximately 24 000 healthcare workers in the United States who had AIDS, representing 5.1% of all AIDS cases in the nation. However, only a small fraction of these workers became infected as a result of workplace exposures. As of December 2001, there were only 57 health care workers with documented HIV seroconversion as a result of occupational exposures and only an additional 138 people with possible HIV infection from occupational exposure.

Community workers who use needles or sharp instruments (tattoo artists, body piercers, and acupuncturists) or who may encounter these devices at work (waste haulers and handlers as well as housekeepers) may be at increased risk of HIV infection through workplace exposures. The prevalence and cumulative incidence of HIV infection among these workers as a result of occupational exposure are unknown.

Causes

Transmission of HIV occurs as a result of contact with infected blood, and can occur through any of three routes: (1) parenteral exposure to infected blood or blood products, such as by intravenous injection using a needle contaminated with HIV; (2) heterosexual or homosexual contact with an HIV-infected person; or (3) perinatally, from an HIV-infected mother to her infant. Although HIV has been isolated from a number of body fluids, only blood, semen, vaginal secretions, and breast milk have been implicated in its transmission. There is no evidence of its transmission through food, insects, or casual contact. It is presumed that HIV-infected people are able to transmit the virus at any time after infection has begun. Transmission risk is increased as the viral load of the HIV-infected person increases.

Occupational risk of infection results from worker exposure to infected blood or blood products, or to body fluids or tissue contaminated with infected blood. This exposure can occur via percutaneous injury, splashes to mucous membranes, and exposure of non-intact skin. In the occupational setting, the vast majority of infection cases are due to percutaneous injuries, through puncture with a needle or other sharp instrument or by a cut injury. In a retrospective case-control study, the risk of HIV transmission to an exposed health care worker appeared to be increased with a deep percutaneous injury, an injury involving a device that had been in an artery or vein, and a hollowed bore needle. Terminal illness of the source patient was also associated with a higher risk of transmission, possibly reflecting a higher viral load. The estimated average risk of HIV infection after a needlestick or cut exposure to HIV-infected blood is about 1 in 300. In addition to percutaneous injuries, there have been documented cases of HIV infection resulting from mucous membrane and/or skin exposure.

Regarding the risk of an HIV-infected health care worker possibly infecting patients, there has been only one identified incident, in which an HIV-infected dentist transmitted the infection to six patients. Other investigations, involving over 22,000 patients of 63 HIV-infected physicians and dentists, have identified no other cases of transmission from healthcare workers to patients. Recommended guidelines for the management of healthcare workers infected with HIV, including any work restrictions, have been published by the Society for Healthcare Epidemiology of America.

Pathophysiology

Like all viruses, HIV can only reproduce inside cells. Using an enzyme called reverse transcriptase, the virus converts viral RNA into DNA, and then incorporates that DNA into the host cell's genome. Once HIV enters the body, it infects a large number of CD4+ cells and replicates rapidly. During this initial or acute phase, the blood contains a great many viral particles, the greatest viral load during the course of infection for that individual. The virus then spreads throughout the body, involving many tissues, including lymphoid tissues or organs. These lymphoid structures include the lymph nodes and spleen.

Within 4 weeks after exposure to the virus, as many as 70% of HIV-infected individuals will experience flu-like symptoms. The patient's immune response to the initial infection is through CD8+ cells and B-cell antibodies. As a result, the viral load is reduced. However, because of many mutations, some of the virus escapes the body's response to the infection. At the same time, the cells involved in controlling the infection (so-called killer T cells) may become depleted or dysfunctional. As a result, the virus will hide within certain cells and serve as a reservoir for future worsening of the infection.

The median time between initial HIV infection, to worsening of the infection ultimately to AIDS, can take 10-12 years in the absence of anti-retroviral therapy. But there is wide variation in the progression of the disease in any one individual. Approximately 10% of infected patients will progress to AIDS within the first 3 years, while up to 5% of individuals will have stable CD4+ counts and have no symptoms, even after being infected for 12 years or longer. Factors such as age, genetic differences, level of virus virulence, and co-infection with other infectious agents will influence the rate and severity of disease progression.

The major immune effect of HIV infection is the disabling and destruction of CD4+ T cells or helper lymphocytes. These cells are involved in the body's protection against other infectious agents, and their loss results in increased susceptibility for the HIV-infected individual to acquire these other infectious agents, leading to severe life-threatening diseases. A healthy, uninfected person usually has 800 to 1200 CD4+ T cells per microliter of blood. During HIV infection, these numbers will decline. When the count for CD4+ T cells falls to less than 200 cells per liter, the person becomes highly vulnerable to opportunistic infections and certain cancers that are associated with AIDS.

Prevention

Many occupations have the potential for exposure to blood and body fluids that may be contaminated with HIV or other bloodborne pathogens, such as hepatitis B virus (HBV) or hepatitis C virus (HCV). Preventive measures are designed to minimize or eliminate contact with blood and body fluids that are contaminated with the virus.

The OSHA Bloodborne Pathogen (BBP) Standard was promulgated in 1991 to provide worker protection from exposure to blood and other potentially infectious material. In 2000, the Needlestick Safety and Prevention Act was signed into law, mandating specific revisions to the OSHA standard that clarified the need for employers to (a) select safer needle devices as they become available and (b) involve employees in identifying and choosing the devices. The Act also requires employers to maintain a log of injuries from contaminated sharps. Many employers mistakenly believe that the BBP Standard is limited to healthcare settings; however, it applies to all covered workplaces where occupational exposures to blood and other potentially infectious material may be reasonably anticipated.

A comprehensive prevention program should consider all aspects of the work environment, and should include employee involvement as well as management commitment. Implementing the use of improved engineering controls, such as safer sharps devices, is one component of such a comprehensive program. Other prevention strategies that should be addressed include modification of hazardous work practices (such as no recapping of needles), administrative changes to address needle hazards in the environment (such as prompt removal of filled sharps disposal boxes), use of personal protective equipment (such as gloves and eye protection), and worker education and awareness.

Improving engineering controls is often the most effective approach to reducing occupational exposures. An example of such an engineering control includes sharp devices that have integrated safety features. Ideal characteristics are reliability and ease of activation and use, which make them safe and effective during patient care and which remain active through the disposal process. Elimination of unnecessary needles also effectively reduces occupational exposures.

Strategies to prevent exposure to blood and body fluids by modifying hazardous work practices and using personal protective equipment are covered by universal precautions. In 1987, the CDC published a set of precautions designed to prevent transmission of bloodborne pathogens when providing first aid or health care. Under universal precautions, all blood and certain body fluids are considered potentially infectious for bloodborne pathogens, and certain precautions should be used in situations where there is potential for exposure to blood. The specific provisions of universal precautions include:

1. Workers should use appropriate barrier precautions to prevent skin and mucous membrane exposure. Protective equipment includes gloves, masks and protective eyewear or face shields, and gowns or aprons. Gloves should be changed after contact with each person.
2. Hands and other skin surfaces should be washed immediately (or as soon as safety permits) if contaminated with blood or body fluids. Hands should be washed immediately after gloves are removed.
3. All workers should take precautions to prevent injuries caused by needles, scalpels, and other sharp instruments or devices during procedures, when cleaning used instruments, during disposal of used needles, and when handling sharp instruments after procedures. Needles should not be recapped by hand, purposely bent or broken by hand, removed from disposable syringes, or otherwise manipulated by hand. Sharp items should be placed in puncture-resistant containers for disposal.

4. Mouthpieces, resuscitation bags, or other ventilation devices should be available for use in areas where the need for resuscitation is predictable.
5. Healthcare workers who have exudative lesions or weeping dermatitis should refrain from all direct patient care, and from handling patient-care equipment, until the condition resolves.
6. Pregnant healthcare workers are not known to be at any greater risk of contracting HIV infection than health care workers who are not pregnant; however, if a health care worker develops HIV infection during pregnancy, the infant is at risk of infection resulting from perinatal transmission. Because of this risk, pregnant healthcare workers should be especially familiar with,and should strictly adhere to precautions in order to minimize the risk of HIV transmission.

The CDC has published precautions for dentistry, autopsy or morticians' services, dialysis, and laboratory personnel. The CDC has also addressed environmental considerations, sterilization and disinfection, housekeeping and clean-up issues, laundry, and handling of infective waste. Facilities involved in producing large volumes of HIV or highly concentrated virus are subject to much more stringent biosafety regulations. Individuals working in such facilities should be acutely aware of the attendant guidelines for such activities.

Precautions for community workers are not as clearly defined as for healthcare workers. Lessons learned from health care exposures may provide guidance in the community setting. Interventions may include improving reporting of exposures, implementing more effective engineering and work practice controls, and providing more effective education. Examples of such preventive measures include the following:

1. Acupuncturists, body piercers, and tattoo artists should be aware of, and use, available sharps devices that have engineered safety features, and follow universal precautions and standard infection control procedures.
2. To prevent needlestick injuries among waste haulers and handlers, housekeepers, janitorial service workers, and other service employees, primary consideration should be given to eliminating used syringes and other sharps from the municipal waste stream, and using engineering controls to isolate contaminated material. This may include locating sharps containers in public areas, such as in hotel rooms, restaurants, public restroom facilities, and airplanes.
3. Police and other public safety workers who may encounter used needles while performing searches or restraining suspects may want to consider using puncture-resistant flexible gloves.

Worker education is an important aspect of prevention, and is required under the OSHA BBP Standard. Employees should be educated about the hazards associated with blood and potentially infectious materials, and the protective measures they should take in order to minimize their risk of exposure. Training should be tailored to the type of work being performed by the employees.

Managing exposures to contaminated blood and body fluids requires a concerted effort among the employer, the employee, and the healthcare provider. The employer must be aware of, and follow, applicable laws and regulations, which includes providing access for employees to prompt evaluation and management after sustaining an occupational exposure to blood or other potentially infectious material. Employees should report all exposure incidents to their employer and follow the established protocols for evaluation. Healthcare providers who evaluate exposed workers should be aware of the most recent guidelines published by the CDC that address the management of occupational exposures to HIV, HBV, and HCV. The guidelines apply to both health care and community worker exposures.

Medical management of occupational exposures to blood is complex and rapidly changing, as new medications are approved and drug resistance increases. Recommendations exist for the risk assessment of workers with occupational exposure to blood and other potentially infectious materials. These include determining the likelihood that the exposure will result in an infection, assessing risk factors for the source patient, testing the source patient when possible, and determining whether preventive therapy with anti-retroviral agents should be recommended. Expert consultation is available to medical providers who are managing occupational exposures, through the National HIV/AIDS Clinical Consultation Center. This toll-free telephone consultation service provides up-to-date HIV, HBV, and HCV clinical information and individualized, expert case consultation.

Further Reading

AIDS/TB Committee of the Society for Healthcare Epidemiology of America (SHEA). Management of health care workers infected with hepatitis B virus, hepatitis C virus, human immunodeficiency virus, or other bloodborne pathogens. *Infection Control and Hospital Epidemiology* 1997; 18: 349-363.

Centers for Disease Control and Prevention. 1993 revised classification system for HIV infection and expanded surveillance case definition for AIDS among adolescents and adults. *MMWR* 1992; 41 (No. RR-17).

Centers for Disease Control and Prevention. Updated U.S. Public Health Service guidelines for the management of occupational exposures to HBV, HCV, and HIV and recommendations for postexposure prophylaxis. *MMWR* 2001; 50(No. RR-11): 7.

Centers for Disease Control and Prevention. Recommendations for prevention of HIV transmission in healthcare settings. *MMWR* 1987; 36 (Suppl .no. 2S).

Huy JM, Ross CS, Boudreau AY, et al. Occupational blood-borne pathogen exposures among community workers. *Clinical Occupational and Environmental Medicine*

2002; 2: 537-556.

National Institute for Occupational Safety and Health. NIOSH Alert: Preventing Needlestick Injuries in Health Care Settings (DHHS [NIOSH] Publication No. 2000-108). Washington, DC: NIOSH, 1999.

National HIV/AIDS Clinical Consultation Center. [2003]. World Wide Web [URL=http://www.ucsf.edu/hivcntr/]. Date accessed: May 10, 2003.

Occupational Safety and Health Administration. *Occupational Exposure to Bloodborne Pathogens* (29 CFR 1910.1030). 1992.

Occupational Safety and Health Administration. *Occupational Exposures to Bloodborne Pathogens; Needlesticks and Other Sharps Injuries; Final Rule.* (29 Federal Register 1910)

Hyperbaric Injury

ICD-10 T70.9

James L. Weeks

Hyperbaric injuries and diseases result from exposure to increased atmospheric pressure. Such pressure occurs during underwater diving, in pressurized underground or underwater caissons, or therapeutically in clinical hyperbaric chambers. Increased atmospheric pressure is used in caissons, sewers, and other tunnel work to purge ground water from the worksite. Diving exposures may include commercial oil field construction, repair work at sea, underwater salvage, oceanographic investigation, scientific research, and some military operations.

Disorders include barotrauma, inert gas narcosis (commonly referred to as "nitrogen narcosis" or, in the United Kingdom, "the narks") or high-pressure nerve syndrome (during very deep diving), and decompression sickness ("the bends" or caisson worker's disease). Noise-induced hearing loss and hypothermia are also associated with work under increased pressure or underwater.

Barotrauma is traumatic tissue damage resulting either from bodily compression or from decompression. Tissue itself, like water, is relatively incompressible and resistant to injury. Air, however, is highly compressible. Thus barotrauma occurs along the boundaries of air-containing structures. The external ear canal, middle ear space, lungs, and sinuses as well as potential spaces, such as dental fillings or intestines, may all be adversely affected by changes in pressure, resulting in loss of hearing, pain, and other forms of traumatic injury.

The most dramatic form of barotrauma is air embolism, in which air, nitrogen, or another gas is introduced into the vascular tree, usually via the pulmonary circulation. Bubbles may migrate to the brain and precipitate acute, severe neurological injury, resulting in death. As little as 0.4 cc of blood air foam delivered to the brain may be fatal if respiratory control centers are damaged.

Inert gas narcosis results from the toxic effects of inert gases on neural cells. Because pure oxygen under pressure is toxic to lungs (where it may cause pneumonitis progressing to fibrosis) and the central nervous system, either compressed air or oxygen diluted with other inert gases (such as helium or neon) is used during diving or other work in a hyperbaric environment. In tunnels and caisson work, where pressure rarely exceeds three atmospheres, compressed air (composed of 20% oxygen, 79% nitrogen, and 1% other gases) is almost always used. At higher pressures or during dives deeper than 30 meters and for longer than 3 hours, nitrogen narcosis may occur. This syndrome is similar to acute ethyl alcohol intoxication; symptoms include numbness, loss of coordination, and mental confusion. Effects are short-term and apparently reversible. Mental confusion while underwater, however, could obviously be disastrous. At 100 meters or more, divers may lose consciousness from nitrogen narcosis.

For dives deeper than about 60 meters, inert gases other than nitrogen are used-primarily helium, but there have been experiments with other mixtures. For even deeper dives, however, a high-pressure nerve syndrome results from the effects of high pressure, aggravated by too rapid ascent. This syndrome can include nausea, vomiting, uncontrollable shaking, and loss of coordination.

Decompression sickness includes both acute effects, which can result in long-term disability, and chronic effects, primarily aseptic osteonecrosis. Decompression sickness, commonly referred to as "the bends," results from too rapid decompression from a hyperbaric environment to sea level. Decompression sickness can also result from ascent to heights as in mountainous regions (see chapter on High-Altitude Illness) or during air travel in unpressurized or inadequately pressurized aircraft, especially after diving. The relative risk of decompression sickness increases geometrically as the time interval between diving and flying decreases. Symptoms of decompression sickness include skin itching, vague to extreme joint pain, vertigo, breathing difficulty, blindness, paralysis, and convulsions leading to death. Onset can be insidious or rapid and can occur during decompression or up to 24 hours later.

Aseptic osteonecrosis (sometimes referred to as dysbarism-related osteonecrosis) may result from a single episode or from repeated episodes of inadequate decompression. The shoulders, the femoral heads, the femur or humerus, and, rarely, the knees may be affected. Occurrence is usually bilateral and at several anatomic sites. Nitrogen bubbles may result in an inadequate blood supply, which can cause gradual destruction of bone and subsequent loss of cartilage and joint function, leading to lifelong disability.

Occurrence

Occurrence of hyperbaric disorders is highly variable by geographic location. In general, it is more common among tunnel workers than among divers. In 1986, there were 507 compensated injuries in 23 states in the United States due to changes in atmospheric pressures. Most (89%) occurred in the transportation and public utilities industries, including sewer services.

In a group of Milwaukee sewer workers studied in 1974, 35% had osteonecrosis. In a group of Gulf Coast commercial divers, the rate of positive x-ray findings for osteonecrosis was 27% in 1972. More recently, the rate of osteonecrosis was reported as 17% among compressed air workers and 4.2% among commercial divers. Prevalence increases with age, work experience, increased pressure (or depth), and with the number of acute episodes of decompression sickness. Osteonecrosis can progress in the absence of further hyperbaric exposure.

Causes

Hyperbaric disorders are caused by increased pressure, excessively rapid decompression, and toxic effects of inert gases. Decompression-related disorders can result from excessive pressure, rapid surfacing following an underwater dive, exposure to decreased atmospheric pressure (such as in an airplane) in a mountainous region, or otherwise at increased elevation, or too rapid decompression from a hyperbaric worksite (tunnel or caisson) or chamber.

Pathophysiology

At sea level, atmospheric pressure is 1 atmosphere (ATA) (= 14.7 pounds/in^2 = 10^5 Nt/m^2 [Pascal]). When a diver descends underwater, the surrounding pressure on the diver is directly proportional to depth and increases by 1 ATA every 10 meters. Thus the total pressure on a diver at 10 meters is 2 ATA, at 20 meters 3 ATA, and so on. Depth gauges used by divers in the United States and the United Kingdom usually measure "fsw" (feet of sea water), which is easily converted into pressure measurements.

As pressure increases, the volume of a fixed quantity of gas decreases, according to Boyle's law:

$$\text{Pressure} \times \text{Volume} = \text{Constant}$$

Therefore, 1000 cc of gas at 1 ATA is compressed to 500 cc at 2 ATA, and 250 cc at 4 ATA. Thus, changes in volume are not linear with depth. The largest changes in gas volume occur during the initial stages of a dive or pressurization-during the first 10 meters. During descent or pressurization, real or potential air spaces within the body must be able to reach equilibrium. Injury occurs when the body's air spaces fail to reach equilibrium during pressurization, resulting in relative vacuums within these air spaces. During decompression, the inability of the air to escape may result in damage from overdistension.

Any potential or real air space may be subject to changes in pressure. Thus, conjunctival hemorrhage may occur from failure to ventilate a sealed facemask, skin injuries may result from sealed dry suits, and intestinal distension or perforation or a hernia may result from swallowing air. Air in the bowels may be sufficient to limit lung expansion.

Rarely, dental pain results from improperly filled teeth where the amalgam incompletely fills the cavity socket. During pressurization, the nerve root and pulp expand into this space resulting in severe pain and, occasionally, loss of an amalgam on descent, sometimes with explosive force.

Changes in pressure may cause ear damage (ear squeezes). External ear canal injury is usually the result of air being trapped in the canal by a tight-fitting hood, foreign body, or, rarely, tortuosity and blockage by osteomas. During pressurization, the relative vacuum in the external ear canal may precipitate pain or injure the tympanic membrane.

The most common form of ear squeeze occurs in the middle ear, which is normally ventilated by the Eustachian tube. During pressurization, a diver usually tries to exhale against a closed glottis (Valsalva maneuver) to force air into the middle ear. Failure to do so may result in implosion of the tympanic membrane (with consequent hearing loss) or hemorrhage into the middle ear with severe pain and, rarely, disruption of the middle ear bones.

Round window blowout is a more profound form of pressure-related ear injury and may occur when a pressurized worker unsuccessfully attempts the Valsalva maneuver. Pressure may be transmitted to the round window of the middle ear, causing rupture with sudden roaring (tinnitus), deafness, and occasionally severe dizziness. This can be a permanent injury.

Alternobaric vertigo, or dizziness secondary to pressure change, is almost always associated with decreases in pressure and is usually transient. It results from unequal clearing of the middle ears, which the vestibular apparatus interprets as spin, resulting in severe dizziness, disorientation, or panic.

The sinuses likewise must be ventilated during changes in pressure. Inability to ventilate can result in hemorrhage, severe localized pain, and subsequent chronic sinus dysfunction. Chronic sinusitis, sinus infection, or allergies may increase the risks of sinus injury.

The pathophysiology of inert gas narcosis is poorly understood but probably results from toxic effects on nerve cells. These toxic effects are apparently a function of the affinity of inert gases for fat. Thus the relative potential for inert gases to cause narcosis can be ranked according to their fat affinity: helium, neon, hydrogen, nitrogen, argon, krypton, and xenon.

Decompression sickness results from inert gas that has been dissolved in tissue under pressure forming bubbles when it comes out of solution. Air emboli thus formed may appear anywhere in the body, blocking blood flow or damaging capillaries or other blood vessels with consequent tissue damage that can lead to death.

Pressure (4.5 ATA or higher) combined with 30% oxygen may also cause myopia.

Prevention

Primary prevention can be achieved by eliminating the need for work under pressure, if possible, or by strict adherence to empirically derived protocols for compression and decompression, commonly referred to as decompression tables. None of the decompression tables presently in use by any navy, commercial oil company, diving agency, construction company, or regulatory agency can eliminate completely the risk of decompression sickness.

The Autodec III-02 Table, employing oxygen for decompression, is recommended as superior to tables for caisson workers adopted by OSHA. Individual day-to-day variation in work conditions and a diver's physical condition result in variable on- and off-gassing of nitrogen during a pressure exposure. Therefore, workers and worksites must be evaluated continuously and individually.

Secondary prevention requires regular medical monitoring of people who work under pressure. Medical examinations should focus on x-ray examination of joints, the chest cavity, sinuses, and teeth, and assessment of the ears, vision, cardiac and CNS functions, and symptoms of osteonecrosis. Emergency treatment of decompression sickness requires on-site decompression chambers, trained personnel prepared for prompt response in case of an emergency, and transportation to a previously identified center for more extensive support if needed.

OSHA regulations applicable to commercial diving are found in the general industry standards at 29 CFR Subpart T Section 401-415. These regulations apply also to construction, marine terminals, and longshoring and are incorporated by reference in regulations applicable to those industries.

Other Issues

People with chronic sinus infection, sinusitis, allergies, asthma, obesity, or improperly filled teeth are at increased risk of injury.

People with right-to-left cardiac shunt are at increased risk of air emboli. More specifically, right-to-left shunt is also associated with an increased incidence of cochleovestibular and cerebral decompression sickness. Gas bubbles form more easily in venous than in arterial circulation, because it is at lower pressure. In normal cardiac function, these bubbles are filtered in the pulmonary circulation. But if there is a right-to-left cardiac shunt, the pulmonary circulation is bypassed and gas bubbles may pass into the arterial tree and form gas emboli.

Work under pressure is frequently associated with higher pay, especially for divers. Therefore, if a worker becomes symptomatic, or is labeled as "bendable," there can be a significant economic penalty, which is a significant disincentive to disclose symptoms. Acceptable alternative employment should be available to symptomatic workers to prevent progression of hyperbaric disorders.

If oxygen is present in a hyperbaric environment, there is a greater risk of fire than at atmospheric pressure because of its higher partial pressure. Therefore, attention must be paid to controlling fuels and sources of ignition.

Noise levels at construction sites-whether on the surface, under water, or in a caisson-are frequently elevated. Coupled with the sound of inrushing air and within the confined environment of a caisson, diving bell, or clinical hyperbaric chamber, noise exposure may be hazardous. For example, when a sound-reducing muffler is not used, the noise level in a clinical chamber may exceed 130 dBA. Even brief exposure to this level of noise will result in transient hearing loss, and long-term exposure will result in permanent loss. Divers rarely experience these noise levels except when they are in a compressed air environment such as a diving bell. However, sound transmission in water is excellent, and concussion injuries may result from underwater blasts and explosions. Noise-induced hearing loss can be prevented by following the guidelines in the section on Hearing Loss, Noise-Induced.

The effects of toxic gases and carbon dioxide (CO_2) under pressure are significantly greater than at atmospheric pressure. For example, a relatively safe concentration of 30 ppm of carbon monoxide (CO) at the surface, resulting from exhaust gases of a compressor venting into an intake, may supply potentially lethal CO levels (equivalent to 150 ppm at 132 feet of seawater [5 ATA]) to a diver breathing this gas. Other gases, such as methane (CH_4) from decaying organic matter, CO_2, and oxides of nitrogen (NO_x) from diesel-powered generators as well as volatile organic compounds, may have serious health effects.

Because of the increased toxicity of these common air contaminants, close attention to the cleanliness and safety of the inflowing gas supply is essential. Injuries and illnesses in divers and other workers have resulted from failure to pay attention to such things as a change in wind direction or the movement of a diving barge during a tide change. Frequent air sampling—usually for CO, CO_2, CH_4, NO_x and oil vapors—of the air supplied to a diver or a clinical chamber followed by prompt corrective action is essential.

There are several websites devoted to hyperbaric work, diving, osteonecrosis, the bends, and related problems.

Further Reading

Cantais E, Louge P, Suppini A, et al. Right-to-left shunt and risk of decompression illness with cochleovestibular and cerebral symptoms in divers: case control study in 101 consecutive dive accidents. *Critical Care Medicine* 2003; 31: 84-88.

Chang CC. Osteonecrosis: current perspectives in pathogenesis and treatment. *Seminars in Arthritis and Rheumatism* 1993; 23: 47-69.

Downs GJ, Kindwall EP. Aseptic necrosis in caisson workers: A new set of decompression tables. *Aviation, Space and Environmental Medicine* 1986; 57: 569-574.

Frieberger JJ, Denoble PJ, Pieper CF, et al. The relative risk of decompression sickness during and after air travel following diving. *Aviation, Space and Environmental Medicine* 2002; 73: 980-984.

Gregg PJ, Walder DN. Caisson disease of bone. *Clinical Orthopedics* 1986; 219: 43-54.

Onoo A, Kiyosawa M, Takase H, et al. Development of myopia as a hazard for workers in pneumatic caissons. *British Journal of Ophthalmology* 2002; 86: 1274-1277.

Hypersensitivity Pneumonitis

ICD-10 J67

Cecile Rose

Hypersensitivity pneumonitis (HP), also known as extrinsic allergic alveolitis, refers to a constellation of inflammatory lung diseases caused by repeated exposure and immunologic sensitization to a variety of organic and chemical antigens. Diagnosis relies on a combination of findings including a history of repeated antigen exposure, characteristic signs and symptoms, pulmonary function abnormalities, radiologic abnormalities, and histologic findings.

There is often considerable overlap in the clinical presentation of HP between acute, subacute, and chronic forms. In acute HP, respiratory symptoms of cough and shortness of breath and systemic symptoms of fever, chills, and myalgias occur 4 to 12 hours after antigen exposure. Physical findings include fever and inspiratory crackles on lung auscultation. The subacute and chronic forms of HP typically present with insidious onset of shortness of breath and cough, accompanied by nonspecific systemic symptoms, such as malaise, fatigue, myalgias, low-grade fever, and weight loss. Physical examination may be normal or reveal basilar crackles. Cyanosis and right-sided heart failure occur in end-stage disease.

The finding of specific IgG precipitating antibodies (precipitins) in the serum of a patient with suspected HP is a helpful diagnostic clue, but is neither sensitive nor specific and is not required for diagnosis. Serum precipitins are markers of antigen exposure and are detectable in 3% to 30% of asymptomatic farmers and in up to 50% of asymptomatic pigeon breeders. False negative results are common. A peripheral leukocytosis is common in the acute forms of HP.

Patients with HP may exhibit obstructive, restrictive, or mixed patterns on pulmonary function testing. In early disease, normal spirometry and lung volumes are not uncommon. Abnormalities of gas exchange, particularly during exercise, are the most sensitive physiologic indicators of early HP. Restrictive changes, usually with decreased diffusing capacity (DLCO), are common in acute HP. Mixed restrictive and obstructive impairments, along with decreased DLCO, occur in subacute and chronic HP. Up to 10% of patients with HP may have obstruction alone, and tests of nonspecific airway hyperresponsiveness, such as methacholine challenge, may be positive.

When disease is detected early, the chest x-ray is often normal. In classic acute HP, chest x-ray findings may include diffuse, ground-glass opacification and a fine reticulonodular pattern. The reticulonodular pattern becomes more prominent in subacute HP. In chronic HP, fibrosis with upper-lobe retraction, reticular opacity, volume loss, and honeycombing are seen.

High-resolution computerized tomography (HRCT) is more sensitive than a routine x-ray for diagnosing HP. Several HRCT abnormalities have been described. Patchy ground-glass opacities, probably representing active alveolitis or fine fibrosis interspersed with areas of air trapping are common in acute illness, but may also be seen in subacute and chronic HP, especially if there is ongoing exposure. Centrilobular nodules with inhomogeneous areas of air trapping, probably reflecting bronchiolitis, are the most frequent HRCT finding in HP. As with ground glass, the nodules may regress with removal from exposure. The irregular linear opacities, traction bronchiectasis, and honeycombing that may be seen in subacute or chronic HP all reflect fibrosis. HRCT may also show emphysema as a sequela of chronic HP, probably due to underlying bronchiolar inflammation and obstruction.

Lung biopsy in patients with HP typically features the triad of bronchiolitis, lymphoplasmacytic interstitial infiltrate, and poorly formed non-necrotizing granulomas. The complete triad is not always present, and pathologic features vary with disease stage. Granulomas are described in 60% to 70% of acute and in less than 50% of chronic HP. Giant cells, airspace foam cells, and bronchiolitis obliterans may occur. A fibrotic pattern similar to usual interstitial pneumonitis (UIP) is frequently seen in chronic HP.

The following diagnostic studies should be considered in a patient with suspected HP: (1) detailed clinical history; (2) physical examination; (3) chest HRCT (unless the plain film is clearly abnormal); (4) complete PFTs, including lung volumes, pre- and post-bronchodilator spirometry, and DLCO; and (5) fiberoptic bronchoscopy with bronchoalveolar lavage (BAL) and transbronchial biopsies. The cornerstone of diagnosis is a detailed history of symptoms plus occupational and/or environmental exposures. Work history should include a chronology of current and previous occupations, with a description of specific work processes and exposures. The work history should also include a list of specific chemical, particulate, and other aerosol exposures; presence of persistent respiratory or constitutional symptoms in exposed co-workers; and the use of respiratory protection. Review of MSDSs for specific chemicals, such as isocyanates, may supplement the work history. The environmental history should explore exposure to pets and other domestic animals, especially birds; hobbies, such as gardening and lawn care, which may involve sensitizing chemical exposures; recreational activities, such as hot tub use and swimming in indoor pools, from which microbial bioaerosols can be generated; use of humidifiers, cool mist vaporizers, and humidified air conditioners, which can be sources of microbial bioaerosols; moisture indicators, such as leaking, flooding, or previous water damage to carpets and furnishings; and visible mold or mildew contamination in indoor environments.

Acute HP can be mistaken for pneumonia, but a pattern of recurring symptoms and signs should suggest the diagnosis. Chronic HP resembles some other interstitial lung diseases; a strong clinical index of suspicion and careful history-taking skills are essential. Clinical signs and symptoms do not readily distinguish HP from sarcoidosis, chronic beryllium disease (CBD), rheumatologic lung diseases, or idiopathic interstitial lung diseases. Unlike sarcoidosis, CBD, and the connective tissue interstitral lung diseases (ILDs), extrapulmonary manifestations do not occur in HP. While the classic histopathologic triad of HP is helpful in the differential diagnosis of granulomatous lung diseases, the histologic findings of HP are nonspecific and should not replace careful history-taking for antigen exposures when a granulomatous lung disease is diagnosed. In cases where physiologic obstruction and airway hyperresponsiveness are prominent, HP may be mistaken for asthma. HP should also be included in the differential diagnosis of emphysema, particularly in nonsmokers.

The clinical course of HP is variable, but permanent sequelae can include persistent airway hyperresponsiveness, emphysema, or progressive interstitial fibrosis. An accelerated decline in lung function with continued antigen exposure has been demonstrated for most forms of HP, underscoring the importance of cessation of exposure. However, the clinical course of HP is variable even with antigen avoidance. Acute HP generally resolves without sequelae. The subacute and chronic forms of HP are frequently recognized later in the disease course; as a result, these patients often have a poorer prognosis than those with acute disease.

Occurrence

The epidemiology of HP is problematic due to variable clinical case definitions and probably frequent underrecognition of disease. Most epidemiologic studies have focused on farmer's lung, a form of HP, and estimates of incidence and prevalence vary significantly by region. Prevalence of HP among bird hobbyists ranges from 0.5% to 21%. Among isocyanate workers, the prevalence of HP is approximately 1%. Attack rates in outbreaks of HP may be quite high. For example, HP occurred in 52% of office workers exposed to a contaminated humidification system, and among 27% to 65% of lifeguards in two sequential HP outbreaks that occurred due to microbially-contaminated aerosols at an indoor swimming pool.

Causes

There are three general categories of antigens known to cause HP: microbial agents, animal proteins, and low-molecular-weight chemicals. Microbial agents include bacteria and fungi. Multiple species of thermophilic bacteria may contaminate decaying vegetable matter and are causally associated with farmer's lung (the prototypical example of HP). Thermophiles as well as other bacterial species may contaminate ventilation systems and humidifiers, and cause HP from exposure to indoor aerosols. Nontuberculous mycobacteria contaminating hot tubs and metal-working fluids have been implicated as causes of HP. Many fungi have been identified as causative antigens in various occupational settings, including woodworking, agriculture, and lumber milling, and as contaminants of air-handling systems or water-damaged indoor settings. Avian antigens are the most common animal proteins associated with HP. A few low-molecular-weight chemicals, including isocyanates, are known to cause HP.

Both environmental and host factors appear to play a role in risk for HP. HP occurs more frequently in nonsmokers, although the prognosis is poorer in smokers. Exposure factors, such as antigen concentration, duration of exposure before symptom onset, frequency and intermittency of exposure, particle size, antigen solubility, use of respiratory protection, and variability in work practices, may influence disease prevalence, latency, and severity. Farmer's lung disease is most common in regions with heavy rainfall, where feed is likely to become damp, and in harsh winter conditions, where this contaminated hay is used to feed cattle in indoor barns with minimal ventilation. Pigeon breeder's lung is most common during the summer sporting season, when exposures are at their highest.

Pathophysiology

HP is characterized by the presence of activated T lymphocytes in bronchoalveolar lavage (BAL) and an interstitial mononuclear cell infiltrate on lung biopsy. Pathogenesis involves repeated antigen exposure, leading to immunologic sensitization and subsequent immune-mediated lung inflammation. The mechanisms underlying this series of events appear primarily to involve cell-mediated immunity, with IL-12 and interferon playing essential immunomodulatory roles.

Prevention

Recognition of an index case of HP may have considerable public health importance, since others exposed to the same environment are at risk for disease. Each case represents a sentinel health event, indicating the need for further investigation of the implicated environment where others may be at risk and where opportunities for prevention may be identified. Interventions may include control of moisture problems by preventing leaks and eliminating aerosol humidifiers or hot tubs. For those with bird breeder's HP, removal of birds from the home should be accompanied by careful cleanup to remove fleecy furnishing, such as carpets, that may be sources of ongoing dust exposure, along with wet-wiping of surfaces and complete avoidance of bird exposure. Changes in work practices to reduce the prevalence of farmer's lung include efficient drying of hay and cereals before storage, use of mechanical feeding systems, and better ventilation of farm buildings. Respirators have been used when removal from exposure was impossible. Although respirators may provide protection in circumstances with limited exposure duration, their effectiveness in preventing new onset or recurrent HP is unproven.

The cornerstone of therapy is removal from exposure to the offending antigen. Elimination of the antigen is the preferred approach and has the added benefit of preventing disease in other exposed individuals. Oral corticosteroids are frequently used to supplement antigen avoidance in severe or progressive HP.

Further Reading

Baur X. Hypersensitivity pneumonitis (extrinsic allergic alveolitis) induced by isocyanates. *Journal of Allergy and Clinical Immunology* 1995; 95: 1004-1010.

Colby TV, Coleman A. The histologic diagnosis of extrinsic allergic alveolitis and its differential diagnosis. *Progress in Surgical Pathology* 1989; 10: 11-26.

Grammar LC. Occupational allergic alveolitis. *Annals of Allergy, Asthma and Immunology* 1999; 83: 602-606.

Lynch DA, Rose CS, Way D, et al. Hypersensitivity pneumonitis: sensitivity of high-resolution CT in a population-based study. *American Journal of Roentgenology* 1992; 159: 469-472.

Rose CS, Martyny JW, Newman LS, et al. "Lifeguard lung": Endemic granulomatous pneumonitis in an indoor swimming pool. *American Journal of Public Health* 1998; 88: 1795-1800.

Hypertension

ICD-10 I10-I15

Kenneth D. Rosenman

The optimal level of blood pressure is less than 120/80 mm Hg. People with blood pressures of 120-139/80-89 mm Hg are considered to have prehypertension and are at high risk of developing hypertension. Hypertension is considered to be a blood pressure of 140/90 mm or greater. Treatment with anti-hypertensive medication and lifestyle modification is indicated for individuals with hypertension. Lifestyle modifications are recommended for individuals with a blood pressure greater than 120/80 and drug therapy for those whose blood pressure is greater than 120/80 mm Hg and who also have kidney disease or diabetes. The diagnosis of hypertension is made from the average of two blood pressure measurements taken while seated at two or more office visits. The patient is typically asymptomatic, unless damage to the eyes, heart, brain, or kidney is present. Increased prevalence of nose-bleeds, unsteadiness, headaches, blurred vision, depression, and nocturia (urination at night) in untreated individuals without organ damage has also been described. Physical examination and laboratory tests are done to assess organ damage and detect the infrequent secondary causes. Blood and urine analyses, a chest x-ray, and an electrocardiogram (ECG) are routinely includ-ed in evaluating a newly-diagnosed individual. Kidney tests, x-rays, and more elaborate blood or urine tests are reserved for individuals suspected of having one of the secondary causes of hypertension.

Occurrence

Twenty-three percent of the U.S. population aged 20 to 74 has hyperten-sion. Prevalence increases with age and is higher among African Americans than among whites.

Hypertension is generally an asymptomatic condition. The most recent data from the National Health Nutrition Examination Survey (1999-2000) found that only 70% of people with hypertension are aware they have it, even less (59%) are under treatment, and even less (34%) were considered ade-quately controlled (systolic blood pressure <140 mm Hg and diastolic <90 mm Hg).

Ambulatory blood pressure measurements are usually lower than those taken in a clinical setting. Blood pressure normally reduces 10% to 20% dur-ing sleep; if it does not, then there is an increased risk of cardiovascular events. Ambulatory blood pressure measurements correlate better with target organ injury than those taken in a clinical setting. Individuals with hyperten-sion have ambulatory blood pressure measurements that average more than 135/85 while awake and 120/75 while asleep.

The percentage of cases of hypertension due to occupational factors is not known.

Causes

Most hypertension is defined as "essential" (cause unknown). Correctable causes, which explain fewer than 10% of all cases of hyperten-sion, include renovascular disorders, primary aldosteronism, pheochromocy-toma, Cushing's syndrome, and coarctation of the aorta. Kidney disease, depending on the type and extent of kidney insufficiency, is associated with hypertension. Kidney disease can affect production of renin, a potent media-tor of the vasoconstrictor angiotensin, as well as salt and water balance in the body. Heavy metals, silica, and organic solvents can cause kidney disease. Lead has been shown to adversely affect renin production. Acute noise exposure causes an increase in catecholamine excretion and serum levels of cholesterol, triglycerides, free fatty acids, and 11-hydroxycortisol; chronic noise exposure has been associated with hyperlipidemia. Hyperlipidemia may be one of the mechanisms responsible for the relationship between workplace noise exposure and elevated blood pressure.

Increased blood lead levels have been associated with high blood pressure: A doubling of the lead level even within normal limits raises blood pressure 1 to 2 mm Hg. Exposure to the heavy metals lead, mercury, cadmium, or arsenic and exposure to silica or solvents also may cause kidney disease, which, in turn, may cause hypertension. When kidney disease and/or gout is present along with hypertension, there is an increased likelihood that lead may be the cause of the hypertension.

Exposure to carbon disulfide, which is associated with ischemic heart disease, may also cause hypertension. Increased noise levels have been associated in most studies with an increase in blood pressure and the clinical diagnosis of hypertension. A meta-analysis has shown that for each 5-decibel increase in noise exposure there is an increase of 0.51 (95% CI, 0.01-1.00) mm Hg in systolic blood pressure and a 1.14 (95% CI, 1.01-1.29) relative risk of hypertension. Animal studies and short-term physiological studies of humans support an association between noise and elevation in blood pressure.

Work in jobs with high demand and low decision latitude have been associated with increased blood pressure, hypertension, and increased left ventricular mass (adverse effect of elevated blood pressure), although some authors have suggested that such jobs represent a marker for low socioeconomic status. (See chapters on Stress and Work Organization for further discussion on stress and hypertension.)

Pathophysiology

Elevated blood pressure, once established, is maintained by increased peripheral vascular resistance, mainly in the small arteries and arterioles, where smooth muscle cell hypertrophy or contraction may account for much of this increased resistance. Stress may lead to structural changes that cause increased vascular resistance. Activation of the sympathetic nervous system, induced by stress, may lead to retention of sodium in the kidney, causing hypertension.

Experiments in animals exposed to lead show increased pressor responses, vascular reactivity, and protein kinase C, a regulator of vascular tone.

Prevention

Reduction in exposure to substances associated with hypertension is necessary for primary prevention. A wellness program that combines reduction in workplace exposures with assistance in nutrition modification to reduce weight, salt intake, and alcohol ingestion can focus on several factors that cause or contribute to hypertension. Modifications of stress factors, such as heavy workloads, a nonsupportive supervisor, and limited job mobility and autonomy, may contribute to prevention. (See chapters on Stress and Work Organization.) The workplace can be an ideal setting for screening for, treating, and monitoring hypertension. (See the chapter on Ischemic Heart Disease for a discussion on preplacement exams.)

Other Issues

Work in a hot environment (readings over 79° F for men and 76° F for women, with a wet bulb thermometer) or exposure to solvents or nitrates may increase the prevalence and severity of adverse effects of medications used to treat hypertension. NIOSH has recommended that individuals taking diuretics ("water pills") not work in hot environments. (See chapter on Heat Stress.)

Further Reading

Campbell KCM, Rybak LP, Khardori R. Sensorineural hearing loss and dyslipidemia. *Journal of the American Auditory Society* 1996; 5: 11-14.

Hu H, Arc A, Payton M, et al. The relationship of bone and blood lead to hypertension. The normative aging study. *JAMA* 1996; 15: 1171-1176.

Nuyts GD, Vlem EV, Thys J, et al. New occupational risk factors for chronic renal failure. *Lancet* 1995; 346: 7-11.

Schnall PL, Kekic K, Landsbergis P, et al. The workplace and cardiovascular disease. *Occupational Medicine: State of the Art Reviews* 2000; 15: 1-334.

Talbott EO, Gibson LB, Burk A, et al. Evidence for a dose-response relationship between occupational noise and blood pressure. *Archives of Environmental Health* 1999; 54: 71-78.

Van Kempen EE, Kruize H, Boshuizen HC, et al. The association between noise exposure and blood pressure and ischemic heart disease: A meta-analysis. *Environmental Health Perspectives* 2002; 110: 307-317.

Infertility and Sexual Dysfunction
ICD-10 N46 (male), N97 (female), F52
SHE/O

Laura Welch

Sexual dysfunction is defined as a decreased libido, a decreased interest in sexual activity, menstrual disorders in women, or erectile dysfunction in men.

An infertile couple is generally defined as a couple that has not conceived after 1 year of unprotected sexual intercourse. Infertility can then be divided into subgroups, using the following World Health Organization definitions:

Primary Infertility

The woman has never conceived despite cohabitation and exposure to pregnancy for a period of two years.

Secondary Infertility

The woman has previously conceived but is subsequently unable to conceive despite cohabitation and exposure to pregnancy for a period of two years.

In many cases, the infertility of the couple can be attributed to either the man or the woman, but in a high percentage of cases, the infertility is due to an interaction of some characteristics of the two. Thus, the term infertile couple is more appropriate than infertile man or infertile woman. For example, a submaximal sperm count in association with some abnormalities of cervical mucus may lead to infertility when neither condition alone would do so.

Occurrence

Approximately 10% to 15% of all couples are infertile. It is estimated that male factors account for 40% of infertility cases, failure of ovulation accounts for 10% to 15%, tubal factors for 20% to 30%, cervical factors for 5%, and 10% to 20% of infertility cases have no known cause. There are no good estimates of the percentage of cases of infertility that are due to occupational factors, or of the rates of sexual dysfunction.

Causes

Sexual dysfunction can interfere with fertility, and several occupational causes of sexual dysfunction exist. Decreased libido and impotence have been reported in male workers in association with exposure to chloroprene, manganese, organic and inorganic lead, inorganic mercury, toluene diisocyanate (TDI), and vinyl chloride, as well as with shift work. Menstrual disorders have been reported in female workers in association with exposure to aniline, benzene, chloroprene, carbon disulfide, inorganic mercury, polychlorinated biphenyls (PCBs), styrene, and toluene.

Research about occupational causes of infertility has concentrated primarily on men; therefore, the remainder of this discussion will be limited to male infertility. Early fetal loss, which may be manifest clinically as infertility, is discussed under Pregnancy Outcome, Adverse.

Two steps are required to demonstrate an occupational cause of male infertility. First, there must be a gonadal disorder as evidenced by at least two of the following abnormal semen parameters: inadequate sperm number,

Table 1. Nonoccupational Causes of Male Infertility

Primary endocrine disorders
Prior nonoccupational testicular injury or testicular surgery
Postadolescent mumps
Gonadotoxic drugs, such as chemotherapeutic cytotoxic drugs and estrogens
Urologic abnormalities:
    Retrograde ejaculation secondary to diabetes,
      neurological disease, or prostatectomy
    Ductal obstruction secondary to tuberculosis, sexually transmitted
      disease, or vasectomy
    Varicocele

abnormal sperm morphology, poor sperm motility, or a decreased ability of the sperm to penetrate the egg. Second, nonoccupational causes must be evaluated; nothing about an abnormal semen analysis per se suggests an occupational etiology. Table 1 lists the nonoccupational causes of male infertility, and Table 2 lists occupational testicular toxins that have been associated with male infertility. Known toxins are those that have been shown to have an effect in humans, often at high doses. Suspected toxins are substances that have been studied in humans with suggestive but inconclusive results, or that have produced positive study results in animals. Table 2 includes a scale for measuring the certainty of the causal association between the occupational exposure and the reproductive effect. The scale does not incorporate details of dose; an association is listed as a strong one even if the data exist only for high doses in animals. In applying the details of these tables in an individual exposure situation, dose-response must be considered in more detail.

Pathophysiology

Reproductive function and sexual interest in both men and women depend, at least in part, on the presence of an intact neuroendocrine system. In men, lutenizing hormone (LH) from the pituitary and testosterone from the Leydig cells of the testes are needed for spermatogenesis and for interest in sexual activity. Pituitary lutenizing hormone and follicle-stimulating hormone (FSH), ovarian and adrenal estrogen, and progesterone are needed for female reproductive function.

This axis can be interrupted by agents that act like steroids or by the neurological input created by stress. Disorders of circadian rhythms, as can occur with some types of rotating work schedules, can also affect the endocrine

Table 2 Definitions (facing page)

*1 = limited positive animal data	2 = strong positive animal data
3 = limited positive animal data	4 = strong positive human data
5 = limited negative animal data	6 = strong negative animal data
7 = limited negative human data	8 = strong negative human data

Table 2. Occupational Testicular Toxins.

Agent	Human Findings	Animal Data	Measure of Association*
Boron	Decreased count Decreased motility	Testicular damage	2,3
Benzene	None	Testicular damage	1
Benzo(a)pyrene	None	Testicular damage	1
Cadmium	Reduced fertility	Testicular damage	2,3
Carbon disulfide	Decreased count, decreased motility	Testicular damage	1,4,7
Carbon monoxide	None	Testicular damage	1
Carbon tetrachloride	None	Testicular damage	1
Carbaryl	Abnormal morphology		3
Chlordecone	Decreased count, decreased motility		2,4
Chloroprene	Decreased motility, abnormal morphology, decreased libido	Testicular damage	1,3,4
Dibromochloropropane (DBCP)	Decreased count	Testicular damage	2,4
Dichlorvos, or dimethyl 2, 2-dichlorovinyl phosphate (DDVP)	None	Decreased count	2
Epichlorohydrin	None	Testicular damage	2,5
Estrogens	Decreased count	Decreased count	2,4
Ethylene oxide	None	Testicular damage	1
Ethylene dibromide (EDB)	Abnormal motility	Testicular damage	2,3,5
Ethylene glycol ethers	Decreased count	Testicular damage	2,3
Heat	Decreased count	Decreased count	2,4
Ionizing radiation	Decreased count		2,4
Lead	Decreased count	Testicular damage, decreased count and motility, abnormal morphology	2,4
Manganese	None	Testicular damage	2
Polybrominated biphenyls (PBBs)	None	Testicular damage	1,5
Polychlorinated biphenyls (PCBs)	None	Testicular damage	1

cycle. The clinical results are menstrual disorders in women and libidinal disorders in both sexes.

Infertility that can be attributed to the male results from some abnormality of semen: an inadequate number of sperm, decreased motility, abnormal morphology, or decreased function. Fertility is a complex process, however, and semen parameters are considered only a surrogate measure. No one test is the best measure, and an abnormal sperm count does not always indicate infertility. Most identified occupational causes of male infertility affect the production or function of sperm in the testes, rather than interfering with the hormonal milieu supporting sperm production, though lead and some other heavy metals have shown some effects on hormonal levels as well.

Sperm begin from diploid cells—the primary spermatocytes—and divide over several cell divisions into haploid cells—the spermatids. They then undergo maturation and develop motility in their transit from the testes through the epididymis. The site of action in the testes has not been identified for most of the known agents and can differ from one agent to another. Ethylene glycol ethers are known to affect cell division at the level of the primary spermatocyte, early in sperm production. Ethylene dibromide (EDB) is hypothesized to act on the maturation of spermatids into mature sperm in the epididymis.

Prevention

Primary Prevention

Reduction of exposure or substitution of products will prevent the development of sexual dysfunction or infertility in the occupational setting. The specific intervention or level of exposure for each of the agents mentioned above differs.

Secondary Prevention

Sperm parameters or fertility status can be monitored in a group exposed to a known or suspected spermatotoxin, with the goal of removing the affected men and reducing exposure if an effect is found. Questionnaire methods of assessing fertility based on the number of live births to exposed and unexposed men have been used; these have not been compared in the same study population to semen analysis, so the relative sensitivities are not known. Repeated semen analyses on the same individuals over time would be a sensitive screening tool; however, voluntary participation could be quite low, while a mandatory company program could be considered an invasion of privacy. Although follicle-stimulating hormone levels rise as sperm counts fall, this is a very insensitive measure in an individual or a small group; oligospermia would need to be severe before it would be detected with FSH as a screening tool.

Tertiary Prevention

There is evidence that male infertility secondary to occupational exposure is reversible if the toxic insult is recognized early in the course of the disease. Abnormal sperm counts secondary to dibromochloropropane (DBCP) exposure have reversed at least partially after the agent was removed unless azoospermia (absence of sperm) was present; recovery took as long as 18 months in men with oligospermia. Data from patients treated with therapeutic radiation suggest that even azoospermia is reversible in some cases, but reversal was not seen until 4 to 5 years had passed.

For this reason, it is desirable to remove an individual with occupational infertility from exposure. Once the diagnosis is made, two semen analyses should be performed, using standardized motility and morphology techniques, before the individual is removed from exposure. The job should be changed, the product substituted, or the exposure eliminated with job controls and protective equipment. If removal is being used as a diagnostic test, it should continue for at least 18 months before the trial can be considered a failure. If monitoring for body burden of the toxin is possible, as it is with lead, these 18 months should begin when the body burden returns to the normal range.

Further Reading

Barlow SM, Sullivan FM (Eds.). *Reproductive Hazards of Industrial Chemicals: An Evaluation of Animal and Human Data.* New York: Academic Press, 1997.

Boekelheide K, Gandolfi AJ, Harris C, et al. (Eds.). *Reproductive and Endocrine Toxicology.* Elmsford, NY: Pergamon Press, 1997.

Boss JM, Hales BF, Robaire B. *Advances in Male Mediated Developmental Toxicity: Advances in Experimental Medicine and Biology,* Vol. 518. Dordrecht, The Netherlands: Kluwer Academic Publishers, 2003.

Frazier LM, Hage ML (Eds.). *Reproductive Hazards of the Workplace.* New York: John Wiley & Sons, 1997.

National Research Council. Subcommittee on Reproductive and Developmental Toxicity, Committee on Toxicology, Board on Environmental Studies and Toxicology, National Research Council. *Evaluating Chemical and Other Agent Exposures for Reproductive and Developmental Toxicity.* Washington, DC: National Academies Press, 2001.

Paul M. *Occupational and Environmental Reproductive Hazards: A Guide for Clinicians.* Lippincott Williams & Wilkins, 1993.

Scialli AR, Zimaman MJ (Eds.). *Reproductive Toxicology and Infertility.* New York: McGraw Hill, 1993.

World Health Organization. *Principles for Evaluating Health Risks to Reproduction Associated with Exposure to Chemicals (Environmental Health Criteria).* Geneva: WHO, 2001.

Injuries, Fatal

ICD-10 501-509, 511-519, 521-529, 531-539, 541-549, 551-559, 561-569, 571-579, 581-589, 591-599, T01-T13, T14.1-T14.9

James L. Weeks

Fatal occupational injuries are caused by acute trauma that occurs because of work. Specific fatal injuries—homicides, suicides, and fatal injuries from falls, from motor vehicle crashes, from burns, and from heat stress—are discussed in more detail in separate chapters.

Occurrence

There are about 6 000 fatal occupational injuries per year in the United States. The greatest number of fatal injuries occurs in construction, transportation, and manufacturing, in that order. The highest rates of fatal injuries, more than five times the rate in the remainder of the labor force, occur in four major industrial groups—mining; construction; agriculture/forestry/ fishing; and transportation/public utilities. All of these rates are declining.

Occupations with the highest rates of fatal injuries are fishers (104.4 deaths per 100 000 workers), most often from drowning; timber cutters (101.0), most from being struck by an object (typically a tree); airplane pilots (97.4) from airplane crashes; and structural metal workers (64.4) from falls. Of the 20 occupations with the highest rates of fatal injuries, seven are construction trades: construction laborers (39.5 deaths per 100 000 workers), structural metal workers (64.4), roofers (29.3), electrical power installers (27.8), electricians (15.9), welders and cutters (12.0), and carpenters (7.6).

A significant minority of occupational fatalities (25% in one study) occurs as multiple fatalities, with more than one person—including fellow workers and nonworkers—involved in a single incident. These incidents include fires, explosions, transportation accidents, entrapment, and building failures. A portion of these injuries occur to fellow workers, some of whom are untrained, or bystanders attempting to rescue others.

The rate of fatal injuries among men is more than ten times that for women (7.4 vs. 0.7 deaths per 100 000 workers). The rate of fatal injuries among Hispanics (5.2 per 100 000) is slightly higher than among whites (4.4), which is slightly higher than that among blacks (4.1). The rate of fatal injuries is strongly related to age; it ranges from 1.6 to 3.5 per 100 000 for workers aged 16-44, to 4.4 among workers aged 45-54, to 6.1 for workers aged 55-64, and to 12.0 for workers over the age of 65.

Causes

The leading causes of fatal occupational injuries for the entire private and public workforce are motor vehicle injuries, homicides, and falls, in that order. The frequency of causes vary by industrial sector. Motor vehicle crashes (on and off the road) are the leading cause of fatal injuries in agriculture, manufacturing, transportation, wholesale trade, and government employment. Contact with objects, such as machines, is the leading cause in mining. Falls are the leading cause in construction (with motor vehicle crashes a close second). Homicide is the leading cause in retail trade and in finance, insurance, and real estate.

Pathophysiology

The pathophysiology of traumatic fatalities is the same as that characteristic of other traumatic injuries. Head and neck injuries are the most common, followed by trauma to multiple body parts and severe crushing injuries. Death usually occurs within hours of serious life-threatening trauma, but for some injuries for which there is delay of treatment or inadequate treatment, death may occur days or more after the initial injury.

Prevention

Preventing occupational fatalities has been achieved in particular industries by focusing attention on specific causes. Preventing fatal injuries caused by motor vehicle crashes, falls from heights, homicides, suicides, and heat stress are discussed in separate chapters and serve as examples of prevention methods that appear to be successful. Fall prevention methods appear to be effective in construction. Generic prevention methods are typically a combination of surveillance, analysis, development of controls, intervention to implement controls, and evaluation to determine whether the intervention was successful. Intervention is occasionally supported by government regulation or the threat of regulation.

Reducing the risk of fatal injuries in coal mining is a notable case study of successful prevention, employing the methods of surveillance, research and development, and implementation. The leading causes were fires and explosions, roof falls, and contact with powered haulage machines. Although research and experience had identified effective controls for these hazards, it was not until controls were made mandatory and enforced, following passage of the Federal Coal Mine Health and Safety Act of 1969, that fatality rates declined-every year for a decade.

Deaths of would-be rescue workers can be reduced with training, education, and supplying workers with appropriate rescue equipment. OSHA's confined space regulations and the experience of underground coal mine rescue teams provide useful insights into successful rescue.

Secondary prevention can be achieved by improvements in the timing and quality of medical care. The time from traumatic injury to death occurs in a trimodal distribution following injury. More than half of deaths occur immediately or very soon (usually less than an hour) after injury. For these victims, even prompt state-of-the-art medical intervention would probably not be able to prevent death. Early deaths, accounting for about one-fourth of all deaths from traumatic injury, occur within a few hours after injury; for these victims, prompt, high-quality medical attention can prevent death in a substantial proportion of cases. Emergency medical services and transportation therefore play a critical role in preventing a significant proportion of traumatic fatalities. Late deaths occur weeks after the initial injury, usually from infection or multiple organ failure. In these cases, the quality of medical care is a critical factor in preventing death.

Further Reading

Bailer AJ, Bena JF, Stayner LT, et al. External cause-specific summaries of occupational fatal injuries. Part I: An analysis of rates. *American Journal of Industrial Medicine* 2003; 43: 237-250.

Bailer AJ, Bena JF, Stayner LT, et al. External cause-specific summaries of occupational fatal injuries. Part II: An analysis of years of potential life lost. *American Journal of Industrial Medicine* 2003; 43: 251-261.

Dong X, Platner JM. Occupational Fatalities of Hispanic construction workers from 1992 to 2000. *American Journal of Industrial Medicine* 2004; 45: 45-54.

Trunkey DD. Trauma. *Scientific American* 1983; 249: 28-35.

Weeks JL, Fox M. Fatality rates and regulatory policies in bituminous coal mining, 1959-1981. *American Journal of Public Health* 1983; 73: 1278-1280.

Injuries, Nonfatal

ICD-10 S00, S10, S20, S30, S40, S50, S60, S70, S80, S90, T00, T74.0

James L. Weeks

This chapter is an overview of nonfatal traumatic occupational injuries. Specific types of injuries—burns, falls, eye injuries, homicide, low back pain, motor vehicle injuries, and skin injuries—are treated in more detail in separate chapters.

Case definitions for reporting, surveillance, and compensation vary from one data source to another. Common sources include regulatory agencies (OSHA and MSHA), the Bureau of Labor Statistics (BLS, which derives estimates from OSHA- and MSHA-required records), state-based workers' compensation systems, the ongoing National Health Interview Survey of households conducted by the National Center for Health Statistics, and the National Electronic Injury Surveillance System (NEISS) based on hospital visits and maintained by the Consumer Product Safety Commission. Private institutions, such as insurance carriers, hospitals, labor unions, and employers, also maintain surveillance systems. These are discussed in more detail in Part I.

Occurrence

Based on its annual national sample of employer reports, the BLS estimated a total of 5.2 million nonfatal occupational injuries in the private sector in the United States in 2001 and a corresponding incidence rate of 5.7 injuries per 100 full-time workers. Both of these estimates are down from prior years. The number of injuries with lost workdays in 2001 was 2.6 million, 1.8 million of which were cases with restricted activity. It is becoming increasingly common for employers to return workers to work before full recovery and, as a result, the number of cases with restricted activity has steadily increased over the past two decades.

The incidence rate in the goods producing sector of the economy (agriculture, mining, manufacturing, and construction) was 7.9 per 100 workers, with the highest rate (8.1) in manufacturing (Figure 1). In the service-producing sector, the overall incidence rate was 5.1 per 100 workers, with the highest rate (6.9) in public utilities. The nationwide incidence rate of injuries with lost workdays based on NEISS incidence data and the Current Population Survey as denominator is approximately the same as that based on BLS data.

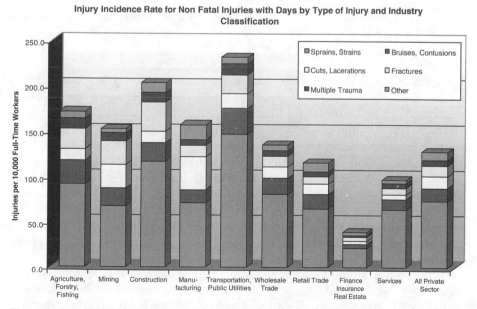

Figure 1. Injury incidence rates, by type of injury and industrial sector.

There is significant evidence of undercounting. In-depth studies of injury occurrence at specific workplaces, using multiple reporting sources, suggest the annual incidence rate of nonfatal injuries may be as high as twice that derived from the BLS Annual Survey. Undercounting probably arises from several factors; some workers injured on the job do not report injuries and some employers do not record injuries due to economic incentives and failure to recognize them. And the BLS survey does not include employers with 10 or fewer employees.

Injury rates and the distribution of types of injuries vary between industrial sectors. The most common types of injuries in all sectors are strains and sprains, followed by contusions. Fractures are common in agriculture, mining, construction, and transportation and public utilities. Cuts and lacerations are common in agriculture and construction. Carpal tunnel syndrome is most common in manufacturing but also appears in most other sectors.

Injury rates vary with experience, age and gender. Younger and less experienced workers generally have higher nonfatal injury rates than other workers. (The rate of fatal injuries is higher in older workers.) The injury rate of workers 18 to 19 is higher than that of other age groups and it declines thereafter. The injury rate of workers aged 15 to 17, after adjusting for hours worked, is about twice that of the remainder of the workforce. Up to about age 35, the injury rate of men is about twice that of women and it remains high into older age groups. Rates also vary with ethnicity and race but most often these associations are secondary to age, experience, and occupation.

The years of productive life lost vary significantly with the type of injury. Estimates based on data in the State of Washington show that back injuries and neck sprains account for significantly more productive years lost than any other type of injury (Table 1).

Causes

Occupational injuries have many causes that differ from one type of event to another. Analysis of specific types of injuries using the methods described in Part I is useful for identifying important causes. Causes common to many types of injuries include inadequate training or supervision, lack of familiarity with a workplace or machine, use of improper or inadequate tools or equipment, and working alone. Stress and distractions to workers—by events on the job or worries from issues away from work—also contribute to the risk of injury.

The most frequent injuries, strains and sprains and not infrequently bruises and fractures often occur from overexertion or slips, trips, and falls to the same level. Falls to lower levels often result in fractures or more serious injuries. Manual materials handling, housekeeping, and slippery floors are common causes. Remedies include substituting mechanical for manual handling, improved housekeeping, and slip-resistant floors and shoes. The causes of lacerations commonly involve sharp objects, such as knives, sheetmetal,

Type of injury	Total YPLL
Back/neck sprains	7003.8
Lower extremity sprains	1058.5
Upper extremity sprains	965.3
Multiple injuries	674.3
Contusions	801.6
Lower extremity fractures	641.9
Lacerations	485.3
Upper extremity fractures	406.7

Table 1. Years of Productive Life Lost (YPLL), by Type of Injury, State of Washington, 1999

broken glass, protruding nails, and similar items. Control of sharps and redesigning work to reduce the risk of contact with sharps can reduce their frequency.

Naturally occurring risk factors, such as the weather, seas, trees, or geologic formations, are important factors in construction, agriculture, fishing, forestry, mining, oil exploration, transportation, and other industries that operate in the natural environment.

Pathophysiology

Injuries are caused by the full spectrum of acute trauma and include strains and sprains, contusions, lacerations, abrasions, fractures, and dislocations. The pathophysiology of burns, eye injuries, low back pain, and skin injury is discussed in more detail in those specific entries.

Prevention

Given many causes, preventing occupational injuries must be multifaceted. Prevention of specific kinds of injuries—falls, burns, eye injuries, homicides, skin injury, and others—is discussed in separate entries. A generic injury prevention strategy, employing the conventional tools of public health, is discussed in Part I; this strategy depends on a health and safety program at work. In brief, such a program consists of material management commitment

Table 2. Estimated Annual Costs (in Billions of Dollars) for Selected Conditions, United States, 1992

	Direct Costs	Indirect Costs	Total
Occupational Injuries	49	96	145
Cardiovascular Disease	67*	97*	164
Cancer			171
Musculoskeletal Disorders (all causes)			149
HIV Infection			30
Alzheimer's Disease	20	47	67

*Extrapolated
Source: Leigh JP, Markowitz SB, Fahs M, et al. Occupational injury and illness in the United States: Estimates of costs, morbidity, and mortality. *Archives of Internal Medicine* 1997, 157: 1557-1568.

to prevention, employee involvement in safety programs, creation of real incentives to support safety, worker and manager education and training, accident analysis and investigation, implementation of preventive strategies, and follow-up to evaluate their effectiveness form the basis of an overall strategy to prevent injuries.

Other Issues

The cost of occupational injuries is substantial, but often inaccurate by failure to measure all costs, both direct and indirect. Direct costs consist of workers' compensation insurance, property damage, police, ambulance, and fire services, medical and rehabilitation expenses, and lost production. Indirect costs include lost earnings and fringe benefits, training and retraining, reemployment, and time delays. More intangible costs include pain and suffering, and effects on children and families. (See Table 2.)

Workers with permanent disabilities are at a significant disadvantage in the labor market and typically have significantly lower lifetime earnings than otherwise similarly situated workers, with associated diminished quality of life for themselves and their dependents. Approximately 4 million workers have permanent disabilities from occupational injuries in the U.S.

When direct and indirect costs are combined, the estimated annual cost of occupational injuries in the U.S. was estimated at $145 billion in 1992. This is slightly less than the annual costs for cardiovascular disease and cancer, about the same as the costs for musculoskeletal disorders (all causes), more than twice that for Alzheimer's disease, and more than three times the cost for HIV infection (Table 2).

It is common, but becoming less so, to classify the causes of accidents (sic) as due to "unsafe acts of persons" or "unsafe conditions." Classifying events in this way lacks utility for preventing injuries. There are usually many causes of injuries, as noted above, and limiting attention to only two broad classes inhibits the search for multiple causes. "Conditions" are created by "acts of persons" in any event.

Further Reading

Centers for Disease Control and Prevention. Nonfatal occupational injuries and illnesses treated in hospital emergency departments-United States, 1998. Morbidity and Mortality Weekly Report 2001; 50: 313-317.

Fulton-Kehoe D, Franklin G, Weaver M, et al. Years of productivity lost among injured workers in Washington State: Modeling disability burden in workers' compensation. *American Journal of Industrial Medicine* 2000; 37: 656-662.

Glazner JE, Borgerding J, Lowery JT, et al. Construction injury rates may exceed national estimates: Evidence from the construction of Denver International Airport. *American Journal of Industrial Medicine* 1998; 34: 105-112.

Jackson LL. Non-fatal occupational injuries and illnesses treated in hospital emergency departments in the United States. *Injury Prevention* 2001; 7 (Suppl, I): i21-i26.

Knight EB, Castillo DN, Layne LA. A detailed analysis of work-related injury among youth treated in emergency departments. *American Journal of Industrial Medicine* 1995; 27: 793-805.

Leavitt RE, Samelson NM. *Construction Safety Management*, (2nd ed.). New York: John Wiley & Sons, 1993.

Leigh JP, Markowitz SB, Fahs M, et al. Occupational injury and illness in the United States: Estimates of costs, morbidity, and mortality. *Archives of Internal Medicine* 1997; 157: 1557-1568.

Miller ME, Kaufman JD. Occupational injuries among adolescents in Washington State. *American Journal of Industrial Medicine* 1998; 34:121-132.

Murphy PL, Sorock GS, Courtney TK, et al. Injury and illness in the American workplace: a comparison of data sources. *American Journal of Industrial Medicine* 1996; 30: 130-141.

Pransky G, Snyder T, Dembe A, et al. Under-reporting of work-related disorders in the workplace: A case study and review of the literature. *Ergonomics* 1999; 42: 171-182.

Shannon HS, Lowe GS. How many injured workers do not file claims for workers' compensation benefits? *American Journal of Industrial Medicine* 2002; 42: 467-473.

Smith GS. Public health approaches to occupational injury prevention: Do they work? *Injury Prevention* 2001; 7 (Suppl. I): i3-i10.

Ischemic Heart Disease

ICD-10 I20-I25

Kenneth D. Rosenman

WHO defines ischemic heart disease as "myocardial impairment due to imbalance between coronary blood flow and myocardial requirements caused by changes in the coronary circulation." The clinical presentation can include repeated chest pain (angina), dyspnea (shortness of breath due to congestive heart failure), palpitations, syncope (fainting), sudden death (due to arrhythmias), and acute chest pain (due to myocardial infarction). (See Congestive Heart Failure and Arrhythmias and Arrhythmia-Induced Sudden Death.) Patients with angina typically develop chest pain with exertion that is relieved by rest and/or nitroglycerin. Different presentations of pain may occur with unstable angina (a changing pattern of pain or pain at rest) or with Prinzmetal's (variant) angina secondary to coronary vasospasm (with myocardial ischemia, also occurring at rest). Results of physical examination, chest x-ray, and electrocardiogram (ECG) are typically normal, unless some other manifestation of ischemic heart disease is present. Exercise tolerance (stress) testing, which may include a radionuclide scan, and cardiac catheterization may be performed, particularly if angioplasty or cardiac bypass surgery is being considered.

Myocardial infarction typically presents with severe acute chest pain, although silent infarcts and infarcts in people not seeking medical care account for up to 20% of all heart attacks. Physical examination is normal unless other manifestations of ischemic heart disease are present. The ECG shows distinct changes, indicating the location and extent of heart muscle destruction. During the acute phase, enzymes, such as creatine kinase-MB isoenzyme and cardiac troponins, are typically elevated in the blood.

Occurrence

In the United States, (a) approximately 13 million people are estimated to have ischemic heart disease; (b) 650 000 individuals a year have a new heart attack; and (c) another 450 000 individuals have a second or further heart attack. Despite the marked reduction in mortality from heart disease since the 1950s, ischemic heart disease (including sudden death) is still the most common cause of death, accounting for 20% of all deaths (of more than 500 000 people) in the United States each year. Approximately 50% of men and 63% of women who die suddenly of coronary heart disease have had no previous symptoms of this disease. Half of coronary heart disease deaths occur outside the hospital. The disease is 1.5 to 3.2 times more common in men than in women, with the ratio decreasing with increasing age. Incidence increases with age. The lifetime risk of coronary heart disease after age 40 is 49% in men and 32% in women. The percentage of cases that are occupationally related is not known.

Causes

Major risk factors for the development of atherosclerosis are dyslipidemia (high low-density lipoproteins and low high-density lipoproteins), cigarette smoking, and hypertension. Other important risk factors include obesity, diabetes, lack of sufficient physical activity, psychosocial factors, and heavy alcohol use. The etiology is multifactorial, and there is an increase in risk of heart disease among people with multiple risk factors. Some authors have estimated that the known risk factors explain only half of the cases of this disease. Other markers being studied include C-reactive protein, homocysteine, coagulation factors, and *Chlamydia pneumoniae* antibody titers.

In addition to risk factors for the development of atherosclerosis, there are risk factors for triggering an acute event, such as a heart attack or arrhythmia. Carbon monoxide and certain solvents are examples of exposures associated with triggering an acute event, but not the underlying atherosclerosis. Other factors, such as stress, are both risk factors for the underlying disease and a trigger for the acute event.

There is good evidence of a causal association between ischemic heart disease and exposure to carbon monoxide (CO), carbon disulfide (CS_2), or nitrates. There is limited evidence that stress, noise, lead, or arsenic causes ischemic heart disease. There is strong evidence from communitywide air pollution studies of an association between exposure to fine particulates (<2.5 µm) and cardiovascular disease morbidity and mortality. There are limited reports on an association between ischemic heart disease and hot or cold temperature extremes, radiation, fibrogenic dusts, cadmium, dioxins, polychlorinated biphenyls (PCBs), and use of ethanol and phenol as solvents. Although review articles have used estimates that 1% to 3% of heart disease is caused by work, there are no actual data to quantify the importance of occupational exposures. CO is ubiquitous and found wherever there is incomplete combustion (fuel-powered engines). If there were a consensus that the association between (a) ischemic heart disease and (b) noise and/or stress was strong, the contribution of occupation to the etiology of ischemic heart disease would be very significant. (See chapters on Stress and Work Organization.) However, given the prevalence of ischemic heart disease in the United States, even if occupational factors were important in only a small percentage of cases, mitigation of such factors would have important public health benefits.

Pathophysiology

Artherosclerosis begins in the second and third decades of life, but typically manifests years later. Initial steps involve the accumulation of lipoproteins, the recruitment and accumulation of leukocytes, and then the proliferation of smooth muscle cells. As an atheromatous plaque grows, more lipids accumulate, smooth muscle cells proliferate and die, and neovasculization and sometimes mineralization occur. Although acute clinical manifestations may occur from gradual blockage of arteries, most acute clinical manifestations appear to be secondary to acute thrombosis, which is commonly secondary to physical disruption of an artherosclerotic plaque.

CO has 200 times the affinity for hemoglobin compared to oxygen and consequently, inhalation of CO displaces oxygen in blood and reduces its transport throughout the body. The percentage of CO bound to hemoglobin (as carboxyhemoglobin, or COHb) is the product of its concentration in air, duration of exposure, and level of physical activity. COHb levels as low as 2.8% have been associated with acute cardiac symptoms. (See portion of Asphyxiants chapter on CO poisoning.)

Chronic CO exposure may be associated with increased artherosclerosis, although data suggesting this association are mixed. Increased vascular permeability and increased platelet adhesiveness have been found in animals exposed to CO.

Suggested mechanisms for the artherosclerotic effect of CS_2 include a role in causing hypertension, hypercholesterolemia, an antifibrinolytic effect, and an effect on heart rate variability. Nitrates cause an acute rebound coronary vasospasm on removal from exposure, but may also be associated with increased artherosclerosis after chronic exposure. Exposure to fine particulates and possibly stress have been shown to change autonomic control which is measured as a reduction in heart rate variability. This is measured by ambulatory monitoring of the electrocardiogram (Holter monitoring). This reduction in heart rate variability would be a factor in precipitating an acute cardiovascular event. Lead, noise, and stress have been associated with blood pressure elevation. Noise and stress have been associated with changes in coagulation factors.

Prevention

Reduction in exposure to substances that cause acute changes, such as CO, or chronic artherosclerotic changes, such as CS_2, is necessary for primary prevention. Whether adequate protection is afforded by current occupational standards has been strongly questioned.

Early diagnosis with removal of symptomatic individuals exposed to CS_2 has been effective in reducing the morbidity and mortality from ischemic heart disease. This is even more likely to be true for people exposed to CO and nitrates, where the evidence for an acute effect is stronger than the evidence for a chronic atherosclerotic effect.

Preplacement and periodic exams potentially would be useful in identifying individuals who, because of the presence of risk factors for ischemic heart disease or ischemic heart disease itself, are at increased risk of accelerating their atherosclerosis or triggering an acute cardiac event from workplace exposures or stress. Unfortunately, none of the current screening tests for coronary heart disease have been shown to have adequate sensitivity and specificity to be used in asymptomatic individuals, including both exercise tolerance (stress) tests and ultra-fast or electron-beam computed tomography, which identifies coronary artery calcification. Stress tests using radioactive technetium tracer or cardiac catheterization are the most common diagnostic tests. The standardized Rose Questionnaire and a resting ECG have historically been used in epidemiology studies of working populations. Measurement of variability of heart rate, which decreases with exposure to substances adversely affecting the heart, has been used in recent studies.

Individuals who have already manifested ischemic heart disease (with a history of a heart attack, congestive heart failure, or arrhythmia) or have cardiac risk factors and are working or plan to work in situations where there are exposures associated with ischemic heart disease would potentially benefit from counseling and workplace modifications. A wellness program that combines reduction of workplace exposures and stress with lifestyle modifications can focus on several factors that cause or contribute to ischemic heart disease.

Other Issues

Symptoms of ischemia, such as angina and claudication, occur with increased frequency and duration as levels of carboxyhemoglobin increase, even within the legal exposure limits for CO. Other evidence suggests that acute heart attacks and death secondary to ventricular fibrillation are more likely to occur in people with underlying ischemic heart disease who are exposed to CO within current legal limits. Therefore, individuals with underlying ischemic heart disease may be particularly susceptible. Such individuals are more likely to have an acute cardiac event during stress or strenuous activity in cold or hot conditions.

Firefighters and police are accorded special compensation rights for ischemic heart disease in approximately 20 states.

Further Reading

Ernst A, Zibrak, JD. Carbon monoxide poisoning. *New England Journal of Medicine* 1998; 339: 1603-1608.

Lustberg M, Silbergeld E. Blood lead levels and mortality. *Archives of Internal Medicine* 2002; 162: 2443-2449.

Nurminen M, Hernberg S. Effects of intervention on the cardiovascular mortality of workers exposed to carbon disulphide: A 15-year follow-up. *British Journal of Industrial Medicine* 1985; 42:32-35.

Rosenman KD. Occupational Heart Disease in Rom WN (Ed.). *Environmental and Occupational Medicine* (3rd ed.). Philadelphia: Lippincott-Raven. 1998; 733-741.

Rudusky BM. Acute myocardial infarction secondary to coronary vasospasm during withdrawal from industrial nitroglycerin exposure—A case report. *Angiology* 2001; 52:143-144.

Schnall PL, Belkic K, Landsbergis P, et al. The workplace and cardiovascular disease. *State of the Art Reviews: Occupational Medicine* 2000; 15:1-334.

Kidney Cancer

ICD-10 C64, C65

Elizabeth Ward

Kidney cancers include renal cell carcinomas, the most common type, and transitional cell carcinomas of the renal pelvis. In early stages, kidney cancer may not produce any signs or symptoms. As a kidney cancer grows, it may produce gross or microscopic hematuria and/or a lump or mass in the flank area. Other less common symptoms, such as fatigue, loss of appetite, weight loss, recurrent fever, a persistent pain in the side, various metabolic disorders, and general malaise, may be caused by local or distant effects of the cancer. A variety of paraneoplastic syndromes may bring the patient to the attention of physicians. Diagnostic tests include computed tomography, magnetic resonance imaging, intravenous pyelogram, ultrasonography, angiography, and thin-needle biopsy.

Occurrence

An estimated 31 900 new cases and 11 900 deaths occurred in the United States in 2003, with a male-to-female incidence ratio of 2:1 and a black-to-white incidence ratio of 1.2:1. The incidence of kidney cancer in the US increased in all population groups from 1973 through the early 1990s and has subsequently stabilized for whites, but not for blacks. There are no formal estimates of the proportion of kidney cancers that are related to occupation.

Causes

The role of occupational factors in the etiology of kidney cancer is poorly understood. Increased risk of kidney or renal cell cancer has been reported for coke oven workers, iron and steel industry workers, chimney sweeps, asbestos-exposed workers, nickel-smelting workers, lead-smelter workers, dry-cleaning workers, textile workers, tailors, firefighters, painters, gasoline station attendants, oil refinery workers, truck drivers, electric power utility workers, farmers, and printers. Several studies have found excesses in kidney cancer associated with solvent exposure, particularly exposure to chlorinated aliphatic hydrocarbons (including trichloroethylene), with suggestion of a gender difference with women being more susceptible than men. Occupational cadmium exposure was associated with renal cell carcinoma in one case-control study published in 1976, but most subsequent studies have found no association. Gasoline came under suspicion as a risk factor for renal cancer when male rats exposed long-term to vapors of unleaded gasoline developed a significant excess of renal cancers. Kidney cancer elevations associated with gasoline-exposed occupations have been observed, but not consistently. Although the data on work-related risks for cancers of the renal pelvis and ureters are limited, risk factors tend to resemble those for bladder cancer. Tobacco smoking and use of phenacetin-containing analgesics (which are no longer available in the U.S.) increase risk of both renal cell carcinoma and cancer of the renal pelvis, while renal dialysis increases risk for renal cell carcinoma only. Diuretics have also been linked to renal cell carcinoma, but so has high blood pressure, a condition they are often used to treat. Familial clustering has been observed. Individuals with a specific genetic disorder, Von Hippel Lindau (VHL) disease, are at increased risk of renal cell carcinoma.

Pathophysiology

There is very little known about the mechanism by which exogenous agents may cause renal cell cancer.

Prevention

Primary prevention includes eliminating or reducing exposure to agents with known or suspected carcinogenicity. Avoidance of tobacco smoking and weight gain will reduce risk due to non-occupational factors. There are no specific recommendations for screening of individuals who may be at increased risk of kidney cancer due to their occupation. Routine urinalysis may detect blood in the urine, but the sensitivity and specificity of this method of screening for kidney cancer has not been evaluated. Five-year survival in the U.S. for the 1992-1998 period was 62.1%, with almost 90% 5-year survival among the 50% of cases diagnosed at a localized stage.

Further Reading

American Cancer Society. All about kidney cancer: detailed guide. Learn about cancer [serial online]. American Cancer Society, Atlanta, GA. [updated 01/04/2001]. Available from URL: http://www.cancer.org/docroot/CRI/CRI_2_3x.asp?dt=22

Dosemeci M, Cocco P, Chow WH. Gender differences in risk of renal cell carcinoma and occupational exposures to chlorinated aliphatic hydrocarbons. *American Journal of Industrial Medicine* 1999; 36: 54-59.

Lewis RJ, Schnatter AR, Katz AM, et al. Updated mortality among diverse operating segments of a petroleum company. *Occupational and Environmental Medicine* 2000; 57: 595-604.

Mandel JS, McLaughlin JK, Schlehofer B, et al. International renal-cell cancer study IV: Occupation. *International Journal of Cancer* 1995; 61: 601-605.

McLaughlin JK, Blot WJ, Devesa SS, et al. In Schottenfeld D, Fraumeni JF Jr. (eds.). *Cancer Epidemiology and Prevention* (2nd ed.). New York: Oxford University Press, 1996, pp. 1142-1155.

National Cancer Institute. What you need to know about kidney cancer. What you need to know about cancer publications [serial online]. National Cancer Institute, National Institutes of Health, Bethesda, MD. [updated 9/16/2002]. Available from URL: http://www.nci.nih.gov/cancerinfo/wyntk/kidney

Pesch B, Haerting J, Ranft U, et al. Occupational risk factors for urothelial carcinoma: agent-specific results from a case-control study in Germany. MURC Study Group. Multicenter Urothelial and Renal Cancer. *International Journal of Epidemiology* 2000; 29: 238-247.

Schulte PA, Ringen K, Hemsteet GP, et al. Occupational cancer of the urinary tract. *Occupational Medicine* 1987; 2: 87-93.

Laryngeal Cancer

ICD-10 C32

Elizabeth Ward

Cancer of the larynx may be asymptomatic initially. With supraglottic (above the glottis) cancers, hoarseness occurs late, and the first symptom may be vague pain or a lump in the neck. Cancers of the glottis lead to hoarseness early. Subglottic cancers rarely cause hoarseness but may lead to shortness of breath due to partial airways obstruction. Diagnosis is made by direct examination and biopsy of the larynx by laryngoscopy.

Occurrence

An estimated 9 500 new cases of laryngeal cancer and 3 800 deaths occurred in the United States in 2003, with a male-to-female incidence ratio of 4:1 and a black-to-white incidence ratio of 2:1. In the United States, incidence over time has been decreasing in men and increasing in women. There are no formal estimates of the proportion of laryngeal cancers that are related to occupation.

Causes

Cigarette smoking and alcohol consumption have both been shown to have a strong association with laryngeal cancer. Studies have revealed a dose-response relationship for each of these factors taken alone and a synergistic (multiplicative) effect of these two factors taken together. Occupational cohort studies have found strong evidence for an association of laryngeal cancer with manufacture of mustard gas, production of ethanol using the "strong acid" process, nickel refining, and metalworking (with straight [mineral] oil exposure), as well as exposure to sulfuric acid in ethanol, steel, and soap production, and to diethyl sulfate in ethanol production. Excess risks observed in brewers, wine makers, and bar workers may be explained by access or exposure to alcohol in their work. Other industries and occupations associated with laryngeal cancer include the railroad and lumber industries, sheetmetal workers, grinding-wheel operators, automobile mechanics and assemblers, metal fitters and assemblers, miners and quarrymen, plumbers and pipefitters, butchers, bakers, painters in construction, production supervisors, "miscellaneous mechanics," construction workers, textile workers, glass formers and potters, transport equipment operators, plastic-working machine operators and handlers, equipment cleaners, workers in grain mills, agricultural and animal husbandry workers, local government services, and laborers. Exposures associated with laryngeal cancer in at least one study include bischloromethyl ether (BCME), woodworking, pinewood dust, fibrous glass insulation production, silica, and insecticides. Some studies have found increased risks of laryngeal cancer associated with asbestos exposure, but this is not consistent.

Pathophysiology

Almost all laryngeal cancers are squamous cell carcinomas. Direct contact with the carcinogenic agent is thought to be responsible for initiating development of this malignancy. Alcohol, tobacco, and occupational exposure to acid mists and other agents can damage the layer of cells lining the larynx, causing mutation and cell proliferation. It has also been suggested that infection with human papillomavirus (HPV), leading to laryngeal papillomas, may play a role in some cases. Supraglottic cancers with late symptoms are the type most likely to metastasize.

Prevention

Prevention rests on eliminating or reducing exposures to specific agents and industrial processes with known or suspected carcinogenicity. The most important measures for primary prevention outside of the occupational setting are avoidance of tobacco and alcohol. Secondary prevention involves early detection and effective treatment of people with this disease. Early detection largely depends on early recognition of symptoms among people with this cancer and on their seeking medical attention. Cancer of the larynx is usually treated with radiation therapy or surgery. Chemotherapy may also be employed. The five-year survival rate for laryngeal cancer in the U.S. for the 1992-1998 period was 64.4%; for the 50% of people whose cancers were diagnosed at a localized stage, the 5-year survival rate was 82.0%.

Other Issues

People who have had cancer of the larynx may have an increased risk of lung and oropharyngeal cancer.

Further Reading

American Cancer Society. All about laryngeal and hypopharyngeal cancers: detailed guide. Learn about cancer [serial online]. American Cancer Society, Atlanta, GA. [updated 01/04/2001]. Available from URL: http://www.cancer.org/docroot/CRI/CRI_2_3x.asp?dt=23

Austin DF, Reynolds P. Laryngeal cancer. In Schottenfeld D, Fraumeni JF Jr. (Eds.). *Cancer Epidemiology and Prevention* (2nd ed.). New York: Oxford University Press, 1996, pp. 619-636.

Eisen EA, Tolbert PE, Hallock MF, et al. Mortality studies of machining fluid exposure in the automobile industry. III: A case-control study of larynx cancer. *American Journal of Industrial Medicine* 1994; 26: 185-202.

Goldberg P, Leclerc A, Luce D, et al. Laryngeal and hypopharyngeal cancer and occupation: Results of a case-control study. *Occupational and Environmental Medicine* 1997; 54: 477-482.

Lynch J, Hanis NM, Bird MG, et al. An association of upper respiratory cancer with exposure to diethyl sulfate. *Journal of Occupational Medicine* 1979; 21: 333-341.

National Cancer Institute. What you need to know about cancer of the larynx. What you need to know about cancer publications [serial online]. National Cancer Institute, National Institutes of Health, Bethesda, MD. [updated 9/16/2002]. Available from URL: http://www.nci.nih.gov/cancerinfo/wyntk/larynx

Ries LAG, Eisner MP, Kosary CL, et al. (Eds.). SEER Cancer Statistics Review, 1973-1999, National Cancer Institute. Bethesda, MD, 2002. Available at: http://seer.cancer.gov/csr/1973_1999/

Steenland K. Laryngeal cancer incidence among workers exposed to acid mists (United States). *Cancer Causes and Control* 1997; 8: 34-38.

Wortley P, Vaughan TL, Davis S, et al. A case-control study of occupational risk factors for laryngeal cancer. *British Journal of Industrial Medicine* 1992; 49: 837-844.

Latex Allergy

ICD-10 T78.4

Lee Petsonk and Barry S. Levy

Allergic reactions to natural rubber latex (NRL) range from rhinitis and conjunctivitis, to contact urticaria (hives), to asthma, to generalized urticaria, and rarely, to anaphylactic shock. Latex allergy results from direct skin (or mucous membrane) contact, as well as from inhalation of particles containing NRL allergens. Continued exposure of sensitized individuals to NRL increases the probability of triggering symptoms.

Latex hypersensitivity is generally documented using skin prick or serologic tests for latex-specific IgE antibodies. Latex allergy is recognized clinically when exposure-related symptoms occur in an individual with latex hypersensitivity. However, not all symptomatic individuals demonstrate hypersensitivity, using currently available test reagents.

Occurrence

Reports of latex hypersensitivity and allergy increased greatly in the 1980s and early 1990s, largely as a result of increased exposure to NRL antigens associated with the use of latex medical gloves. In the health-care setting, NRL is used in many products and NRL allergy can be a major occupational health problem. If protective latex gloves are used, workers in other industries also may become sensitized, such as during greenhouse work, housekeeping, food service, hairdressing, and emergency services. Latex products, such as balloons and condoms, are also frequently encountered outside the workplace. As a result, career and lifestyle options may be substantially limited for people with severe latex allergy. A number of studies have indicated that from 1% to 6% of the general population demonstrate laboratory evidence of latex hypersensitivity. Among health care workers and others who have frequent occupational contact with NRL allergens, the rate of latex hypersensitivity has averaged from 8% to 12%. Atopic individuals have significantly higher rates of latex hypersensitivity, generally rates 2- to 4-fold higher, than non-atopic individuals. The proportion of latex-sensitive individuals who are symptomatic depends upon the degree of allergen exposure. Among hospital employees, asthma triggered by latex exposure has been documented in up to half of sensitized workers. Recent evidence suggests latex-related health problems may be declining in the face of measures taken to reduce NRL antigen exposures by glove users, health-care institutions, and manufacturers.

Causes

"Latex allergies" are not actually reactions to the rubber polymer, but to one or more of the several hundred polypeptides and proteins in the milky fluid from the rubber tree (*Hevea brasiliensis*) that is used to make natural rubber latex. Exposures to less than a dozen of these proteins have been implicated as causing latex hypersensitivity and clinical latex allergy. In contrast to natural rubber latex, a number of commercial products are derived exclusively from synthetic rubber, such as latex paints. Because they do not contain NRL proteins, these products do not represent a hazard for the development of NRL allergy.

Pathophysiology

Latex hypersensitivity is generally mediated by the production of specific IgE, which can be triggered when proteins from a latex product are transferred to the skin or mucous membranes, such as during the wearing of gloves or the handling of other products made from natural rubber latex. Use of NRL gloves in the presence of eczema or rashes caused by detergents or by chemicals used in glove processing appears to increase the likelihood of developing latex allergy. In the past, most examination and surgical gloves were coated with starch powder. It is now recognized that the allergenic NRL proteins adhere to these powders, and can remain in contact with the skin after the gloves are removed. Additionally, protein-powder particles are aerosolized when powdered latex gloves are removed. These particles can be inhaled by nearby workers and patients, exposing mucous membranes to NRL allergens.

Prevention

Prevention of occupational latex allergy is achieved by reducing potential exposure to NRL allergens. In the health-care setting, this involves the replacement of powdered latex gloves with either latex-free gloves or reduced-protein powder-free latex gloves, while maintaining effective glove barrier protection against microorganisms. Individuals at increased risk, such as those working in health care settings where latex gloves are used, should be adequately informed about the potential hazards of latex gloves and the personal and institutional measures that can be taken to prevent latex allergy.

A number of preventive measures have been endorsed by public health agencies and professional organizations:

- Employers should provide latex-free gloves for activities that are not likely to involve contact with infectious materials, such as food preparation, routine housekeeping, and maintenance.
- In workplaces where latex gloves and other products continue to be used, employers should routinely inform workers about signs and symptoms of latex allergy, and risk factors associated with its development, as well as the importance of early medical evaluation and intervention.
- All workers whose job requires exposure to natural rubber latex should be periodically evaluated for symptoms of latex allergy, and employers should develop policies related to latex allergy, providing for appropriate medical evaluation and care, and reasonable accommodations when necessary.

Other Issues

Workers should avoid contact with NRL gloves in the presence of eczema or other active skin disorders. Individuals may also develop latex allergy while receiving medical care, such as during surgical, obstetric, or gynecologic procedures resulting in skin or mucous-membrane contact with latex gloves or other medical supplies or equipment. Persons who have undergone multiple surgical procedures in which latex gloves were used, such as children with spina bifida or congenital urological anomalies, are at high risk of developing latex allergy. These children generally have been shown to have NRL allergy prevalence in the range of approximately 25% to 65%. Individuals with NRL allergy may also manifest allergic reactions to some fruits, including avocado, banana, chestnut, and kiwi, possibly due to allergen cross-reactions.

Further Reading

American College of Occupational and Environmental Medicine (ACOEM). Evidence Based Statements: Guidelines for Employee Health Services in Health Care Facilities, V.1.0 , ACOEM Section on Medical Center Occupational Health. April 1998. available at: http://www.acoem.org/guidelines/article.asp?ID=3 (January 2, 2004)

Allmers H, Schmengler J, Skudlik C. Primary prevention of natural rubber latex allergy in the German health care system through education and intervention. *Journal of Allergy & Clinical Immunology* 2002; 110: 318-323.

Dillard SF, Hefflin B, Kaczmarek RG, et al. Health effects associated with medical glove use. *AORN Journal* 2002; 76: 88-96.

Hamilton RG, Peterson EL, Ownby DR. Clinical and laboratory-based methods in the diagnosis of natural rubber latex allergy. *Journal of Allergy & Clinical Immunology* 2002; 110 (2 Pt. 2): S047-S056.

Kaczmarek RG, Silverman BG, Gross TP, et al. Prevalence of latex-specific IgE antibodies in hospital personnel. *Annals of Allergy, Asthma and Immunology* 1996; 76: 51-56.

Kibby T, Akl M. Prevalence of latex sensitization in a hospital employee population. *Annals of Allergy, Asthma and Immunology* 1997; 78: 41-44.

Liss GM, Sussman GL. Latex sensitization: occupational versus general population prevalence rates. *American Journal of Industrial Medicine* 1999; 35: 196-200.

National Institute for Occupational Safety and Health. *NIOSH Alert: Preventing Allergic Reactions to Natural Rubber Latex in the Workplace* (DHHS [NIOSH] Publication No. 97-135). Cincinnati, OH: NIOSH, 1997.

Petsonk EL. Couriers of asthma: antigenic proteins in natural rubber latex. *Occupational Medicine* 2000; 15: 421-430.

Phillips VL, Goodrich MA, Sullivan TJ. Health care worker disability due to latex allergy and asthma: a cost analysis. *American Journal of Public Health* 1999; 89: 1024-1028.

Saary MJ, Kanani A, Alghadeer H, et al. Changes in rates of natural rubber latex sensitivity among dental school students and staff members after changes in latex gloves. *Journal of Allergy & Clinical Immunology* 2002; 109: 131-135.

Vandenplas O, Delwiche JP, Evrard G, et al. Prevalence of occupational asthma due to latex among hospital personnel. *American Journal of Respiratory and Critical Care Medicine* 1995; 151: 54-60.

Lead Poisoning

ICD-10 T56.0

James Keogh* and Barry S. Levy

Lead poisoning is a syndrome of intoxication caused by absorption of metallic lead, inorganic lead compounds, or, in rare instances, organic lead

* James Keogh (deceased) wrote the Lead Poisoning chapter for the first edition of this book. The chapter for this edition was revised and updated by Barry S. Levy, with the assistance of Howard Hu and Rick Rabin.

compounds. Centuries after the dangers of lead intoxication were recognized, lead poisoning remains one of the most common environmental diseases. Because lead poisoning can masquerade with a wide variety of clinical presentations, diagnosis depends on a high degree of clinical suspicion. Patients may present with predominantly neurological, gastrointestinal, or rheumatologic complaints. Diagnosis is generally confirmed by demonstration of elevated levels of lead in blood and/or bone.

The useful purpose of lead, its widespread availability, and the ease with which it can be manipulated by even simple technology contribute to the widespread occurrence of lead poisoning. Because lead can be melted at relatively low temperatures (327.4°C), is malleable, is water resistant, and makes brightly colored pigments, it has many uses throughout the world.

Sophisticated modern technology does not necessarily provide a margin of safety from this "old" disease. State-of-the-art "mass-burn" solid-waste incinerators produce ash with a high content of lead that poses a potential threat to operators. Sophisticated air pollution control devices and complex chemical plants rely on the time-honored skills of lead burning for their maintenance.

About 3 million tons of lead are produced each year worldwide, mainly from mines in Australia, China, the United States, Peru, Canada, and Mexico, and many more millions of tons are recycled. In the U.S., approximately 1.5 million tons are consumed annually, about 30% from new production of lead and about 70% from recycling scrap (mainly from used batteries). Two-thirds of the US consumption is used in manufacturing lead-storage batteries. Because batteries have a limited life span, about 80% of this lead is returned to scrap to be recycled. Sheet lead is used for waterproofing, noise and vibration reduction, and lining chemical reaction vessels.

Large amounts of lead compounds are used in paints for color, for rust prevention, and as drying agents. Although the use of indoor paint containing significant amounts of lead has been banned in the U.S., red lead paint is still used extensively as a primer on structural steel. Lead is also used in the plastics and ceramics industries, and in both bullets and primer for firearms.

Organic lead compounds, mainly tetraethyl lead, have been used since the 1920s as anti-knock additives in fuels. Substantial atmospheric pollution can result from the inorganic lead compounds produced by the combustion of these fuels. Leaded gasoline has been phased out totally in the United States and approximately 50 other countries, but it is still being used elsewhere.

Thirty years ago, 80 µg/dL was considered an acceptable limit for lead in blood, and many exposed workers had blood lead levels (BLLs) just below this level. Industry-sponsored mortality studies have since shown that these levels of exposure caused increased mortality from renal and cardiovascular disease. The OSHA lead standard requires medical evaluation at a BLL of 40 µg/dL and requires that physicians conducting surveillance be provided with a copy of guidelines describing signs and symptoms of intoxication.

Studies of "subclinical" and "asymptomatic" lead poisoning have shown that workers with BLLs between 40 and 60 µg/dL not only show neurological abnormalities, but also have symptoms when asked.

Occurrence

Many states have implemented surveillance systems for laboratory reporting and subsequent investigation of cases. In 1998, the 25 states with required reporting of BLLs of 25 µg/dL or higher identified 10 501 adults with elevated BLLs at or above this level. It is thought that these data grossly underestimate the number of cases of work-related lead poisoning in these states.

In 1991, Massachusetts established an occupational lead registry to systematically identify workers and workplaces with overexposure to lead in order to target and evaluate prevention efforts. Massachusetts regulations require laboratories performing BLLs to report all those of 15 µg/dL or greater in individuals aged 15 or older. The registry attempts to perform follow-up on all persons with a BLL of 40 µg/dL or greater in order to obtain data on source of exposure, health and safety training, and participation in blood lead monitoring programs. During the 1991-1995 period, the registry received reports of 664 cases with at least one BLL of 40 µg/dL or greater, the level at which federal and state regulations require increased medical surveillance. Of these cases, 85% were in workers who were occupationally exposed to lead. More than two-thirds (70%) of these workers were employed in the construction industry, primarily as de-leaders and bridge and house painters. In addition, 23% of these cases were in manufacturing workers, primarily in foundries, smelters, and plastic and glass products manufacturing facilities. Only cases in which the diagnosis was suspected and the BLL was measured are counted in this surveillance system. Because an accurate diagnosis is difficult to obtain, many cases undoubtedly go unrecognized and uncounted.

Causes

Occupational lead poisoning can occur in a variety of settings. Perhaps the most common is where work is done with lead as a paint or coating. Although exposure can occur while paint is being applied, the worst exposures occur when existing paint is disrupted. Workers removing lead paint from homes during renovation or remodeling can produce much fine lead dust, especially during burning or powered sanding operations. Such processes many not only poison workers but also render a home much more dangerous to occupants. Similarly, workers removing old paint from steel structures by grinding or abrasive blasting prior to repainting may also be exposed. Demolition of such structures by the use of oxyacetylene torches, an extremely common practice, has resulted in many documented epidemics of lead poisoning.

Because lead poisoning has long been recognized in the battery industry, hygienic conditions in many plants have been improved. Nevertheless, dangers still exist in production and especially recycling old batteries.

Lead used for waterproofing can be a hazard both for those installing it and for those involved in repair or demolition. Several episodes of lead poisoning have occurred in small plants reclaiming telephone cable that has a protective lead sheath, and overexposure of cable splicers has been documented. The use of lead as a cleaning and coating agent in the manufacture of steel and wire may also be a hazard.

The use of lead solder in manufacturing and repairing automobile and truck radiators exposes many individuals, usually in very small and poorly ventilated shops. Additionally, lead in bullets and primer presents a significant hazard on indoor target ranges. Discharge of pistols and rifles produces much lead fume and particulate, a significant hazard to firing range employees and regular users.

Lead poisoning may also occur in the arts with the use of lead-containing pigments and glazes for painting and ceramic work. Stoneware made with lead glazes may allow lead to leach into acidic foods and beverages. Lead is also used in stained-glass work and in making metal objects, figurines, and fishing sinkers.

Lead remains in the paint of many older homes, representing a potential hazard to young children. It is also present in water pipes leading into many homes. Some traditional medicines from Mexico and other countries contain high concentrations of lead. Some calcium supplements from health food stores are produced from animal bones and may be heavily contaminated with lead. Hair coloring and some cosmetics may also contain lead.

Pathophysiology

Lead as a fume or very fine particulate is readily deposited in and absorbed through the lungs. Inhalation is probably the most significant route of absorption in most occupational cases. Larger particles that are inhaled will usually be trapped by the mucociliary defensive system; unfortunately, most lead that is cleared from the tracheobronchial tree is swallowed. Lead dust that contaminates hands, clothing, food, and tobacco products may also be ingested. Absorption from the gastrointestinal tract is particularly efficient in children. Although organic lead compounds, such as tetraethyl lead, can be absorbed through the skin, inorganic lead cannot.

Once in blood, lead is largely bound in red cells. With circulation, it is distributed to soft tissues and bone. Soft-tissue concentrations are highest in the liver and kidneys. Lead is largely excreted in the urine, and to a small extent, in the feces. Excretion is slow, with estimates of a half-life as long as 10 years. Lead also crosses the blood-brain barrier and the placenta.

There is relatively slow equilibration of lead among body compartments, so that levels of lead in blood may decrease after an acute exposure as lead is distributed to soft tissue and bone. After equilibration, more than 95% of total body lead is present in bone. Although bone may act acutely as a metabolic "sink," trapping lead that would otherwise be available to cause greater harm elsewhere, the gradual accumulation of lead in bone that occurs with chronic exposure may maintain soft-tissue concentrations years after occupational exposure ceases. Under circumstances that cause calcium mobilization, lead that has been stored for some time in bone may be remobilized and may elevate blood lead levels. The mobilization of stored lead with pregnancy, menopause, and aging is a cause of concern.

Lead is a potent central nervous system poison. Subtle CNS effects occur at BLLs previously thought to be safe. Severe episodes of intoxication may cause permanent cognitive and emotional changes. The earliest symptoms of intoxication are usually moodiness, irritability, change in sleep patterns, and difficulty with concentration. Patients typically develop headaches, a general decrease in well-being and energy, and a loss of libido. In many cases, these symptoms develop gradually, and patients and their family members tend to attribute them to an emotional or psychological cause rather than to a physical one. Medical attention may not be sought for some time, and if the patient is unable to alert the physician to the possibility of lead poisoning, diagnosis is often delayed. The insidious development and slow recognition of CNS manifestations of lead poisoning can cause great family and social disruption, perhaps the most devastating aspect of this illness.

Lead may cause a peripheral neuropathy, usually manifested by motor weakness and, in severe cases, by wrist drop or foot drop. Lead also predisposes to nerve entrapment syndromes, especially in workers who use their arms vigorously and repetitively.

Gastrointestinal symptoms can occur, including abdominal pain and constipation.

The renal effects of lead rarely cause symptoms, but are a major cause of long-term morbidity and mortality. Lead principally affects the renal tubular cells. Aminoaciduria may develop, and abnormalities of uric acid excretion may lead to episodes of gout. With severe poisoning, end-stage renal disease may occur.

Individuals with even modest elevations in BLL may have persistent renal dysfunction and hypertension. Mortality studies of battery workers show increased deaths from hypertensive cardiovascular disease, renal disease, and stroke.

In men, lead is associated with a reduction in sperm count and an increase in the number of abnormal sperm. In women, lead exposure has been associated with spontaneous abortion and stillbirth. Studies of the relationship between increased umbilical cord blood lead and delayed cognitive development suggest that lead can be a significant transgenerational toxin.

Mobilization of lead from maternal bone occurs during pregnancy. Even modest lead exposures can result in maternal bone lead burdens that can be mobilized and affect fetal development many years later, even if current exposures are low.

The impact of lead on heme synthesis is typically measured in blood, but heme is critical not only to hemoglobin but also to cytochromes and other enzyme systems in many other tissues. Measurement of the various porphyrin markers in blood serves as a marker for the biochemical disruptiveness of lead in other organs.

Lead causes a normochromic normocytic anemia, in which red cells may show characteristic basophilic stippling. Even when patients are not frankly anemic, their hemoglobin values are usually depressed and typically rise slightly during recovery. When patients are frankly anemic, it is often due, in part, to hemolysis related to lead effects on the red cell membrane.

It is likely that subpopulations of people who are genetically susceptible to greater toxicity, given any specific amount of lead exposure, will be identified. (The policy and ethical ramifications of this issue are complex.)

Prevention

Primary prevention can be achieved by substitution of other materials, by elimination of exposure through process changes, and by good work practices.

Substitution has led to elimination of tetraethyl lead as an anti-knock additive, elimination of much of the lead soldering in steel-can manufacture, use of epoxy paints in place of red lead paint on some exterior metal surfaces, use of fluidized silica beds to replace traditional lead baths in wire manufacture, and changes in automobile design to remove the hazard posed by solder grinding.

Where lead is still used, processes can be changed to reduce fumes dramatically or to reduce the concentration and increase the size of lead particulate. Removing lead paint from the interior of homes traditionally involves burning and sanding surfaces, methods that can leave a house more dangerous to inhabitants than before and can poison those doing the work. State regulations banning the use of powered sanders and open flames and requiring postabatement measurement of surface dust lead levels can dramatically reduce this risk. To comply with these regulations, contractors and property owners can replace woodwork, perform off-site stripping, and chemically remove paint on-site.

Other process changes can reduce the risks in painting exterior steel surfaces. Careful inspection and use of sanders with HEPA vacuum attachments on problem areas, or the use of wet methods, can reduce the enormous amount of leaded dust that is generated when surfaces painted with old lead paint or primer are prepared by dry abrasive blasting. Such methods can also reduce environmental pollution.

Jacketed ammunition and low-lead primers have been used to reduce the burden on ventilation systems at indoor pistol ranges.

When exposure cannot be eliminated by changes in process, enclosure and ventilation can be highly effective. To avoid turning an occupational exposure into a community exposure, filtration and extraction of lead fume and particulate from exhaust air may be necessary.

In industry, and even on construction and demolition sites, critical thinking about process and industrial hygiene can prove cost-effective by reducing the burden placed on personal protective equipment as a final line of defense. The use of wet methods of cleaning, HEPA vacuums, and particle accumulators or precipitators can reduce dustiness. Longer torches can enable demolition or scrapping workers to be farther back from the lead-containing plume.

Wherever dust is being generated or fume may be settling out, surfaces must be cleaned frequently. Providing facilities for frequent washing at work and forbidding eating, drinking, and smoking at the workplace can reduce the likelihood of lead ingestion. The use of protective clothing, coupled with a shower before leaving work, can eliminate the danger of lead particles being transported to workers' homes.

Good respiratory protection is a major part of primary prevention in settings where other methods have not been sufficient. Both workers directly involved in the generation of fume or dust and those working nearby need respiratory protection. Positive-pressure respirators are preferred; the OSHA lead standard provides that they be made available at the employee's request. Especially in construction and demolition work, where exposures can vary greatly from day to day and can be extremely high at times, negative-pressure respirators may be inadequate.

Because of the insidious nature of lead poisoning, worker education is critical to primary prevention. Worker training should include specific information about the danger of lead to workers and their families, symptoms of lead poisoning, importance of engineering controls and good work practices, and the provisions of the OSHA standard and other applicable regulations.

The OSHA occupational lead standard (CFR 1910.1025), promulgated in 1978, remains the cornerstone of regulatory prevention of occupational lead poisoning. It established a permissible exposure limit of 50 $\mu g/m^3$ of lead in air and an action level of 30 $\mu g/m^3$, and it prescribed that engineering controls be used whenever feasible to meet these goals. The standard requires removal of workers from exposure when BLLs exceed 60 $\mu g/dL$ (or average over 50 $\mu g/dL$). Medical evaluation is required whenever a worker has a BLL over 40 $\mu g/dL$, complains of possible symptoms, or expresses concern about adverse reproductive effects. The standard also directs affected workers' physicians to design suitable restrictions and remedies, relying on their medical judgment.

In interpreting surveillance data, physicians should be sensitive to early symptoms of lead intoxication and not rely on the numerical requirements of the OSHA lead standard. Although the idea of defining specific levels that call for immediate action by employers is sound, BLLs that mandate worker removal in the current OSHA standard represent outdated thinking about "safe" levels. BLLs well below the current level of concern of 40 $\mu g/dL$ are associated with toxicity, and some experts have advocated that the level of concern be revised downwards to, for example, 15 or 20 µg/dL. Beyond absolute values, a rising BLL is a signal that preventive measures have broken down and that the individual is at risk.

Although the standard was based on the now untenable idea that it was appropriate to allow working populations to endure an average BLL of 40 µg/dL, enforcement of the standard has achieved much better-than-anticipated results in many industries. The elaborate requirements for sampling, surveillance, employer-paid second medical opinions, protective equipment, and income maintenance while on medical removal have stimulated many employers to implement preventive measures to achieve exposures well below the action level.

In addition to OSHA regulation, some states and cities have adopted regulations linked to efforts to abate lead in housing. Some of these are more proscriptive and prescriptive than OSHA regulations and treat all facets of the problem, from identification of children at risk to disposal of lead-contaminated waste.

Construction and demolition work were exempted from the federal OSHA standard. In most of the U.S., construction workers have little regulatory protection from lead exposure. Twenty years ago, Maryland extended a somewhat modified version of the standard to construction, and has since demonstrated that such regulations can be effectively enforced in the construction industry. As another example, Massachusetts has adopted a law that establishes minimum work standards to protect the health and safety of inspectors, de-leaders, residential renovators, and the general public, and standards of competency for persons or entities engaged in or performing de-leading work.

While never a substitute for the primary preventive measures described above, careful medical surveillance can detect evidence of lead absorption and intoxication in time to identify gaps in primary prevention and prevent permanent damage. Health professionals conducting medical surveillance must be alert for the subtle symptoms of lead poisoning elicited by careful history taking. Equally important is repeated reinforcement of the health education the worker has received about lead intoxication. Without these two elements, monitoring of laboratory tests may have little lasting benefit.

Measurements of intermediates in the heme synthesis pathway can serve as simple screening tests and can improve interpretation of blood lead measurements. Protoporphyrin IX, bound to zinc in vivo, can be measured directly and instantaneously with a hematofluorometer. Because zinc protoporphyrin (ZPP) can also be elevated by iron deficiency, results must be interpreted carefully, especially in screening children. Iron deficiency is so infrequent in working men, however, that an elevated ZPP nearly always indicates lead intoxication. Protoporphyrin is now recognized as being fairly insensitive.

Cumulative lead exposure, reflected either by integrating BLLs over time (the cumulative blood lead index) or K-XRF (x-ray fluorescence)-measured bone lead levels, is now thought to be a much more useful measure than measures of acute exposure. These measures of cumulative lead exposure have been found to be more valid than BLLs in predicting many forms of chronic toxicity, such as hypertension, cognitive dysfunction, and adverse reproductive outcomes. (Bone lead level measurement, only available at a small number of health-care centers, is therefore not recommended for clinical practice.)

Measurement of BLL itself may be used to confirm an abnormal ZPP or to serve as the primary screening test for lead absorption. Careful choice of an experienced laboratory and good quality control are important. Determination of BLL can be difficult, and confidence in the laboratory being used is critical. Even with CDC efforts at laboratory certification, much variability and inaccuracy still plague commercial clinical labs. State health departments may provide the most reliable services or give advice about selecting a dependable lab.

Because there are many nonoccupational sources of lead poisoning and because many occupational cases occur in settings where the potential for lead exposure has been ignored, sporadic cases are likely to continue despite the best efforts at organized control. Public health workers have begun to address this problem in some states by educating primary medical care practitioners and by requiring clinical laboratories to report elevated lead levels. Surveillance systems allow health departments to respond to each identified case by providing information and support to the patient and physician and, when appropriate, by involving OSHA.

Whenever a case of lead poisoning occurs, the nature of the patient's exposure should be characterized, family members or coworkers who may have also been exposed should be identified, and interventions should be made to stop exposure to lead.

The principal therapy of occupational lead poisoning is removing the worker from exposure and investigating the source. In cases in which BLLs or body burdens are high, where patients are symptomatic, or there is evidence of end-organ damage, chelation therapy is often indicated. Chelation therapy should never be given prophylactically or administered to a patient when lead absorption may be continuing.

Further Reading

Chia KS, Jeyaratnam J, Lee J, et al. Lead-induced nephropathy: relationship between various biological exposure indices and early markers of nephrotoxicity. *American Journal of Industrial Medicine* 1995; 27: 883-895.

Hertz-Picciotto I, Croft J. Review of the relation between blood lead and blood pressure. *Epidemiologic Reviews* 1993; 15: 352-373.

Hertz-Picciotto R. The evidence that lead increases the risk for spontaneous abortion. *American Journal of Industrial Medicine* 2000; 38: 300-309.

Hu H, Watanabe H, Payton M, et al. The relationship between bone lead and hemo-
   globin. *JAMA* 1994; 272: 1512-1517.
Kaufman JD, Burt J, Silverstein B. Occupational lead poisoning: can it be eliminated?
   American Journal of Industrial Medicine 1994; 26: 703-712.
Loghman-Adham M. Renal effects of environmental and occupational lead exposure.
   Environmental Health Perspectives 1997; 105: 928-939.
Silbergeld EK. The international dimensions of lead exposure. International *Journal of
   Occupational and Environmental Health* 1995; 1: 336-348.
Vork KL, Hammond SK, Sparer J, et al. Prevention of lead poisoning in construction
   workers: a new public health approach. *American Journal of Industrial Medicine*
   2001; 39: 243-253.
Wedeen RP, Ty A, Udasin I, et al. Clinical application of in vivo tibial K-XRF for mon-
   itoring lead stores. *Archives of Environmental Health* 1995; 50: 355-361.
White RF, Diamond R, Proctor S, et al. Residual cognitive deficits 50 years after lead
   poisoning during childhood. *British Journal of Industrial Medicine* 1993; 50: 613-622.

Leukemia

ICD-10 C90.1, C91-C95
SHE/O

Howard Kipen

Leukemias are malignant neoplasms of the blood-forming system. They include myeloproliferative and lymphoproliferative disorders. The myeloproliferative disorders are clonal neoplasms arising in a pluripotent hematopoietic stem cell and are characterized either by excessive production of phenotypically normal mature cells (chronic myeloproliferative disorders) or by impaired or aberrant maturation of hematopoietic precursor cells (acute myeloproliferative and myelodysplastic disorders). They include chronic myelogenous leukemia (CML) and acute myelogenous leukemia (AML).

AML is a hematopoietic stem cell disorder resulting in lethal overgrowth of incompletely differentiated bone marrow precursor cells. Various subtypes are identified by predominant cell type (myelomonocytic, monoblastic, promyelocytic, and megakaryocytic leukemias, and erythroleukemia). CML is also a clonal stem cell disorder characterized clinically by overproduction of mildly defective granulocytes. Ninety percent of CML patients have the Philadelphia marker chromosome. In contrast to the myelogenous leukemias, acute and chronic lymphocytic leukemias (ALL and CLL) are lymphoproliferative disorders of committed lymphopoietic stems cells, originating in lymphoid precursors of bone marrow, thymus, and lymph nodes.

Symptoms of leukemia include fatigue, weakness, weight loss, repeated infections, enlarged lymph nodes, bruising, and bleeding. Diagnosis is based on symptoms, on signs such as an enlarged spleen and enlarged lymph nodes, and on laboratory tests, including complete blood counts and smears (which reveal abnormally low or high blood cell counts and abnormal cells) and examination of the bone marrow.

Occurrence

The most recent SEER data indicate that there are approximately 31 500 new cases of leukemia each year in the United States, divided about equally between acute and chronic cases. Most occur in adults. Leukemia is approximately 20% more common in men than women. Overall, leukemia accounts for 3% of cancer incidence in the United States. The incidence of acute leukemia has remained relatively steady in recent decades. Age-adjusted incidence is somewhat higher among whites than among blacks.

From 1995 to 1999, the overall U.S. death rate from leukemia was 7.8 per 100 000 (10.4 per 100 000 for males, and 7.8 per 100 000 for females). Death rates from acute leukemia rise sharply with age. Geographical variation of death rates from leukemia in the United States has been observed, with elevated death rates occurring in several central states (Nebraska, Minnesota, Oklahoma, Colorado, Montana, Kansas, and Iowa). Maine, Oregon, and California also have relatively high death rates from leukemia. Reliable rates of occurrence have not been calculated for occupational groups known to be at risk.

Causes

Occupational causes of leukemia include the following:

Based on numerous cohort studies, benzene is causally associated with AML (especially erythroleukemia), less strongly to CML, Hodgkin's and non-Hodgkin's lymphomas, ALL, CLL, and multiple myeloma. There is a dose-response relationship between cumulative benzene exposure and leukemia risk, although other dose aspects have not been completely explored.

Several retrospective cohort studies have indicated excess mortality from leukemia. Studies in several different countries have demonstrated excess leukemia in benzene-exposed workers in various industries, in workers exposed to ionizing radiation, in workers exposed to ethylene oxide, and in some workers exposed to electromagnetic radiation. In addition, studies of agricultural workers and residents of agricultural areas have indicated increased risk of leukemia, although the specific agent or combination of causal factors have not been identified consistently. Finally, health care workers who are exposed to anti-neoplastic drugs may be at increased risk for leukemia, although epidemiologic studies have not demonstrated this to date.

Nonoccupational causes of leukemia include alkylating agents, particularly melphalan, cyclophosphamide, chlorambucil, and busulfan. Hodgkin's disease patients receiving such anti-neoplastic drugs have a 5- to 10-fold increase in risk for AML, and multiple myeloma patients have a 100-fold increase in risk for this disorder. Chlorambucil has been associated with an excess incidence of acute non-lymphocytic leukemia (ANLL).

Chloramphenicol-induced aplastic anemia patients have subsequently developed leukemia. Excess incidence of ANLL has been reported among patients with ovarian cancer, multiple myeloma, and breast cancer following therapy with alkylating agents.

Individuals with Down syndrome are reported to have a higher risk of myelogenous leukemia. Genetic factors, which include the occurrence of chromosomal damage or unstable chromosomal patterns, are associated with ANLL.

Pathophysiology

All leukemias are thought to be clonal disorders and are often characterized by chromosomal translocations and other aberrations. The known leukemogens, ionizing radiation and benzene, are potent clastogens, and ethylene oxide has been shown to cause sister chromatid exchange in a dose-dependent manner in exposed workers, suggesting a possible unifying mechanism. Chemotherapeutic agents are also clastogenic. Induction of leukemia by radiation or chemotherapy peaks at about 5 to 10 years after exposure, compared with the 20- to 30-year latency for radiation-induced solid tumors.

Prevention

The absence of effective secondary or tertiary prevention makes reduction of exposures to radiation and benzene critically important. Exposure to ionizing radiation in the workplace, including health care facilities, should be minimized, as all exposures are believed to confer some degree of increased cancer risk, including leukemia. The reduction of the OSHA benzene exposure standard from 10 ppm to 1 ppm reflected concern for reducing increased leukemia risk from an estimated 95 deaths per 1 000 workers to fewer than 10 deaths for a lifetime working exposure, based on the risk assessment favored by OSHA. Screening populations or individuals who are exposed to leukemogens for hematologic abnormalities is not likely to indicate significant deviation from exposure standards or to provide clinically meaningful early detection of neoplasia.

Further Reading

Committee on the Possible Effects of Electromagnetic Fields on Biologic Systems, National Research Council. *Possible Health Effects of Exposure to Residential Electric and Magnetic Fields*. Washington, DC: National Academy Press, 1997.

Goldstein BD. Clinical hematotoxicity of benzene. In Mehlman (Ed.), *Carcinogenicity and Toxicity of Benzene*. Vol 4. Princeton, NJ: Princeton Scientific Publishing, 1983, pp. 51-61.

Hayes RB, Songnian Y, Dosemeci M, et al. Benzene and lymphohematopoietic malignancies in humans. *American Journal of Industrial Medicine* 2001; 40: 117-126.

Linet MS. *The Leukemias: Epidemiologic Aspects*. New York: Oxford University Press, 1985.

Linet MS, Cartwright RA. The leukemias. In Schottenfeld D, Fraumeni FR Jr., (Eds.). *Cancer Epidemiology and Prevention*. New York: Oxford University Press, 1996, pp. 841-892.

Low Back Pain Syndrome

ICD-10 M54.5

Laura Punnett

Low back pain (LBP) syndrome is a common presentation for many independent pathological conditions affecting the bones, tendons, nerves, ligaments, and intervertebral discs of the lumbar spine. It is categorized as (a) regional low back pain, with or without pain radiating into the legs (sciatica); or (b) back pain due to systemic disease, such as cancer, osteomyelitis, the spondylo-arthropathies, and Paget's disease. The second category is very uncommon in most workers but should be considered in people over age 60 and in workers with backache at rest, or with backache and fever regardless of age.

Regional LBP may result from a specific incident, injury, or fall; chronic or repetitive trauma or strain; or no definable precedent. Localized pain, decreased range of spinal motion, and often muscle spasm are the predominant features. Lumbar spine x-ray findings in non-traumatic LBP are usually normal or show minimal degenerative changes consistent with the patient's age and are of little diagnostic value.

Whether LBP is work-related or not, it is often difficult to identify the particular tissue or structure that has been damaged. Typically, the mechanism of disease cannot be determined for up to half of all cases. Thus, an individual's pain, limitation of movement, and disability in performing daily tasks at home or at work are often used as criteria to determine the existence and severity of a condition. Quality and duration of symptoms are the two most important classification criteria.

Sciatic-type LBP includes mild to severe pain with radiation down one or both legs. Normal walking is difficult, and mobility of the lumbar spine is limited. Pain is aggravated by jarring movements, such as riding in a car, coughing, or sudden changes in position. The straight-leg raise test and neurological signs are often used clinically to detect nerve root tension. CT scans, MRI, and electromyography may be useful in diagnosing sciatica but are not always conclusive and are not indicated unless back pain has continued for more than 6 weeks or if there are nerve root symptoms.

Occurrence

In the United States, Canada, Japan, and the Nordic countries, more people are disabled from work as a result of musculoskeletal disorders, especially back pain, than from any other group of diseases. Back pain is especially common, affecting as many as 80% of all adults during their lifetime. It is relatively infrequent before age 20 but may be increasing in this age group; first episodes most often occur before the age of 40. After age 65, the prevalence of LBP decreases.

Estimates of prevalence or incidence vary widely, reflecting differences in case definitions, age and gender distribution, and occupation and industry distribution. One-year period prevalence for the total U.S. working popula-tion is approximately 15%. Sciatica occurs in only about 1% of cases. Men and women are equally likely to be affected, but men are much more likely to receive workers' compensation. LBP is associated with lower socioeconomic status, possibly reflecting the higher physical demands these individuals face at work.

In the U.S., National Health Interview Survey data suggest that over 50% of all back pain in the working population is attributable to work. Occupationally related LBP cases with time lost from work number between 5 and 6 million annually. The Bureau of Labor Statistics estimates the annual incidence of work-related back injuries in the private sector to be about 5.0 cases per 1 000 person-years.

Back disorders are more often reported in occupations that require fre-quent heavy lifting, especially if combined with other stressors, such as awk-ward posture or exposure to whole-body vibration. Truck and tractor drivers, nurses and nurses' aides, construction laborers and carpenters, and janitors and cleaners have the highest rate of work-related back injuries. Truck driv-ers are thought to incur their injuries most often while loading or unloading trucks rather than while driving, although chronic exposure to whole-body vibration may be an important antecedent. Also at high risk are warehouse workers and other materials handlers, operators of cranes and other large vehicles, and other health-care workers.

In general, the prognosis for an isolated episode of acute regional low back pain is excellent. With conservative therapy (such as rest and exercise) alone, 80% of those affected will be better in 2 weeks, and 90% in 1 month. However, the rate of recurrence has been estimated to be about 35% to 40% within 3 years following an initial work-related episode and higher in physi-cally stressful jobs, such as nursing and trucking.

About 75% of workers who lose time from work because of LBP return to work within 1 month; however, 5% to 10% are disabled for more than 6 months, accounting for 70% of lost workdays and LBP workers' compensa-tion costs. Estimates of the annual direct cost of compensable LBP in the United States are as high as $16 billion, or one-third of all workers' compen-sation claims costs. About two-thirds of this amount results from lost work time.

Those with chronic LBP may suffer a variety of economic and social con-sequences, including lost work time and income, medical care utilization and expenses, and reduced household and recreational activities. High physical and psychological job demands and low supervisory support have been asso-ciated with lower return to work among compensated LBP cases. Among

Table 1. Summary of epidemiologic studies with risk estimates of null and positive associations of physical risk factors at work and the occurrence of back disorders

Risk Factor	Number of epidemiologic studies		
	No Associations found	Positive Associations found	Range of attributable fractions (in percent)
Manual material handling	4	24	11 - 66
Frequent bending and twisting	2	15	19 - 57
Heavy physical load	0	8	31 - 58
Static work posture	3	3	14 - 32
Repetitive movements	2	3	41
Whole body vibration	1	16	18 - 80

Source: National Research Council, 2001 (Table 4.2, p. 99)

people with LBP, pain and activity limitation are associated with low socioeconomic status, suggesting that appropriate job accommodation on return to work is particularly important for lower-status individuals. As many as 60% of injured workers have multiple episodes of work disability over an extended period.

Causes

The causes of LBP vary among jobs and may be either immediate or prolonged. Lifting objects or persons is the most cited factor; the risk increases both with load weight and with the reach distance to the load center of gravity. Other important occupational exposures are bending and twisting of the trunk and whole-body vibration (Table 1), particularly in prolonged driving of motor vehicles. Professional drivers also experience vehicle acceleration and deceleration, inability to alter body position, lack of back support, and static leg position. However, evidence is inconclusive regarding prolonged sitting. Acute-onset LBP also results from slips and falls associated with poor walking or working surfaces and related safety hazards.

The best predictor of future back pain is having suffered it in the past, which may simply represent continuing exposure to the original injurious agent or personal factors related to risk.

LBP also has been associated with psychosocial stressors, such as repetitive and dissatisfying work. While psychological factors affect reporting and recovery from LBP, the evidence for a primary etiologic role is much less clear.

Nonoccupational associations cited by some studies but contradicted by others include height, obesity or increased body weight, full-term pregnancy, depression, and cigarette smoking. The association with recreational activities is generally weak. One study found that the combined effects of seniority, age, gender, smoking, height, weight, and leisure time physical activity explained only 4% of the variability in workers' LBP occurrence.

Pathophysiology

The pathophysiology of regional LBP is thought to be due to injury or microtrauma to the lumbar spine or its supporting structures. Injury may arise from direct trauma, such as a fall; a single overexertion, such as lifting a particularly heavy or bulky object; or repetitive loading of soft tissues. The variety of conditions that can cause LBP include muscular or ligamentous strain, facet joint arthritis, spinal stenosis, segmental instability, inflammatory lesions, and disc pressure on the annulus fibrosis, vertebral end-plate, or nerve roots.

The pathophysiology of sciatic-type back pain is thought to be through either nerve root compression or other mechanisms of nerve root irritation. Nerve root compression can be due to a herniated disc or to arthritic changes in the spine in the region of the neural foramen, where the nerve root exits the spinal canal. Herniation of the intervertebral discs is diagnosed in only a small proportion of LBP cases. Irritative causes of sciatic-type back pain include arachnoiditis, active rheumatoid arthritis, and the so-called facet syndrome. The most common type of back pain occurs with nonspecific symptoms. The pathophysiology of LBP in a given individual usually cannot be determined.

Prevention

Primary prevention measures should concentrate on the ergonomic design of workstations, equipment, tools, and work organization relative to the size and strength capabilities and tissue tolerances of the human body. Engineering controls involve designing the job to reduce known ergonomic stressors associated with increased risk. In high-risk jobs, often two or more of these factors are present and synergistically exert adverse effects. To be effective, control measures usually need to address all existing exposures. Examples of effective ergonomic design to reduce LBP include:

1) Mechanical aids. Effective mechanical lifting devices will reduce or eliminate the load on the supportive structures of the lumbar spine during handling of heavy loads, important for health-care workers lifting patients and materials handlers. Lifting tasks should be free from any obstruction that forces a worker to lift over or around it or to jerk the load free. Excessive loads can often be reduced by purchasing supplies in packages of appropriate size and weight.

2) Good workstation design. This reduces unnecessary bending, twisting, and reaching by bringing the work within the standard reach envelope and directly in front of the worker. Work locations that are below waist height, farther forward than arm's length, or behind the midline of the body are undesirable for regular activities. Chairs should have lower-back support and adequate seat padding; furniture should be adjustable in height and other relevant dimensions so as to conform to the range of body sizes of workers. Prolonged sitting and standing can be modified by providing flexible workstations that allow workers to perform their tasks in a variety of positions.

3) For drivers, vehicle design. Good design includes improved vehicle suspension, padding and mechanical isolation of the vehicle seat from the vibrating components, cab design for good visibility and accessibility of controls, and improved loading and unloading facilities at truck terminals.

4) Administrative controls. These include reducing the duration or frequency of manual handling tasks by providing longer or more frequent rest breaks, permitting operators to set their own work pace, and including non-handling tasks in the job description to provide more variety of motion patterns. Similar principles apply to reducing the duration of exposure to vibration.

The NIOSH Lifting Equation summarizes the research on the human body's tolerance of lifting demands and incorporates different lifting conditions that interact to determine acceptable load weight.

In the unlikely event that a job cannot be redesigned to reduce the hazardous exposures, job placement strategies may be considered. These may be appropriate for certain high-risk jobs that are difficult to design and control, such as firefighting and police work. Selection techniques can be grouped into medical examination, strength and fitness testing, and job-rating programs.

There is no good evidence to support utilizing preplacement medical examinations to identify workers susceptible to back pain or to prevent future back pain. Pre-employment low back x-rays are ineffective as a preventive tool and potentially harmful to workers due to unnecessary x-ray exposure. Although some studies have shown a relationship between strength, flexibility, and/or fitness and the incidence of low back pain, prospective data are few. Evidence is weak regarding the effectiveness of preplacement use of strength or fitness screening criteria to reduce the incidence of future LBP at work, except perhaps for very high-risk jobs. In addition, for high-risk jobs, it may be useful to screen for a history of previous low back conditions, since a worker with prior LBP is more likely to experience a recurrence. In otherwise asymptomatic people, available methods for screening workers for LBP have serious technical, ethical, and legal limitations and should not be used in lieu of ergonomic controls as a way of preventing LBP.

Job-rating programs are structured attempts to evaluate and create a good match between the job and the worker. Rather than screening out applicants, job-rating programs identify appropriate work for people with varying abilities. These programs appear promising in principle, but no studies have demonstrated the success of any job-rating program in modifying the incidence or severity of work-related LBP syndrome.

Training workers to perform job tasks safely and effectively is the oldest and most commonly used approach to preventing back problems in industry. Equally important is educating management, unions, and clinicians who deal with back pain in the workplace. Worker training programs have concentrated on safe lifting and on strength and fitness. A supervised training program at work to improve strength and fitness was shown to reduce the subsequent occurrence of back pain in those who had recovered from prior acute LBP, although effectiveness as primary prevention has not been documented. Similarly, back schools, as coordinated ways of retraining and rehabilitating injured workers, may facilitate earlier return to work, but there have been few controlled studies to document their effectiveness for the asymptomatic worker.

Tertiary measures, which overlap secondary preventive measures, include medical treatment and rehabilitation, work hardening, and, most important, job modification. Job modifications for the impaired worker include changing the tools, position, or manner in which a worker does a job, thus allowing the worker to accomplish the essential tasks despite physical impairment. Modified duty ("light duty") is a related concept in which the worker is assigned to a new job with requirements that will not cause aggravation of pain or progression of disease. In the ideal situation, the job modifications will be individualized based on the degree and nature of the individual's impairment. For the worker with LBP, the modifications usually include decreasing the frequency and degree of spinal flexion, decreasing weight loads lifted by the worker, and providing for flexible sitting or standing postures. The same ergonomic changes that have been associated with decreased incidence of LBP are effective in enabling workers with mild LBP to continue to work.

Modified duty may not be available in many workplaces. In these situations, workers and their physicians may be told that a worker needs to be "100% better" before being allowed to return to work. The purpose of work-hardening programs is to allow such workers an opportunity to return gradually to full working capacity in a supervised setting that is similar to the workplace but provides for modifications and worker autonomy while the worker completes the healing process. Recent studies that integrate work-hardening techniques with intensive physical and psychological rehabilitation programs have shown promising results in patients with chronic disabling back pain.

Surgical operations are available for herniated spinal discs. However, surgery, cortisone injections, and use of other drugs are last resorts and should be avoided if at all possible. Surgery may be only temporarily effective if the worker is returned to an ergonomically stressful job that has not been modified; possible loss of strength and flexibility and increased vulnerability to mechanical insult following surgery make it essential that job assignments be selected carefully to avoid recurrence. Use of transitional workshops for affected workers and graded retraining under the supervision of an experienced physical therapist may also reduce the risk of recurrence.

Other Issues

One of the most difficult problems in placing a worker with a history of LBP is determining a safe job assignment that will protect the worker from serious future injury without limiting his or her work and economic opportunity. If a worker has a history of recurrent, severe, and disabling back pain in relation to tasks and requirements similar or identical to those anticipated in the job placement in question, modifications of the job or restrictions from known exacerbating factors are clearly indicated. Absolute refusal of job opportunities based on the history, routine physical exam, or x-ray is rarely justified, given the prevalence of LBP in the general population and lack of evidence that such restrictions are effective in preventing future episodes of back pain.

Further Reading

Bernard BP (Ed.). *Musculoskeletal Disorders and Workplace Factors: A Critical Review of Epidemiologic Evidence for Work-Related Musculoskeletal Disorders of the Neck, Upper Extremity, and Low Back.* Washington, DC: NIOSH, 1997.

Cailliet R. *Low Back Pain Syndrome* (5th ed.). Philadelphia: F.A. Davis, 1995.

Guo H-R, Tanaka S, Cameron LL, et al. Back pain among workers in the United States: National estimates and workers at high risk. *American Journal of Industrial Medicine* 1995; 28: 591-602.

Hagberg M, Silverstein BA, Wells RP, et al. (Eds.). *Work-related Musculoskeletal Disorders (WMSD): A Handbook for Prevention.* London: Taylor and Francis, 1995.

Himmelstein JH, Andersson G. Low back pain: Fitness and risk evaluations. *State of the Art Reviews in Occupational Medicine* 1988; 3: 255-269.

Jayson M. Back pain and work. In: Baxter PJ, Adams PH, Aw T-C, et al. (Eds.) *Hunter's Diseases of Occupations* (9th ed.). New York: Oxford University Press, 2000.

Krause N, Dasinger LK, Deegan LJ, et al. Psychosocial job factors and return-to-work after compensated low back injury: A disability phase-specific analysis. *American Journal of Industrial Medicine* 2001; 40: 374-392.

National Institute for Occupational Safety and Health. *Low Back Atlas of Standardized Tests/Measures.* Morgantown, WV: NIOSH, 1988.

National Research Council and the Institute of Medicine. *Musculoskeletal Disorders and the Workplace: Low Back and Upper Extremities.* Washington, DC: National Academy Press, 2001.

Pransky G, Benjamin K, Hill-Fotouhi C, et al. Work-related outcomes in occupational low back pain: A multidimensional analysis. *Spine* 2002; 27: 864-870.

Waters TR, Baron SL, Piacitelli L, et al. Evaluation of the revised NIOSH lifting equation. A cross-sectional epidemiologic study. *Spine* 1999; 24: 386-394.

Lung Cancer

ICD-10 C34

Barry S. Levy

Occupational exposures can cause all types of lung cancer. Approximately 10 percent of lung cancer cases in men and approximately 2 percent of lung cancer cases in women the United States are, at least in part, due to workplace exposures. Arsenic, asbestos, bis-chloromethyl ether (BCME), beryllium, cadmium, hexavalent chromium (chromium VI), coke oven and coal gasification fumes, ionizing radiation (including radon), nickel, silica, and soot are among the IARC Group 1 (definite) carcinogens that cause lung cancer, and acrylonitrile and diesel exhaust are among the IARC Group 2A (probable) carcinogens that cause lung cancer. In about half of patients with lung cancer, the initial symptoms are respiratory, such as cough, sputum production, and chest discomfort; however, many lung cancer patients have coincident pulmonary disease (chronic bronchitis or emphysema), which makes it difficult to identify the new symptoms of cancer.

Occurrence

Lung cancer incidence and mortality have been increasing for the past 50 years in the United States, first among males and, more recently, among females. Annually, there are more than 175,000 cases of, and more than 160,000 deaths from, lung cancer in the United States. In both men and women, it is the second most frequently occurring malignancy and the most frequent cause of cancer deaths. Slightly more than half of these deaths occur in men; however, at the current rate of increase, women are expected to have equivalent death rates to those of men within the next several years. It has been difficult to determine exactly how much lung cancer is directly attributable to occupation because lung cancer is strongly associated with cigarette smoking and some cases of lung cancer are due to a combination of occupational exposures and smoking. In addition, the latency period between initial exposure and recognition of lung cancer has a very wide range (averaging approximately 20 years), which makes identification of causal associations difficult.

Causes

The best-recognized cause of lung cancer is cigarette smoking. In addition, there is increasing evidence that exposure to ambient air pollution and to environmental tobacco smoke in indoor air are causally associated with lung cancer. Research findings suggest a genetic predisposition to lung cancer in individuals with altered ability to metabolize hydrocarbons (with inducible aryl hydrocarbon hydroxylase). Some of the causes of lung cancer that are known or suspected to be due to occupation are listed below, accompanied by a brief review of the current understanding of the association between each agent or environment and lung cancer.

Arsenic: The best-described association between occupational arsenic exposure and lung cancer has been documented among smelter workers in Japan, Sweden, and the United States; whether particulates and sulfur dioxide in smelters may play a role in causation is under investigation. Excess risk of lung cancer among arsenical pesticide manufacturers and applicators has also been documented. (See Arsenic, Adverse Effects.)

Asbestos: One of the earliest and best-described workplace agents associated with lung cancer is asbestos, which causes increased risk of lung cancer regardless of fiber type. (See Asbestos-Related Diseases.) In a review of several studies, elevated occurrence of lung cancer was found to be highest among insulators and textile workers, intermediate for asbestos product manufacturers, and lowest among asbestos miners. With high exposure, even exposures of a few months' duration have been shown to increase lung cancer risk. Of particular importance is the well-documented synergistic (multiplicative) effect of asbestos and cigarette smoking in causing lung cancer. In some studies, asbestos workers who smoke cigarettes may have a 60-fold increased risk of lung cancer compared with workers who are not exposed to asbestos and do not smoke. Asbestos workers who do not smoke are also at increased risk of lung cancer.

BCME: The alkylating agent BCME, a known carcinogen, is a frequent contaminant of chloromethyl ether (CME). As such, BCME is strongly associated with lung cancer in workers who encounter it as a contaminant of CME. Small cell carcinoma, in particular, occurs more frequently among BCME-exposed workers than in the general population, but the occurrence of other cell types of lung cancer have been reported among BCME-exposed workers.

Beryllium: Beryllium and several beryllium compounds cause lung cancer in experimental animals. Recent cohort mortality studies show elevated lung cancer rates in workers exposed to beryllium that were unlikely to be accounted for by smoking.

Chromium: Hexavalent chromium salts are human carcinogens. Chromium salts have been associated with lung cancer in several different types of employment. Causal associations have been found in chromate production workers and in workers using chromate pigments. In addition, there is suggestive evidence of excess risk of lung cancer among chrome platers.

Coke Oven Emissions: An excess of lung cancer has been documented among coke oven workers, especially those who were exposed on the top of coke ovens. Environmental controls have dramatically reduced exposure to polyaromatic hydrocarbons (PAHs), the suspected carcinogens.

Diesel Exhaust: Diesel exhaust is a complex mixture of substances, with polycyclic aromatic hydrocarbons surrounding an elemental carbon core. The gas phase includes oxides of nitrogen and carbon monoxide. Diesel exhaust is a lung carcinogen in animals. Two meta-analyses of human epidemiologic studies of worker populations have found relative risks for lung cancer of 1.47 and 1.33.

Ionizing Radiation: Workers exposed to ionizing radiation, particularly alpha-emitting radioactive isotopes in the form of radon progeny, have an increased risk of lung cancer. This was first documented among uranium miners, but now is also suspected to occur in hematite (iron-ore) and other metal-ore miners. There is an apparent excess of small-cell lung cancer among these workers and, among uranium miners, a multiplicative association with cigarette smoking. Excess risk has also been demonstrated in nonsmokers.

Nickel: A substantial excess risk of lung cancer has been documented among nickel production workers. The risk was apparently associated in the past with early stages of the refinery processes; is likely to be expected to be controlled in modern production operations.

Silica: Various work environments with silica exposures have been studied, including mines, quarries, granite industry, foundries, ceramics, and stone work. There is a clear excess of lung cancer in workers exposed to silica. In one analysis of 15 cohort and case-control studies, the relative risk was 2.80 (95% CI, 2.50-3.15). It appears that (a) relatively high doses of silica that are necessary to induce silicosis cause lung cancer, and/or (b) some aspect of the silicosis disease process increases the risk of lung cancer.

Vinyl Chloride: Vinyl chloride monomer (VCM) is thought to be a lung carcinogen, in addition to its other cancer risks. The excess risk appears to be limited to environments where exposures have been substantial, often where polyvinyl chloride (PVC) polymerization has been done, rather than where VCM has been manufactured or where PVC has been put to final use.

A number of other industrial exposures have been associated with lung cancer excess in working populations. The studies in support of these associations have not been frequently or widely replicated. These materials should be considered as suspect lung carcinogens. Illustrative examples are the following:

Acrylonitrile: Acrylonitrile has been studied experimentally and found to be mutagenic and an animal carcinogen. Several human epidemiology studies have shown lung cancer excess, but only in the aggregate is the evidence strong enough to be convincing.

Aluminum Industry: Four mortality studies of aluminum workers, when combined, suggest that there is an excess of lung cancer among primary aluminum industry workers. A specific agent has not been identified.

Foundry Workers: Epidemiological studies consistently suggest that foundry workers are at a greater risk of dying from lung cancer than the general population and that this risk varies with job performed, calendar time employed, duration of exposure, and type of foundry. Current studies should assist in refining the location and source of excess risk.

Oil Mists (Cutting or Machining Fluids): Studies of oil mist exposures in various settings have determined these agents to be carcinogenic. Whether there is a specific risk for lung cancer, however, is still under study. When the major studies of machining operators and metal industry workers are taken together, the evidence supports such an association. Because industrial processes require a variety of machining fluids (straight mineral oil, semi-synthetic oils, and synthetic oils) and the composition of any of these fluids varies from lot to lot and over time, it has been difficult to specify the agent(s) that might be causative. Cigarette smoking may also be confounding the understanding of this risk.

Printing Industry: An excess of lung cancer has been shown repeatedly in studies of this industry. The excess risk is not high, and agents have not been identified. It is possible that oil mists, common in the industry, are a source of the risk.

Synthetic Vitreous Fibers (SVFs): Suspicion of lung cancer risk as a result of SVF exposure results from evidence that the physical dimensions of asbestos fibers are key to the mechanism of their carcinogenicity. Studies of SVFs, however, have been confounded by a continuing development of smaller and thinner fibers and by many employees having previous exposure to asbestos. IARC has aggregated studies from Europe and the evidence, when added to that found in U.S. studies, suggests that some fibers are associated with excess lung cancer risk. (See Synthetic Vitreous Fiber-Related Disease.)

Transportation Workers: Case-control and cohort studies of truck drivers and other transportation workers have, with reasonable consistency, shown an excess risk of lung cancer. Because the work requires prolonged driving time, it is difficult to separate cigarette smoking from exhaust emissions as agents that might explain the risk. When controlled for smoking, recent studies have suggested an increased risk from exposure to diesel particulates.

Pathophysiology

There are three stages in the natural history of lung cancer. The development stage occurs with exposure to a carcinogen. Basal cell hyperplasia, stratification, and squamous metaplasia develop in areas of high exposure. As the tumor enlarges, the production of signs and symptoms signals the second stage. Tumor spread to contiguous areas marks the end of this stage, and the third stage occurs coincident with metastases. Tumor growth rates vary by cell type. Small-cell cancers grow most rapidly, and adenocarcinoma grows least rapidly. Residence time between estimated onset and manifestation of symptoms is estimated to be 2 to 3 years for small cell carcinoma, 7 to 8 years for squamous cell (epidermoid) carcinoma, and 15 to 17 years for adenocarcinoma.

Prevention

Lung cancer has a high case-fatality rate and low 5-year survival. CT scans of the chest and sputum cytology exams may detect cases at an early treatable stage. Studies of the impact of these medical surveillance tests are under way to determine if these tests can, in addition to detecting some cases of lung cancer at an early stage, reduce lung cancer mortality among those at increased risk. (See last section of Further Reading.)

The most important approach to prevention, however, is through primary prevention, including reducing or eliminating exposures to occupational lung carcinogens and cessation of cigarette smoking. Further study of the interaction between cigarette smoking and occupational exposures, along with a better understanding of genetic, nutritional, and immunologic factors, is needed before host-factor susceptibility is likely to be identified or alternative therapies developed.

Other Issues

There are no means to distinguish occupational lung cancer from lung cancer due to other causes. Cigarette smoking interacts with asbestos and radon progeny to multiply the risk of lung cancer in exposed workers. Similar effects have been inconsistently reported for workers exposed to arsenic and chloromethyl ethers.

Further Reading

General

Axelson O. Alternative for estimating the burden of lung cancer from occupational exposures-Some calculations based on data from Swedish men. *Scandinavian Journal of Work Environment and Health* 2002; 28: 58-63.

Feinstein MB, Bach PB. Epidemiology of lung cancer. *Chest Surgery Clinics of North America* 2000; 10: 653-661.

Ruano-Ravina A, Figueiras A, Barreiro-Carracedo MA, et al. Occupation and smoking as risk factors for lung cancer: A population-based case-control study. *American Journal of Industrial Medicine* 2003; 43: 149-155.

Steenland K, Loomis D, Shy C, et al. Review of occupational lung carcinogens. *American Journal of Industrial Medicine* 1996; 29: 474-490.

Taioli E, Gaspari L, Benhamou S, et al. Polymorphisms in CYP1A1, GSTM1, GSTT1 and lung cancer below the age of 45 years. *International Journal of Epidemiology* 2003; 32: 60-63.

Williams MD, Sandler AB. The epidemiology of lung cancer. *Cancer Treatment Research* 2001; 105:31-52.

Specific Causes

Bhataia R, Lopipero P, Smith AH. Diesel exhaust exposure and lung cancer. *Epidemiology* 1998; 9: 84-91.

Ding M, Shi X, Castranova V, et al. Predisposing factors in occupational lung cancer: Inorganic minerals and chromium. *Journal of Environmental Pathology, Toxicology and Oncology* 2000; 19: 129-138.

Ferreccio C, González C, Milosavjlevic V, et al. Lung cancer and arsenic concentrations in drinking water in Chile. *Epidemiology* 2000; 11: 673-679.

Goodman M, Morgan RW, Ray R, et al. Cancer in asbestos-exposed occupational cohorts: a meta-analysis. *Cancer Causes and Control* 1999; 10: 453-465.

Greenberg M. A study of lung cancer mortality in asbestos workers: Doll, 1955. *American Journal of Industrial Medicine* 1999; 36: 331-347.

Lee PN. Relation between exposure to asbestos and smoking jointly and the risk of lung cancer. *Occupational and Environmental Medicine* 2001; 58: 145-153.

Lipsett M, Campleman S. Occupational exposure to diesel exhaust and lung cancer: a meta-analysis. *American Journal of Public Health* 1999; 89: 1009-1017.

Nurminen MM, Jaakkola MS. Mortality from occupational exposure to environmental tobacco smoke in Finland. *Journal of Occupational and Environmental Medicine* 2001; 43: 687-693.

Pohlabeln H, Wild P, Schill W, et al. Asbestos fibreyears and lung cancer: a two phase case-control study with expert exposure assessment. *Occupational and Environmental Medicine* 2002; 59: 410-414.

Soutar CA, Robertson A, Miller BG, et al. Epidemiological evidence on the carcinogenicity of silica: Factors in scientific judgment. *Annals of Occupational Hygiene* 2000; 44: 3-14.

Viren J, Silvers A. Nonlinearity in the lung cancer dose-response for airborne arsenic: apparent confounding by year of hire in evaluating lung cancer risks from arsenic exposure in Tacoma smelter workers. *Regulatory Toxicology and Pharmacology* 1999; 30: 117-129.

Wells AJ. Lung cancer from passive smoking at work. *American Journal of Public Health* 1998; 88: 1025-1029.

Lung Cancer Screening

Henschke CI. Early lung cancer action project: overall design and findings from baseline screening. *Cancer* 2000; 89: 2474-2482.

Lam S, Lam B, Petty TL. Early detection for lung cancer: New tools for casefinding. *Canadian Family Physician* 2001; 47: 537-544.

Melamed MR. Lung cancer screening results in the National Cancer Institute New York Study. *Cancer* 2000; 89: 2356-2362.

Parkin DM, Moss SM. Lung cancer screening: improved survival but no reduction in deaths-The role of "overdiagnosis." *Cancer* 2000; 89: 2369-2376.

Strauss GM, Dominioni L. Perception, paradox, paradigm: Alice in the Wonderland of lung cancer prevention and early detection. *Cancer* 2000; 89: 2422-2431.

Lyme Disease

ICD-10 A69.2

Linden Hu and Steven Hatch

Lyme disease is a multisystem disease caused by the spirochete *Borrelia burgdorferi sensu latu*. It is transmitted by the bite of Ixodes ticks. Humans are incidental hosts who become infected when bitten by a tick carrying the bacteria.

Although Lyme disease can have protean manifestations, the vast majority of cases have very characteristic findings. Lyme disease can be divided into early and late stage disease. The most characteristic manifestation of early Lyme disease is the erythema migrans (EM) rash, which typically begins at the site of the tick bite. Although classically described as a "bull's eye" lesion, the rash can appear homogeneously erythematous, vesicular or maculo-papular. The rash is generally not painful or pruritic. Over days to weeks, the spirochetes will disseminate widely from the original EM lesion. At this stage, patients often have systemic symptoms, including fever, myalgias and arthralgias, and headache. More specific findings during this early, disseminated phase include cardiac and neurologic changes. Cardiac findings include atrio-ventricular conduction delays, myocarditis/endocarditis, and, rarely, congestive heart failure. Common neurologic findings are radiculitis, cranial nerve palsies (especially facial palsy), and meningitis. Treatment during the early phase of the disease speeds recovery for many of these symptoms and leads to an excellent prognosis.

Left untreated, all of the manifestations of early Lyme disease will resolve spontaneously over weeks to months. However, late manifestations of the disease will occur in 60% to 80% of patients. The most common late-stage manifestation of Lyme disease in the United States is oligoarticular arthritis. Other late-stage manifestations include continued peripheral nervous system involvement (radiculitis or mono-neuritis multiplex) and a subtle encephalopathy that predominantly affects mood and memory. Two additional skin abnormalities are seen almost exclusively in European patients: (a) lymphocytomas, which are collections of lymphocytes that appear as bluish masses, most commonly on ear lobes or nipples; and (b) acrodermatitis atrophicans, which is a chronic violaceous rash that appears late in Lyme disease, often at the site of the original tick bite.

The diagnosis of Lyme disease is typically made on clinical grounds in patients living in endemic areas. Serologic tests are frequently used in patients with intermediate risk or nonclassical clinical findings. To reduce the incidence of false-positive testing, both the CDC and the American College of

Physicians have advocated a two-step screening process utilizing (a) a less specific enzyme-linked immunosorbent assay (ELISA) or indirect fluorescent antibody (IFA) test, which, if positive, is followed by (b) a more-specific confirmatory western blot test. All three tests may be performed for IgM and/or IgG antibody responses. All of these tests, especially tests for IgM antibody, have been plagued by lack of appropriate sensitivity and specificity and poor intra- and inter-laboratory reliability. IgM testing, in particular, should be reserved for patients with acute (<1 month) duration of disease. A newly FDA-approved test, the C6 peptide antibody test for IgG antibody to a single borrelial peptide, may represent a substantial improvement in Lyme disease testing. Early results have shown that IgG antibody responses to C6 develop as early as IgM antibody in traditional tests and that sensitivity and specificity are as good or better than traditional two-step testing. Testing in late stage Lyme disease with either a C6 antibody test or a traditional two-tiered approach results in excellent sensitivity (>90%) and specificity (95% to 100%). As expected, sensitivity of both the C6 antibody test and two-tiered testing is substantially lower in early disease (30-50% at 1 week). The most appropriate use of the C6 antibody test and other newer generation, peptide-based antibody tests remains to be determined. Polymerase chain reaction (PCR) testing has been used to detect *B. burgdorferi* DNA from skin, blood, CSF, and synovial fluid samples, but remains mostly investigational.

Occurrence

Lyme disease is the most common vector-borne disease in the United States, with approximately 18,000 cases reported in 2000. Because reporting of Lyme disease is passive, it has been estimated that there are 5 to 10 unreported cases for each reported case of the disease. The incidence of reported cases of Lyme disease in the U.S. doubled from 1991 to 2000. The proportion of cases that is occupationally related is not known.

Lyme disease is endemic in both Northern Europe and the United States. In the U.S., there have been reports of Lyme disease from 44 states. However, the vast majority of cases occur in two primary areas of disease-the Northeast (from Maine to Maryland) and the Upper Midwest (Wisconsin and Minnesota).

Two other diseases, babesiosis and human granulocytic ehrlichiosis (HGE), can be co-transmitted with *B. burgdorferi* during feeding of an Ixodes tick. The incidence of co-infection varies by geographic region, but has been reported to be as high as 15%. There does not appear to be an increase in the incidence or severity of late-stage manifestations of Lyme disease in patients who are infected simultaneously with more than one of these organisms.

Common sense would predict that outdoor workers who are exposed to Ixodes ticks in regions endemic for Lyme disease would be at increased risk for contracting the disease. There are multiple case reports of Lyme disease in workers in various outdoor occupations, such as forest workers, railway and road clearing or maintenance crews, construction workers in recently defor-ested areas, farmers, land surveyors, utility workers, and park and wildlife management personnel. However, despite the increased exposure in outdoor workers, there have been few studies that have quantitated the amount of increased risk. Studies have shown wide variation in the relative risk associ-ated with outdoor work, which is not unexpected, given the variations in geo-graphic prevalence of the disease and the difference in risk associated with different professions. Most studies have shown an increase in seroprevalence (indicating exposure to *B. burgdorferi*) in outdoor workers compared with matched groups of non-outdoor workers. However, most studies have found either only small or no increases in cases of symptomatic Lyme disease.

Causes

In both Europe and the U.S., Ixodes ticks serve as the vector for trans-mission. Small rodents, particularly the white-footed mouse, are the wild reservoir of the disease. Transmission of the disease is closely tied to the life cycle of the Ixodes tick. Ticks feed once during each stage of their life cycle from larvae to nymph to adult. Because there is little to no vertical transmis-sion, ticks typically become infected during the larval stage while feeding on infected animals. Infected ticks can transmit the disease during their subse-quent feedings. Nymphal ticks appear to be responsible for the majority of human transmission. Because larvae feed in late summer, nymphs feed in spring to early summer and adults feed in late fall to early winter, most human cases occur in late spring and summer.

All borrelia that cause Lyme disease are from the genus *Borrelia burgdor-feri sensu latu*. Different genomic subgroups of *B. burgdorferi sensu latu* pre-dominate in different geographic regions. All Lyme disease in the U.S. is caused by *B. burgdorferi sensu stricto*. In Europe, three separate strains have been found: *B. afzelii*, *B. garinii*, and *B. burgdorferi sensu stricto*. Differences between the strains are likely responsible for subtle differences in manifesta-tions between European and U.S. disease.

Infection occurs by inoculation of *B. burgdorferi* directly into the skin by an Ixodes tick. Because *B. burgdorferi* remains dormant in very low numbers in the midgut of an infected tick between blood meals, transmission of *B. burgdorferi* to a new host does not occur until the tick has been attached for at least 24 hours. Although *B. burgdorferi* can be found in the blood and urine of infected mice, there are no documented cases of transmission of Lyme disease following contact with infected animals or their secretions.

Pathophysiology

The vast majority of symptoms occurring in both early and late Lyme dis-ease are directly attributable to the effects of the host immune response to the presence of *B. burgdorferi*. Eradication of the organism results in resolution of most of the manifestations of disease.

Multiple antibiotic regimens are effective in the treatment of Lyme disease. Early disease is generally treated with oral antibiotics. In patients without contraindications, doxycyline is the treatment of choice due to its good oral absorption and CNS penetration. In addition, doxycyline is active against *Anaplasma phagocytophila*, which causes human granulocytic ehrlichiosis and can be co-transmitted with *B. burgdorferi*. Other oral options include amoxicillin, cefuroxime, and newer generation macrolides. CNS disease, high degree A-V block, and arthritis unresponsive to oral antibiotics are treated with intravenous ceftriaxone, cefotaxime, or penicillin.

A small percentage of patients with late-stage arthritis do not respond to repeated courses of antibiotic therapy. It is thought that these patients may have an autoimmune syndrome triggered by infection with *B. burgdorferi*. Arthritis in these patients typically responds to immunosuppressive medications such as methotrexate. Tumor necrosis factor-α antagonists also appear to be effective anecdotally.

In the U.S., a minority of treated patients may continue to have a syndrome of persistent fatigue as well as mood and memory disturbance indistinguishable from fibromyalgia. Although it remains somewhat controversial, it appears that most of these patients do not have detectable persistence of the bacteria and do not respond to long courses of antibiotic therapy. It is thought that this syndrome may represent a post-infectious syndrome following Lyme disease. This syndrome is rare among European patients.

Prevention

For workers in tick-infested forests, contact avoidance constitutes the primary strategy for risk reduction. Key measures include the clearing of leaves, tall grass, and brush from work areas; wearing hats, high boots, and long-sleeved shirts; tucking trouser legs into socks; and wearing light-colored clothing in order to make ticks more easily visible. Because Ixodes ticks cannot transmit Lyme disease before 24 hours of attachment, "tick checks" at the end of work shifts are an extremely effective measure for prevention. Ticks should be removed with tweezers by applying gentle, tugging pressure without crushing the tick. After several minutes, the tick's mouth parts fatigue and the tick releases intact from the skin.

Insecticides containing either DEET or IR3535 are effective tick repellants. Permethrin, a compound toxic to ticks, can be sprayed on clothing but should not be applied directly to skin.

A vaccine for Lyme disease, LYMErix, was approved for human use in patients with high risk of exposure to Lyme disease. Vaccine efficacy was approximately 85%. Unfortunately, this vaccine was withdrawn from the market in 2002 due to poor sales. No other vaccine is currently available.

For patients in an endemic area who find an Ixodes tick that has been attached for more than 24 hours, antibiotic prophylaxis can be considered. A single dose of doxycycline (200 mg) has been shown to be 87% effective in preventing the development of Lyme disease. Although antibiotic prophylaxis is commonly practiced in many areas endemic for Lyme disease, cost-benefit analyses have generally not supported its use and careful observation for symptoms after a tick bite is an appropriate alternative.

Further Reading

Steere AC. Lyme disease. *New England Journal of Medicine* 2001; 345: 115-125.
Piacentino JD, Schwartz BS. Occupational risk of Lyme disease: An epidemiological review. *Occupational and Environmental Medicine* 2002; 59: 75-84.

Lymphoma

ICD-10 C81-C85

Richard W. Clapp

Typically lymphoma presents with enlarged lymph nodes that are not painful and can be palpated by the patient or health-care provider. Patients may have no other symptoms or may have fever, night sweats, or weight loss. The two major types of lymphoma, Hodgkin's disease (or Hodgkin's lymphoma) and non-Hodgkin's lymphoma, may present somewhat differently, although the importance of distinguishing between the two types is becoming less evident. Hodgkin's disease is typically more localized to contiguous nodes in the neck or axilla, whereas non-Hodgkin's lymphoma patients may have more disseminated or extranodal spread. Lymphoma cells can be detected in blood. CT scans, lymphangiograms, and other radiologic procedures can document the extent of disease. There are genetic markers and a wide variety of morphologic and behavioral subtypes of lymphoma, with different treatments and prognoses.

Occurrence

Each year, about 60 000 new cases of lymphoma are diagnosed in the United States, and approximately 25 000 deaths occur from this disease. It is not known what percentage of lymphoma cases or deaths is occupationally related. Lymphoma is currently the sixth most common cause of cancer death in the US. It occurs more commonly among men than among women, and more often among whites than in blacks. The risk of Hodgkin's disease is bimodal, with incidence rates rising from childhood to the early adult period, then declining in the 30- 50-year age group, and then rising again among those age 60 and above. Prevalence of Hodgkin's disease in the U.S. in early 2000 was estimated to be about 120 000 cases. Incidence of non-Hodgkin's lymphoma rises steadily to age 80 and above. Overall incidence of lymphoma has been increasing at about 3% to 4% per year during the past three decades in the U.S., although recent trends indicate a decline in this rate of increase since 1995. Lymphoma prevalence in the U.S. was estimated at slightly more than 300 000 cases in early 2000. Rates of non-Hodgkin's lymphoma have risen particularly rapidly in young men in areas with many HIV/AIDS patients, such as the San Francisco-Oakland area, due to the increase in non-Hodgkin's lymphoma of the brain in this population. Burkitt's lymphoma is rare in the U.S., but relatively common in parts of West Africa.

Causes

Studies of workers in a variety of occupations and with a variety of exposures have demonstrated increased risk of lymphomas. For example, farmers and workers in the chemical industry have been shown to be at risk for non-Hodgkin's lymphoma in Scandinavia, Europe, North America, Australia, and New Zealand. Furthermore, studies of rubber workers, pulp and paper workers, meat workers, chemists, printing pressmen, firefighters, and other groups of workers have demonstrated increased risks of lymphoma. Benzene and other solvents, phenoxyacetic acid herbicides, and dioxins have been implicated as causes of lymphoma. Vietnam veterans have been compensated if they were diagnosed with Hodgkin's disease or non-Hodgkin's lymphoma after service in the Vietnam War. Several viruses (Epstein-Barr virus, HTLV-1, and HIV-1) are associated with specific types of lymphoma, including Burkitt's lymphoma, Hodgkin's disease, and non-Hodgkin's lymphoma of the brain. Patients who have received immunosuppressive therapy and/or organ transplants are at increased risk for non-Hodgkin's lymphoma, suggesting that there may be a combination of chemical immunosuppression and viral infection resulting in lymphomas in some individuals.

Pathophysiology

The precise mechanism by which lymphoid cells become malignant is not known. At least some of the steps in this process involve a combination of lesions to proto-oncogenes and tumor suppressor genes in a cascading series of steps leading to uncontrolled growth. Viruses and viral DNA may play a role in some of these steps.

Prevention

Primary prevention of lymphoid malignancies includes prevention of exposure to organic solvents, certain pesticides, and viruses, including HIV. Secondary prevention is possible by examining individuals at risk for evidence of lymphadenopathy, although large-scale studies of the efficacy of this approach for early detection of lymphoma are not available.

Further Reading

Franceschi S, Serraino D, LaVecchia C, et al. Occupation and risk of Hodgkin's disease in north-east Italy. *International Journal of Cancer* 1991; 48: 831-835.

Kogevinas M, Kauppinen T, Winkelmann R, et al. Soft tissue sarcoma and non-Hodgkin's lymphoma in workers exposed to phenoxy herbicides, chlorophenols, and dioxins: two nested case-control studies. *Epidemiology* 1995; 6: 396-402.

Lynge E, Antilla A, Hemminki K. Organic solvents and cancer. *Cancer Causes and Control* 1997; 8: 406-419.

Zheng T, Blair A, Zhang Y, et al. Occupation and risk of non-Hodgkin's lymphoma and chronic lymphocytic leukemia. *Journal of Occupational and Environmental Medicine* 2002; 44: 469-474.

Memory Impairment

ICD-10 R41

Robert G. Feldman, Patricia A. Janulewicz, and Marcia H. Ratner

Memory impairment is often a symptom of chronic exposure to neuro-toxic chemicals. Memory depends on the specific physiological and psychological processes of perception of a stimulus, encoding of information, and of recall or reproduction of this stored material. Anterograde memory impairment is the inability to assimilate new material. Retrograde memory impairment is the inability to recall previously learned material. Clinically, a distinction is made between deficits in long-term memory (in which retrieval is usually hours, days, or years after presentation), short-term memory (in which retrieval in several minutes to an hour after presentation), and immediate memory (in which retrieval is seconds to a few minutes after presentation). Distinguishing retrograde from anterograde memory disorders is easier if there is a specific time of injury, as there is with head injury or an acute, one-time high-level chemical exposure. It is more difficult if memory loss emerges insidiously due to chronic exposure to neurotoxic chemicals.

Memory impairment may lead to an inability to adapt to situations requiring that new information be learned; therefore, for efficient work practices to continue, it is critical that health professionals pay attention to signs of memory impairment. When questioned about recent events, memory-impaired workers may confabulate, either because they are embarrassed about their inability to remember or because they believe their words to be true. Generally, workers suffering from memory disorders do not identify themselves as such. It is therefore, difficult to describe memory loss objectively as a symptom. Interviewing friends and family members about the patient's recall ability is helpful. Memory loss may be disguised as a patient with behavioral problems or personality changes.

Deficits in short- and long-term memory may be more rigorously assessed by means of a neuropsychological evaluation that includes subtests of the Wechsler Adult Intelligence Scale and Wechsler Memory Scale (Information, Orientation, Mental Control, Memory Passages, Digit Spans for immediate recall, Visual Reproductions, and Associated Language). The California Verbal Learning Test (CVLT) is a test of short-term memory function that is sensitive to deficits that might interfere with activities of daily living. Likewise, Albert's Famous Faces Test is a specific test sensitive to loss of information stored in remote or long-term memory.

Neuropsychological testing may quantify deficits in memory by comparing results with preexposure test data, which implies that there is a worker surveillance system in place. When preexposure information is not available, tests of vocabulary ability that measure old, established verbal skills are of use in estimating baseline ability. Depressive symptoms concomitant with memory difficulties may be a response to the cognitive impairment experienced by the worker in exposure cases. Other neuropsychological signs that are often seen in workers with memory impairment include impaired initiative, reaction time, intellect, and concentration; easy fatigability; mood changes; anxiety; nervousness; emotional lability; and irritability.

Occurrence

Memory impairment caused by occupational exposure is infrequently seen in isolation and is often a component of the symptom complex associated with toxic encephalopathy, which may include attention and executive function deficits as well. It occurs as one of several symptoms of head injury or systemic poisoning by heavy metals, organic solvents, or pesticides. A review of clinical case reports showed that out of 220 chemicals, 149 had caused documented neurotoxicity in humans.

Causes

Impaired memory can result from head trauma and from acute or chronic exposure to (a) heavy metals, such as arsenic, lead, manganese, and mercury; (b) solvents, such as carbon disulfide, perchloroethylene, toluene, and trichloroethylene; or (c) organophosphates. Reversibility of symptoms depends on the dose of neurotoxic chemical and, thus, extent of damage. Memory problems may be due to acute impairments of cognitive function, such as disturbance of neurotransmission, associated with chemical exposure. In such cases, the memory problems resolve immediately with cessation of exposure although the impact of impaired encoding of novel materials that occurred during the exposure period may persist indefinitely. By contrast, memory problems due to neuronal loss are typically irreversible. Short-term memory deficits, especially of visual memory, in the presence of tremor and psychomotor disturbances have been associated with inorganic mercury intoxication. Workers exposed to lead demonstrate short-term memory problems, depression, and poor problem-solving abilities; long-term memory recall is relatively intact. As with metal exposure, solvent exposure causes mainly short-term and immediate recall impairment, along with impaired visuospatial functioning and affective problems such as depression. With immediate recall problems, deficits in learning ability are evident.

Nonoccupational causes of memory impairment include alcoholism, senility, depressive pseudodementia, Alzheimer's disease, and cerebral concussion. Korsakoff's syndrome consists of anterograde amnesia, variable retrograde amnesia with intact early memories, and confabulation. This syndrome has been associated with closed head injury, trauma, vascular lesions, surgically-induced lesions, metabolic and nutritional deficiencies, central nervous system infections, and exposure to toxic agents. Korsakoff's disease or psychosis, an alcohol-related syndrome of amnesia, often presents in patients with Wernicke's encephalopathy, which is associated with thiamine deficiency. People with this condition are in a confusional state and have ocular motor dysfunctions, memory disturbances, perceptual and conceptual function disturbances, ataxia, and, commonly, peripheral neuropathy. Cortical dementias, such as in Alzheimer's disease, involve loss of both anterograde and retrograde memory, impaired language function, and diminished ability to execute motor tasks due to cognitive deficits.

Pathophysiology

Memory can be affected at several levels, including encoding of new information, retrieval of encoded information, ability to inhibit interference during learning and retrieval, and retention of encoded information. Many patients exposed to neurotoxicants have relative deficits on tests of short-term or anterograde memory function compared with retrograde memory or long-term memory function. This dichotomy reflects the sensitivity to neurotoxicant exposure-induced brain damage of the complex dynamic processes involved in short-term memory function, particularly encoding processes, and the relative resilience of the previously stored information tapped by tests of retrograde memory function which is considerably more dependent upon retrieval mechanisms. Because of the divergent patterns of memory impairment possible following toxic exposure, we use a rather extensive memory battery in assessing these patients.

Memory deficits in the presence of impaired verbal fluency, reduced psychomotor speed, and depression suggest bilateral frontotemporal dysfunction with lesions involving the amygdala and hippocampus. This is the suspected target in lead and organotin exposure. Short-term memory deficits in solvent-exposed workers have been attributed to diffuse brain damage. Memory is mediated by a complex interaction of neurotransmitters, such as acetylcholine, and genetics regulation of receptor protein synthesis that collectively modulates neuronal excitability. Thus, toxic chemical exposure may affect memory at various levels to produce acute and persistent deficits in function.

Prevention

Risk of head trauma and exposure to neurotoxicants should be minimized. Effective industrial hygiene monitoring and control of respiratory, gastrointestinal, and dermal routes of chemical entry are discussed in Part I. Biological monitoring of workers for specific chemical exposures may be used as a guide to check on individual exposure levels, but careful attention must be paid to the interpretation of individual results. ACGIH has issued threshold limit values (TLVs) and biological exposure indices (BEIs) for many chemicals.

Periodic neuropsychological testing to assess memory problems in workers by trained neuropsychological personnel may be useful.

Other Issues

People affected by alcoholism and vitamin B_1 or B_{12} deficiency often develop memory disorders, which may confound the diagnosis of occupational toxic exposure. Other confounding influences, fatigue and other attentional deficits, and learning disabilities can affect the diagnosis of memory impairment as a result of toxic chemical exposure. Medications can also confound neuropsychological testing and be the root of a memory problem.

Further Reading

Feldman RG, Travers PH. Environmental and occupational neurology. In: Feldman RG (Ed.) *Neurology: The Physician's Guide.* New York: Thieme-Stratton, 1984, pp. 91-212.

Feldman RG. *Occupational and Environmental Neurotoxicology.* Philadelphia: Lippincott-Raven, 1999, pp. 10-11.

Juntunen J. Alcoholism in occupational neurology: Diagnostic difficulties with special reference to the neurological syndrome caused by exposure to organic solvents. *Acta Neurologica Scandinavica* 1982; 66 (suppl 92): 89-108.

Lindstrom K. Behavioral effects of long-term exposure to organic solvents. *Acta Neurologica Scandinavica* 1982; 66 (suppl 92): 131-141.

Ratner MH, Feldman RG, and White RF: Neurobehavioral Toxicology. In: Ramachandran V.S. (Ed); *Encyclopedia of the Human Brain.* New York, Elsevier Science, Vol. 3, pp 423-439, 2002.

White RF. Differential diagnosis of probable Alzheimer's disease and solvent encephalopathy in older workers. *Clinical Neuropsychologist* 1987; 1: 53-160.

World Health Organization. Joint/WHO Nordic Council of Ministers Working Group. Chronic Effects of Organic Solvents on the Central Nervous System and Diagnostic Criteria. Copenhagen, Denmark: World Health Organization, 1984, pp. 166-184, 219-242, and 243-262.

Mercury, Adverse Effects

ICD-10 T56.1

Rose H. Goldman

Mercury (Hg) poisoning can occur from exposure to either inorganic mercury (elemental mercury or mercury salt compounds) or organic mercury compounds. Diagnosis is based upon the combination of symptoms, the signs on physical examination, and the history of exposure. Laboratory tests to determine the concentration of mercury in blood and urine can be helpful in supporting the diagnosis. The signs and symptoms vary by the type of mercury or mercurial compound, and by the dose and length of exposure.

Acute poisoning by elemental or inorganic mercurial compounds is rare, and often presents with respiratory tract or skin irritation, kidney damage, stomatitis (inflammation of the mouth and gums), and gastrointestinal complaints. Renal injury is of particular concern after exposure to mercuric chloride. Neurological symptoms can develop subsequently, and are similar to those seen from chronic overexposure. Chronic exposure is much more common and is associated with the classic triad of tremor, psychological disorder (extreme shyness and emotional labilty termed "erethism"), and stomatitis. An initial fine resting tremor may progress to an intention tremor. A sensory peripheral neuropathy may develop with distal paresthesias. Early cognitive changes can eventually progress to dementia as a later manifestation.

Organic mercury is felt to be more toxic than inorganic mercury. Exposure to organic compounds is usually chronic, and results in the insidious onset of progressive neurological symptoms. Early symptoms include numbness and tingling of the extremities and around the lips. Later symptoms include decreased motor coordination, with gait ataxia, tremor, and difficulties with fine motor movements. Other symptoms of severe exposure and poisoning include constriction of the visual fields, central hearing loss, spasticity with increased deep tendon reflexes, and cognitive changes. Children are more sensitive to the effects of mercury, and thus are affected at much lower doses, raising issues for pregnant workers. Most of the research done concerning developmental effects of mercury on children relates to methylmercury, rather than inorganic mercury compounds that are more common in the workplace. Nevertheless, this research offers some insight on possible acceptable levels of mercury exposure to women of child-bearing age in the workplace. In 2000, a National Research Council (NRC) committee recommended a benchmark dose level of 58 µg/L mercury in cord blood, based on adverse developmental effects from in utero methylmercury exposure. In order to account for uncertainties and individual variability in response to toxic effects, the NRC recommended dividing by an uncertainty factor of 10 and calculated a reference dose of 5.8 µg/L mercury in cord blood. ATSDR defines the minimum risk level (MRL) as 13.6 µg/L based upon the results of another study. Urinary mercury is the preferred biomarker for low-level exposures to elemental and inorganic mercury. In terms of inorganic work place compounds, urinary mercury is the preferred biomarker for low-level exposures to elemental and inorganic mercury.

Clinical diagnosis is supported by elevation of the concentration of mercury in blood and/or urine. Inorganic mercury can be detected in both blood and urine. Organic mercury is predominately excreted through the gastrointestinal tract (feces), so it is best detected in blood. The finding of an elevated blood mercury, but not detectable urine mercury, is very suggestive of exposure to organic mercury, particularly methylmercury. There does not appear to be a precise correlation between the biological monitoring results and toxic manifestations. For persons nonoccupationally exposed, urinary mercury is generally less than 5 µg Hg/gram creatinine and usually less than 1 µg/g. Subclinical neurological effects for elemental mercury have been reported at urinary mercury levels between 20 and 100 µg/L. ACGIH recommends that the blood inorganic mercury of workers not exceed 15 µg/L and that urine values not exceed 35 µg/g creatinine. Organic mercury is rarely detected in urine unless exposures are enormous. Neurotoxicity thresholds for blood mercury in adults have been suggested to be between 50 and 200 µg/L. Some recent data have been suggestive of increased risks of cardiovascular disease with mercury levels that are much lower.

Occurrence

Reliable data concerning the actual incidence of mercury poisoning are unavailable. The number of cases of adverse effects of mercury and the proportion that is occupationally related are not known.

Causes

Mercury is a silvery white metal that naturally occurs in the earth's crust and in the ocean. It is found in numerous rocks and is recovered primarily from cinnabar ore (HgS). Elemental mercury has been used in scientific instruments such as thermometers and barometers. Mercury vapor has also been used to illuminate street lights, fluorescent lamps, and advertising signs. Mercury can also form alloys (or amalgams) with other metals, such as gold, silver, zinc, and cadmium. Amalgams can then be used to help extract gold, create dental fillings (of silver-mercury amalgam), and help extend the life of batteries (with nickel and cadmium) and to be a component in some cosmetics. Use of mercury in the tanning and taxidermy felt (hat) industries in the 19th century led to psychosis among exposed workers and the phrase "mad as a hatter."

Inorganic mercury exists in two oxidative states (mercurous and mercuric) and combines with other elements, such as chlorine (mercuric chloride), sulfur, and oxygen to form inorganic mercury compounds or salts. Inorganic mercury enters the air from mining of ore, burning of coal, and incineration of medical wastes. Some of these inorganic compounds have been used for medicinal purposes, such as mercuric chloride ($HgCl_2$) to disinfect wounds and mercurous choride (an antiseptic called calomel [Hg_2Cl_2]) to kill bacteria. Mercuric sulfide (HgS) has been used to make a red paint pigment, vermilion, and mercuric oxide (HgO) to make mercury batteries.

Mercury can also combine with organic compounds to form methylmercury, phenylmercury, and merthiolate. The most common population exposure to mercury is to methylmercury, through fish consumption. Methylmercury forms when inorganic mercury is released into the air or water through industrial pollution, accumulates in aquatic environments, and is acted upon by microbes. Methylmercury bioaccumulates through the food chain so that concentrations are highest in large predatory fish. Other forms of organic mercury include ethylmercury (thimersol), which has been used as a preservative in vaccines; phenylmercurics, used as fungicides; and toxic dimethylmercury for chemistry laboratory processes.

Most workplace occupational mercury poisonings in the United States are due to exposure to elemental mercury or inorganic mercury compounds. Inorganic mercury exposure has been described in many occupations, including miners; workers manufacturing or repairing mercury-containing instruments, fluorescent lamps, batteries, and pharmaceuticals; workers in chlor-alkali mercury cell operations for production of chlorine and caustic soda; platers of jewelry; and dentists, dental technicians, and other laboratory workers. In South American nations and some other developing countries, metallic mercury has been used to extract gold during the mining process. Frequently, there is little or no control of exposure, resulting in overexposures among workers, as well as contamination of land and waters and exposure to residents of mining areas. Organic mercury poisoning has occurred as a result of environmental contamination; and eating of contaminated fish (fisherman and community members in Minamata, Japan) and ingestion of seeds and grains treated with mercury fungicide (Iraq). Workers handling organic mercury fungicides may be at risk for exposure. Rare cases of fatal poisoning in laboratory workers exposed to the highly toxic dimethylmercury have been described.

Pathophysiology

Elemental mercury is absorbed after the inhalation of mercury vapor. There is negligible gastrointestinal absorption. Soluble mercurial salts can also be absorbed after inhalation, and to a limited extent after ingestion. Alkylmercurials such as methylmercury are highly absorbed from the gastrointestinal tract, and also from inhalation and skin contact. In the blood, most mercury (especially alkylmercury) is found within red blood cells. Mercury compounds are distributed to many tissues, particularly the kidney and the brain. Mercury binds to sulfhydryl groups and may interfere with many cellular enzyme systems. The exact mechanism of action for neuronal damage is not known.

Both organic and elemental mercury compounds can cross the blood-brain barrier and the placenta, and are secreted in breast milk. Mercury compounds are eliminated in the feces, saliva, and sweat. Some inorganic mercury, but almost no methylmercury, is excreted in the urine.

Prevention

The first step in prevention is to reduce exposures. Substitution of less toxic substances will eliminate exposure. For example, substitutes can be found for mercury-containing instruments, such as mercury-containing sphygmomanometers and thermometers. These changes decrease not only mercury that enters the environment and community, but also exposures to workers who would otherwise be involved in manufacturing mercury-containing instruments.

When it is not possible to substitute, then use of engineering controls, often with enclosure or ventilation of processes that use mercury or mercurial compounds, is often the most effective means of reducing exposures. Use of appropriate protective clothing and gloves is important because mercury can be readily absorbed through the skin. Proper respiratory protection needs to be provided in situations in which other engineering methods are not possible or sufficient. Workers who handle mercury and those working nearby need respiratory protection.

Proper housekeeping and clean-up of any spilled mercury is essential in order to avoid ongoing exposures. So is providing locations to wash hands, and to eat and smoke in areas separate from where mercury is handled. Changing clothes and showering before leaving work decreases ongoing worker exposure, and the likelihood of contaminating the home environment.

Worker training should include specific information about the health effects of mercury and the need to use proper protective measures and work processes.

Preplacement examinations should be geared to identifying persons with any risk factors for increased sensitivity to mercury, such as renal disease or pregnancy. Periodic urinary mercury levels (spot samples corrected for creatinine) are useful for monitoring exposure to inorganic mercury, and, if levels are high, for triggering review of working conditions. Use of urine monitoring will avoid confusion with exposure from fish, since methylmercury is not excreted in urine (but can be detected in blood).

In order to protect the developing fetus, careful control of mercury exposure is particularly important for women workers of childbearing age, especially pregnant workers. Women who are pregnant, or in the process of conceiving, should also limit environmental exposures to methylmercury by avoiding eating high-mercury-containing fish, such as swordfish, tile fish, shark, and king mackerel. EPA advises keeping daily intake to about one 7-ounce can of tuna per week.

Further Reading

Agency for Toxic Substances and Disease Registry. *Toxicological Profile for Mercury*. Atlanta: ATSDR, 1999.

Clarkson T, Magos L, Myers GJ. The toxicology of mercury: current exposures and clinical manifestations. *New England Journal of Medicine* 2003; 349: 1731-1737.

Kales SN, Goldman RH. Mercury exposure: current concepts, controversies, and a clinic's experience. *Journal of Occupational and Environmental Medicine* 2002; 44: 143-154.

National Research Council. *Toxicological Effects of Methylmercury*. Washington, DC: NRC, 2000.

Schober SE, Sinks TH, Jones RL, et al. Blood mercury levels in US children and women of childbearing age, 1999-2000. *JAMA* 2003; 289: 1667-1674.

Mesothelioma

ICD-10 C45

James Leigh

Malignant mesothelioma is a cancer with unique features of etiology, diagnosis, management, and prevention. It most commonly occurs in the pleura, less commonly in the peritoneum, and much more rarely in the pericardium or tunica vaginalis. Nearly all cases are related to asbestos exposure. There has been a dramatic epidemic increase in incidence throughout the world during the past 40 years. Mesothelioma is a disease of long latency, generally 20 to 60 years from first exposure to the time of diagnosis.

Previously considered a rare disease, it is now as common in some countries as kidney cancer, liver cancer, cervical cancer, and uterine cancer. Unlike lung cancer, it is not related to tobacco smoking.

The clinical presentation of the pleural tumor is often a pleural effusion (in 95% of patients), chest pain, and dyspnea (in 40% to 70% of patients). The peritoneal tumor commonly presents with ascites or bowel obstruction. A detailed history of occupational and environmental asbestos exposure is vital to diagnosis. Chest radiography and tomography demonstrates pleural opacity. Tissue diagnosis can be made on pleural or peritoneal fluid cytology or pleural biopsy obtained via needle biopsy, thoracoscopy, or thoracotomy. Peritoneoscopy or laparotomy may be required to obtain tissue in peritoneal mesothelioma.

Occurrence

The ILO and WHO have estimated incidence in North America, Western Europe, Japan, Australia, and New Zealand at 10 000 cases per year, and a greater number of cases per year elsewhere in the world. Incidence in the United States is estimated at 3 000 cases per year. More than 90% of cases are pleural; less than 10%, peritoneal. Incidence in males is 5 to 6 times that of females. Incidence is increasing in both sexes and is predicted to reach a maximum in 2020 in Western Europe, in 2010 in Australia, and much later in Japan, Eastern Europe, and developing countries. Incidence may be leveling out or decreasing in the U.S. and Scandinavian countries, which restricted asbestos exposure earlier than other countries. National incidence bears a strong ecological relationship to historical national per-capita asbestos consumption. Australia has the world's highest incidence rate: 35 per million in 1999 (male: 51, female: 9), compared with almost 13 per million in 1982. In comparison, the U.S. male rate reached a maximum of 19 per million in 1992, and declined to approximately 11 per million in 1998.

Despite active research into new treatments, mesothelioma is nearly always fatal, with no effective curative therapy available. Thus, mortality equals incidence. Median survival from initial diagnosis is generally about 6 months. Peritoneal tumors have a worse prognosis than pleural tumors; sarcomatous and mixed-cell tumors have a worse prognosis than epithelial cell tumors.

The total global burden of asbestos-related mesothelioma in the 20th and 21st centuries has been estimated by the ILO at 2 million cases, with most cases yet to occur.

Asbestos exposure in mesothelioma cases has been found to have occurred in a wide range of occupational and nonoccupational settings. The Australian Mesothelioma Register is the only complete national register which has systematically monitored incidence and exposure histories over a long period (1980 to present). According to this register, some common exposure histories during the 1986-2002 period were: repair and maintenance of asbestos materials (13%); shipbuilding (3%); asbestos cement production (4%); railways (3%); power stations (3%); boilermaking (3%); crocidolite mining (5%); wharf labor (2%); para-occupational, hobby, and environmental (4%); carpenter (4%); builder (6%); navy (3%): plumber (2%); brake linings (2%); and multiple exposure (12%). A similar pattern of exposure has been found in the U.S. The Australian occupational groups exposed in the 1950-1980 period with the highest lifetime risks are shown in Table 1.

Cases have occurred as a result of very low exposures (a few days) and a result of exposure to all commercial asbestos fiber types (chrysotile, crocidolite, amosite, anthophyllite, and tremolite). In Australia, most patients (55%) have had mixed amphibole-chrysotile exposure, 13% amphibole only, 7% amphibole and possibly chrysotile, 6% chrysotile and possibly amphibole, and 4% chrysotile only (with 15% exposed to fibers of unknown type). Mean

Table 1. Mesothelioma Risks in Occupational Groups, Australia, 1950-1980

Occupation	Percent Lifetime Risk of Mesothelioma
Wittenoom crocidolite mine or mill worker	16.6
Power station worker	11.8
Railway laborer	6.4
Navy/merchant navy	5.1
Wittenoom town	3.1
Waterside worker	2.1
Plasterer	2.0
Boilermaker/welder	1.9
Bricklayer	1.8
Plumber	1.7
Electrical fitter, mechanic, electrician	0.7
Vehicle mechanic	0.7
All Australian men	0.39
All Australian women	0.07

latency from first exposure to presumptive diagnosis has been 37.4 years (range, 4 to 75 years). Three percent of cases have resulted from exposure of less than 3 months.

A history of asbestos exposure can be obtained in approximately 90% of cases. In Australia, 88% of those with no known history of exposure had transmission electron microscopy (TEM) fiber counts of >200,000 per gram of dry lung tissue, and 30% had fiber counts of >1 million, suggesting that nearly all mesothelioma patients have been exposed to asbestos. Past exposure, especially in women, is not always recognized as such. In addition, even the absence of fibers in the lungs does not negate exposure and asbestos causation of mesothelioma, as fibers may have been cleared before post-mortem examination of the lungs.

Accounting for about 5% of cases is nonoccupational exposure to bystanders, in domestic or hobby situations, or environmental exposure near a mine, factory, or natural or human-made surface deposits of asbestos.

Causes

Nearly all human mesothelioma cases result from exposure to asbestos or erionite (a fibrous zeolite). Other proposed risk factors or co-factors, such as ionizing radiation and inflammatory lung conditions, are supported only by anecdotal evidence. The case for SV40 virus as a co-factor is still unproven. While there is a dose-response relationship with asbestos exposure, no threshold has been identified. Recent studies have shown that if a threshold exists, it is less than 0.15 fiber-yr per ml. If there is a "background rate" in the complete absence of asbestos exposure, it has been estimated to be much less than 1 per million per year.

The relative carcinogenic potency of crocidolite, amosite, and chrysotile is still uncertain, but there is consistent evidence that crocidolite and amosite are more likely to cause mesothelioma than chrysotile.

Pathophysiology

While there are still many outstanding research questions in the pathogenesis of asbestos-induced mesothelioma, the current consensus view is that asbestos is involved in both the initiation and the promotion/proliferation phases of mesothelioma development. Therefore, all of an individual's cumulative asbestos exposure must be considered in causation.

Prevention

In view of the almost unique causation of mesothelioma by any type of asbestos, the absence of a demonstrated threshold below which there is no risk, and the availability of safe substitutes for the industrial uses of asbestos, the logical preventive measure is elimination of all asbestos exposure. This goal is being achieved by bans or phase-outs on new use of any form of asbestos, with few technical exemptions, in more than 40 countries, including all EU countries, Australia, New Zealand, Switzerland, Argentina, Chile, some Brazilian states, all Scandinavian countries, most Eastern European countries, Saudi Arabia, and the United Arab Emirates. Crocidolite and amosite use has been banned in most countries since the mid-1980s. The U.S., which is considering federal legislation to ban new use of all forms of asbestos, uses about 12,000 metric tons per year of asbestos, mainly in gaskets, roofing material, and friction products, such as brake linings.

New chrysotile use continues in some developing countries, and chrysotile is still mined in and exported from Russia, Brazil, Canada, and Zimbabwe. Chrysotile mining for domestic use occurs in Russia, Brazil, China, and Kazakhstan. There is a strong international movement advocating a global ban, but the asbestos industry maintains that safe controlled use is possible in developing countries, where asbestos has been claimed to have economic advantages over substitute products.

Even in developed countries, there is still much asbestos in place and very strict controls are implemented on exposure in removal, renovation, and general operations involving asbestos and asbestos products.

Other Issues

Compensation for asbestos-induced mesothelioma is available under statutory schemes and under civil law in most countries, including common-law countries.

It is recommended that mesothelioma sufferers seek immediate legal advice, in view of the poor prognosis and lack of provision for survivor actions in many jurisdictions.

Criminal law sanctions apply in some jurisdictions. Jail sentences have been given in the UK, France, and Italy to directors of companies who recklessly exposed persons to asbestos.

Mesothelioma is a very distressing and feared disease. Palliative care, psychosocial support, and assistance by associations of asbestos disease patients are very important in management.

Further Reading

British Thoracic Society. Statement on malignant mesothelioma in the United Kingdom. *Thorax* 2001; 56: 250-265.

Butnor KJ, Sharma A, Sporn TA, et al. Malignant mesothelioma and occupational exposure to asbestos: An analysis of 1445 cases. *Annals of Occupational Hygiene* 2002; 46 (Supp 1): 150-153.

Leigh J, Davidson P, Hendrie L, et al. Malignant mesothelioma in Australia, 1945-2000. *American Journal of Industrial Medicine* 2002; 41: 188-201.

Robinson BWS, Chahinian AP (Eds.) *Mesothelioma*. London: Martin Dunitz (Taylor & Francis), 2002.

Metal Fever

ICD-10 T56

David C. Christiani

Metal fever, previously called metal fume fever, is an acute and usually self-limited syndrome that follows exposure to metal fumes or very fine metal dust particles. It is characterized by a complex of symptoms that includes fever, chills, excess sweating, nausea, weakness, fatigue, throat irritation, cough, headache, myalgias, and arthralgias. The first symptoms are often thirst and a metallic taste in the mouth. Onset usually occurs after a delay of several hours from time of exposure to the offending agent(s). Peripheral leukocytosis often occurs.

The diagnosis is made by taking a careful history and making special note of recent exposures. Repeated episodes may occur. There is no chronic form of the disease *per se*, although the occurrence of wheezing has reportedly been associated with a prior history of metal fume fever, and a history of repeated episodes is associated with decreased lung function.

When, in the absence of an accurate occupational history, attention is focused on the most recent exposure, metal fume fever is often incorrectly identified as the flu.

Occurrence

No population-based studies are known to have been conducted to establish the frequency of this condition adequately. However, given the common occurrence of metal fume exposure in the workplace, the number of workers at risk is large, and the syndrome is considered common.

Causes

Metal fever is caused by exposure to metal fumes (very small condensation products of metal vapor) and by fine particles of metal oxides. The most common metal causing this syndrome is zinc. Other metals such as magnesium, aluminum, antimony, cadmium, copper, iron, manganese, nickel, selenium, silver, and tin are thought by some to cause the condition. Metal fever commonly occurs following welding, such as on galvanized steel, brazing soldering, or other exposure to molten metal. Exposure in a poorly ventilated workplace may be a contributing factor.

Exposure to cadmium or manganese may also result in pneumonitis, and exposure to nickel may also result in occupational asthma. (See Asthma.) There are also case reports of asthma following exposure to zinc fumes.

Pathophysiology

Understanding of the pathophysiology of this syndrome is incomplete. Metal fever can be provoked in laboratory animals. It has been postulated that the pathophysiological mechanism is as follows: minute particles (0.05- to 0.5-μm particles in 5- to 10-μm aggregates) of metal fume or dust penetrate deeply into the respiratory tract and easily reach alveoli; they then activate cells, such as macrophages, with subsequent release of pyrogenic mediators. Unless accompanied by bronchiolitis obliterans or pneumonitis, they cause no permanent structural damage. Curiously, tolerance is acquired after continuous exposure but lost after several days. This suggests some form of unknown immunologic mechanism.

Prevention

Primary prevention involves reducing exposure to offending agents through engineering controls or local exhaust ventilation. Respirators are available but should be selected to filter particles in the submicron-size range. Early identification and diagnosis of this syndrome, documentation of its occurrence, and follow-up evaluation are also important. Workers potentially exposed to molten metal, such as welders, solderers, and foundry and smelter workers, or to very fine metallic aerosols should be alerted to this condition and urged to take precautionary measures such as use of local exhaust ventilation. Often, experienced workers know the syndrome and are a useful source of information for less experienced workers.

Further Reading

Gordon T, Fine JM. Metal fume fever. *Occupational and Environmental Medicine* 1993; 8: 504-517.

Kales SN and Christiani DC. Progression of chronic obstructive pulmonary disease after multiple episodes of an occupational inhalation fever. *J Occup Med* 1994; 36:75-78.

Nemery B. Metal toxicity and the respiratory tract. *European Respiratory Journal* 1990; 3: 202-219.

Mold, Adverse Effects

ICD-10 T65-8

William B. Patterson

Fungi are widely distributed eukaryotic organisms that are occasionally unicellular, but more commonly multicellular. When fungi grow as a fuzzy mat of microscopic filaments (hyphae), they are often referred to as mold. Fungi lack chlorophyll and commonly spread by dissemination of spores. Health effects related to mold and the implications of mold exposure have been attracting increasing attention. Many argue that there is little evidence that mold causes frequent or serious health effects other than allergic responses in a small number of people. Others suggest that mold may cause a wide variety of symptoms and health problems involving many body systems, the exact pathophysiology being thus far undefined. This entry focuses on the indoor environment, which is the setting for most mold-related complaints.

The best documented health effects related to mold exposure are allergic in nature and primarily involve the upper and lower respiratory tracts. Most commonly, these reactions present as type 1 immediate hypersensitivity reactions, usually rhinitis and asthma. Patients may be unaware of their mold exposure, and a detailed occupational and environmental history is critical to recognizing the causal relationship between exposure and disease.

Less commonly, mold disease may present as a type 3 hypersensitivity reaction such as hypersensitivity pneumonitis. Molds also play a role in organic toxic dust syndrome, usually seen in the agricultural setting. In rare cases, usually not related to occupational exposures, mold may cause allergic fungal sinusitis and allergic bronchopulmonary mycosis. There is some evidence that unexplained interstitial lung disease may be associated with mold exposure; if confirmed, the spectrum of mold toxicity will be more broadly defined.

Mold-related invasive infections are usually opportunistic in nature and would not be common in the occupational setting, except in workers with underlying immunocompromised systems and substantial exposure. Superficial skin infections due to mold are a concern for workers in high-humidity environments who have impaired access to facilities for personal hygiene. Constant heat and moisture may predispose to intertriginous infection, which can be easily managed with routine measures and patient education.

Other presenting problems related to mold are less well supported by current medical research. In many cases, patients will present to their physicians complaining of various symptoms that they attribute to mold exposure. Commonly, patients will report central nervous system symptoms, such as headache, nausea, cognitive impairment, and fatigue. Mucous-membrane irritation may be prominent, and patients will often attribute their complaints to sick building syndrome. Assertions that occupational mold exposure may cause neurological effects, immunosuppression, and other serious or disabling conditions are currently supported only by circumstantial evidence and are not proven.

Given widespread publicity about "toxic mold," alleged systemic disease from mold, and destruction of whole buildings to control mold, many health professionals are called upon to answer the question "What does this mold exposure mean to me?" The identification of mold exposure may be visual, following water penetration and/or accumulation in a building from a wide variety of sources, such as flooding, maintenance deficiencies, building integrity violation, indoor accidents, and heating, ventilation, and air-conditioning (HVAC) problems.

Mold exposure is also commonly identified by smell, characteristically musty and dank. The smell itself is often more persistent and troublesome than other smells, since there is less olfactory fatigue with the microbial volatile organic compounds that are often secreted by mold and the odor threshold of these substances is quite low. Industrial hygiene evaluation of a building may identify unrecognized mold exposure, especially if it is being performed for suspected air quality complaints. Conditions that increase the likelihood of mold growth in a building include standing water and high relative humidity that leads to persistent condensation on surfaces. If there is a sharp temperature differential, such condensation can occur at very low ambient relative humidity, as demonstrated by window condensation in cold-winter climates. Air sampling for mold is complex and subject to both false negatives and overestimation of risk. It should be performed only by experienced personnel and the results interpreted with caution.

Occurrence

The numbers of all cases and of occupationally related cases of mold-related disorders is not known. Mold, however, is ubiquitous as a commensal organism (*Candida*) and as a common organism in almost any environment. Fungal growth is subject to variability due to geographic, environmental (such as temperature and growth media), seasonal, and temporal factors. Human environmental interventions, such as continuous and heavy-duty air conditioning in the southern United States, also play an important role in mold proliferation, primarily by cooling surfaces without removing or preventing the intrusion of hot humid air. As a rule of thumb, any indoor environment that is subject to standing water for more than 24 hours is a candidate for fungal growth.

Causes

Although there are more than 100 000 species of fungi, only a few comprise those identified in routine air sampling and in air quality investigations. Common genera include *Cladosporium* and *Alternaria*, which may play a role in allergy. *Trichophyton* species, which cause tinea, and *Candida* are usually not occupational problems, except when they cause skin infections.

Any fungus probably can cause allergic disease if high enough exposures occur. Common opportunistic fungi include *Aspergillus fumigatus*, and *Cryptococcus*. *Histoplasma, Blastomyces,* and *Coccidioides* are virulent fungi that can cause illness in many inhalation-exposed people, although most infections are subclinical. Many fungi produce a wide variety of toxic substances, such as mycotoxins. The most notorious of these is *Stachybotrys chartarum* (syn. *S. atra*). However, from a global public health perspective, ingestion of *Aspergillus flavus*, which produces aflatoxins, and *Aspergillus ochraceous*, which produces ochratoxins, is probably more important. Although outbreaks of ingestion-related disease from mycotoxins are well documented, inhalational outbreaks of disease have yet to be documented, except in agricultural environments.

Pathophysiology

The growth and dispersion of mold are complex and are the subjects of much current research. The allergic effects of mold appear to be related primarily to enzymes released from the spores. Irritant and toxic effects may be related to cell wall components or secondary metabolites. Volatile organic compounds released by fungal metabolism are typically low-molecular-weight alcohols, ketones, and aldehydes. Mycotoxins are relatively low-molecular-weight organic compounds that are produced during growth of some types of fungi. Mycotoxins are found in spores and in the growth medium. In animals, there are well-documented adverse effects of ingestion and inhalation of mycotoxins, but in humans, there is much less evidence that meaningful disease is common, except in specific situations, such as aflatoxin ingestion leading to liver cancer. Glucans, which form part of the cell walls of most fungi, have been associated with indoor air-quality symptoms.

There are many factors that will determine whether workers are affected by mold growth. Exposure factors include the duration of exposure, occupant activity, and whether the fungal growth is disrupted by maintenance, construction, air movement, or deterioration. For example, schools are a common source of complaints related either to perceived adverse health effects or exposure-related concerns. Lack of funding for school maintenance, use of temporary classrooms, design considerations, and school expansion and remodeling are all contributors to water incursion or accumulation and subsequent mold growth.

There are also host factors that influence the likelihood of adverse effects related to mold. These include co-morbid diseases, such as leukemia and AIDS; pre-existing IgE-mediated allergy, such as asthma; and airways disease, which increases the risk of allergic bronchopulmonary aspergillosis. IgG-mediated allergy is associated with hypersensitivity pneumonitis, and deep wounds may be prone to infection in certain situations.

Prevention

Moisture control is the primary approach to limit and control unintended mold growth. The first consideration is building construction. Attention placed onto building design will have a substantial impact in determining whether or not the building envelope remains impervious to water penetration and accumulation. There are many small factors in the design of a building and specification of materials that will affect its permeability to the elements, including water dams for window and door frames, vapor barriers, treatments for the building exterior, and contact of insulation with outside moisture. Many people may be unaware that masonry absorbs and transmits water, requiring appropriate design to prevent penetration of the building envelope. Potential moisture sources, such as bathrooms, washing areas, kitchens, basements, and crawl spaces must be appropriately ventilated or isolated. For example, mold spores may migrate from a crawl space to a classroom when gaps around pipes and conduits are not properly sealed and the crawl space is positively pressured in comparison to the occupied space.

Effective building maintenance is a critical preventive measure. The integrity of the roof, the functioning of unit ventilators (especially in schools), and the reliability of the HVAC system are all critical factors in preventing water accumulation. In general, the humidity of buildings should be maintained below 50%, and using humidifiers should be discouraged, since they often involve standing water and may easily become contaminated and spread spores in a building or occupied space. Buildings that have low air turnover may experience moisture build-up; increasing the ventilation may be an important preventive measure. In selected situations, dehumidifiers may be very useful in reducing the likelihood of mold growth. Properly controlling the temperature of cold surfaces where moisture may condense by the use of insulation, heating coils, or isolation, may prevent condensation and water accumulation in hard-to-reach areas. Finally, those responsible for buildings should pay special attention to carpet that is installed on concrete or basement flooring and the prompt drying or removal of wet rugs.

Preventive measures in any building should include the identification of standing water and mold as part of regular inspections. Ceiling tiles and other building materials may show signs of water damage, which should lead to prompt evaluation and remediation of the unintended water incursions.

Certain manufacturing processes and operations may be associated with standing water, such as machining operations that use water-based coolants. Storage of damp organic materials provides a food source for fungal growth. Measures should be taken to minimize mold overgrowth and to provide engineering and appropriate protective equipment for exposed workers in these situations.

A critical preventive measure for mold-related disease is the implementation of an effective response when mold growth or standing water is identified. For example, water incursion or accumulation should be addressed immediately and dried within 24 hours.

Identified mold should be cleaned up or removed promptly, and contaminated building materials should be replaced.

The prevention of mold-related health effects also includes providing appropriate advice to patients who are or may be susceptible to adverse consequences of mold, including advising those with allergies and immunocompromise to minimize mold exposure. Workers in hot environments may be advised to wear clothing that wicks away sweat and to wash and dry carefully in order to decrease fungal overgrowth on the skin.

Further Reading

Burge HA. Fungi: toxic killers or unavoidable nuisances? *Annals of Allergy, Asthma, and Immunology* 2001; 87: 52-56.

Committee on Damp Indoor Spaces and Health, Institute of Medicine. *Damp Indoor Spaces and Health.* National Academies Press, Washington DC. 2004

Flannigan B, Miller JD. Health implications of fungi in indoor environments: an overview. In: Samson RA, et al, (Eds.) *Health Implications of Fungi in Indoor Environments.* Amsterdam: Elsevier, 1994, pp. 3-28.

Hardin BD, Kelman BJ, Saxon A. Adverse Human Health Effects ssociated with Molds in the Indoor Environment: American College of Occupational and

Environmental Medicine Position Statement. Arlington Heights, IL: ACOEM, 2002.

Jaakkola MS, Nordman H, Piipari R., et al. Indoor dampness and molds and development of adult-onset asthma: A population-based incident case-control study. *Environmental Health Perspectives* 2002; 110: 543-547.

Menzies D, Bourbeau J. Building-related illnesses. *New England Journal of Medicine* 1997; 337: 1524-1531.

Rao CY. Toxigenic fungi in the indoor environment. In: Spengler JD, Samset JM, McCarthy JS, (Eds.). *Indoor Air Quality Handbook.* New York: McGraw Hill, 2001, pp. 46-2 and 46-4.

Roth VS, Muniz MA. Special report: controversy surrounding environmental mold exposure-Current scientific knowledge and future directions for investigational studies. *OEM Report* 2002; 16: 89-96.

Sorenson WG. Fungal spores: hazardous to health? *Environmental Health Perspectives* 1999; 107: 469-472.

Trout D, Bernstein J, Martinez K, et al. Bioaerosol lung damage in a worker with repeated exposure to fungi in a water-damaged building. *Environmental Health Perspectives* 2001; 109: 641-644.

Motor Vehicle Injuries

ICD-10 S00, T14, T20-T32, T79. T90-T95, T98.2

Leon Robertson

Injury while an occupant of a motor vehicle, or being struck by a motor vehicle during work, should be considered and recorded as an occupational injury. Persons killed or injured by vehicles driven by workers who survive unscathed have traditionally not been counted as occupational injuries. Both on- and off-highway vehicles are involved in occupational injuries. Although the National Highway Traffic Safety Administration (NHTSA) has a code for "occupational" in its file of on-road fatal crashes, the data are not recorded in most cases.

Occurrence

Even with underreporting, it can be inferred that motor vehicles account for more occupationally-related deaths than exposure to any other single hazard. Of the 5900 deaths estimated to have occurred to persons while engaged in an occupation in 2001, the BLS attributes 23.8% to vehicle occupants and 6.5% to persons struck by vehicles. Motor vehicles account for more than half the deaths in agriculture, trucking and warehousing, wholesale trade, and government work. These numbers substantially undercount those killed or injured coincident with occupationally-related crashes. In 2000, the NHTSA counted 741 large-truck occupant deaths, but 4060 deaths to occupants of vehicles struck by large trucks and 410 deaths to pedestrians struck by such trucks. While large trucks accumulate only 8% of miles driven, they are involved in 12% of deaths to occupants of passenger vehicles. In addition, in the same year, approximately 140 000 people were nonfatally injured in large-truck crashes.

Causes

The causes of injuries from motor vehicle crashes can be classified by characteristics of the vehicle operators, the vehicles involved, and environmental factors. The severity of injuries is a function of energy exchange in the crash. Energy is measured by multiplying the vehicle's mass times its velocity squared, and dividing the product by two. Energy increases with the square of speed, but vehicles with greater energy-absorbing properties are less likely to injure severely at the same speeds.

Major driver risk factors are gender, age, length of time driven without rest, and alcohol consumption. Vehicle risk factors include incompatible height and weight among vehicles that collide, poor maintenance, high center of gravity, and energy management in the event of a crash, such as with seat belts, air bags, and reinforcing and energy-absorbing structures. Environmental risk factors include conditions of roadways, shoulders and berms, weather, and time of day.

Men experience twice the injury rate per mile driven as women. The rate of injuries declines exponentially with age of driver and increases exponentially with blood alcohol concentration. Alcohol is primarily found in male drivers 20 to 45 years old, although alcohol is present much less frequently in drivers of large trucks in crashes than in drivers of other vehicles. Truckers with more than 6 hours on the road in a given trip are at one-third higher risk. Risk increases as a function of convictions for moving offenses in the previous 3 years.

Although double-trailer trucks are 2-3 times more likely to be involved in a fatal crash, use of these trucks has been expanded to all states by federal regulation, overriding laws in some states banning them. These trucks have greater mass and are less controllable than single-trailer trucks. Brake defects have been found about 1.5 times as often in trucks involved in crashes as in trucks not involved in crashes in the same driving environment. Steering defects have been found more than twice as often in the crashed trucks.

In 2000, motorcyclists were 21 times more likely to die per mile traveled than occupants of passenger cars. Occupants of cars with wheelbases less than 100 inches average twice those with wheelbases 115 inches or longer, due to reduced distance for occupant deceleration.

Pathophysiology

Motor vehicle deaths are usually from blunt trauma and occur within minutes or hours. Death and permanent disability occur in relation to damage to the central nervous system or vital organs, internal hemorrhage, or asphyxiation from damage to the larynx. Later deaths are more related to complications of infection and immobility.

Prevention

Known risk factors lead to preventive strategies. Rules limiting the hours of operation of large trucks in a single day should be better monitored and enforced.

High-risk vehicles, such as motorcycles, small utility vehicles, small cars, and double-trailer trucks, should be avoided. Where possible, shipment by rail or barge should be preferred to shipment by truck. When utility vehicles are necessary, the stability coefficient, the ratio of the distance between the center of the tires to twice the height of center of gravity from the ground, should be 1.2 or greater. Antilock brakes should be installed in all vehicles. Headlamps should be kept on at all times. Reflective tape should be placed around the outline of large trucks to enhance visibility at night. Speed limits should be lowered on roads with higher crash rates and should be enforced, such as by governors or black boxes on large trucks that record speeding.

Underride barriers on the rear of truck trailers should be replaced with lower, energy-absorbing structures. Front bumpers on trucks and utility vehicles should be low enough and energy absorbent in order to engage the bumpers of passenger cars in a collision. Seatbelt use laws should be strictly enforced. Vehicles with sharp points or edges on their front structures should be avoided to reduce damage to pedestrians. Deceleration lanes for trucks with overheated brakes should be available on sections of highway with downhill grades.

Other Issues

Many occupational injury prevention programs are based on education. However, there is no good experimental evidence that driver training reduces crashes. Studies of high-school driver education, defensive driving courses, and professional driver courses indicate that they do not reduce the risk of a crash or injury. Most studies that claim that such courses are effective do not control for selectivity of lower-risk drivers into the courses. This does not preclude the possibility of courses being effective, but they should be proven to be so in controlled trials before being adopted.

Further Reading

Bureau of Labor Statistics. *Fatal Occupational Injuries By Industry and Event or Exposure, All United States, 2001.* Available at www.bls.gov, 2003.

Ervin RD. Injury risks posed by incompatibilities between trucks and other road users. *Bulletin of the New York Academy of Medicine* 1988; 64: 816-834.

National Highway Traffic Safety Administration. *Traffic Safety Facts, 2000.* Washington, DC: U.S. Department of Transportation, 2001.

Robertson LS. *Injury Epidemiology: Research and Control Strategies.* New York: Oxford University Press, 1998.

Robertson LS, Maloney A. Motor vehicle rollover and static stability: An exposure study. *American Journal of Public Health* 1997; 87: 839-841.

Stein HS, Jones IS. Crash involvement of large trucks by configuration: A case-control study. *American Journal of Public Health* 1988; 78: 491-498.

Stein HS, Jones IS. Defective equipment and tractor-trailer crash involvement. *Acc. Anal. Prev.* 1989; 21: 469-481.

Multiple Chemical Sensitivities

ICD-10 T65

Mark Cullen

Multiple chemical sensitivities (MCS) is an incompletely understood syndrome whose hallmark is symptoms affecting multiple organs and occurring after exposure to small amounts of diverse chemicals. Although the spectrum of symptoms attributed by some practitioners to chemical exposure is broad and not very distinctive, a subgroup of patients with the following suggestive historical features can be recognized and the possession of several of these features have been proposed as a case definition:

1) Symptoms first appear in a previously healthy person after a typical occupational or environmental illness, such as intoxication or injury due to a high-level exposure to a known toxin. This precipitating illness may be isolated or recurrent, mild or severe, and usually becomes known to a physician in the course of taking a patient's history.

2) Reexposure to decreasing amounts of the same or a similar toxin causes symptoms resembling those of the precipitating illness.

3) Symptoms become generalized to other organ systems, almost invariably including the central nervous system.

4) The triggers become generalized. Increasingly diverse and chemically unrelated classes of substances at decreasing concentrations evoke symptoms. Responses typically occur at exposures that are orders of magnitude below accepted industrial exposure limits.

5) Common tests of organ function cannot explain the constellation of symptoms.

6) Psychosis or major medical conditions are absent.

MCS may be a complication of a well-characterized occupational or environmental disease, but it is distinct from any toxic or allergic reaction yet described because of the diversity of symptoms and triggers and because of very low exposure. The inability of organ function tests to explain symptoms and the absence of major medical conditions or overt psychosis have suggested to some observers a milder psychiatric disorder or neurosis. More severely disabled people may, in fact, become anxious or depressed, as do those disabled by chronic illness or injury. However, the features of full-blown MCS are sufficiently distinct to warrant this diagnosis until its pathogenesis and means of control are understood.

This distinctiveness is less clear when not all the above cardinal features are met. Current knowledge is far too limited even to predict the likelihood that people with only some of the features have the same underlying disturbance. In such cases, other diagnoses should be considered, especially when longstanding symptoms predate known occupational or environmental disease or when symptoms vary without relation to environmental reexposures, however small.

Occurrence

Because no uniform definition has been accepted, data on the occurrence of MCS are limited. In a random-digit dial telephone survey of several thousand adults conducted by the California Department of Health in the 1990s, 16% self-identified as "allergic or very sensitive" to chemicals, 6% said a physician had diagnosed MCS or "environmental disease," and 1% said they had changed jobs because of their sensitivity to chemicals. These prevalences were similar in wealthy and poor areas and in rural and urban areas of the state. Since there is evidence to suggest that a portion of the former military population participating in the Persian Gulf War in 1991 developed symptoms consistent with this diagnosis, included among the group characterized as suffering from "Gulf War Syndrome," a survey of military personnel was conducted in Iowa. Among these individuals who were in uniform at the time but did not go to the war region, approximately 2% met survey criteria for MCS; the rate was almost twice as high among those who did participate in the war. These same data suggest female predominance of about 2 or 3 to 1; neither racial nor other social factors predisposed.

Clinically significant MCS with impairment of function is far less common based on clinical series. Such series also suggest that the most common setting in the workplace is in the setting of epidemic nonspecific building-related illness (see below). Workers in their 30s and 40s are most likely to seek care, although patients with onset at all ages have been described.

Most cases occur in isolation, suggesting host idiosyncrasy is important, though risk factors have not been well delineated. A higher than expected fraction of patients have history of prior anxiety and depression. Clusters of cases have been reported in occupational groups and in families, though the significance of this observation regarding pathogenesis is uncertain. Additionally, a higher than expected fraction of MCS patients have prior or associated fibromyalgia or chronic fatigue syndrome, suggesting some interrelationship among these disorders. In the California survey, the prevalence of self-identified chemical sensitivity was higher in those who reported hay fever or asthma.

Causes

Although the cause or causes are not known, enough experience with MCS has now accrued to allow discussion of some environmental and occupational factors associated with it. These exposure factors appear to be a necessary component of the illness sequence; hence, they may appropriately be considered as "causes" even while it remains evident that other contributors, including host factors, must also be causally linked.

Virtually any toxic substance in the workplace or ambient environment, when present in sufficient concentration to cause predictable and acute toxicological effects, appears capable of precipitating MCS in some individuals. Overwhelmingly, most important among these substances are organic solvents, pesticides, and respiratory tract irritants. It is unknown whether these large classes of compounds have unique features or whether these agents are prominent merely because of their ubiquitous presence in the environment, coupled with their high potential to cause perceptible, albeit often mild, clinical effects.

A very important environmental precipitant is the constellation of agents that collectively make up the air of "sick buildings." Among individuals adversely affected in large outbreaks of sick building syndrome, some develop MCS, often with little hiatus or clear demarcation between the responses that are shared by large proportions of exposed workers and the insidious development of more generalized responses in a few. Sick building syndrome and MCS must be distinguished, although the latter may complicate the former in some individuals. Unlike most of their co-workers, those who develop MCS will not predictably improve when control measures are adopted to improve indoor air quality.

Once the illness is established, patients with MCS will react adversely to an extraordinary array of environmental agents, both toxic and nontoxic. Substances with low thresholds for irritation and/or odor, as well as those with various properties similar to those of the original offending agent, seem to be the most potent stimuli. This may reflect the ease with which the patient recognizes such substances rather than any special importance the substances have in precipitating symptoms as compared with other widely distributed chemicals. Exhaust fumes, aerosols, and indoor contaminants, such as formaldehyde and tobacco smoke, often appear on the long lists of intolerable substances.

Pathophysiology

The mechanism(s) by which a toxic exposure and subsequent disorder precipitates MCS in a susceptible worker is unknown. The pathways that lead from low-level exposure to the provocation of often disabling symptoms are also obscure. Moreover, it is unclear whether MCS is due to a single disease process.

One or more of several theories may explain MCS. Most important from a societal point of view is that MCS and its variations represent a form of global, cumulative poisoning of the immune system resulting in immune dysregulation. This proposal, espoused for many years by a group of practitioners known as clinical ecologists or environmental physicians, explains MCS as a result of an excessive total body burden of toxic compounds-largely human-made chemicals of petrochemical origin. According to this view, collective toxicity of these prevalent chemicals in hosts who are perhaps predisposed because of nutritional, infectious, or other stressors creates an effect that exceeds the potential of any toxin alone. The result is the development of "allergy" to a broad range of substances, which can be mitigated only by extreme chemical avoidance coupled with modification of other stressors. Although there is little scientific information to support this theory, it has been widely disseminated, many MCS patients probably have learned of it, and some have clung to it as a plausible basis for an otherwise inexplicable illness. Some clinical ecologists believe that the condition can only be diagnosed by placing the patient in a chemical-free environment for several days and then challenging them with chemicals in a specially designed chamber in an allegedly blind fashion. This technique provides MCS labels to some patients who do not give clear histories of initiating events or ongoing triggering. There are controversies in the literature about the adequacy of the blinding techniques used and the subjectivity of the symptoms triggered.

The three major alternative theories entail psychological mechanisms: (a) that MCS is a somatization disorder adapted to the culture of postdevelopment society; (b) that MCS is a variant of post-traumatic stress disorder (PTSD), in which the precipitating illness fulfills the role of the trauma with subsequent exposures serving to recapitulate the trauma, expressed as somatic symptoms rather than as nightmares or flashbacks that are more typical of PTSD; and (c) that MCS is a form of "psychological sensitization" to workplace and/or environmental chemicals, with symptoms serving as conditioned responses to stimuli that evoke an initial unpleasant experience. None of these views has been substantiated.

Other biologic hypotheses include the possibility that MCS may represent an atypical host response to neurotoxins or airway irritants, which are the most common initial precipitants. Meggs and colleagues have focused attention on altered olfactory and irritant responses in the nasal mucosa. A theory involving limbic "kindling" has been promulgated by Bell, with some experimental support. Miller and colleagues more recently have proposed a related theory, "toxicant-induced loss of tolerance." This explanation involves loss of tolerance developing after a single high-level chemical exposure, or after repeated lower-level exposures, followed by subsequent triggering of symptoms by everyday exposure to chemicals, certain foods, caffeine, alcohol, and drugs. What keeps the patients and their doctors from recognizing these chemical triggers, according to this theory, are (a) the overlapping effects of multiple exposures, and (b) habituation phenomena that may mask the effects of particular chemicals and symptoms.

Only open-minded scientific inquiry appears likely to resolve these issues.

Prevention

No established primary prevention strategy can be offered to reduce the occurrence of MCS. Because acute occupational diseases, such as intoxication by solvents and pesticides or injury due to airway irritants, appear to predispose to MCS, reduction in the incidence of these illnesses would likely be somewhat effective. On the other hand, global reductions in chemical use in our culture, recommended by those who postulate an immune basis for MCS due to "total body burden" of xenobiotics, would have uncertain benefits based on present confirmed knowledge of the condition.

Because victims of acute occupational and environmental disorders appear to be at risk for MCS, it would be prudent to focus secondary preventive efforts on these individuals. Unfortunately, it remains unclear what to do for these individuals to alter the likelihood of MCS occurring as a complication. If psychological formulations prove to be correct, interventions aimed at uncoupling the trauma of the acute illness from the experience of less noxious exposure to chemicals might be justified, although no effort of this kind has been systematically undertaken or evaluated. If limbic kindling or loss of tolerance theories prove correct, avoidance of the initiating or triggering agents may reduce the severity of the illness. In any event, it certainly makes sense to follow patients who have experienced acute illnesses more closely and expectantly than the level of reversible injury might otherwise demand. In particular, sensitive and early follow-up of all cases of acute occupational disease can be justified, because this intervention alone may be of value; at a minimum, early recognition and treatment of MCS could ensure avoiding the crises in care reported by most patients.

Regarding tertiary prevention, once MCS is established, the goal of therapy must be to maximize patient function and minimize suffering. For most patients, this begins with establishing a therapeutic relationship with a clinician who accepts the illness, even though the nature of the underlying disorder is obscure, and requires abandoning the tireless search for the cause that will vindicate their suffering and prove someone has harmed or poisoned them. For clinicians, this relationship entails receiving the patient reports with full seriousness, even though test results do not reveal pathology.

In practical terms, the biggest issues become (a) the degree to which avoidance of chemicals should be undertaken; (b) determination of appropriate benefits, such as workers' compensation; and (c) the use of specific treatment modalities, including the experimental ones now available based on immunologic theory. It seems best to minimize extreme avoidance by emphasizing to the patient that there is little basis for the "total body burden" view; it is proposed that the harm of each offending exposure is limited to the symptomatic consequences, however unpleasant. If this is accepted, patients are encouraged to remain in contact with offending environments to the extent that these environments are important for continued functioning. Although the patient should not be directly returned to the work environment in which the precipitating illness occurred, a rapid return to appropriately modified work, with job retraining if necessary, is strongly encouraged.

To allow the usually obligatory job modification, legitimate claims for compensation benefits should be supported, viewing MCS as a complication of occupational disease. Without such benefits, it is inevitable that clinical symptoms will be severely aggravated by material loss, although rarely is such loss sufficient to press the patient back into appropriate work. Although therapy should encourage prompt return to some work, rarely will the financial benefit of compensation provide a strong disincentive to do so.

Radical therapies may be difficult to endorse. Often, however, the patient has placed considerable emotional investment in one or more of these, so the value of criticism is unlikely to outweigh the cost. On the other hand, since many experimental therapies are very expensive and are frequently coupled with admonitions to avoid all chemicals, it is often necessary to discuss the harm of these modalities.

The mainstay of therapy, given present knowledge, is support. The goals of therapy should be realistic, focusing on treatable co-pathology and short-term, achievable life objectives.

Further Reading

Ashford NA, Miller CS. *Chemical Exposures: Low Levels and High Stakes* (2nd ed.). New York: John Wiley and Sons, 1998.

Black DW, Doebbeling BN, Voelker MD, et al. Multiple chemical sensitivity syndrome. *Archives of Internal Medicine* 2000; 160: 1169-1176.

Bell IR. White paper: Neuropsychiatric aspects of sensitivity to low-level chemicals: a neural sensitization model. *Toxicology and Industrial Health* 1994; 10: 277-312

Buchwald D, Garrity D. Comparison of patients with chronic fatigue syndrome, fibromyalgia, and multiple chemical sensitivities. *Archives of Internal Medicine* 1994; 154: 2049-2053.

Meggs WJ. Neurogenic inflammation and sensitivity to environmental chemicals. *Environmental Health Perspectives* 1993; 10: 234-238.

Miller CS. Toxicant-induced loss of tolerance. *Addiction* 2001; 96: 115-137.

Sparks PJ, Daniell W, Black DW, et al. Multiple chemical sensitivity syndrome: a clinical perspective. *Journal of Occupational Medicine* 1994; 36: 718-730.

Sparks PJ (Ed.). Multiple chemical sensitivity/idiopathic environmental intolerance. *Occupational Medicine: State of the Art Reviews* 2000; 15: 497-675.

Multiple Myeloma

ICD-10 C90

Richard W. Clapp

Multiple myeloma is a B-cell malignancy of plasma cells that often presents with symptoms of bone pain and fractures, weakness, and anemia, and is sometimes accompanied by infections. X-ray or MRI evaluation typically shows patchy lesions in skeletal and vertebral bone structures. Renal failure may be evident and monoclonal immunoglobulin peaks (M-components) can be seen in serum or urine electrophoresis studies. Bone marrow biopsy reveals plasmacytoma or increased plasma cells; this malignancy may arise from altered lymphoreticular stem cells. Prognosis, which is strongly related to stage and renal function, is generally poor, although there are some less aggressive cases with better long-term survival.

Occurrence

Each year, approximately 15 000 cases of multiple myeloma are diagnosed in the United States and about 11 000 people die of this disease. It is not known how many cases and deaths are occupationally related. The disease affects males and females approximately equally, but, in the U.S., rates in blacks are typically double those in whites or Latinos. Asians have very low rates. Incidence rates increase sharply after age 50. Prevalence was estimated to be about 47 000 cases in the U.S. in early 2000. Five-year survival has remained at about 25% to 30% for the past three decades.

Causes

Family history of multiple myeloma; exposure to ionizing radiation, benzene and other solvents, and dioxin; and farming and forestry work all appear to increase risk. Chronic immune stimulation seems to modify or promote the risk of disease. Male and female residents of the zone with the second heaviest dioxin exposure in Seveso, Italy, had significantly increased incidence and mortality from multiple myeloma (although there were no cases in the most-exposed zone). Vietnam veterans have been compensated if they were diagnosed with multiple myeloma after service in Vietnam. Some studies have suggested increased risk of multiple myeloma in workers exposed to diesel-engine exhaust.

Pathophysiology

Multiple myeloma appears to arise from a single transformed cell, which then forms clones and metastatic sites. The typical aggressive course and varied clinical presentation of this malignancy makes treatment challenging. Recently, thalidomide has been shown to be useful in treating late-stage disease.

Prevention

Primary prevention of multiple myeloma involves avoiding exposure to ionizing radiation, benzene and other solvents, and dioxin. Patient education about risk factors is also warranted. Screening for early detection of disease has not been shown to be effective.

Further Reading

Bertazzi PA, Pesatori AC, Consonni D, et al. Cancer incidence in a population accidentally exposed to 2,3,7.8-tetrachlorodibenzo-para-dioxin. *Epidemiology* 1993; 4: 398-406.

Cardis E, Gilbert ES, Carpenter L, et al. Effects of low doses and low dose rates of external ionizing radiation: Cancer mortality among nuclear workers in three countries. *Radiation Research* 1995; 142: 117-132.

Eriksson M, Karlsson M. Occupational and other environmental factors and multiple myeloma: a population based case-control study. *British Journal of Industrial Medicine* 1992; 49: 95-103.

Wing S, Richardson D, Wolf S, et al. A case control study of multiple myeloma at four nuclear facilities. *Annals of Epidemiology* 2000; 10: 144-153.

Multiple Sclerosis

ICD-10 G35

Robert G. Feldman, Seth Nelson, and Marcia H. Ratner

Multiple sclerosis (MS) is a disorder of neurological dysfunction characterized by damage to the myelin sheaths (white matter) of CNS axons. It is defined pathologically by randomly occurring areas of inflammation, demyelination, axon loss, and gliosis, that appear in spinal cord nerve pathways and in large myelinated structures, such as the centrum semiovale, corpus callosum, internal capsules, and pyramidal bundles. Periventricular lesions are also common. The areas of demyelination impair conduction of nerve impulses within the CNS. The persistent neurological deficits seen MS are most likely related to neuronal loss. The size and distribution of the areas of demyelination and plaques change over time and, thus, can be documented with serial MRI studies. These changes in the lesions result in a consequent variability of clinical manifestations. Clinically, MS is characterized by two different symptoms or signs of neurological dysfunction reflecting distinct pathological loci that are separated by space and time. For example, a patient with an episode of transient eye blindness who later presents with an episode of left leg weakness or cerebellar dysfunction. Cerebrospinal fluid laboratory results often support MRI and clinical findings. Thus, the diagnosis is based on clinical, pathological, radiological and laboratory data.

Patients experience exacerbation and remission of subtle neurological symptoms, such as unsteadiness of gait, slurred speech, blurred vision, and mood changes. If left untreated, patients may begin to experience more severe symptoms, such as motor weakness, paresthesias, diplopia, nystagmus, dysarthria, intention tremor, ataxia, impairment of deep sensation, bladder dysfunction, and paraparesis. Progression of the disorder becomes more evident as many of the neurological features—which may be hardly noticeable to the patients—begin to accrue, functional impairments become persistent, and objective findings are found on neurological examination.

The diagnosis of MS depends on documenting the distribution of symptoms and signs over time and body areas, reflecting the development of new lesions in different CNS areas as the disease progresses. It requires serial clinical examinations with documentation of periods of exacerbation and remission of neurological symptoms and signs. Typically, patients initially present with episodes of focal disorder in the optic nerve, spinal cord, or brain. An MRI is necessary to document the diffuse white-matter lesions throughout the CNS, often more concentrated in periventricular regions. CSF examination helps to confirm the diagnosis of demyelination by finding moderately-increased total protein and increased numbers of oligoclonal bands on CSF protein electrophoresis. As the disease progresses and more lesions begin to manifest themselves in specific areas of the cerebrospinal axis, a diagnosis of MS becomes more apparent. After initial symptoms manifest, a long period of latency—sometimes as long as 10 years or more—can be seen, followed by either the recurrence of symptoms seen previously or the appearance of new symptoms in other areas of the nervous system with additional lesions. More than half of MS patients have either an intermittent or a steady progression of the disease. Early symptoms include limb weakness and/or numbness in at least half of all patients. Numbness and tingling in the extremities is common. Tight, band-like sensations around the trunk or limbs may occur, possibly due to the involvement of the posterior spinal cord columns, that carry the large myelinated fibers for proprioception to the lower extremities. There is a lower chance of remission each time an attack occurs. Relapse rates average 30% to 40% per year, but the mean time from onset to first relapse is extremely variable. Those who have more attacks, a shorter interval between the first two attacks, and relapse more frequently have more disability.

MS must be differentiated from occupationally-induced demyelination from exposure to solvents, such as toluene, and heavy metals, such as mercury. MS patients and people exposed to solvents may experience subtle neurological symptoms, such as unsteadiness of gait, slurred speech, blurred vision, and mood changes, requiring that the differential diagnosis between early MS and chemical toxicity rely on careful documentation of clinical and laboratory findings. For example, unsafe levels of toluene can cause diffuse white-matter demyelination, but without the predilection for the periventricular region seen in MS. In addition, toluene toxicity does not manifest periods of exacerbation and remission after cessation of exposure, and the optic nerve is not typically involved.

There are slight differences in pathology between MS and heavy metal toxicity. MRI may help differentiate between MS and heavy metal toxicity as well as other demyelinating disorders of the white matter.

Occurrence

There are an estimated 250 000 to 300 000 people in the United States suffering from MS. Occurrence has not been consistently associated with particular occupations. A latitudinal association has been convincingly associated with MS occurrence, suggesting the importance of environment in determining its causes. The typical age at onset is between 20 and 40. The diagnosis of MS should be questioned in an older person with white matter lesions and a history of occupational chemical exposure.

Causes

Symptoms of MS may emerge after a period of physical or emotional stress or may appear insidiously.

The cause of MS has not been fully elucidated. Speculation has led to various hypotheses. Many studies associate numerous factors with prevalence. There is a slightly amorphous gradient by latitude. Prevalence is 30 to 80 per 100 000 people in Canada, 6 to 14 in the U.S. and Southern Europe, and less than 1 in equatorial areas. The gradient is less clear south of the equator. Migrants from areas of high to lower risk generally retain the risk level of their countries of origin, suggesting the exposure to an environmental factor plays an important role in the cause of MS.

There is strong evidence for familial and hereditary factors. In the U.S., people of color are less likely to develop MS than whites. Fifteen percent of patients have a relative who is also affected. There is 26% concordance rate between monozygotic twins, but much less between dizygotic twins, where concordance rate is similar to that of non-twin siblings. Histocompatability (HLA) antigens are present in MS patients in higher frequencies than in control subjects. Overrepresented HLA antigens can be seen most prominently in the D or DR locus on chromosome 6, which is thought to be an immune response gene. If one of these markers is present, the risk of developing MS is 3- to 5-fold.

Other causal hypotheses focus on a yet-to-be-found virus and an autoimmune response triggered by a virus.

Exposure to industrial chemicals, such as organic solvents, may be a contributing factor in MS development. A study of automobile assembly plant workers found that exposures to lead and organic solvents were related to an increased incidence of MS. Occupational and environmental exposure to zinc has also been implicated as a causal factor. A study of welders, however, showed no difference in MS occurrence between those exposed and those not exposed to solvents. In another study involving twins, findings indicated no causal link between occupational exposure to industrial chemicals and MS. A study of Danish men afflicted with MS whose background included exposure to industrial chemicals found no association between exposure and MS. Although no specific chemical has been associated with a significant increase in the incidence of MS, the results of studies do suggest that exposure to chemicals may unmask latent MS.

Pathophysiology

Demyelination shows a predilection for periventricular locations, specifically around the subependymal veins that line the lateral ventricle. Lesions are also seen in the optic chiasm, optic nerves, and in the spinal cord near sites of white matter bordered by pial veins. The lesions do not appear to progress along specific tracts of fibers within the brain. Lesions can vary in size, from smaller than a millimeter to as large as a few centimeters. In the early stage of the disease, diagnosis is difficult because there is normally only one locus of lesioning present in the CNS.

Younger lesions will show partial or complete demyelination while axons are left relatively unharmed around a zone of many small perivenuous foci. Older lesions may be associated with marked axonal loss. At the perivenous sites, there is slight degeneration of oligodendroglia and prominent infiltration of mononuclear cells and lymphocytes around the perivascular and para-adventitial areas. Older lesions show an increase in size and number of astrocytes in and around the lesion, and infiltration of macrophages. The oldest lesions may show thickly matted fibroglial tissue that is acellular in nature, with few paravascular lymphocytes and macrophages. Some axons within lesions may remain intact, possibly having been partially remyelinated and possibly accounting for shadow patches within lesions.

Prevention

Damage to CNS myelin can be prevented by avoiding exposures to those chemicals possibly associated with MS. The use of precautionary measures, including gloves, proper ventilation, and respirators, can also reduce the risk of CNS demyelinating disorders due to chemical exposure. Those persons with MS who may come in contact with chemicals should take extra precautions.

Among other common factors that evoke or exacerbate MS are trauma and infection. Since trauma is known to exacerbate disease progression, those with MS should take extra precautions when undertaking any activity or occupation that can increase the likelihood of head injury. Likewise, persons working in jobs with exposure to infectious agents should also take precautions to minimize exposures. Health-care workers who have MS should wear surgical masks and gloves when possibly exposed to infectious patients.

Other Issues

Prognosis for MS patients is best when treatment is initiated as early in the clinical course as possible. Therefore, clinicians need to accurately diagnose MS and chemically-induced demyelination in order to initiate appropriate therapeutic interventions that may improve the prognosis.

Further Reading

Filley C, Kleinschmidt-DeMasters BK. Current concepts: Toxic leukoencephelopathy. *New England Journal of Medicine* 2002; 345: 425-432.

Gronning M, Albrektsen G, Kvale G, et al. Organic solvents and multiple sclerosis: a case-control study. *Acta Neurologica Scandinavica* 1993; 88: 247-250.

Juntunen J, Kinnunen E, Antti-Poika M, et al. Multiple sclerosis and occupational exposure to chemicals: A co-twin control study of a nationwide series of twins. *British Journal of Industrial Medicine* 1989; 46:417-419.

Landtblom AM, Flodin U, Soderfeldt B, et al. Organic solvents and multiple sclerosis: A synthesis of the current evidence. *Epidemiology* 1996; 7: 429-433.

Landtblom AM, Flodin U, Karlsson, et al. Multiple sclerosis and exposure to solvents, ionizing radiation and animals. *Scandinavian Journal of Work Environment & Health* 1993; 19: 399-404.

Langauer-Lewowicka H, Zajac-Nedza M. Changes in the nervous system due to occupational metallic mercury poisoning. *Neurol Neurochir Pol* 1997; 31: 905-913.

Mortensen JT, Bronnum-Hansen H, Rasmussen K. Multiple sclerosis and organic solvents. *Epidemiology* 1998; 9: 168-171.

Nelson NA, Robins TG, White RF, et al. A case control study of chronic neuropsychiatric disease and organic solvent exposure in automobile assembly plant workers. *Occupational and Environmental Medicine* 1994; 51: 302-307.

Victor M, Ropper AH. Multiple Sclerosis and Allied Demyelinative Diseases. In M Victor, AH Ropper, RD Adams (Eds.). *Adams and Victor's Principles of Neurology* (Seventh Edition). New York: McGraw-Hill, 2001, pp. 954-982.

Nasal/Sino-Nasal Cancer

ICD-10 C11.3, C30.0, C31

Elizabeth Ward

Cancer of the nose and paranasal sinuses, often referred to as sino-nasal cancer, includes malignancies of the internal nose, the middle and inner ear, and the maxillary, ethmoid, frontal, and sphenoid sinuses. Presenting features can include pain, bleeding, sores that do not heal, chronic sinusitis, and cranial nerve abnormalities. Diagnosis is confirmed by biopsy and microscopic examination. The two most common histological types in the United States are squamous and undifferentiated carcinoma, and adenocarcinoma.

Occurrence

In the United States, the age-adjusted incidence rates are 0.9 per 100 000 for males and 0.6 per 100 000 for females. Rates are similar for whites and blacks. Incidence increases with age, with the highest rates occurring for the group aged 75 years and older. There is substantial international variation in incidence rates, with high rates in parts of Asia and low rates in North America. There are no formal estimates of the proportion of sino-nasal cancers that are related to occupation. However, occupational factors seem to play an important role in the etiology of sino-nasal cancer.

Causes

Well documented occupational causes of sino-nasal cancer include exposure to nickel compounds in refining processes, wood and other organic dusts in the furniture-making industry, boot and shoe dust, hexavalent chromium in pigment manufacturing, and radioisotopes, based on studies of radium dial painters. A recent pooled analysis of wood dust and sino-nasal cancer reported a 13-fold increased risk of adenocarcinoma among men with employment in wood-related occupations and a 45-fold increased risk among men with the highest exposure. Formaldehyde was also associated with adenocarcinoma in the pooled analysis, with a 3-fold increased risk among exposed men and a 6-fold increased risk among exposed women. Although many formaldehyde-exposed jobs involve concomitant exposure to wood dust, there is some evidence suggesting that formaldehyde has an independent effect. Cumulative exposure to textile dust was associated with adenocarcinoma in women only, and cumulative exposure to asbestos was associated with squamous cell carcinoma in men only. A significantly elevated risk for adenocarcinoma in women was found at the highest level of cumulative silica exposure. With respect to non-occupational causes, there is limited evidence to suggest an association with cigarette smoking and intranasal snuff use. Persons with recurrent nasal polyps and chronic sinusitis may be at increased risk.

Pathophysiology

It is presumed that direct inhalation accounts for exposure of the target tissues. Known carcinogenic agents may be trapped in the nose and sinuses because of large particle size. Delayed mucociliary transport may be another factor.

Prevention

Prevention relies heavily on reducing or eliminating worker exposure to the carcinogenic agents. Exposure can be reduced by engineering measures, changed work practices, substitution, and personal protective equipment. Education and training of workers and managers is also important. It may be useful to educate workers as well as physicians and dentists to recognize symptoms early, which would lead to earlier and more effective treatment. Screening of asymptomatic individuals is not of proven value. Five-year survival rates in the U.S. for the 1992-1998 period were 57% for whites and 40% for blacks.

Further Reading

American Cancer Society. All about nasal cavity and paranasal cancer: detailed guide. Learn about cancer [serial online]. American Cancer Society, Atlanta, GA. [updated 01/04/2001]. Available from:
http://www.cancer.org/docroot/CRI/CRI_2_3x.asp?dt=75

Demers PA, Kogevinas M, Boffetta P, et al. Wood dust and sino-nasal cancer: Pooled reanalysis of twelve case-control studies. *American Journal of Industrial Medicine* 1995; 28: 151-166.

Luc D, Leclerc A, Begin D, et al. Sinonasal cancer and occupational exposures: A pooled analysis of 12 case-control studies. *Cancer Causes and Control* 2002; 13: 147-157.

Roush GC. Cancers of the nasal cavity and paranasal sinuses. In: Schottenfeld D, Fraumeni JF Jr. (Eds.). *Cancer Epidemiology and Prevention* (2nd ed.). New York: Oxford University Press, 1996, pp. 587-602.

Organic Dust Toxic Syndrome

ICD-10 T52.9

David C. Christiani

Organic dust toxic syndrome (ODTS) is characterized by acute onset of symptoms following exposure to concentrations of organic dust during agricultural work or the handling of agricultural products. Symptoms include a flu-like complex of fever, chills, arthralgias, myalgias, cough, shortness of breath, and sometimes chest constriction. Notable laboratory features include normal findings on both chest x-ray and usual pulmonary function tests (no demonstrable bronchial construction). Occasionally, the diffusion capacity is reduced. The syndrome is distinct from silo-filler's disease, which is bronchiolitis obliterans after exposure to oxides of nitrogen. Recovery from ODTS is similar to that from metal fever and polymer fever and is usually complete within 24 hours. Interestingly, after repeated exposures there is tolerance to at least the febrile effects of exposure.

Occurrence

The prevalence of ODTS is not known, but the syndrome has been described among a number of populations. The varieties of ODTS have been reported using descriptors that refer to the particular work environment. These varieties include mill fever in cotton textile workers; grain fever in grain handlers; humidifier fever among workers exposed to humidified air; pulmonary mycotoxicosis among farmers exposed to moldy silage; and fowl fever among workers in poultry confinement buildings. Because of tolerance to the febrile effects, the syndrome is probably underreported.

Cause

Various organic dusts have been associated with this syndrome. An agent common to most exposures is gram-negative bacterial endotoxin. Threshold exposure and dose-response relationships for endotoxin-induced ODTS have not been defined.

Pathophysiology

Activation of pulmonary macrophages and release of mediators are probably the responsible mechanisms for the responses seen after inhalation of organic dusts. Activated macrophages are capable of releasing a number of mediators that are biologically active, locally and systemically. These include interleukin-1, interleukin-8, prostaglandins, platelet-activating factor, and polymorphonuclear cell chemotactic factors. In addition, some organic dusts have been shown to activate the complement system directly, resulting in inflammation.

Prevention

The syndrome occurs after high levels of exposure and after initial exposures. Thus, primary prevention involves reducing exposures to organic dusts. This is the most feasible way of controlling the problem, because elimination of bacterial endotoxin is impossible in most of these environments. In the case of moldy silage, modification of storage procedures to minimize mold growth would constitute primary prevention.

Further Reading

Blanc PD. Husker days and fever nights: counting cases of organic dust toxic syndrome. *Chest* 1999; 116: 1157-1158.

Seifert SA, Von Essen S, Jacaobitz K, et al. Organic dust toxic syndrome: a review. *Journal of Toxicology and Clinical Toxicology* 2003; 41: 185-193.

Spurzen JR, Romberger DJ, Von Essen SG. Agricultural lung disease. *Clinics in Chest Medicine* 2002; 23: 795-810.

Osteoarthritis (Degenerative Joint Disease)
ICD-10 M19.9

Laura Punnett

Arthritic disorders include osteoarthrosis (OA), autoimmune diseases such as rheumatoid arthritis (RA) and ankylosing spondylitis, gout, other metabolic disorders, and other less common diseases. Like "cumulative trauma disorders," the term "arthritis" has often been used as a catch-all diagnosis for non-specific joint pain and other hard-to-diagnose musculoskeletal disorders. Thus, some general population surveillance data are lacking in specificity.

Osteoarthrosis involves erosion of the joint cartilage to the extent that the bone surfaces below the cartilage are no longer adequately protected from the mechanical forces of articular movement. The bone can respond to these forces by thickening and growth of bone spurs, or osteophytes, at the edges of the joints. This new bone tissue does not correspond to the original shape of the articulating surfaces and thus causes joint stiffness and pain on motion, especially after periods of inactivity (including overnight). Eventually the pain and stiffness are constant, range of motion is reduced, and the joints may become visibly deformed. Key screening criteria include early morning stiffness of more than 30 minutes; discomfort aggravated by joint motion or load and relieved by rest; reduced range of motion and difficulty walking or with other activities of daily living. Other physical examination findings may include joint effusion and crepitus and altered joint attitude or angular deformation.

In the spine, osteophytes may develop at the facet joints or at the vertebral margins, secondary to disc degeneration (spondylosis). Disc degeneration often occurs without facet joint OA and may be a precursor. At other joints, OA may coexist with soft tissue conditions, such as tendonitis and peripheral nerve compression syndromes.

Diagnosis of OA is by x-ray, with or without symptoms. However, radiographic changes, especially in the spine, are common without accompanying symptoms or loss of function. On the other hand, early changes in the cartilage and synovium are not visible radiographically until thinning of the cartilage takes place, which will be seen as narrowing of the joint space. As the disease progresses, the thickened bone tissue and bone spurs will be visible. Four radiographic grades of OA (1=doubtful, 2=minimal, 3=moderate, and 4=severe), based on joint space narrowing and osteophytes, are in virtually universal use. Although some have reported high variability among examiners, this grading system shows a strong correlation with findings, such as loss of range of motion in the knee. Even grade 1 findings have some predictive value for progression to more severe stages. Arthroscopy also has utility for diagnosis and grading.

RA is distinguished in its clinical presentation from OA in that typically multiple joints are affected, often in a bilaterally symmetric pattern and often in combination with other signs, such as synovitis. In the fingers, the proximal interphalangeal (IP) and metacarpophalangeal (MCP) joints are affected, whereas in OA the distal IP and thumb carpometacarpal (CMC) are more usually involved. An elevated serum rheumatoid factor and the presence of subcutaneous rheumatoid nodules help make the diagnosis.

Occurrence

Degenerative joint disease is common in the general population, especially in older ages. The spine, especially the low back and neck, is particularly involved. In the extremities the knees are more affected than hips, hands, or feet (Table 1), although it is somewhat difficult to compare findings across studies because of differences in case definitions and age distributions.

Overall, the prevalence of OA increases with age and is higher among women, although this varies by joint. The prevalence of degenerative changes in the spine in the United Kingdom is about 50% in men and 40% in women aged 35 to 44, increasing to 90% in men and 80% in women 65 years or older. Knee OA is much more common in women, while hip OA shows less of a gender effect (Table 2). Arthritis is elevated in African-American and lower in

Table 1. Prevalence of symptomatic OA with radiographic or clinical findings in the adult general population, by body site, gender, and country*

Site	Country	Men	Women	Total
Hands	England	1.7%	2.9%	2.4%
Feet	England	1.6%	3.6%	2.0%
Hips	Finland	4.6%	5.5%	-
Knees	England	6.2%	12.2%	9.4%
	Finland	5.5%	14.5%	-

* Data for age ranges starting at 25 to 35, up to 74 or older, adapted from Lawrence (1998) and Riihimäki (1995).

Table 2: Prevalence of hip and knee OA in the general population of Finland*, by age and gender. (OA defined by symptoms and findings on clinical examination, but not x-ray.)

Age	Hip		Knee	
	Men	Women	Men	Women
45-54	1.9%	2.8%	4.0%	9.7%
55-64	6.6%	7.9%	9.0%	24.2%
65-74	11.9%	11.7%	12.8%	33.0%
75 and older	16.4%	21.2%	15.7%	38.4%

* Data from Mini-Finland Health Survey (in Finnish), adapted from Riihimäki (1995).

Hispanic individuals, and higher in those who have not completed high school, are physically inactive, or are overweight.

In a 2001 survey of the U.S. adult population, about 10 million adults with chronic joint symptoms (22% of 47.5 million adults affected) had never seen a health-care provider for their symptoms. This included 2 million people who were limited in their activities by joint symptoms. Not seeking medical attention is more common among younger persons, males, Hispanics, those with a high school education or less, lack of health insurance, and having no personal physician.

High-risk occupations are somewhat joint-specific, although prevalences are higher overall in blue-collar and agricultural work. Miners were among the earliest occupational groups described; high prevalence of back OA was linked to heavy lifting and prolonged periods of working in a stooped position in low coal seams. OA in the hands and fingers is more common among women employed in papermaking, textile manufacturing, and food preparation. Risk of knee and hip OA is elevated among agriculture and forestry workers, male firefighters and construction workers, and female cleaners and mail carriers. Knee problems, including, but not limited to, joint degeneration, are also in excess among floor and carpet layers, who often experience direct impact to the joint as well as kneeling and squatting.

The severity of OA has been relatively well-studied because it has a high prevalence in the general population and accounts for a high proportion of disability and lost productivity among working-age (18 to 64 years) adults in many countries. In Canada, this disease group has consistently ranked among the top three causes of restricted activity days, seeking medical attention, and use of pain and other medication. About 14 million adults in the United States general population are disabled by arthritis and related conditions, with difficulty climbing stairs, walking three city blocks, or lifting or carrying 10 pounds. Persons with hand OA have reduced grip strength and difficulty writing and handling small objects. Among working-age adults, 17.7 million (10.5%) have been found to have a limitation in their ability to work outside the home, representing potentially high economic and social costs. In Finland, two-thirds of people with hip or knee OA had functional impairment (reduced range of joint motion) on clinical examination, and OA was considered responsible for 18% of all severe functional impairment in the general population.

Pain and activity limitation among patients with arthritis are inversely associated with measures of socioeconomic status, such as employment grade, education, and household income. Thus, appropriate job accommodation in return-to-work is particularly important for lower status individuals.

Causes

OA is attributed to microfractures in the bone just beneath the cartilage, resulting from repetitive impulse loading and stresses that are distributed unevenly. The epidemiologic associations are generally consistent with the effect of physical joint usage patterns. For example, hand OA is associated with manual "overuse at work" and "repetitive tasks." Repetitive motion may induce inflammatory changes that could promote or aggravate joint tissue damage. In the shoulder, acromioclavicular OA has been linked to heavy manual work. Associations have been reported between segmental vibration and radiographic OA findings in the shoulder, elbow, and wrist, especially in working groups with high exposure to physical loads.

Cervical (neck) OA is not usually thought to be work-related, although one study reported much higher prevalence of degenerative changes in the cervical spine and shoulder among dentists than farmers. Another reported associations with frequent lifting and possibly operation of vibrating equipment.

With regard to the back, an extensive literature has linked spinal disc degeneration (typically defined as radiographic disc space narrowing) with mechanical loading on the spine. Physically heavy work, especially heavy lifting and frequent or prolonged stooping and bending, are strongly associated with increased risk. In addition, onset occurs 5 to 10 years earlier, on average, among exposed persons, such as miners and construction workers. There is no literature that specifically addresses the relationship of facet joint OA to physical demands at or outside work.

Kneeling, squatting, and knee bending; striking with the knee; stair climbing; heavy lifting and other "heavy physical work" affect OA of both the knees and hips. There are no studies on the effect of vibration on knee or hip OA.

Non-occupational risk factors of probable importance for knee OA include obesity, history of acute knee injury (sports or occupational), and activities that load the knee (such as kneeling and squatting) when they occur in housework, athletic activities, and elsewhere. Higher body weight results in higher biomechanical forces on the joints during physical activity. No association with cigarette smoking has been found. Women have different occupational and non-occupational exposure profiles than men but sex hormones may also play a role.

OA may occur either as systemic disease or as a consequence of local joint usage. For example, generalized OA and some forms of knee OA are associated with family history. Different mechanisms may exist for unilateral and bilateral disease, since acute injury is a stronger risk factor for the former and obesity for the latter. There may be ethnic differences in predilection or susceptibility to OA of different joints; however, the association of knee OA with obesity does not appear to be mediated by genetic factors.

Prevention

The entries on Upper Extremity Musculoskeletal Disorders and Low Back Pain Syndrome discuss general principles of ergonomic exposure reduction that also apply here. OA is a chronic and often progressive condition that cannot be treated, although adaptive behavior and exercise therapy can maintain and even improve people's functional capacity. Exposure to risk factors at work usually occurs for many years before symptoms occur. In light of its high prevalence, progressive nature and functional outcomes, primary prevention through ergonomics should be of high priority.

For the lower extremity in particular, powered carpet-stretching tools have been developed that permit carpet-laying without using the knee as a hammer, as is done with the traditional "knee kicker" device. However, the increased cost and time required to operate such equipment appear to mitigate against its widespread use to date.

Orthotic wedged-shoe insoles are a good conservative treatment for medial OA of the knee in its early stages and for somewhat reducing the amplitude of the shock waves resulting from climbing stairs, jumping off platforms, and other activities that produce high accelerations and energy absorbed by joint tissues in the leg. However, the value of insoles for reducing the risk of OA has not been demonstrated. Knee braces and canes may also help to reduce joint forces and pain during daily activities. Non-surgical treatment includes regular application of ice to the joint for brief periods and analgesics orally or in topical creams. Chondroitin sulfate and glucosamine may prove effective in the future.

Further Reading

Allan DA. Structure and physiology of joints and their relationship to repetitive strain injuries. *Clinical Orthopedics and Related Research* 1998, pp. 32-38.

Andersson GBJ. The epidemiology of spinal disorders. In: JW Frymoyer (ed.) *The Adult Spine: Principles and Practice* (2nd ed.) Philadelphia: Lippincott-Raven, 1997, pp. 93-141.

Felson DT. Occupation-related physical factors and osteoarthritis. In: Kaufman LD, Varga J (Eds.). *Rheumatic Diseases and the Environment*. New York: Oxford University Press, 1999, pp. 189-195.

Lavender SA, Andersson GBJ. Job accommodation with respect to the lower extremity. In: Nordin M, Andersson GB, Pope MH (Eds.) *Musculoskeletal Disorders in the Workplace: Principles and Practice*. Philadelphia: Mosby-Year Book, 1997, pp. 528-535.

Milne AD, Evans NA, Stanish WD. Nonoperative management of knee osteoarthritis. *Women's Health in Primary Care* 2000; 3: 841-846.

Ranney D. Chronic musculoskeletal injuries in the workplace. *Elbow, Forearm, Wrist, and Hand*. Philadelphia: WB Saunders, 1997.

Riihimäki H. Back and limb disorders. In: McDonald C (Ed.) *Epidemiology of Work Related Diseases*. London: BMJ Publishing Group, 1995, pp. 207-238.

Rossignol M, Leclerc A, Hiliquin P, et al. Primary osteoarthritis and occupations: A national cross sectional survey of 10412 symptomatic patients. *Occupational and Environmental Medicine* 2003; 60: 882-886.

Yelin E, Callahan LF, National Arthritis Data Work Group. The economic cost and social and psychological impact of musculoskeletal conditions. *Arthritis and Rheumatism* 1995; 38: 1351-1362.

Pancreatic Cancer

ICD-10 C25

Elizabeth Ward

The early stages of pancreatic cancer are generally asymptomatic. The classic triad of pain, weight loss, and jaundice usually results from advanced disease. Diabetes mellitus, acute pancreatitis, and thrombophlebitis are other associated findings that may lead to diagnosis. Abdominal CT scanning may demonstrate a pancreatic mass. Fine-needle biopsy or laparotomy allows a histological diagnosis.

Occurrence

An estimated 30,700 new cases of pancreatic cancer and 30,000 deaths were predicted to occur in the United States in 2003, with a male-to-female ratio of 1.3:1, and a black-to-white ratio of 1.5:1. Pancreatic cancer is the fourth most common fatal cancer in both men and women. Risk increases with age; the disease rarely occurs before age 40 and has the highest incidence rates after age 75. In a recent meta-analysis, the etiologic fraction of pancreatic cancer due to occupational exposures was estimated at 12%.

Causes

There is accumulating evidence for the role of occupational exposures in pancreatic cancer, largely from case-control studies. A recent meta-analysis examined the role of 23 chemical or physical agents present in the work environment and found associations for exposure to chlorinated hydrocarbon solvents and related compounds, nickel and nickel compounds, chromium and chromium compounds, polycyclic aromatic hydrocarbons (PAHs), organochlorine insecticides, silica dust, and aliphatic and aromatic hydrocarbons. Independent meta-analyses concluded that there is an elevated pancreatic cancer risk among dry cleaners and among workers exposed to cadmium. Occupational cohort studies have identified elevated pancreatic cancer mortality among workers processing vinyl resins and polyethylene, producing chlorohydrin with potential exposure to ethylene dichloride and bis-chloroethyl ether (BCEE), and manufacturing DDT. Increased pancreatic cancer has also been reported in cohorts of leather tannery workers.

Recent case-control studies have found increased risks of pancreatic cancer among men in a variety of occupational and industrial groups, including electricians, stone miners, textile workers, cement and building materials, manufacturing workers, pharmacists and sales associates in pharmacies, wood machinists, gardeners, and transport inspectors and supervisors, metalworkers, tool makers, plumbers, welders, glass manufacturers, potters and construction workers, science technicians, painters and varnishers, and machinery mechanics. Among women, a number of studies report associations with employment in the textile industry and one, an increased risk associated with employment in agriculture. Increased risks have also been reported for exposure to ionizing radiation, electromagnetic fields, nonchlorinated solvents, pesticides, inorganic dust containing crystalline silica, heat stress, and rubber chemicals including acrylonitrile. Evidence for association with occupational exposure to asbestos and ionizing radiation is inconsistent.

Exposure to cadmium has been hypothesized to cause pancreatic cancer for mechanistic reasons, with limited support from epidemiologic data.

Non-occupational risk factors include smoking, diabetes mellitus, heavy alcohol use, and chronic pancreatitis. The attributable risk of pancreatic cancer due to smoking is estimated to be 23%. Dietary factors that may increase risk include consumption of meats, grilled meats, and fat. Consumption of fruits and vegetables decreases risk. Combinations of risk factors may lead to substantially more disease than the presence of individual risk factors.

Pancreatic cancer has been induced experimentally by polycyclic aromatic hydrocarbons, nitrosamines, and aromatic amines, which are present in cigarette smoke, in the diet, and in many of the occupations where increased pancreatic cancer risk has been observed.

Pathophysiology

Extensive evidence indicates that mutations in codon 12 of the K-ras protooncogene play a role in pancreatic cancer, with a high frequency of these mutations in human pancreatic adenocarcinomas. Activation of codon 12 of the K-ras protooncogene has been found in heavy smokers and among pancreatic cancer patients with higher serum concentrations of p,p´-DDT, p,p´-DDE, and di-ortho-chlorinated polychlorinated biphenyls (PCBs). Activation of codon 12 of K-ras is also seen in hamsters treated with a pancreatic carcinogen (*N*-nitroso-bis-[2-oxypropyl] amine). The increased risk of pancreatic cancer associated with pancreatitis suggests that inflammation may play a role.

Prevention

Primary prevention includes eliminating or reducing exposure to agents with known or suspected carcinogenicity. In addition, avoidance or cessation of smoking, reduction of alcohol use, and a diet rich in fruits and vegetables may lead to a lower probability of developing disease. No methods for screening and early detection of pancreatic cancer are available at this time. Five-year survival in the United States for the 1992-1998 period was 4.3%.

Further Reading

American Cancer Society. All about pancreatic cancer: detailed guide. Learn about cancer [serial online]. American Cancer Society, Atlanta, GA. [updated

01/04/2001]. Available from URL:
http://www.cancer.org/docroot/CRI/CRI_2_3x.asp?dt=34

Anderson KE, Potter JD, Mack TM. Pancreatic cancer. In Schottenfeld D, Fraumeni JF Jr. (Eds.). *Cancer Epidemiology and Prevention* (2nd ed.). New York: Oxford University Press, 1996, pp. 725-771.

Dell L, Teta MJ. Mortality among workers at a plastics manufacturing and research and development facility: 1946-1988. *American Journal of Industrial Medicine* 1995; 28: 373-384.

Ji BT, Silverman DT, Dosemeci M, et al. Occupation and pancreatic cancer risk in Shanghai, China. *American Journal of Industrial Medicine* 1999; 35: 76-81.

Kauppinen T, Partanen T, Degerth R, et al. Pancreatic cancer and occupational exposures. *Epidemiology* 1995; 6: 498-502.

National Cancer Institute, What you need to know about cancer of the pancreas. What you need to know about cancer publications [serial online]. National Institutes of Health, Bethesda, MD. [updated 9/16/2002]. Available from URL: http://www.nci.nih.gov/cancerinfo/wyntk/pancreas

Ojajarvi IA, Partanen TJ, Ahlbom A, et al. Occupational exposures and pancreatic cancer: a meta-analysis. *Occupational and Environmental Medicine* 2000; 57: 316-324.

Schwartz GG, Reis IM. Is cadmium a cause of human pancreatic cancer? *Cancer Epidemiology, Biomarkers, and Prevention* 2000; 9: 139-145.

Weiderpass E, Partanen T, Kaaks R, et al. Occurrence, trends and environment etiology of pancreatic cancer. *Scandinavian Journal of Work, Environment and Health* 1998; 24: 165-174.

Yassi A, Tate R, Fish D. Cancer mortality in workers employed at a transformer manufacturing plant. *American Journal of Industrial Medicine* 1994; 25: 425-437.

Parkinsonism

ICD-10 G20-G22

Robert G. Feldman, Patricia A. Janulewicz, and Marcia H. Ratner

Idiopathic Parkinson's disease (IPD) is a progressive neurodegenerative movement disorder characterized by slowness of movement, rigidity, and tremor. Although an interaction between genetic and environmental factors has been reported, the etiology of IPD remains unknown. In contrast, the association between parkinsonism and chemical exposure, including manganese, has been well established. Since IPD and parkinsonism are the result of dysfunction in common neuroanatomical structures, it is not surprising that the two syndromes also share many of the same clinical expressions. IPD is differentiated from parkinsonism induced by exposure to neurotoxicants by the clinical manifestations and the therapeutic response to levodopa, which reflect the differences in the neuropathology of IPD and chemical-exposure-induced parkinsonism.

Occurrence

The prevalence of IPD has been estimated to be 18 to 328 per 100,000 people. Exposure to 1-methyl, 4-phenyl, 1,2,3,6-tetrahydro-pyridine (MPTP), methamphetamine, manganese, carbon disulfide, carbon monoxide, and toluene may be associated with symptoms of parkinsonism. Studies indicate a younger age of onset of parkinsonism than Parkinson's disease.

Causes

Chemical Exposures

MPTP

Exposure to MPTP has been associated with parkinsonism in humans and non-human primates. The first reports of an association of MPTP with parkinsonism occurred after a group of young unsuspecting drug abusers injected themselves with illicitly produced meperidine analogues contaminated with MPTP. The clinical picture that resulted was rapidly progressive Parkinson's disease. Researchers studying the effects of MPTP in animals are at risk for occupational exposures to this neurotoxicant, which induces a marked loss of neurons in the substantia nigra and a clinical picture very similar to that of IPD.

Amphetamine

Exposure to amphetamines, such as methamphetamine, has been associated with parkinsonism. Human and animal studies have shown that methamphetamine decreases markers of dopaminergic neuron terminal integrity in the basal ganglia. Research suggests that the brain injury associated with exposure to methamphetamine is due to an increase in free-radical formation and mitochondrial damage, which leads to a failure of cellular energy metabolism followed by a secondary excitotoxicity. Positron emission tomography (PET) studies reveal that methamphetamine users have significant decreases in dopamine transporter density in the caudate nucleus and putamen as compared with controls, suggesting that methamphetamine users may be at increased risk for developing parkinsonism as well as IPD.

Manganese

There have been reports of higher incidence of tremor and parkinsonism among manganese smelter workers and welders who use welding rods containing manganese alloys. Exposure to manganese is one of the identifiable occupational causes of parkinsonism. Exposures to manganese have been associated with disturbances of dopaminergic neurotransmission and neuronal loss in the substantia nigra and basal ganglia that are manifested clinically as insomnia, emotional lability, hallucinations, tremor, dystonia, bradykinesia, incoordination, propulsive gait, postural reflex impairment, soft speech, and reduced facial expression. Clinical features typically stabilize or may even improve when the exposure to manganese ceases. Damage to neurons is thought to occur through free-radical formation and mitochondrial damage, with subsequent cell death via apoptosis. Neuropathological studies reveal damage is primarily to the basal ganglia (putamen and globus pallidus) but the hypothalamus, cerebellum, and the substantia nigra are also affected by manganese toxicity. Manganese-induced parkinsonism is differentiated from IPD by clinical features and its lack of clinical response to levodopa therapy. Positron emission tomography (PET) studies using [18F]-6-fluorodopa as a tracer indicate that parkinsonism due to manganese exposure occurs without a profound reduction in putamenal dopamine uptake, as is seen in IPD, indicating that this particular imaging technique may be useful in the differential diagnosis of IPD and manganese-induced parkinsonism. The T-1 weighted MRI images of patients exposed to manganese may reveal hyperintensities in the globus pallidus during exposure. Although these MRI findings typically normalize with cessation of exposure, the symptoms of parkinsonism associated with these findings may persist indefinitely. Exposure to manganese has also been associated with younger age at onset of IPD, suggesting that neuronal loss in the substantia nigra mediated by manganese may hasten the clinical course of this idiopathic neurodegenerative disease.

Carbon Disulfide

Carbon disulfide (CS_2) is an industrial solvent that can be inhaled, swallowed, or absorbed through the skin. Workers in the viscose rayon and cellophane industries are at risk for exposure to toxic levels of carbon disulfide. Fat-soluble carbon disulfide readily accumulates in the lipid-rich content of brain tissue. Individuals with toxic exposures present initially with behavioral alterations, such as pressured speech, insomnia, violent fluctuations in mood, excessive sexual behavior, paranoia, hallucinations, and nightmares. These changes often appear within weeks to months of high-level exposure. Depression, tremor, parkinsonism, incoordination, memory loss, peripheral neuropathy, and cerebrovascular disease follow the initial mania. With chronic low-dose exposure, the symptoms most frequently seen include depression, weakness, gait disorder, spasticity, ataxia, parkinsonism, and dementia. Studies in primates reveal that exposure to carbon disulfide produces necrosis of neurons in the globus pallidus and substantia nigra and a depletion of dopamine in the putamen and globus pallidus. The role of carbon disulfide in contributing to oxidative stress in an exposed individual has been demonstrated and may account, in part, for its effects on the central nervous system and its ability to produce symptoms of parkinsonism. Carbon disulfide exposure may also hasten the progression of IPD.

Carbon Monoxide

Exposure to carbon monoxide (CO) occurs in poorly-ventilated areas where fumes from incompletely combusted organic materials, such as gasoline, wood, coal, or natural gas, accumulate. Acute carbon monoxide exposure results in the formation of carboxyhemoglobin (COHb), which occurs when carbon dioxide displaces the oxygen that normally binds to the hemoglobin in red blood cells. The formation of carboxyhemoglobin reduces the oxygen-carrying capacity of the red blood cells and leads to tissue hypoxia. Carbon monoxide also interferes with the activity of cytochrome oxidase. These toxic effects on oxygen transport and utilization collectively lead to cell death. Acute symptoms of carbon monoxide toxicity include dizziness, headache, visual changes, changes in level of consciousness, convulsions, coma, and cardiorespiratory arrest. Within days or weeks after recovery from coma, survivors often develop delayed symptoms of memory loss and personality change. Signs of parkinsonism also emerge at this time and include flat facial expression with decreased eye blink frequency, increased muscle tone, and short and shuffling gait, with impairment of postural reflexes.

Abnormalities in the density of the putamen and globus pallidus are seen on MRI following carbon monoxide poisoning. Rapid replacement of oxygen supply can reverse the acute effects of carbon monoxide poisoning as evidenced by the less profound abnormalities on MRI and better clinical prognosis. The longer a patient remains comatose the worse the overall prognosis. Hyperbaric oxygen chambers may reduce the severity of CNS effects of exposure and are often used in acute carbon monoxide poisoning when carboxyhemoglobin levels are significantly elevated. Residual effects, however, may persist in the form of parkinsonism with slow movement and cognitive impairments even when considerable efforts are made to minimize effects. These parkinsonian features are not responsive to levodopa therapy. In such cases, there may be permanent scarring or gliotic changes in the globus pallidus, seen pathologically and on MRI. Symptoms of cognitive dysfunction often accompany the parkinsonism associated with carbon monoxide poisoning and depending upon the severity of the acute intoxication these may be reversible or may persist indefinitely.

Hydrocarbons

Toluene is a hydrocarbon that is often used as an organic solvent. It is a constituent of gasoline and is found as an ingredient in paint thinners, inks, and glues. The effects of acute exposure to toluene are those of narcosis due to inhibition of GABAergic neurotransmission. These acute symptoms include blurred vision, headache, fatigue, attention deficits, confusion, lightheadedness, inebriation, memory difficulties, disturbed equilibrium, and incoordination. With occupational exposures to toluene, these acute symptoms may persist between repeated episodes of acute exposure due to accumulation of this lipid-soluble solvent in lipid-rich tissues of the body, which serve as an internal source of exposure between acute events.

Persistent neurological abnormalities emerge insidiously with chronic and repeated exposures to toluene and include features of parkinsonism and chronic toxic encephalopathy. These symptoms of parkinsonism may include disturbances in posture, and fine motor control along with a marked tremor of the hands, feet, and sometimes the trunk. Chronic exposure to toluene affects white matter and must be differentiated from the similar features of other diseases, including the effects of hypoglycemia, hyperglycemia, transient ischemic attacks, head injury, hysteria, and multiple sclerosis.

Head Trauma

Closed head injury due to a single severe blow to the head or a series of more subtle head injuries, such as repeated torsion injury to nerve cells as the result of blows to the head during boxing, can result in signs and symptoms of parkinsonism. Clinically, post-traumatic parkinsonism presents with rigidity of limbs (legs more than arms), slowed movements, and slurred speech. These findings may appear asymmetrical in severity but usually affect both sides of the body. Typical parkinsonian gait, posture, and postural instability are eventually seen as the syndrome progresses, even after repeated head trauma has ceased. Tremor, while a frequent finding, often has an irregular and jerky (myoclonic) quality. Other findings include visual-spatial disorientation, amnesia, and slowed cognitive performance.

Some cases of trauma-induced parkinsonism respond well to levodopa therapy, suggesting that the head injury served to precipitate IPD. In other cases of head-injury-induced parkinsonism, the response to levodopa is less pronounced, suggesting that the clinical features are due to the actual traumatic damage to the brain areas concerned with movement and posture.

Pathophysiology

Loss of neurons in the substantia nigra is the pathological hallmark of IPD. Therefore, any factor that induces diffuse neuronal loss in the CNS can be expected to induce some neuronal loss in the substantia nigra and therefore can be expected to contribute to the progression of IPD. Factors which may lead to neuronal loss in the CNS include free radicals, disruption of cellular respiration, and head trauma. The role of free radicals is appreciated in both IPD and parkinsonism. The earliest studies of MPTP revealed the role of free radicals in neurotoxicant-induced parkinsonism. Many of the chemicals that have been associated with occupational-exposure-induced parkinsonism, such as manganese, are also metabolized to free radicals and/or potentiate the formation of free radicals.

Parkinsonism due to chemicals other than MPTP is differentiated neuropathologically from IPD by the loss of neurons in the substantia nigra, which is typically greater in IPD than in parkinsonism. In addition, these syndromes are typically associated with relatively greater neuronal loss in structures of the basal ganglia, such as the globus pallidus. These neuropathological differences between these syndromes and IPD account for the differences in the clinical presentations.

Prevention

Effective industrial hygiene monitoring should minimize exposure to neurotoxicants. Biological monitoring of workers and personal and air sampling for chemicals may be used to check exposure levels. ACGIH has recommended Biological Exposure Indices (BEIs) for a number of chemicals. Administration of monoamine oxidase-B (MAO-B) inhibitors, such as deprenyl, to workers with acute accidental exposure to MPTP may prevent the metabolism of MPTP to its toxic metabolite MPP+, and therefore may reduce the risk of neurotoxicity and parkinsonism. Prevention of blunt trauma to the head is the first line of defense against trauma-induced parkinsonism. All workers at risk for closed head injuries should wear protective headgear, such as hardhats.

Other Issues

Exposure to chemicals, including pesticides, has been reported in some studies to increase the incidence of IPD. However, there is still no incontrovertible evidence that chemicals cause IPD or related neurodegenerative movement disorders. Nevertheless, chemical exposure should still be considered when dealing with a patient with IPD, as exposure may hasten the clinical course of the disease and lead to a younger age at onset.

Further Reading

DiMonte DA, Lavasani M, Manning-Bog AB. Environmental factors in Parkinson's disease. *Neurotoxicology* 2002; 23: 487-502.

Engel CS, Checkoway H, Deifer MC, et al. Parkinsonism and occupational exposure to pesticides. *Occupational and Environmental Medicine* 2001; 58: 582-589.

Feldman RG. Manganese. In Vinken PJ, Bruyn GW, De Wolff FA (Eds.), *Handbook of Clinical Neurology, Volume 20, Intoxications of the Nervous System, Part I*. New York: Elsevier, 1994, pp. 303-322.

Feldman RG, Ratner MH. The pathogenesis of neurodegenerative disease: Neurotoxic mechanisms of action and genetics. *Current Opinion in Neurology* 1999; 12: 725-731.

Hageman G, van der Hoek J, van Hout M, et al. Parkinsonism, pyramidal signs, polyneuropathy, and cognitive decline after long-term occupational solvent exposure. *Journal of Neurology* 1999; 246: 198-206.

Kim Y, Kim KS, Yang JS, et al. Increase in signal intensities on T1-weighted magnetic resonance images in asymptomatic manganese-exposed workers. *Neurotoxicology* 1999; 20: 901-907.

Levy BS, Nassetta WJ. Neurological effects of manganese in humans: A review. *International Journal of Occupational and Environmental Health* 2003; 9: 153-163.

Lockwood AH. Pesticides and parkinsonism. *Current Opinion in Neurology* 2000; 13: 687-690.

McCann UD, Wong DF, Yokoi F, et al. Reduced striatal dopamine transporter density in abstinent methamphetamine and methcathinone users: Evidence from positron emission tomography studies with [11C]WIN-35,428. *Journal of Neuroscience* 1998; 18: 8417-8422.

Pezzoli G, Canesi M, Antonini A, et al. Hydrocarbon exposure and Parkinson's disease. *Neurology* 2000; 55: 667-673.

Racette BA, McGee-Minnich L, Moerlein SM, et al. Welding-related parkinsonism: clinical features, treatment, and pathophysiology. *Neurology* 2001; 56: 8-13.

Sanchez-Ramos JR. Toxin-induced parkinsonism. In: Stern M, Kuller W (Eds.). *Parkinsonism Syndromes*. New York: Marcel Dekker, 1993, pp. 155-171.

Peripheral Nerve Entrapment Syndromes

ICD-10 G61-G62

Laura Punnett and Judith E. Gold

Peripheral nerve entrapment is a general term for compression or pinching of any nerve in the arm or leg. The peripheral nerves may be injured by internal sources of pressure, such as contact with bone, tendon, or ligaments, or by external sources of pressure, such as the sharp edge of a desk or tool handle or the hard surface of a chair or knee pedal.

The symptoms and signs of peripheral nerve entrapment depend on whether the affected nerve is sensory, motor, and/or autonomic. Compression of sensory nerve fibers usually results in tingling, numbness, or pain in the supplied areas. These symptoms often begin gradually and intermittently. They may progress to burning, painful numbness, deep dull aching, or a sensation of swelling without objective signs; sometimes they spread up or down the limb. Compression of motor nerve fibers often causes weakness or clumsiness in the supplied muscles, as evidenced by difficulty in holding small objects without dropping them. Compression of an autonomic nerve interferes with normal sweating and blood flow.

One of the best-known peripheral nerve entrapments is carpal tunnel syndrome (CTS) (see entry). Other examples include compression of the median nerve under the muscle of the forearm (pronator teres syndrome); the ulnar nerve at the wrist (Guyon's canal syndrome) or at the elbow (cubital tunnel syndrome); the sciatic nerve and its branches, such as the peroneal nerve, in the hip area or just below the knee; the radial nerve in the upper arm; and any or all of the nerves that travel from the neck through the shoulder and into the arm, such as in thoracic outlet syndrome (TOS), which may also involve compression of the accompanying blood vessels. A single nerve may be compressed at more than one location, as in the "double-crush" syndrome.

Neurologic TOS is a controversial topic, with claims of both underdiagnosis and overdiagnosis by medical professionals. According to those who endorse the possibility of compression at the thoracic outlet, the distribution of pain and numbness can range over the neck, shoulder, arm, or hand (usually in the little finger and the outside of the fourth finger that are supplied by the ulnar nerve, but potentially in the entire hand). Symptoms may occur at night or after prolonged sitting and may be aggravated by overhead arm activities. Headaches and range of motion limitations in the shoulder and neck may be present. Because there may be simultaneous compression of blood vessels, other symptoms can occur, such as blanching, discoloration, coldness, or swelling. Symptoms may be less localized and less intermittent than those in a simple nerve entrapment.

Signs of sympathetic nerve fiber compression include extremely dry and shiny skin and blanching associated with poor circulation. Sensory nerve damage may lead to decreased sensitivity to touch, pain, temperature, or vibration; two-point discrimination may be poor. With motor nerve damage, there may be objective loss of strength or dexterity in the affected muscles, with atrophy (muscle wasting) in severe cases. Tendon reflexes may be weak or absent with either sensory or motor nerve involvement, depending on the severity. In general, however, early stages of nerve entrapment may not be apparent upon physical examination without specific provocation.

Sensory symptoms may be provoked or worsened by specialized tests for specific compression syndromes. Usually such a test consists of either percussion over the site of entrapment, such as Tinel's sign over the ulnar nerve at the wrist crease for Guyon's canal syndrome, or maneuvers that compress or place tension on a nerve. Two examples of the latter, for forearm median-nerve compression, are: (a) positioning the arm in full supination with the wrist in neutral and elbow in extension, with pressure applied to the pronator teres muscle; and (b) resisted motion involving muscular contraction at the site of entrapment, such as by resisted pronation with clenched fist and flexed wrist. For thoracic outlet syndrome, hand symptoms may be precipitated or exacerbated by raising the arms above the head or out to the sides with palms upward. While such maneuvers may be used in clinical practice, there is scant scientific confirmation of their reliability or predictive value.

The diagnosis may sometimes be confirmed objectively, more often when the condition is well-established and more severe. For the more distal compression syndromes, nerve conduction studies may be performed to measure whether the nerve impulse is conducted more slowly (decreased nerve conduction velocity) or less efficiently (decreased amplitude and prolonged residual latency) through the entrapped section of the nerve than through other segments of the same nerve or, if only one limb is affected, the corresponding nerve on the other side. These electrodiagnostic studies, as well as sensory tests, including two-point discrimination, may detect more severe stages of compression. Magnetic resonance imaging and ultrasound have been proposed for diagnostic testing; however, although useful for equivocal cases of some conditions (such as cubital tunnel syndrome), these are unlikely to be routinely useful.

Diagnosis of nerve entrapment syndromes tends to be problematic. Neither examination methods nor diagnostic criteria have been fully standardized. The reliability and validity of almost all the available tests remain to be demonstrated. There is also a lack of data on how well results of examinations in the early stages of a nerve compression disorder predict progression.

The differential diagnosis includes local tendon disorders, compression of the same nerve at another location, nerve root problems, and various systemic diseases. As with upper extremity disorders in general, nerve compression disorders may coexist with other types of musculoskeletal disorders. Many patients needing surgery for cubital tunnel syndrome have carpal tunnel, pronator teres, and/or thoracic outlet syndromes simultaneously.

Occurrence

Except for CTS, accurate data are lacking on the incidence and prevalence of nerve entrapment syndromes, both for the general population and for exposed occupational groups, partly because of the diagnostic difficulties noted above. Although cases have been described in the medical literature of entrapments other than CTS, few of these reports include sufficient information for adequate statistical analysis of the potential role of occupational or non-occupational factors. The prevalence of any single syndrome is probably low, except perhaps in populations highly exposed to factors that produce nerve compression at a particular body site, such as localized external compression or postural load.

In one study of three occupations, 18% of the workers had findings consistent with TOS on physical examination. The highest rate (32%) was observed in female cash register operators who had to keep their right arm elevated continuously, producing high static load on the muscles of the shoulders, neck, and back. The rate was 17% among men and women working on a heavy assembly line and 10% in office workers using video display terminals.

The frequencies of these disorders by age or gender are not well known. In the TOS study cited above, the prevalence was higher among women than among men in both office work and heavy industry. Gender differences in body size and resulting worker-machine fit may have accounted for some of this discrepancy.

Of 160 consecutive cases who required surgery for cubital tunnel syndrome, the recurrence rate was 13%. Most recurrences occurred in those with concomitant disorders, such as carpal tunnel syndrome or TOS.

Causes

The specific cause of a nerve entrapment depends on the particular nerve and where it is compressed. A peripheral nerve may be entrapped by either an internal or an external source of pressure. Internal entrapments may occur at a point where the nerve passes between two anatomic structures; they are most common near the joints. Certain body postures can cause increased direct pressure or tension on a nerve. Local shortened or tight muscles may act to further compress the nerves. External sources of pressure include mechanical force concentrations created by contact of the body with hard surfaces and sharp edges of tools, workstations, chairs, and equipment, especially where the nerve is close to the surface and vulnerable to external pressure.

For example, with repeated pronation (palm-down rotation) of the forearm, and especially with repetitive or forceful wrist or finger flexion, the pronator muscle becomes hypertrophied and compresses the median nerve as the nerve travels through the forearm (pronator teres syndrome). Repetitive elbow flexion, especially in combination with wrist flexion, may increase pressure within the elbow joint and compress the ulnar nerve as it passes through the elbow (cubital tunnel syndrome). Frequent, prolonged, or forceful exertions with the arm raised to or above shoulder height, or with the arm extended back behind the midline of the body, compresses the nerves and blood vessels for the entire arm at the thoracic outlet between the shoulder muscles, the rib cage, and the collarbone. Nerve compression may also be secondary to muscular contraction, muscle hypertrophy, or tendon inflammation and swelling.

External mechanical compression of a nerve by hard or sharp edges may occur at a site where the nerve is close to the surface; an example would be resting the elbow on an unpadded table surface or corner (ulnar nerve compression in cubital tunnel syndrome) or repetitive use of a thigh-operated pedal (compression of the peroneal nerve). Mechanical pressure applied to the palm (using the hand as a hammer or using a tool whose handle presses into the center of the hand) may compress branches of the median or ulnar nerve. Molded tool handles with indentations too wide or too narrow for an individual's fingers may compress the digital nerves along the sides of the fingers. Carrying a weight on the shoulders, such as a mailbag or other type of knapsack, may cause mechanical compression of the soft tissues at the shoulder. The sciatic nerve may become compressed from prolonged sitting on a small or inadequately padded seat.

In addition to ergonomic causes, occupational or non-occupational exposure to solvents or heavy metals, such as lead or mercury, damages the peripheral nerves and may cause symptoms and signs resembling those of nerve entrapment. This damage may also make the nerves more susceptible to the effects of mechanical pressure, although there are few epidemiologic data on the effects of such combined exposures. Occupational or non-occupational acute trauma, such as fracture of a bone near the site of the nerve damage, may leave scar tissue or a bone fragment pressing on a nerve.

Non-occupational causes include alcoholism, diabetes mellitus, thyroid problems, rheumatoid arthritis, kidney failure, malnutrition, possibly obesity, and other systemic conditions causing swelling that may lead to nerve entrapment. These conditions are relatively common in the general population and should not be permitted to distract attention from the workplace and easily preventable causes, especially when a high rate of one or more specific nerve compression syndromes is observed. Uncommon non-occupational associations include Paget's disease of bone, neoplasms including multiple myeloma, gout, myxedema, acromegaly, renal disease anemia, Guillain-Barré syndrome, and toxic shock syndrome.

Pathophysiology

The mechanism of damage is mechanical compression or irritation of a peripheral nerve between two body structures or between an internal structure and an outside source of pressure. This compression causes restricted movement of the nerve and probable disruptions in transport of nutritive materials for the nerve. In the early stages of nerve compression, there is impaired blood flow to the nerve, followed by an increase in scar tissue in the area. Prolonged compression may result in slowed or incomplete transmission of sensory and motor nerve signals through the limb to the part supplied by that nerve, causing discomfort and loss of nervous and muscular function of the innervated body part. Simultaneous compression of accompanying blood vessels may result in loss of blood supply to the nerve and aggravate the nerve damage.

TOS is posited to involve compression or irritation at the shoulder of the entire neurovascular bundle (the major artery and vein that supply the arm, and the nerve fibers that give rise to most of the peripheral nerves of the arm). Symptoms vary depending on whether nerves, blood vessels, or both are compressed. Blood vessel compression in TOS, unlike in brachial plexus nerve compression, may be identified through objective vascular tests. However, the existence of TOS is contested by some, and the diagnosis is problematic.

Prevention

Primary prevention involves the use of engineering controls (ergonomic design and selection of tools, tasks, and workstations) to minimize the force and repetition of manual exertions, mechanical compression, sources of excessive cold, and frequency and amplitude of vibration. Ergonomic design of the workstation and tools should eliminate the necessity of working with the body in an awkward or non-neutral posture, such as bent wrist, pronated forearm, or elevated upper arm (see Upper Extremity Musculoskeletal Disorders).

For secondary prevention, health care providers should be trained in interview and clinical examination procedures necessary to identify occupational peripheral nerve entrapments at an early stage. Once reported, cases should be treated conservatively, and jobs should be analyzed for ergonomic features that may be modified. Treatments include nocturnal splinting, and occupational or physical therapy, with postural re-education and stretching and strengthening of affected areas. Removing a worker from exposure is critical to prevent the disorder from progressing to irreversible nerve damage. Follow-up is important to ensure that job modifications have been effective, that "light-duty" jobs have been correctly selected to avoid continuing ergonomic stress, and that symptoms and signs have not progressed. People with extant or previous peripheral nerve conditions should be cautious in considering new employment in jobs with repetitive manual demands.

Except for CTS, surgery to release the compressed nerve is rarely used and should be considered extremely carefully. While surgery may be needed to reduce pain and preserve nerve function, patients are often insufficiently advised regarding possible loss of function as a consequence of surgery. Moreover, surgery may be only temporarily effective if the worker is returned to an ergonomically stressful job that has not been modified; possible loss of muscle strength or build-up of scar tissue following surgery makes it imperative that job assignments be selected carefully to avoid recurrence. Use of transitional workshops for affected workers and graded retraining under the supervision of an experienced physical therapist may also reduce risk of recurrence.

Other Issues

There are no epidemiologic data on the effects of combined exposures to ergonomic stressors and peripheral neurotoxins, such as heavy metals or solvents. It is also unclear whether smoking might play a possible role as an aggravating factor because of its adverse effect on peripheral circulation.

Further Reading

Bernard BP. (Ed.). *Musculoskeletal Disorders and Workplace Factors: A Critical Review of Epidemiologic Evidence for Work-Related Musculoskeletal Disorders of the Neck, Upper Extremity, and Low Back.* Washington, DC: NIOSH, 1997.

Cailliet R. *Neck and Arm Pain* (3rd ed.). Philadelphia: F.A. Davis, 1991.

Cailliet R. *Soft Tissue Pain and Disability* (3rd ed.). Philadelphia: F.A. Davis, 1996.

Feldman RG, Goldman RH, Keyserling WM. Peripheral nerve entrapment syndromes and ergonomic factors. *American Journal of Industrial Medicine* 1983; 4: 661-681.

Keogh JP, Gucer PW, Gordon JL, Nuwayhid I. The impact of occupational injury on injured worker and family: Outcomes of upper extremity cumulative trauma disorders in Maryland workers. *American Journal of Industrial Medicine* 2000; 38: 489-506.

Novak CB, Mackinnon SE. Nerve injury in repetitive motion disorders. *Clinical Orthopaedics and Related Research* 1998; 351: 10-20.

Piligian G, Herbert R, Hearns Met et al. Evaluation and management of chronic work-related musculoskeletal disorders of the distal upper extremity. *American Journal of Industrial Medicine* 2000; 37: 75-93.

Rempel DM, Dahlin L, Lundborg G. Biological response of peripheral nerves to loading: Pathophysiology of nerve compression syndromes and vibration-induced neuropathy. *Work-Related Musculoskeletal Disorders: The Research Base.* Washington, DC: National Academy Press, 1998, pp. 98-115.

Seradge H, Owen W. Cubital tunnel release with medial epicondylectomy: factors influencing the outcome. *Journal of Hand Surgery (American)* 1998; 23: 483-491.

Sluiter JK, Rest KM, Frings-Dresen M Criteria document for evaluation of the work-relatedness of upper extremity musculoskeletal disorders. *Scandinavian Journal of Work, Envirnoment, and Health* 2001; 27 (Suppl 1): 1-102.

Peripheral Neuropathy

ICD-10 G61-G62

Robert G. Feldman and Marcia H. Ratner

Peripheral neuropathy presents with insidious onset of symptoms, such as intermittent tingling and numbness, and may progress to dysesthesias and an inability to perceive sensation. Muscle weakness and eventual atrophy result from damage to the motor nerve fibers. Based on the structural components of the nerve primarily involved, toxic polyneuropathies may be subdivided into (a) axonopathies, which present as distal sensorimotor loss (most evident in the lower extremities, where the axons are the longest); (b) myelinopathies, such as spotty segmental demyelination; and (c) neuronopathies. The most common pattern seen in metabolic or toxic neuropathies is "dying back," or distal, axonopathy with segmental demyelination occurring as a secondary effect. Objectively, insensitivity to pinprick or touch indicates peripheral neuropathy. Two-point discrimination, position, vibration, and temperature sensation may also be impaired.

There are two main occupational causes of peripheral neuropathy. The more obvious cause involves laceration or traumatic injury such as nerve compression by repetitive motion and ergonomic inadequacies. The more difficult causal relation to establish is that of peripheral neuropathy induced by exposure to the neurotoxic effects of chemicals found in certain occupational settings (toxic neuropathies). The clinical manifestations of peripheral neuropathy include (a) numbness and tingling of the distal fingers and toes, often followed by more peripheral loss of sensation in the extremities, usually affecting the lower extremities first; and (b) eventually onset of weakness of hands and feet and muscles. On examination, there is decreased perception of the pinprick, temperature, joint position, and vibration. Deep tendon reflexes are reduced in the extremities. Such findings are bilateral in toxic neuropathies. Clinical findings in compression neuropathies are localized to the laterality and area affected by the particular nerve that is compressed.

Depending on the type or severity of neuropathy, electrophysiological examination of nerves can reveal slowed conduction velocities, reduced sensory or motor action potentials or amplitudes, and/or prolonged latencies. Prolonged sensory action potentials serve as a complementary test to confirm clinically-observed diminished sensation. Slowed motor or sensory conduction velocity is generally associated with demyelination of the nerve fibers, while preservation of normal values in the presence of muscle atrophy indicates axonal neuropathy. However, exceptions occur when there is (a) progressive loss of motor/sensory nerve fibers in axonal neuropathy that affects the maximal conduction velocity (which results from the dropping out of the large-diameter, fast-conduction fibers); and (b) immature regeneration of nerve fibers. Regenerated fibers are known to conduct slowly, so conduction may be slowed early in the recovery period of axonal neuropathy; this usually occurs only in the distal nerve segments, but occurrence depends on the severity of damage.

Occurrence

Knowledge about occurrence of peripheral neuropathy or any other occupational neurological disease is limited to case series and a partial listing of compensation awards. A review of clinical case reports showed that out of 220 different chemicals, 149 had caused documented neurotoxicity in humans. In 1986, a total of 8723 workers' compensation awards for occupational conditions of the nervous system among both public and private sector employees were reported to the BLS from 23 states. The vast majority of these awards (98%) were for "diseases of the nerves and peripheral ganglia." The remainder were for conditions that either were unclassified or affected the central nervous system. Most (63%) were for conditions that occurred in the manufacturing sector. The percentages attributable to either exposure to workplace toxins or other causes are unknown.

Causes

Clinical manifestations of neuropathy have been associated with exposure to heavy metals, organic solvents, pesticides, and other chemicals (Table 1). Arsenical neuropathy follows acute and chronic exposures and initially presents with dysesthesias and diminution of sensation in the feet and hands. Neuropathy induced by thallium exposure is progressive, beginning with parasthesias of the extremities; pain increases in severity as weakness and atrophy of muscles develops. The skin of the trunk is painful to touch, and the muscles and nerves are tender to pressure.

Neuropathy caused by organic solvents, such as methyl butyl ketone (MBK) and *n*-hexane (which are both metabolized to the neurotoxicant 2,5-hexanedione), initially manifests as tingling paresthesias in the fingers and toes, with loss of sensation to pinprick, temperature discrimination, and touch. Organic solvent intoxication results in slowed peripheral motor conduction velocities and prolonged terminal latencies because of secondary changes in the myelin sheath. Cranial neuropathies have been observed following exposure to trichloroethylene and its environmental breakdown product dichloroacetylene; the effects are primarily on the trigeminal nerve, causing motor and sensory loss in the face.

Certain monomers, such as acrylamide, cause a peripheral neuropathy that is similar to that associated with exposure to the solvents *n*-hexane and MBK. Clinical manifestations include initial sensory changes, which are followed by disturbances of peripheral motor nerve fiber function.

Table 1. Exposures Associated with Peripheral Neuropathy

Neurotoxin	Major Uses or Sources of Exposure
Metals	
Arsenic	Pesticides
	Pigments
	Antifouling paint
	Electroplating industry
	Semiconductors
	Smelters
	Chromium-arsenate treated lumber
Lead (Motor nerves only)	Solder
	Lead shot
	Illicit Whiskey
	Insecticides
	Auto body shops
	Foundries
	Storage battery manufacturing plants
	Lead-stained glass
	Smelters
	Lead pipes
Mercury	Scientific instruments
	Photography
	Pigments
	Chlorine production
	Electroplating industry
	Amalgams
	Taxidermy
	Electrical equipment
	Felt making
	Textiles
Thallium	Rodenticides
	Manufacturers of special lenses
	Infrared optical instruments
	Fungicides
	Mercury and silver alloys
	Photoelectric cells
Solvents	
Carbon disulfide	Manufacturers of viscose rayon
	Preservatives
	Rubber cement
	Electroplating industry
	Paints
	Textiles
	Varnishes

(continued)

Table 1 *(cont.)*. Exposures Associated with Peripheral Neuropathy

Neurotoxin	Major Uses or Sources of Exposure
Solvents *(cont.)*	
Methyl butyl ketone	Paints
	Quick-drying inks
	Metal-cleaning compound
	Varnishes
	Lacquers
	Paint removers
n-Hexane	Lacquers
	Rubber cement
	Pharmaceutical industry
	Stains
	Printing inks
	Glues
Perchloroethylene	Paint removers
	Extraction agent for vegetable and mineral oils
	Dry-cleaning and textile industries
	Degreasers
Trichloroethylene	Degreasers
	Adhesive in shoe and boot industry
	Process of extraction of caffeine from coffee
	Paints
	Painting industry
	Varnishes
	Dry-cleaning industry
	Rubber solvents
	Lacquers
Monomers	
Acrylamide	Paper, pulp industry
	Water, waste treatment facilities
	Grouting materials
	Photography
	Dyes
	Laboratories using gel electrophoresis
Gases	
Ethylene oxide	Gas sterilization processes
Insecticides	
Organophosphorus compounds	Agricultural industry
	Commercial pesticide applicators

Occupational exposures to the gas ethylene oxide have been associated with peripheral neuropathy as well. Persons involved in sterilization processes are at increased for neurotoxic effects of this gas, which include peripheral sensory and motor disturbances.

Organophosphorus compounds, such as malathion, chlorpyrifos, mipofox, leptophos, and tricresyl phoshate, can produce a rare type of delayed-onset polyneuropathy, known as organophosphate ester-induced delayed neuropathy (OPIDN). The delayed neurotoxic effects are due to the inhibition of the enzyme neuropathy target esterase (NTE), which occurs within hours following exposure and is gradually restored; however, clinical onset occurs 3 to 4 weeks later. Symptoms present primarily as a distal symmetrical motor polyneuropathy, characterized as a distal weakness in the limbs. Sensory loss may occur, depending on dose and duration. Electrophysiological studies of subjects with organophosphate-induced neuropathy reveal small or absent amplitudes of sensory nerve action potentials.

Nonoccupational causes of peripheral neuropathy may be genetic, nutritional, infectious or post-infectious, paraneoplastic (associated with carcinomas or leukemia), metabolic (due to diabetes or thiamine deficiency), or physical (exposure to cold or radiation). Nonoccupational neuropathy may also be due to alcoholism, uremia, paraproteinemia, amyloidosis, or sarcoidosis. Alcoholic neuropathy is characterized by axonal degeneration and regeneration; electrophysiological testing reveals fibrillation potential, positive sharp waves, increased mean motor unit potential duration, reduced interference pattern, and reduced sensory and motor nerve conduction velocities.

Pathophysiology

Many of the chemicals associated with peripheral neuropathy are electrophilic agents that can bind with cellular macromolecules and disrupt axonal transport. Axonal degeneration and segmental demyelination of nerve fibers can occur. Most organic solvents, acrylamide monomer, and ethylene oxide gas cause mainly an axonal neuropathy mediated by the electrophilic metabolites of these compounds. Accumulation of neurofilamentous protein aggregatisms within axons is responsible for the neuropathy caused by diketones, such as *n*-hexane, 2,5-hexanedione, acrylamide, and MBK. Lead-induced peripheral neuropathy is generally of mixed segmental demyelination with some axonal changes. Pathological studies of thallium-exposed persons has revealed axonal degeneration. Trichloroethylene and its environmental breakdown product dichloroacetylene also damage myelin directly. In organophosphate-induced delayed neuropathy, the longer nerve fibers to the legs undergo axonal degeneration, as do the long spinal cord dorsal columns. Due to involvement of the spinal cord long tracts, organophosphate-induced delayed neuropathy is generally irreversible.

Prevention

Exposure to neurotoxins should be minimized. Effective industrial hygiene monitoring and control of respiratory, gastrointestinal, and dermal routes of chemical entry are discussed in Part I. Biological monitoring of workers for specific chemical exposures may be used as a guide to check on individual exposure levels, but careful attention must be paid to the interpretation of individual results. For more information, the ACGIH has issued TLVs and biological exposure indices (BEIs) for a number of chemicals.

New portable machines are being used to screen workers for some of the early sensory signs and symptoms that are sometimes the only manifestation of early peripheral nervous system dysfunction. These instruments can measure vibrotactile performance thresholds as well as sensory and motor function latencies and F-waves. They may provide possible objective evidence of worsening or improvements in sensation. Careful workstation and tool design will reduce the risk of physical injury to the nerves. Surgical release of entrapped nerves will correct peripheral neuropathy of physical origin.

Other Issues

Disease of the spinal cord can cause problems in differential diagnosis. Workers who have been exposed to lead may later develop a clinical picture indistinguishable from the amyotrophic lateral sclerosis-motor neuron disease forms seen in people without lead or other metal exposures. The diagnosis of toxic neuropathy due to workplace hazards must include a comprehensive examination that takes into account confounding presentations. The patient's medications should be evaluated, as some may produce neuropathic symptoms. Individuals with metabolic or hereditary disorders that manifest with peripheral neuropathy present variables that can confound the diagnosis of toxic peripheral neuropathy due to exposure to workplace hazards. Workers with neuropathy due to diabetes, alcohol, previous injury, or post-infectious syndromes are more susceptible to compression neuropathy.

Toxic neuropathy may be considered in cases of unexplained neuropathy in patients with environmental or occupational exposures. These are often diagnoses of exclusion. Glucose tolerance tests, however, are important to consider prior to making such a diagnosis since subclinical diabetes is the most common cause of unexplained abnormalities on NCV and EMG.

Further Reading

Feldman RG. *Occupational and Environmental Neurotoxicology.* Philadelphia: Lippincott-Raven, 1999, pp. 7-29.

Feldman RG, Jabre JF. Analysis of electrophysiological studies in arsenic exposure. In Chappell WR (Ed.), *Arsenic Exposure and Health.* London: Thomas Science Publishing, 1997, pp. 145-158.

Feldman RG. Niles C, Proctor S, et al. Blink reflex measurement of effects of trichloroethylene exposure on the trigeminal nerve. *Muscle & Nerve* 1992; 15: 460-495.

Grandjean P, Sandoe SH, Kimbrough RD. Non-specificity of clinical signs and symptoms caused by environmental chemicals. *Human & Experimental Toxicology* 1991; 10: 167-173.

Spencer PS, Schaumburg HH, Ludolph AC (eds.). *Experimental and Clinical Neurotoxicology*, 2nd Edition. New York: Oxford University Press, 2000.

Pesticides, Adverse Effects

ICD-10 X48

Patrick O'Connor-Marer and Marc B. Schenker

A pesticide is any substance used to prevent, destroy, repel, or mitigate insects, rodents, nematodes, fungi, weeds, or other organisms considered pests. Plant growth regulators, defoliants, desiccants, and insect repellents are also among the many substances regulated as pesticides. Pesticides are ubiquitous, being used in homes, schools, workplaces, and communities. Pesticides play an important pest management role in most conventional and organic agricultural operations worldwide. The widespread dissemination of pesticides, however, creates potential for related illness and injury, especially among pesticide handlers and agricultural workers.

Formulated pesticide products consist of active ingredients (those chemical compounds that have pesticidal action) and inert ingredients (chemicals such as soaps, adjuvants, and carriers having no pesticidal action, but nevertheless possibly being harmful to humans). The chemical components of some pesticide materials cause poisoning by interfering with biochemical and physiological functions. The nature and extent of injury depends on (a) the hazards of the inert components, (b) the toxicity of active ingredients, and (c) the dose. Some pesticide compounds are very toxic and cause poisoning at low doses; ingesting a few drops of these has the potential to cause severe illness or even death. Other pesticide compounds are so mildly toxic that large quantities of the active ingredient must enter the body before signs of illness or injury can be detected. Exposure to certain chemicals used as pesticides correlates to specific chronic disorders, although it is difficult to accurately predict what effects can result from long-term repeated exposures to most types of pesticidal chemicals.

There are several ways workers are exposed to pesticides, most commonly when they mix or apply them or when they enter, or perform work in, treated areas soon after pesticide application. People also may be exposed if they work near areas where pesticides are used, eat contaminated produce, or touch recently treated foliage, livestock or poultry, stored products, clothing, or furnishings. Spills or accidents may result in high levels of exposure. Sometimes the circumstances of exposure reveal the approximate amount absorbed. For example, exposure to drift from a spray application that has been properly diluted for field application is not likely to convey a large dose, unless exposure has been prolonged. In contrast, spills of concentrated material onto the skin or clothing may well represent a large dose of pesticide, unless the pesticide is promptly and properly removed.

In some cases, poisoning or injury may result from a single exposure to a specific pesticide. In other cases, symptoms do not occur until a person has been exposed repeatedly to small doses of a single compound or combination of materials over a period of time. Individuals commonly vary in their reaction to a pesticide exposure dose. Some people may exhibit no noticeable effect to a dose that causes severe illness in others. A person's age, body size, and health status often influence response to a given dose; thus, infants and young children are normally affected by smaller doses than are adults. In addition, women often are affected by lower doses than men.

Usually the most harmful levels of pesticide exposure result from occupational accidents. Workers in the agricultural industry have the highest number of pesticide-related accidents, most occurring during mixing or application, when workers handle concentrated pesticides.

Spills, explosions, or similar accidents during manufacturing and packaging of pesticide products have the potential to seriously injure plant employees and people living or working nearby. Persons involved in transporting pesticides risk possible illness and injury if pesticide containers rupture and spill their contents or are involved in a fire. Pesticide spills during transport also endanger the public.

The type and severity of pesticide poisoning depends on the toxicity and mode of action of the pesticide, the amount absorbed (dose), and the speed of absorption, metabolism, and excretion. Severity of poisoning may sometimes be reduced by prompt decontamination and medical treatment. Very small doses usually produce no symptoms. Depending on the toxicity of the pesticide, larger doses may cause severe illnesses. Effects of exposure may be localized, such as irritation of the skin, eyes, nose, or throat, or generalized, when pesticides are absorbed. Pesticides may adversely affect several organ systems concurrently.

Symptoms vary among classes of pesticides and among pesticides within a class. Presence and severity of symptoms are usually proportional to dose. Common symptoms include rashes, headaches, and irritation of the eyes, nose, and throat, all of which symptoms may resolve within a short period of time and are often difficult to distinguish from similar symptoms produced by allergies and respiratory infections. Other more-specific symptoms, which might indicate exposure to certain classes of pesticides, include blurred vision, dizziness, excessive sweating, weakness, nausea and vomiting, abdominal pain, diarrhea, extreme thirst, and blistered skin. Serious poisoning may also lead to apprehension, restlessness, anxiety, unusual behavior, shaking, convulsions, or unconsciousness. Certain individuals exhibit allergic reactions when exposed to some types of pesticides; the substance causing the reaction may be the pesticide or one of the ingredients in the pesticide formulation. Symptoms often include breathing difficulties, sneezing, eye watering and itching, rashes, apprehension, and general discomfort.

Most pesticide poisoning is reversible, although irreversible damage can occur and result in a chronic illness, disability, and/or death.

Occurrence

Surveillance for adverse health effects due to pesticide exposure tends to focus on acute effects, and is not generally designed to identify chronic health effects. There are few population-based pesticide surveillance programs, and little data on the completeness or representativeness of those that exist. The California Pesticide Illness Surveillance Program, operated by the California Environmental Protection Agency, is one major source of information about pesticide effects on human health. In the decade from 1993 through 2002, it collected 12,275 case reports of pesticide illnesses. These were classified as definitely (21%), probably (47%), or possibly (32%) related to pesticide exposure based on investigation by county agricultural commissioners. Of the 12,275 cases, 82% resulted from exposures during work, and 41% from exposure to pesticides used for agricultural purposes. Field worker exposures constituted 17% of the total, with 56% due to residue, 41% to drift, and 3% to other or unknown causes.

Causes

Table 1 lists many occupations in which pesticide exposure may occur. Although workers who apply, mix, and/or load pesticides are most at risk, farm field workers, tractor drivers, and irrigators may receive injurious exposures if they are not protected from pesticide residues. The establishment of restricted-entry intervals following application of highly toxic or otherwise hazardous pesticides has been an important step in reducing farm worker injury. Techniques, such as reducing drift and applying pesticide spray when workers are not present in adjacent fields, are other regulatory controls designed to reduce potential for exposure of workers in nearby areas.

Workers who perform maintenance or repair on pesticide application or handling equipment may come in contact with pesticide residues on that equipment. Oil-soluble pesticides are a major concern because they accumulate more in grease deposits and on oily surfaces and may be difficult to remove. Employees in packing sheds and food processing plants may unknowingly be exposed to contaminated produce or soil-borne residues, especially when persistent pesticides have been used during crop production. The activities of greenhouse and nursery workers often expose them to treated foliage due to the density of plants and the narrow aisles. Greenhouses have limited ventila-

Table 1. Occupations in Which There May Be Risks of Pesticide Exposure

agricultural field workers
agricultural irrigators
agricultural pesticide applicators
agronomists
aquatic area workers
building maintenance workers
crop duster maintenance workers
emergency responders
entomologists
farmers and farm workers
field scouts
firefighters
flaggers
forestry workers
formulators of pesticide products
fumigators
golf course workers
greenhouse, nursery, mushroom house workers
hazardous waste workers
institutional maintenance crews
interior plantscapers
landscapers
livestock handlers
maintenance gardeners
manufacturers of active ingredients
marina workers
medical personnel
mixers and loaders of pesticides
park and campground workers
pest management consultants
pesticide application equipment maintenance workers
pesticide applicators
plant pathologists
produce packing-house workers
public health pest control workers
recreational area workers
research chemists
right-of-way workers
rodent control specialists
sewer line workers
storage and warehouse workers
structural pest control operators
transporters of pesticides
treaters of contaminated workers
veterinarians
weed specialists
wood treatment workers

tion, increasing the potential for respiratory, dermal, and ocular exposures. Similar conditions exist for pest-control operators working in enclosed areas of buildings, homes, warehouses, factories, and offices. Residues remaining after fumigation of buildings and other work areas are a source of possible exposure to people if they enter these areas too soon after application. Fabrics, furniture, and carpeting are sometimes treated to prolong useful life by protecting them from insect damage or reducing the build-up of fungi or bacteria, and, in some instances, may subject people to low levels of exposure. Livestock and pets treated with pesticides may be a low-level source of exposure if people work in close contact with recently treated animals. Lawns, shrubs, and other parts of residential, industrial, and public landscape that have been treated for pests sometimes can be sources of pesticide exposure.

Certain types of pesticides, such as some chlorinated hydrocarbons, persist in the environment for as long as 50 to 100 years or more. Other pesticides, organophosphates for example, break down rapidly (within a few hours to several weeks or months) under normal environmental conditions. The persistence of a pesticide is often referred to as its half-life-the measure of how long it takes for the material to be reduced to half of the amount originally applied. Besides the chemical nature of the pesticide, there are other factors that influence persistence. The types of adjuvants in a pesticide formulation affect persistence. Pesticides dissolved in oils or petroleum solvents may volatilize more slowly than water-soluble materials, and therefore persist longer. The pH of the water used for mixing pesticides affects the speed of degradation. In addition, the pH of plant or animal tissues may have a similar influence. High temperatures and humidity cause chemical changes in some compounds, accelerating breakdown. Sunlight produces photochemical reactions that decompose many pesticides.

Pesticides enter the body via skin or eye absorption, ingestion, or inhalation. Dermal contact is the most frequent route of entry. Pesticides on the skin may cause a mild dermatitis or more severe skin injury. The ability of a pesticide to be absorbed dermally depends on chemical characteristics of the pesticide and its formulation components. Pesticides that are more soluble in oil or petroleum solvents penetrate the dermis more easily than those that are more soluble in water. In addition to injury to ocular tissues, the eyes provide another route for pesticide absorption. Ingestion may occur by accidentally drinking a pesticide (usually because it has been improperly stored in a beverage or food container), by accidentally splashing liquids or dusts into the mouth during mixing or application, or by eating foods or drinking beverages that have become contaminated during pesticide handling. Intentional ingestion during suicide attempts also accounts for many illnesses and deaths in some countries. Smoking while handling pesticides may also increase the potential for ingestion. In the lungs, some types of pesticides are quickly absorbed and transported to other sites. Certain chemicals used as pesticides or adjuvants in a pesticide formulation directly damage lung tissue.

Prevention

Pre-Market Testing and Registration: Federal law requires that before selling or distributing a pesticide in the United States, a person or company must obtain a registration of this product from the EPA. Before registering a new pesticide or new use for a previously registered pesticide, the EPA must ensure that the pesticide, when used according to label directions, can be used with reasonable certainty of no harm to human health and without posing unreasonable risks to the environment. To make such determinations, the EPA requires more than 100 scientific studies and tests from applicants. Where pesticides may be used on food or feed crops, the EPA also sets tolerances (maximum pesticide residue levels) for the amount of the pesticide that can legally remain in or on foods.

Most states conduct a review of each label of pesticides sold in order to ensure that they comply with federal labeling requirements and any additional state restrictions of use. States may require the registration of pesticides and inert ingredients that are exempt from the requirements of federal registration.

The EPA is reregistering existing pesticide products in order to ensure that older pesticides meet current safety standards. Changes in the way a pesticide is used may be necessary to protect consumers, workers, or the environment.

Surveillance and Biological Monitoring: Medical surveillance of pesticide-exposed workers includes several components common to proper risk management in the workplace. There is a need for both broad-based efforts to reduce exposure and prevent disease, as well as focused efforts to monitor those workers at the highest risk. Each surveillance system has different strengths and weaknesses, and may be appropriate in different situations.

Some recent efforts have focused on individual susceptibility to the adverse effects of pesticide exposure. For example, individual differences in "toxifying" and "detoxifying" metabolic rates for alkylating and arylating pesticides may explain inter-individual susceptibility to their adverse effects. However, such individual susceptibility information is not well understood and generally has no known role in the development and management of risk among individuals with possible pesticide exposure.

Medical surveillance programs can be offered at the population level or the individual level. At the population level, monitoring of pesticide illnesses provides data on overall rate of acute pesticide illnesses, indicates possible localized epidemics, and identifies risk factors for acute illnesses in the population being monitored. Such surveillance can be performed at the national, state, county, or any other level for which data are collected. The most well-known example is the reporting requirement for pesticide illnesses in California. This system has been used to identify individual cases or clusters of pesticide illness for evaluation, to monitor the overall number and distribution of pesticide illnesses in the state, and to investigate risk factors for acute pesticide illnesses. Using a surveillance system as a case-finding database, one can also evaluate short- and long-term health outcomes of individuals.

Surveillance of pesticide illnesses at the national level has also been successfully achieved. For example, three national registers (of occupational accident and disease reports, hospitalizations, and deaths) were analyzed in Costa Rica in order to characterize pesticides associated with the highest risk of poisoning and population characteristics associated with increased occupational poisonings. A surveillance system in Malaysia identified the most common pesticides associated with poisonings in that country. An epidemic of acute pesticide poisonings was identified in Nicaragua by use of a ministry of health surveillance system for detecting pesticide poisonings.

Medical surveillance for pesticide illness at the individual level can be used with various data sources to detect probable cases. Surveillance may also be conducted by measurement of biological markers of exposure to pesticides. Such markers may detect either pesticide metabolites or the effects of pesticide exposure.

The Sentinel Event Notification System for Occupational Risks (SENSOR), sponsored by NIOSH, uses targeted sources of information to recognize and report selected occupational disorders to a state surveillance center. This system is a cooperative federal-state effort designed to develop local capability for preventing selected occupational disorders. It has been used to monitor various occupational diseases, including pesticide poisoning. Other surveillance can be conducted through use of other databases of mortality, hospital discharge, and poison control center data.

Recent advances have improved the ability to estimate long-term, chemical-specific pesticide exposures in epidemiologic studies. Detailed questionnaires can be used to collect intensity-related exposure information, such as maintenance or repair of mixing and application equipment, work practices, and personal hygiene. While such methods are useful for epidemiologic studies of chronic effects from pesticide exposure, they are not appropriate for surveillance of acute pesticide illness.

Exposure assessment for pesticides may be conducted on environmental samples or by measurement of the pesticide or its metabolites in body tissues (biologic monitoring). Biologic markers exist for several pesticides and pesticide metabolites. Biologic monitoring measures the actual dose absorbed by any route, thus reflecting actual exposure conditions, work practices, and use of personal protective equipment. In some instances, biologic markers may reflect both occupational and nonoccupational exposures. Methods for biologic monitoring of pesticides have been reviewed, reflecting biologic exposure indices for several classes of pesticides, including phenoxyacids, quaternary ammonium compounds, coumarin rodenticides, synthetic pyrethroids, organochlorine pesticides, and chlortriazines.

One advantage of biologic monitoring of pesticides or their metabolites is that exposures can be measured at very low concentrations, reflecting exposure not known to be associated with adverse health effects. For example, urinary alkylphosphates reflect exposure to organophosphate pesticides (OPs) and can be measured to evaluate acute effects of OP exposure. Paraquat can be measured at extremely low concentrations in the urine, reflecting exposures not known to be associated with adverse acute effects. Exposure to high concentrations of paraquat results in respiratory toxicity, and urinary monitoring may be useful to identify workplace factors associated with paraquat exposure so they can be minimized. Biologic markers of pesticide exposure are useful for epidemiologic studies, but have significant limitations for medical surveillance programs, the major one being the short half-life of most currently used pesticides (usually 1 to 2 days). The cost of these assays also limits their use for routine surveillance.

The most well-established biologic marker for monitoring OP exposure is blood cholinesterase. Cholinesterases hydrolyze esters of choline; for example, acetylcholinesterase hydrolyzes the neurotransmitter acetylcholine. Acetylcholinesterase is an extremely useful biomarker for evaluation of acute illness possibly due to cholinesterase-inhibiting pesticides. It has also been used for monitoring exposure to a broad range of OPs and carbamate pesticides. Monitoring of blood cholinesterase is required in California for workers using OPs and carbamates. Recently-developed automated kits for cholinesterase measurement are not recommended for routine use without standardization against a laboratory standard. Because of poor correlation of cholinesterase assays among commercial laboratories, there is a need for better standardization of assay methods.

There are several components of surveillance programs for workers exposed to OPs or carbamates, including: (1) identification of high-risk populations requiring surveillance, (such as mixers, loaders, applicators, flaggers, and others who directly handle pesticides); (2) baseline cholinesterase determination (red cell cholinesterase is the preferable measurement, although problems with standardization of this assay persist); (3) periodic surveillance (the repeat testing interval is variable, and may depend on the type of system used and the duration of pesticide application); and (4) criteria for medical removal from work (although exact criteria are difficult to establish because of variability in blood cholinesterase measurements, depressions of 15% to 30% from baseline should be medically evaluated).

Appropriate Diagnosis and Treatment: Recognition of acute pesticide poisoning and prompt treatment, essential elements in the prevention of serious adverse effects, are particularly important because many symptoms of mild acute pesticide poisoning are nonspecific and may be misdiagnosed as other health problems, such as systemic viral illness, gastroenteritis, or asthma. It is important to consider possible pesticide poisoning in any individual with an altered state of consciousness or with findings that suggest OP poisoning. A history of working with pesticides should increase suspicion of this diagnosis, but pesticide poisoning can also occur from nonoccupational exposures.

The signs and symptoms of acute pesticide poisoning due to OPs and/or carbamates are sometimes remembered by the mnemonic MUDDLES (miosis, urination, defecation, diaphoresis, lacrimation, excitation, and salivation). OPs and carbamates inhibit acetylcholinesterase at nerve endings, resulting in an accumulation of acetylcholine. Severe OP poisoning must be distinguished from an acute cerebrovascular accident and, especially in agricultural workers, heat stroke, heat exhaustion, and infection. Clinical manifestations reflect the organ systems where acetylcholine is the neurotransmitter (Table 2). Severity of the symptoms may reflect dose; pre-existing medical conditions, such as asthma, and localized effects.

Treatment of possible OP poisoning should never be delayed pending results of blood cholinesterase testing. If the clinical diagnosis of OP poisoning is suspected, samples should be drawn for laboratory analysis and a test dose of atropine administered. The required dose of atropine depends on the severity of the poisoning. Other considerations in treatment include: (1) decontamination, including bathing of skin and gastric lavage, as appropriate; (2) increasing doses of atropine every 15 minutes to reverse the muscarinic effects; (3) pralidoxime chloride (2-PAM), only for OP poisoning; and (4) clearance of secretions, artificial ventilation, oxygenation, and other supportive measures, as clinically indicated. More complete descriptions of the recognition and management of pesticide poisonings can be found in a book available free of charge

Table 2. Signs and Symptoms of Acute Organophosphate Poisoning

System	Receptor Type	Organ	Sign or Symptom
Parasympathetic	Muscarinic	Eye, iris, and ciliary muscles	Miosis
Sympathetic		Glands: lacrimal, salivary, respiratory, gastrointestinal, sweat	Tearing, salivation, bronchorrhea, pulmonary secretion, nausea, vomiting, diarrhea, frequent urination, perspiration
		Heart: sinus node, AV node	Bradycardia, arrhythmias, heart block
		Smooth muscle: bronchial, gastrointestinal	Bronchoconstriction, cramps, vomiting, diarrhea
		Bladder	Frequent urination, incontinence
Neuromuscular	Nicotinic	Skeletal muscle	Fasciculation, cramps, followed by weakness, loss of reflexes and paralysis
Central nervous system		Brain	Headache, dizziness, malaise, confusion, convulsions, and, as a late effect, loss of consciousness

from the EPA (see Further Reading) and in standard medical texts on diagnosis and treatment.

There are many other pesticides, used for control of diverse pests ranging from insects to weeds to rodents (Table 3). Acute health effects vary with many factors, including dose, route, duration of exposure, and chemical structure. Adverse health effects for specific agents should be ascertained from more detailed reference texts.

Protecting People through Pesticide-Use Decisions: Often several pesticides can be used for control of the same weed, insect, pathogen, nematode, or vertebrate pest in a particular situation. One of the ways that employers can reduce hazards to employees is to avoid or reduce the use of specific pesticides, and when needed, select the least hazardous materials for the pest control situation. Integrated pest management (IPM) is a method of controlling pests that reduces pesticide use and therefore reduces human health and environmental concerns. IPM uses life history information and extensive monitoring to understand a pest and its potential for causing economic damage. Control is achieved through multiple approaches including prevention, cultural practices, exclusion, natural enemies, host resistance, and limited pesticide applications. The goal is to achieve long-term suppression of target pests with minimal impact on nontarget organisms and the environment. Selecting a pesticide that is effective and that does not present hazards to people can be a difficult task. In order to make decisions, pest managers often gather information about pesticide choices from pesticide labels, county agents, consultants, university publications, and other sources. Factors such as cost, hazards to users, label-imposed plantback restrictions (restrictions that limit the type of commodity that can be grown in an area for a designated period of time after a certain pesticide has been used), persistence characteristics, ease of use, ability of the pesticide to combine with other materials, its effects on natural enemies and beneficial insects, and required restricted-entry intervals and harvest limitations all have an influence on which pesticide a pest manager chooses.

Each pesticide has a toxicity category rating that suggests the relative hazard of the pesticide to people, animals, and plants in the environment. The signal words "danger," "warning," or "caution" printed on product labels reference the hazard categories. Hazards are modified by such factors as formulation type (for example, microencapsulated formulations are safer for applicators to use than wettable powders), persistence in the environment, and amount of pesticide used.

Regulation: Registration of pesticide products and establishment and enforcement of regulations for transporting, handling, applying, and disposing of these products play important roles in preventing harmful exposures. Regulation of pesticides does not focus solely on assessing toxicity, but also on managing risk by controlling exposure. State and federal pesticide regulatory programs focus not on eliminating all pesticides, but on protecting people and the environment from harmful exposures. If a pesticide product cannot be used

Table 3. Major Categories of Pesticides and Associated Acute Adverse Health Effects*

Type of Pesticide	Use	Examples	Major Acute Adverse Health Effects
Organophosphate	Insecticide	Azinphos-methyl, chlorpyrifos, tribufos (DEF), demeton, diazinon, dichlorvos, dimethoate, ethyl 2,4-dichlorophenyl thionobenzene phosphorate (EPBP), famphur, fenthion, malathion, mevinphos, parathion, phenthoate, phosalone, tetraethyl pyrophosphate, trichlorfon	Salivation, incontinence, convulsions, sweating, diarrhea, meiosis, rhinorrhea, bronchorrhea, pulmonary edema (see Table 2)
Carbamate	Insecticide	Aldicarb, bendiocarb, carbaryl, carbofuran, dioxacarb, isoprocarb, methiocarb, methomyl, pirimicarb, propoxur, thiodicarb	Symptoms are similar to those for organophosphates.
Organochlorine	Insecticide	Aldrin, benzene hexachloride, chlordane, chlordecone, DDT, dicofol, dieldrin, endosulfan, endrin, heptachlor, hexachlorobenzene, lindane, methoxychlor, mirex, toxaphene	Hyperesthesias, paresthesias, headache, dizziness, nausea, vomiting, mental confusion, convulsions, respiratory depression.
Biological	Varies with agent	azadirachtin, nicotine, pyrethrum, rotenone	Varies with agent.
Chlorophenoxy acids	Herbicide	2,4-D (2,4-dichlorophenoxyacetic acid); 2,4-DP (2,4-dichlorophenoxy propionic acid); 2,4-DB (2,4-dichlorophenoxy butanic acid); 2,4,5-T (2,4,5-trichlorphenoxyacetic acid); MCPA (4-chloro-2-methylphenoxy acedtic acid); MCPB (4-chloro-2-methylphenoxy butanoic acid); MCPP (2-methyl-4-chlorophenoxy propionic acid); 2-methyl-3,6-dichlorobenzoic acid	Skin, mucous membrane and respiratory irritation, vomiting, diarrhea, headache, confusion.
Paraquat and diquat	Herbicide	Paraquat, diquat	Vomiting, dysphagia, oropharyngitis, restlessness, jaundice, cyanosis, hemoptysis.
Arsenicals	Pesticide	Arsenic acid, arsenic trioxide, cacodylic acid, copper arsenite, lead arsenate, methane arsonic acid, sodium arsenate, zinc arsenate	Garlic odor, vomiting, abdominal pain, bloody-watery stools, headache, dizziness, drowsiness, confusion, muscle weakness and spasms, convulsions, hematuria, cardiac arrhythmia, jaundice.
Fungicides	Fungicide	Substituted benzenes (chloroneb, chlorothalonil, dicloran); thiocarbamates (metam-sodium, thiram, ziram); EBDC compounds (mancozeb, maneb, nabam, zineb); thiophthalimides (captafol, captan, folpet); copper compounds; organomercury; organotin; cadmium compounds	Variable with agent. Dermatitis, mucous membrane irritation common. Systemic toxicity with ingestion.
Fumigants	Multiple organisms	Halocarbons, hydrocarbons, nitrogen compounds, oxides and aldehydes, phosphorus and sulfur compounds	Varies with agent. Highly toxic. Irritants, CNS and respiratory effects common.
Coumarins, indandiones	Rodenticides	Coumachlor, warfarin, chlorophacinone, diphacinone	Nosebleed, hematuria, melena, ecchymoses.

*Acute adverse health effects depend on many factors, including dose, route, and duration of exposure.

safely, then its use will be banned; however, the initial step is to impose strict controls on the use.

Regulatory provisions include requiring pesticide manufacturers to submit studies that document the potential health and environmental effects of their products; regulatory scientists then evaluate these data to ensure that the chemicals can be used safely. Risks are mitigated by imposing restricted-entry intervals of varying length for treated areas and harvest limitations on specific crops. Depending on the hazards of the specific product, the regulatory agencies may also require handlers to wear certain types of personal protective equipment, such as chemical-resistant gloves and footwear, eye protection, respiratory protection, and full-body protection.

Engineering Controls: In order to reduce the risks associated with handling and applying pesticides, engineering controls can be used that help to reduce or practically eliminate exposure to these toxic chemicals. Certain types of packaging of pesticide products, such as pre-weighed water-soluble bags, reduce the risk of exposure to workers mixing powdered pesticides; these bags dissolve in a spray tank, thereby removing the need for a worker to open large containers and measure specific quantities to put into the tank.

Closed transfer systems allow accurate and safe measuring of pesticides being put into a spray tank; concentrated liquid pesticides are moved from original shipping containers to a sprayer mixing tank with minimal or no worker contact and without the need to pour these materials into measuring vessels. Pesticide injection systems allow pesticides to be mixed directly with water in the sprayer plumbing system, usually at the sprayer pump or manifold, rather than in the water tank of the sprayer; only clean water is held in the sprayer tank, reducing spill and leak hazards and eliminating interior tank cleaning. Liquid pesticide containers that have been drained still contain amounts of concentrated pesticides; container rinse systems provide a way to thoroughly wash the inside of these containers and transfer the rinse water into the spray tank. Enclosed cabs installed on tractors protect operators from exposure to pesticides. Some types block spray droplets and mists while providing the operator with a comfortable air-conditioned environment. Certain types of enclosed cabs are acceptable for respiratory protection since their air-conditioning systems are equipped with multiple-stage filters that include a pre-filter, a high efficiency particulate air (HEPA) filter, and an activated carbon filter.

Training and Education: Most workplace exposures to pesticides result from carelessness, accidents, or failure to follow instructions. Reducing exposures and pesticide-related illnesses among workers who handle pesticides or who work in areas where pesticides have been applied requires training on proper and safe handling methods. It also involves educating workers on how to avoid contact with pesticide residues on treated surfaces and contaminated equipment. Workers should be instructed on proper emergency decontamination procedures as well as routine washing during and after work activities.

Further Reading

Karalliedde L, Feldman S, Henry J, et al. (Eds.). *Organophosphates and Health*. London: Imperial College Press, 2001.

O'Connor-Marer PJ. *Pesticide Safety: A Reference Manual for Growers*. (DANR Publication No. 3383), Davis, CA: University of California (1997.

O'Connor-Marer PJ. *The Safe and Effective Use of Pesticides* (2nd ed.) (ANR Publication No. 3324), Davis, CA: University of California, , 2000.

Routt Reigart J, Roberts JR. *Recognition and Management of Pesticide Poisonings* (5th ed.) (EPA 735-R-98-003). Washington, DC: U.S. Environmental Protection Agency, 1999. Available at: http://www.epa.gov/pesticides/safety/healthcare

Pigment Disorders

ICD-10 L80-L81

Michael E. Bigby

Pigment disorders can be identified by lightening or darkening of the skin. Hypopigmented or hyperpigmented macules often occur in areas of previous dermatitis or trauma to the skin or following exposure to certain chemicals.

Occurrence

Post-inflammatory hyperpigmentation and hypopigmentation are the most common pigment disorders. Any dermatitis or trauma to the skin, such as thermal or chemical burns, may lead to an increase or decrease in pigmentation in that area. Heavily pigmented individuals have postinflammatory pigment abnormalities most notably, and these abnormalities are slower to resolve. (See Contact Dermatitis for data on incidence rates for occupational skin diseases.)

Causes

Monobenzyl ether of hydroquinone, an antioxidant used in rubber manufacturing, was the first chemical implicated in inducing work-related loss of pigment (chemical leukoderma). Certain phenolic compounds used as antioxidants or germicidal disinfectants also produce pigment loss. Hypopigmented macules that are not in sites of direct contact may also occur after exposure to these chemicals. Hyperpigmentation can also be caused by exposure to aerosols of metallic silver, mercury compounds, and arsenic.

Pathophysiology

Chemically induced leukoderma occurs because the chemical either blocks the synthesis of melanin (the pigment responsible for skin color) or, in rare instances, causes direct injury to melanocytes (cells responsible for melanin synthesis). In post-inflammatory hypopigmentation, the transfer of melanosomes (pigment-containing granules) from melanocytes to keratinocytes is blocked. Post-inflammatory hyperpigmentation results from an increase in melanin synthesis and the release of large pigment granules into the dermis.

Prevention

Chemical leukoderma can be prevented by implementing engineering controls and by using protective clothing and gloves to prevent skin contact with hydroquinones and phenolic compounds. Monobenzyl ether of hydroquinone is rarely used now in rubber gloves. Post-inflammatory hyperpigmentation and hypopigmentation are best prevented by the prompt recognition and treatment of inflammatory skin diseases.

Other Issues

Currently, phenolic germicides, present in detergents, may be the most common cause of occupational leukoderma. Workers at risk are those who clean and sanitize floors and equipment, such as hospitals. Workers at risk for developing leukoderma due to the hydroquinones have often been in the cosmetics industry, where bleaching creams are made. Heavily pigmented workers (with olive-colored or darker skin) are at greatest risks for developing post-inflammatory hyperpigmentation or hypopigmentation.

Further Reading

Adams RM. *Occupational Skin Disease* (3rd ed.). Philadelphia: W.B. Saunders, 1999.
Taylor SC. Skin of color: Biology, structure, function, and implications for dermatologic disease. *Journal of the American Academy of Dermatology* 2002; 46: S41-S62.

Pleural Diseases, Asbestos-Related

ICD-10 J92.0

David C. Christiani

Three kinds of pleural disease result from exposure to asbestos fibers: (a) pleural plaques/discrete pleural thickening, (b) diffuse pleural thickening, and (c) benign exudative pleural effusion.

Pleural Plaques/Discrete Pleural Thickening

The presence of bilateral pleural thickening is a strong marker of previous asbestos exposure. The individual is usually asymptomatic unless other respiratory abnormalities are present or the plaques are extensive. Many pleural plaques are not detectable on routine radiograph. On chest x-ray, most plaques occur in the middle of the diaphragm, in the posterolateral chest wall between the seventh and ninth ribs. Plaques can calcify and can be seen along the diaphragm and lateral wall margins. They can also calcify along the mediastinum and left heart border. Plaques that are not seen on routine chest x-ray can be detected on computed tomography (CT), which is also helpful in distinguishing between pleural and parenchymal location of nodules. Pleural plaques due to asbestos exposure often appear alone, in the absence of underlying interstitial disease. The differential diagnosis of discrete pleural plaques is limited; these plaques may be seen with mesothelioma (malignant or benign fibrous mesothelioma), metastatic cancer, lymphoma, or myeloma.

Asbestos exposure frequently, though not always, causes bilateral plaques; hence, if the plaque formation is unilateral, other etiologies should be considered. The most frequent nonasbestos etiology of plaques is previous trauma. Calcified pleural plaques can occur with trauma or infection, either of which results most often in unilateral discrete pleural thickening. They also can occur with a rib fracture, x-ray therapy, scleroderma, and chronic mineral oil aspiration.

Diffuse Pleural Thickening

The clinical importance of distinguishing diffuse pleural thickening from discrete pleural plaque formation is not well established. Diffuse pleural thickening results from visceral pleural thickening, as opposed to the parietal pleural plaques noted above. The individual with mild diffuse pleural thickening is usually asymptomatic. If the pleural thickening is extensive, dyspnea and chest discomfort on inspiration may be experienced. On chest x-ray, diffuse thickening generally involves the costophrenic angle and has ill-defined borders on all x-ray views. It is also much less likely to calcify than circumscribed plaques. Pulmonary function tests can yield normal results or can demonstrate a restricted lung capacity. The diffusion capacity is normal when corrected for the actual lung volume. Diffuse pleural disease in the absence of parenchymal asbestosis can cause significant pulmonary impairment and has reportedly caused respiratory failure.

Diffuse thickening is less specific for asbestos exposure than discrete pleural plaques. Other conditions that can result in diffuse thickening include infection (tuberculosis or other bacterial infections), connective tissue diseases (especially scleroderma, rheumatoid arthritis, and systemic lupus erythematosus), and, rarely, sarcoidosis, uremia, or drug reactions.

Benign Exudative Pleural Effusion

A benign exudative effusion can occur suddenly often within the first 10 years after initial exposure to asbestos. The volume of effusions is usually small but may be several liters. In general, effusions resolve within 1 year without residual abnormalities, but they may leave blunted costophrenic angles or diffuse visceral pleural thickening. Fluid examinations are always sterile exudates that may be serous or seronsanguinous. Pulmonary function will be abnormal if the pleural effusion is large enough to create areas of atelectasis and shunting, or if there is concomitant parenchymal disease. The diagnosis is made by establishing exposure to asbestos, performing thoracentesis, ruling out other causes of effusion (usually requiring a pleural biopsy), and not finding evidence of malignancy for 3 years. The main differential diagnoses are malignant mesothelioma , lung cancer, and metastatic cancer from another primary site.

Occurrence

Some occupational groups with 40 or more years of exposure to asbestos have a prevalence of pleural plaque formation as high as 80%. The presence of plaques depends on the concentration and duration of exposure. The mean latent period is about 20 years from the time of first exposure to radiographic evidence of a plaque. There is a relationship between pleural thickening and cumulative asbestos dose, independent of the time elapsed since first exposure, but this relationship is less well established than that for parenchymal fibrosis. Plaques tend to calcify with time, although calcification is not always readily apparent on the plain chest x-ray. The prevalence of diffuse pleural thickening in occupationally exposed groups has not been well described.

In contrast to pleural plaques and diffuse thickening, asbestos-induced benign exudative effusions often occur early, many within 10 years from first exposure to asbestos. The prevalence of this disorder is not well known; in one study, it was dose-related.

All occupational groups exposed to asbestos (directly, in handling it, or indirectly, by being close to workers handling it) are at risk of developing asbestos-related pleural diseases. In the general population, surveys have revealed x-ray evidence for pleural plaques ranging up to 3% among those surveyed. In the United States, the vast majority of pleural plaques occur in people who are occupationally exposed, directly or indirectly, to asbestos.

Recent studies have demonstrated reduced pulmonary function in workers with circumscribed plaques in comparison with workers similarly exposed but without plaques. The relationship between development of plaques and subsequent risk for development of parenchymal fibrosis or lung cancer has not been well quantified. Other asbestos-related diseases can occur in the absence of pleural diseases.

Causes

Asbestos-related pleural diseases are caused by inhalation of asbestos fibers of any type. (See Asbestos-Related Diseases.)

Pathophysiology

The exact pathogenesis of pleural disease due to asbestos exposure is not known. Pleural plaques are discretely elevated, gray-white areas on the parietal pleura (rarely occurring on the visceral pleura). Microscopically, they are composed of relatively acellular strands of collagen interposed between normal tissue with intact lamellae and a covering layer of mesothelial cells. Asbestos bodies are not visible, but fibers can be demonstrated by electron microscopy and microchemical analysis. The mechanism by which pleural plaques undergo calcification is not clear.

The pathology of diffuse asbestos-related pleural thickening results from the thickening and fibrosis of the visceral pleura, often fusing with the parietal pleura. The pathogenesis is unclear. Diffuse pleural thickening appears to be a common sequela of old asbestos-induced pleural effusion, with costophrenic angle blunting and diffuse visceral fibrosis. It is not known how often diffuse visceral pleural fibrosis occurs after an asbestos-induced pleural effusion.

Prevention

Strategies necessary to prevent asbestos-related pleural diseases are the same as those needed to prevent asbestos-related parenchymal disease. (See Asbestos-Related Diseases.)

Further Reading

Chapman SJ, Cookson WO, Musk AW, et al. Benign asbestos pleural diseases. *Current Opinion in Pulmonary Medicine* 2003; 9: 266-271.

Hillerdal G. Pleural malignancies including mesothelioma. *Current Opinion in Pulmonary Medicine* 1995; 4: 339-343.

Post-Traumatic Stress Disorder

ICD-10 F43.1

Evan Kanter

Post-traumatic stress disorder (PTSD), as defined in the Diagnostic and Statistical Manual of Mental Disorders, Fourth Edition (DSM-IV), is an anxiety disorder that develops following exposure to a traumatic event characterized by actual or threatened death or serious injury, or a threat to the physical integrity of oneself or others. The response to the trauma must involve intense fear, helplessness, or horror. A diagnosis of PTSD then requires the presence of symptoms in each of three categories.

Reexperiencing symptoms include intrusive memories, nightmares, flashbacks, triggered psychological distress, and triggered physiological reactivity. *Avoidant* symptoms include avoidance of reminders of the trauma, amnesia for important aspects of the trauma, diminished interest in activities, isolation, emotional numbing, and a sense of a foreshortened future. *Hyperarousal* symptoms include difficulty sleeping, irritability or anger outbursts, difficulty concentrating, hypervigilance, and an exaggerated startle response. Symptoms must be present for at least one month and there must be clinically significant distress, such as impairment in social or occupational functioning.

Occurrence

The consensus estimate for the lifetime prevalence of PTSD in the United States is 5 to 6% in men and 10 to 21% in women. PTSD is at least twice as prevalent in women, despite a higher incidence of trauma exposure in men, and it persists longer. Twenty million Americans have had a lifetime episode of PTSD; 5.2 million currently meet clinical criteria.

Increasing attention is being given to industrial and workplace accidents as precipitants of PTSD. Moreover, a variety of occupations are by nature associated with risk for PTSD. Police officers, firefighters, rescue workers, disaster workers, emergency medical technicians, and those in military service all may face significant trauma exposure. Bank tellers and armored couriers may become victims of violent crime. PTSD has also been recognized in journalists who cover war, disaster, and social violence.

The rate of comorbidity of other psychiatric disorders, especially depression and substance abuse, is extremely high. Stress-related medical conditions are also highly comorbid. PTSD should be considered in the differential diagnosis of patients presenting with unexplained medical symptoms.

Forty percent of people with PTSD recover within the first year after trauma exposure. One-third to one-half of those with PTSD continue to be symptomatic, even after many years, and the disorder becomes a chronic, often disabling, illness. Traumatic exposure and PTSD have a significant impact on overall health, functional level, and health-care utilization.

Causes

Most trauma-exposed individuals do not develop PTSD. What causes some individuals to be susceptible is not well understood, though a variety of risk factors have been identified that increase the likelihood of developing PTSD and the severity of PTSD symptoms. The most important of these is the intensity and duration of the trauma. In effect, there is a dose-response relationship between trauma exposure and PTSD.

Interpersonal traumas of intentional human design are more likely to result in PTSD than accidents or "acts of God." Other aspects of the trauma itself that increase the likelihood of subsequent PTSD include physical injury, significant object loss (such as death of a loved one), and real or perceived responsibility for a tragic outcome. Predisposing factors related to the individual and the environment include age at the time of trauma, gender, social support, family history, childhood experiences (especially abuse), personality variables, and preexisting mental disorders. Genetics also plays a role, and twin studies have demonstrated a genetic component of as much as 30% for some PTSD symptoms.

Pathophysiology

Neuroendocrine studies have identified chronic physiologic alterations of both major stress-response systems. Evidence of sympathetic nervous system alteration includes increased circulating catecholamines and increased heart rate in response to trauma-specific cues. Alterations of the hypothalamic-pituitary-adrenal axis include low basal cortisol levels and increased glucocorticoid negative feedback.

Neuroimaging studies have demonstrated hyperresponsiveness of limbic structures, including the amygdala, that are associated with conditioned fear. Decreased hippocampal volume has been reported in people with PTSD. Hypofunction of anterior cingulate and prefrontal cortical regions associated with working memory, executive function, and emotional regulation has also been observed.

Prevention

Psychotherapies with established efficacy in the treatment of PTSD include cognitive behavioral therapy and various forms of exposure therapy. Normalization and validation of feelings is vital, as is education about the psychological effects of trauma. Serotonergic antidepressants and antiadrenergic agents are among the drug classes commonly used in psychopharmacologic treatment of PTSD. Pharmacologic prophylaxis by administration of such agents immediately after the trauma is an intriguing theoretical possibility.

Advance preparation and training may enhance the ability to cope with trauma exposure. Stress inoculation training attempts to simulate conditions as close as possible to what one expects to encounter on the job. Public education about psychological trauma and PTSD has been increasing. This should facilitate early recognition and treatment, leading to prevention of significant morbidity and disability.

Much attention has been given to "critical incident debriefing," with the assumption that allowing one to talk about the experience can help alleviate stress. A single session of debriefing immediately after the trauma has actually been shown to increase PTSD symptoms, however. The timing of the intervention is likely to be the crucial factor. Presumably there is a critical period for consolidation of traumatic memory. Retelling of traumatic events during this time window might cause overconsolidation. More research is needed to define the most effective parameters for debriefing interventions.

Other issues

PTSD is a prototypical syndrome; however, it is not the only possible pathological outcome of psychological trauma. Overwhelming traumatic stress may produce a variety of effects in different individuals, including mood disorders, other anxiety disorders, dissociative disorders, and somatization disorders. In addition, other clinical entities, such as fibromyalgia and chronic fatigue syndrome, are highly associated with trauma histories.

Further reading

Carlier IV, Lamberts RD, Gersons BP. Risk factors for posttraumatic stress symptomatology in police officers: a prospective analysis. *Journal of Nervous and Mental Disease* 1997; 185: 498-506.

Corneil W, Beaton R, Murphy S et al. Exposure to traumatic incidents and prevalence of posttraumatic stress symptomatology in urban firefighters in two countries. *Journal of Occupational Health and Psychology* 1999; 4: 131-141.

Feinstein A, Owen J, Blair N. A hazardous profession: war, journalists, and psychopathology. American Journal of Psychiatry 2002; 159: 1570-1575.

MacDonald HA, Colotla V, Flamer S, et al. Posttraumatic stress disorder (PTSD) in the workplace: a descriptive study of workers experiencing PTSD resulting from work injury. *Journal of Occupational Rehabilitation* 2003; 13: 63-67.

Ursano RJ, Fullerton CS, Vance K, et al. Posttraumatic stress disorder and identification in disaster workers. *American Journal of Psychiatry* 1999; 156: 353-359.

van der Ploeg E, Kleber RJ. Acute and chronic job stressors among ambulance personnel: predictors of health symptoms. *Occupational and Environmental Medicine* 2003; 60: i40-i46.

Pregnancy Outcome, Adverse

ICD-10 O20-O29

Laura Welch

Adverse pregnancy outcome includes spontaneous abortion, stillbirth, congenital defects, prematurity, and low birth weight. The World Health Organization defines spontaneous abortion as any nondeliberate interruption of an intrauterine pregnancy before the 28th week of gestation, timed from the date of the last menstrual period, in which the fetus is dead when expelled. After 28 weeks, if the fetus is dead at birth, it is considered a stillbirth. The term *congenital defects* refers to abnormalities of appearance or function that are present at birth. Prematurity is a birth before 37 weeks of gestation. The term *low birth weight* is used if a baby weighs less than 2500 grams at birth, regardless of gestational age.

Occurrence

The cause of most spontaneous abortions is fetal death due to fetal genetic abnormalities which are acquired rather than inherited. Other possible causes for spontaneous abortion include infection, conditions the mother may have, hormone factors, immune responses, and serious diseases of the mother, such as diabetes or thyroid problems).

It is estimated that up to 50% of all fertilized eggs die and are lost (aborted) spontaneously, usually before the woman knows she is pregnant. Among known pregnancies, the rate of spontaneous abortion is approximately 10% and usually occurs between the 7th and 12th weeks of pregnancy.

No surveillance system for spontaneous abortion exists in the United States. In Finland, all birth outcomes are registered, including early fetal loss of a recognized pregnancy, and this registry can be linked to registries of occupation to look for increased risk of miscarriage. Much literature in this field comes from use of these registries.

About 4% to 5% of all newborns have a detectable birth defect, 7.5% of newborns are of low birth weight, and 11% of all births are pre-term; and another group, less well defined, has developmental or functional deficits in childhood. Surveillance systems exist at the state and national levels to monitor changes in birth defects. The national system is not population-based, and 25% of births are not covered by state-based systems. It is possible that these systems would fail to detect birth defects secondary to occupational exposure; neither is able to track birth defects by industry, so an increase in one geographic area or industry might be lost in the aggregate data. Because the etiology of most birth defects is not known, it is also possible that an increase in environmental causes would be masked by a decrease in infectious causes, such as congenital rubella.

Causes

Table 1 lists the possible nonoccupational causes of spontaneous abortion, but for many of these there is no firm evidence nor is the degree of risk understood. Congenital defects are caused by genetic factors, environmental factors, or both. Most congenital defects of known cause are genetic, and although many environmental agents have been implicated as teratogens, relatively few have been associated with a specific defect.

Table 2 lists the nonoccupational causes of congenital defects. Table 3 lists the known and suspected occupational causes of adverse pregnancy outcome; among these are agents that have been well studied to date and have not been clearly linked with adverse reproductive outcome. Table 3 also includes a scale for measuring the certainty of the causal association between the occupational exposure and the reproductive effect. The scale does not incorporate details of dose; an association is listed as a strong one even if the data exist only at high doses in animals. In applying the details of these tables to an individual exposure situation, dose-response must be considered in

Table 1. Possible Nonoccupational Causes of Spontaneous Abortion

Strong Evidence	Weaker Evidence
Previous spontaneous abortion	Gravidity
Parental age	Oral contraceptives
Maternal smoking	Spermicidal agents
Maternal alcohol consumption	Prior induced abortion
Conception with intrauterine	Stress
device (IUD) in place	Nutrition
Ionizing radiation of gonads	Hormonal factors
Maternal medication/drug use	
Maternal illness, such as diabetes	
Uterine abnormalities, other	
gynecologic conditions	

Table 2. Nonoccupational Causes of Congenital Defects

Inherited disorders
Drugs and chemicals (exposure during pregnancy)*:
    Alcohol
    Androgens
    Anticoagulants
    Antithyroid drugs
    Chemotherapeutic drugs
    Diethylstilbestrol (DES)
    Isotretinoin
    Lithium
    Phenytoin
    Ribavirin
    Streptomycin
    Tetracycline
    Thalidomide
    Trimethadione
    Valproic acid
Infections during pregnancy*:
    Cytomegalovirus (CMV)
    Rubella
    Syphilis
    Toxoplasmosis
    Varicella
Ionizing radiation exposure during pregnancy*

*Exposure could occur in the occupational setting. For the drugs listed, the data are available from high-dose animal experiments or therapeutic use in humans; a threshold has not been defined but may exist for some adverse reproductive outcomes.

more detail. Table 4 lists industries that have been associated with adverse reproductive outcome, without identifying a specific agent.

Pathophysiology

After fusion of the sperm and ovum, the embryo undergoes growth and development. Weeks 1 and 2 are the period of rapid division of the zygote,

Table 3. Occupational Agents Causing Adverse Pregnancy Outcome

Agent	Outcome Seen in Human Studies	Outcome Seen in Animal Studies	Strength of Association*
Anesthetic gases	Spontaneous abortion	Birth defects	1,3,5,7
Antineoplastic drugs	Spontaneous abortion	Birth defects, fetal loss	2,3
Arsenic	Spontaneous abortion, low birth weight	Birth defects, fetal loss	2,3
Benzo(a)pyrene	None	Birth defects	1
Cadmium	Low birth weight, skeletal disease	Fetal loss, birth defects	2,3
Carbon disulfide	Menstrual disorders, spontaneous abortion		
Carbon monoxide (CO)	None specific to CO	Birth defects, low birth weight	2
Chlordecone	None	Fetal loss	2,5
Chloroform	Eclampsia	Fetal loss	1
Chloroprene	None	Birth defects	2,3,5
Ethylene glycol ethers	None	Fetal loss	2
Ethylene oxide	Spontaneous abortion	Fetal loss	1,3
Formamides	None	Fetal loss, birth defects	2
Inorganic mercury	Spontaneous abortion	Fetal loss	1,3
Lead	Spontaneous abortion, prematurity, neurological dysfunction in child	Birth defects, fetal loss	2,4
Organic mercury	Cerebral palsy, CNS malformation	Birth defects, fetal loss	2,4
Ozone	None	Fetal loss	1
Physical stress	Prematurity	None	4
Polybrominated biphenyls (PBBs)	None	Fetal loss	2
Polychlorinated biphenyls (PCBs)	Low birth weight	Low birth weight	2,4
Mixed solvents	Spontaneous abortion, CNS defects	Spontaneous abortion, CNS defects	2, 3, 7
Radiation	CNS defects, skeletal abnormalities	Fetal loss, birth defects	2,4
Selenium	Spontaneous abortion	Low birth weight	2,7
Tellerium	None	Birth defects	2
2,4-D	Skeletal defects	Birth defects	1,3,8
Video display terminals	Spontaneous abortion	Birth defects	1,8
Vinyl chloride	CNS defects	None	3,6
Xylene		Fetal loss	

*1 = limited positive animal data   2 = strong positive animal data
3 = limited positive human data   4 = strong positive human data
5 = limited negative animal data   6 = strong negative animal data
7 = limited negative human data   8 = strong negative human data

implantation, and formation of the bilaminar embryo. There are critical periods of sensitivity for major organ systems, during which an environmental exposure could cause damage to the embryo or fetus.

Table 4. Industries with a Reported Increased Risk of Adverse Reproductive Outcome in Exposed Women, without Linkage to Specific Exposures

Industry	Reported Outcome
Rubber industry	Spontaneous abortion
Leather industry	Spontaneous abortion
Chemical industry	Spontaneous abortion
Electronic industry (in solderers)	Spontaneous abortion
Metal work	Spontaneous abortion
Laboratory work	Spontaneous abortion, birth defects
Construction	Birth defects
Transportation	Birth defects
Communications	Birth defects
Agriculture and horticulture	Birth defects
Jobs with mixed solvent exposures	Birth defects, spontaneous abortion
Textiles	Spontaneous abortion

Because of these critical periods, the timing of an exposure largely determines the effect of a toxin. Delivering an insult to an organism just before or during the early stages in the development of an organ can cause a defect, whereas the same organ would not be susceptible to injury later in fetal development. A variety of agents given at the same developmental stage could cause the same defect, and the same agent given at different periods could cause different defects.

In the first 3 weeks, the most probable effect of significant exposure is severe damage and death of the embryo; organogenesis has not yet begun. The period from 4 to 9 weeks is the time when classic birth defects might be induced. After 9 weeks, organogenesis is basically complete, although the central nervous system continues to develop through the prenatal period. Exposures after 9 weeks may cause postnatal functional abnormalities. Carcinogens could potentially exert an effect at any stage in development.

The effect of exposure also varies with the dose of a toxin, with the degree of abnormal development increasing as the dose increases. At lower exposures, damages may be repairable; high doses may lead to death of the embryo; and doses in between may cause a congenital defect.

In addition, a toxic agent may affect the embryo even if one of the parents is exposed before conception. An exposure can damage germ cells in the ovary or the testis. Damage to the ova can persist for years, while damage to the sperm can be repaired by cell turnover if the germinal epithelium remains intact. Exposure prior to conception can affect the development of the fetus when the toxin persists in the maternal body. For example, PCBs are known to be stored in fat for a significant period of time, and lead is stored in bone. The storage of toxins exists in a steady state with the blood, so the fetus can be exposed to these body stores through the maternal circulation.

Prevention

Primary Prevention: Two basic approaches have been used to prevent exposure of a pregnant woman to an identified teratogen: reduction or elimination of her exposure to the toxin, or transfer of the woman to a different job without that exposure. Engineering control of exposure or product substitution are the preferred solutions (see Part I). Because there is a time lag between conception and the recognition of a pregnancy—a time when the embryo or fetus is especially vulnerable to the effects of exposure—significant exposure may occur if medical removal is the only preventive strategy.

In the face of uncontrollable exposure, medical removal may be needed. A job transfer can be used effectively for prevention, given that pregnancy lasts for a limited time and the woman can return to her job after delivery. In addition, if a woman becomes pregnant in a high-risk job, a transfer can be accomplished more quickly than product substitution or the institution of appropriate engineering controls. However, the implementation of "fetal protection" policies has often resulted in discrimination against women in hiring and job placement, and decisions about job placement for pregnant or "potentially pregnant" workers have been made without adequate knowledge or counsel. This issue is discussed in more detail below.

After the child is born, occupational exposures of the parents can continue to affect the child's growth and development. An understanding of these as well as of prenatal exposures can afford opportunities for primary prevention. Hazards can occur from substances that are (a) brought home on the clothes of a parent, (b) used in the homes of self-employed people, or (c) excreted in breast milk. As a general rule, good work practices, which include a change of clothes and care not to contaminate street clothes, will control the first two possibilities. Breast-feeding is discussed in more detail below.

Secondary Prevention: After a hazardous exposure has occurred, a congenital defect or developmental disorder could theoretically be prevented by terminating the pregnancy. Each time this question arises, the physician must analyze the circumstances and details of the exposure, providing appropriate consultation and advice in each case. The risk of an exposure must be evaluated against the background risk in any pregnancy. In most cases, an occupational exposure does not add significantly to the background risk.

Other Issues

Job Transfers: Suggesting a job transfer may raise difficult social and economic issues. As discussed above, a man may recover testicular function if he is removed from exposure. Removal from exposure must be negotiated with the employer in each case. A clinician who suggests this step needs to be sensitive to the dilemma that can potentially be created, because job transfer may mean income loss. As in similar situations during pregnancy, issues of benefits and legal rights should be referred to a lawyer with expertise in workers' compensation and employment law. A recommendation to a woman to attempt a job transfer or to leave her job for the duration of her pregnancy must be discussed in great detail. The woman should be informed of the magnitude of the risk, insofar as it is known, and of her legal rights. Some employers have a policy regarding pregnancy, and many will follow the recommendations of a physician regarding job placement.

In the United States, pregnant workers have limited legal rights. However, an employer is not required to transfer a pregnant worker to a safe job. The federal Pregnancy Discrimination Act (Public Law No. 95-555) does not require any employer to offer specific benefits, but it does require that employers provide coverage for pregnancy equal to their coverage of any other temporary medical disability. For example, if an employer guarantees any employee that a job will be available when the employee returns from sick leave, then the same accommodation must be made for leave due to pregnancy.

The law also prohibits discrimination against a pregnant worker in ways unrelated to fringe benefits. An employer cannot refuse to hire or promote a woman because she is or may become pregnant. The law does not require that unpaid maternity leave be available, so a pregnant woman is not protected from loss of her job if she opts to leave work because of a potentially dangerous exposure. Some states have mandated additional benefits for pregnant women.

The Family and Medical Leave Act (FMLA) requires an employer to provide up to 12 weeks of medical leave, some or all of which can be unpaid leave. If a doctor states that a pregnant woman cannot perform her job due to hazardous exposures, the FMLA may require the employer hold her job for this 12-week period.

The U.S. Supreme Court has declared a "fetal protection" policy to be unconstitutional. A woman may not be denied employment because she may become pregnant in a job that may pose a risk to the fetus.

Breast-Feeding: Although obstetricians and pediatricians encourage breast-feeding based on data showing that breast milk provides immunoglobin IgA, an amino acid profile, and fats essential for the developing infant, these advantages could be outweighed by the transmission of toxic chemicals. Chemicals make their way into breast milk by a passive transfer that depends on three major characteristics: polarity of the chemical, lipid solubility, and molecular weight. A compound with a low molecular weight that is polarized and lipid soluble can be transferred readily.

The infant's dose depends on the metabolic fate of the substance in the mother. If a substance is rapidly metabolized or excreted by the mother, the dose in breast milk will decrease rapidly when exposure stops. Solvents, although very fat soluble, are excreted through the lung, liver, and kidneys as well as through the breast milk, so the maternal body burden decreases rapidly after exposure has ceased. Excretion through breast milk can be a major route for toxins that the body has no other way of handling; this is the case for PCBs, PBBs, DDT, and related halogenated hydrocarbons. The baby's dose may be high even after the mother has been removed from exposure.

Only for a very few substances have the acute health effects of a given dose to an infant been defined; chronic effects of low-dose exposures are virtually unknown. This uncertainty makes it difficult to decide whether to recommend against breast-feeding, given the well-documented benefits. Details of pharmacokinetics and specific recommendations can be found in the Mattison publication in Further Reading.

Further Reading

Barlow SM, Sullivan FM (Eds.). *Reproductive Hazards of Industrial Chemicals: An Evaluation of Animal and Human Data.* New York: Academic Press, 1997.

Boekelheide K, Gandolfi AJ, Harris C, et al. (Eds.). *Reproductive and Endocrine Toxicology.* Elmsford, NY: Pergamon Press, 1997.

Clarkson TW, Sager PR, Nordberg GF. *Reproductive and Developmental Toxicity of Metals.* New York: Plenum Press, 1983.

Frazier LM, Hage ML (Eds.). *Reproductive Hazards of the Workplace.* New York: John Wiley & Sons, 1997.

Mattison D. Physiological variations in pharmacokinetics during pregnancy. In Fabro S, Scialli AR (Eds.). *Drug and Chemical Action in Pregnancy.* New York: Marcel Dekker, 1986, pp. 37-102.

National Research Council. Subcommittee on Reproductive and Developmental Toxicity, Committee on Toxicology, Board on Environmental Studies and Toxicology, National Research Council. *Evaluating Chemical and Other Agent Exposures for Reproductive and Developmental Toxicity.* Washington, DC: National Academies Press, 2001.

World Health Organization. *Principles for Evaluating Health Risks to Reproduction Associated with Exposure to Chemicals (Environmental Health Criteria).* Geneva: WHO, 2001.

Radiation, Ionizing, Adverse Effects

ICD-10 T66

Arthur C. Upton

Ionizing radiation consists of short-wavelength, high-frequency electromagnetic radiations (x-rays and gamma rays) as well as various particulate radiations (electrons, protons, neutrons, alpha particles, and other atomic particles). Both types of radiation are produced when atoms disintegrate, whether naturally or in such devices as nuclear reactors or cyclotrons. Ionizing radiation differs from nonionizing radiation in its capacity to disrupt atoms in the absorbing medium, thereby giving rise to ion pairs and free radicals and breaking chemical bonds.

Ionizing radiation can cause many types of adverse health effects, depending on the dose of radiation that is absorbed, the rate at which it is absorbed, the quality (linear energy transfer) of the radiation, and the conditions of exposure. Most of the adverse health effects of ionizing radiation, including various types of tissue injury, such as erythema of the skin, cataract of the lens, impairment of fertility, and depression of hematopoiesis, are produced only when the relevant threshold doses are exceeded. Certain other health effects, however, are assumed to have no threshold and, therefore, to increase in frequency with any increase in dose; included in this category are mutagenic and carcinogenic effects.

The four main types of adverse health effects

Effects on genes and chromosomes

The frequency of mutations and chromosome aberrations appears to increase linearly with the dose of radiation in the low-to-intermediate dose range. Such changes are not pathognomonic of radiation exposure, however, since they can result from other causes. Furthermore, heritable effects of radiation have yet to be demonstrated in humans, and the dose required to double their frequency in human germ cells is estimated to be at least 1 sievert.

Effects on tissues

Mitotic inhibition and other cytological abnormalities are detectable almost immediately after exposure to a large dose of ionizing radiation. In tissues such as the gastrointestinal mucosa and bone marrow, the killing of stem cells may interfere with the normal replacement of differentiated cells and thereby cause impairment of organ function. Thus, a dose that is large enough to kill a major percentage of progenitor cells in the marrow or intestinal mucosa, such as >2 sieverts [Sv] absorbed in a single brief exposure, will cause the acute radiation syndrome. The typical prodromal symptoms of this syndrome include anorexia, nausea, and vomiting during the first few hours after exposure, followed by a symptom-free interval until the main phase of the illness. The main phase of the intestinal form of the illness typically begins 2 to 3 days after exposure, with abdominal pain, fever, and increasingly severe diarrhea, dehydration, toxemia, and shock, leading to death within 7 to 14 days. In the hematopoietic form, the main phase typically begins in the second or third week after exposure, with a reduction in white blood cells and platelets; if damage to the bone marrow has been severe enough, death from septicemia or exsanguination may occur between the fourth and sixth week after irradiation.

Effects on the growth and development of the embryo

Embryonal and fetal tissues are unusually radiosensitive. Thus, intensive exposure in utero can cause prenatal death, various malformations, or other disturbances in growth and development, including mental retardation.

Effects on the incidence of cancer

The incidence of many, but not all, types of cancer may be increased by exposure to ionizing radiation. The induced cancers include leukemias and cancers of the thyroid, lung, female breast, respiratory tract, digestive tract, and bone. Such cancers do not appear until years or decades after exposure, however, and they have no distinguishing features identifying them individually as having been caused by radiation. For the most part, moreover, they have been detected only after relatively large doses (0.5 to 2.0 Sv). Therefore, the carcinogenic risks from low doses can be estimated only by interpolation or extrapolation from higher doses, based on assumptions about the dose-response relationship.

Occurrence

Because of the relatively large doses that are required to elicit acute tissue reactions, the occurrence of such reactions is limited primarily to radiotherapy patients and heavily exposed radiation accident victims. Smaller doses, however, have been observed to increase the risks of lung cancer in underground hardrock miners, leukemia in early radiologists, and cancers of additional types in various other exposed populations. The weight of existing evidence implies, moreover, that although less than 3% of all cancers in the United States can be attributed to natural background irradiation, up to 10% of lung cancers in the U.S. may result from exposure to radon in indoor air.

No U.S. population-based data on the occurrence of radiation effects now exist. It has been estimated, however, that more than 150 000 U.S. workers are exposed to ionizing radiation in the nuclear energy fuel cycle; more than 70 000 are exposed in various defense-related activities; more than 400 000 are exposed in the healing arts; more than 100,000 are exposed in research; and more than 1 million are exposed, at substantially lower average exposure levels, in manufacturing and other industrial sectors. The types of workers who may be exposed occupationally to ionizing radiation include atomic energy plant workers; dentists and dental assistants; electron microscopists; fire-alarm makers; industrial fluoroscope operators and radiographers; inspectors using, and workers located near, sealed gamma ray sources; nuclear submarine workers; petroleum refinery workers; physicians (including radiologists); nurses; medical technologists (including x-ray technicians); thorium ore producers and workers with thorium alloys; uranium workers; and uranium mill workers.

Causes

Next to radon in indoor air, x-irradiation for diagnostic medical imaging represents the major source of exposure for the general population. Among workers who are exposed occupationally, the largest doses are received, on average, by those employed in nuclear power plants or other parts of the nuclear fuel cycle.

Pathophysiology

The injuries caused by ionizing radiation result primarily from damage to genes and chromosomes that kill or alter the affected cells, especially dividing cells, which are highly radiosensitive as a class. To kill enough cells in a given organ to seriously impair its function, however, generally requires a relatively large dose. Mutations, chromosome aberrations, and some forms of cancer, on the other hand, appear to increase in frequency as linear, non-threshold functions of the dose at low-to-intermediate levels of exposure; however, the possibility that their induction may be greatly inhibited by adaptive responses at lower levels of exposure cannot be excluded.

Prevention

The guiding principle in radiation protection holds that the dose should be kept as low as is reasonably achievable. In addition, absolute limits have been set on the permissible doses to radiation workers and other members of the population. The dose equivalent (DE) limit that has been recommended to prevent the impairment of organ function has been set at 0.50 Sv (50 rem) per year for all organs other than the lens of the eye, for which the recommended annual DE limit has been set at 0.15 Sv (15 rem). To restrict the risks of mutagenic and carcinogenic effects to levels that are acceptably low, additional DE limits have been recommended for the various organs of the body, as well as for the body as a whole-which is 50 mSv (5 rem) in any one year, but no more than 20 mSv per year, on average.

In order to restrict radiation exposure to acceptable levels, radiation sources, facilities where they are produced or used, and work procedures must be designed accordingly. This requires thorough training and supervision of involved personnel, implementation of a well-conceived radiation protection program, and systematic health physics oversight and monitoring. Also needed are careful provisions for dealing with radiation accidents, emergencies, and other contingencies; systematic recording and updating of each worker's exposures; thorough labeling of all radiation sources and exposure fields; appropriate monitoring of facilities and interlocks to guard against inadvertent irradiation; and various other precautionary measures.

General principles to be observed in every radiation protection program include:

1. A well-developed, well-rehearsed, and updated emergency preparedness plan to enable prompt and effective response in the event of a malfunction, spill, or other radiation accident, including appro-

priate provisions of isolation and decontamination of contaminated objects and people.

2. Appropriate use of shielding in facilities, equipment, and work clothing, such as aprons and gloves.
3. Appropriate selection, installation, maintenance, and operation of all equipment.
4. Minimization of exposure time.
5. Maximization of distance between personnel and sources of radiation, recognizing that the intensity of exposure decreases inversely with the square of the distance from the source.
6. Appropriate training and supervision of workers to accomplish routine tasks with minimal exposure and to cope safely with unanticipated events.

Further reading

Mettler FA, Upton AC. *Medical Effects of Ionizing Radiation.* New York: W.B. Saunders, 1995.

National Council on Radiation Protection and Measurements. *Limitation of Exposure to Ionizing Radiation. NCRP report No. 116.* Bethesda, MD: National Council on Radiation Protection and Measurements, 1993.

National Council on Radiation Protection and Measurements. *Operational Radiation Safety Program. NCRP report No. 127.* Bethesda, MD: National Council on Radiation Protection and Measurements, 1998.

Shapiro J. Radiation Protection: *A Guide for Scientists and Physicians* (3rd ed.). Cambridge, MA: Harvard University Press, 1990.

United Nations Scientific Committee on the Effects of Atomic Radiation. *Sources and Effects of Ionizing Radiation: Report to the General Assembly, with Annexes.* New York: United Nations, 2000.

Upton AC. Ionizing radiation. In Levy BS, Wegman DH, (Eds.). *Occupational Health: Recognizing and Preventing Work-Related Disease and Injury (4th ed.).* Philadelphia: Lippincott Williams & Wilkins, 2000, pp. 355-366.

Radiation, Nonionizing, Adverse Effects

ICD-10 T66

Robert J. Nordness

Throughout the electromagnetic spectrum, as wavelength increases, the energy level of emitted radiation decreases. Nonionizing radiation refers to the longer wavelength portion of the electromagnetic spectrum that does not possess enough energy to ionize, or eject an electron from, target atoms or molecules.

Forms of nonionizing radiation, listed in order of increasing wavelength (and thus decreasing energy) include: ultraviolet, visible light, infrared, and microwave/radiofrequency. Laser is an artificial form of nonionizing radiation with special properties of monochromacity and low divergence.

Body tissue injury can occur through many mechanisms, and is a function of the amount of energy transferred from the source to the target tissue. The amount of energy transferred, and thus the risk of injury, depend on the energy of the radiant source, the distance to the target, and the time of exposure.

Strategies to reduce the risk of injury are focused on reducing the amount of energy transferred by decreasing the source energy, utilizing shielding (or personal protection equipment), increasing the distance from the source, and/or reducing the total time of exposure.

Ultraviolet Radiation (UVR)

UVR exhibits poor penetration into tissues overall, thus the most common acute effects from UVR exposure are to the skin and eyes. Sunburn occurs within several hours of UV exposure and is the result of local vasodilation with subsequent increased blood flow. An individual's skin response to UVR exposure is highly variable and depends on prior exposure, natural skin pigmentation, and other characteristics. The minimal erythema dose (MED) to provoke sunburn is approximately 200 J/ms for wavelengths between 250 and 300 nm (predominately UV-B). With ocular UVR exposure, the cornea absorbs the bulk of the transferred energy, though the conjunctival tissues can also be damaged affected, leading to keratitis and/or conjunctivitis. Ocular symptoms of pain, foreign body sensation, increased tearing, and spasm of the eyelids often present after a latent period of 6 to 12 hours and are typically associated with UV-B and UV-C exposures. This condition is commonly known as "welder's flash" (or "ground glass eyeball" or "flash burns") and is commonly associated with arc welders, or those observing welders, who are not wearing adequate eye protection.

Chronic injury associated with UVR exposure to the skin causes elastosis, resulting in a deeply wrinkled, leathery appearance to the skin caused by loss of natural elasticity of the skin tissue. Chronic exposure to UVR has been strongly associated with cataracts. The link between sunlight and skin cancers is well known, with the UV-B wavelengths believed to be more carcinogenic than the UV-A portion of the spectrum. Most UVR-associated skin cancers are epithelial in nature, though some studies have shown links between malignant melanomas and UVR. Research also suggests that UVR exposure depresses immune system function.

Occurrence

The sun is the major source of UV exposure for most workers. Artificial sources include welding arcs, fluorescent lights, plasma jets, and ultraviolet lights ("black lights"). The amount of UVR reaching the earth varies greatly as a function of time of day, season, atmospheric condition, and geographical position. Most UVR reaching earth occurs during the summer months with approximately 60% of UVR occurring between the hours of 10:00 a.m. and 2:00 p.m. UVR-induced injury can also be potentiated by many natural and human-made photosensitizers. Many medications, in a variety of drug classes, are photosensitizing agents and workers exposed to UVR who are on any medication should discuss this possibility with their physician. Common natural photosensitizers include food plants containing psoralens and furocoumarins, such as celery, mango, lime, and figs. Some common industrial photosensitizers include coal tar, pitch, naphthalene, and chemicals containing phenols. The proportion of adverse effects that are occupationally related is not known.

Causes

UVR is the portion of the electromagnetic spectrum that falls between x-rays and visible light, with wavelengths between 160 and 400 nanometers. The ultraviolet spectrum is further subdivided into UV-A (320-400 nm), UV-B (280-320 nm), and UV-C (160-280 nm). UV-C radiation with the lowest wavelength, and thus the highest energy within the UVR spectrum, actually has sufficient energy to ionize target molecules.

Pathophysiology

Eye injury is most commonly the result of thermal injury, increasing with proximity to the source and arc strength. Skin injury is predominantly photochemical, with injury caused by the products of photochemical reaction with the cell, and potentiated by photosensitizing agents. Sunburn is caused by a resultant vasodilation of exposed skin tissues with a subsequent increase in blood flow.

The chronic skin effects of UVR are a result of repeated injury leading to collagen destruction and loss of elasticity. With chronic exposure, the skin atrophies and many abnormal disorderly patterned cells are evident histologically. These areas may be considered "premalignant," at high risk for skin cancer development.

Prevention

Artificial UV sources should be shielded to the greatest extent possible with materials appropriate for the source. Interlocking access panels are advisable with high-UV-producing sources. Workers operating near these sources must wear clothing with appropriate UV protection factors (UPFs) and goggles with protective qualities matched to the spectral distribution and intensity of the source. Overexposure to natural UVR should be avoided by wearing protective clothing and sunglasses. Most commercial sunscreens offer protection in the UV-B range, though only benzophenone and anthranilate sunscreens also provide limited protection in the UV-A range. Opaque sunblock, such as zinc oxide and titanium dioxide, offer the most complete protection, reflecting up to 99% of the radiation in the UVR and visible spectrums, and may be essential if there is photosensitization. Exposure guidelines have been established by ACGIH, but not by OSHA.

Other Issues

The decrease in the earth's ozone layer has heightened concern over UVR injury, especially skin cancer. Ozone is an effective filter of the sun's higher energy UVR transmissions, and a small loss of ozone can lead to significant increase in skin cancer incidence. Having light skin and working at high altitude (where solar radiation is more intense) are risk factors for sun-induced skin cancer.

Further Reading

Goettsch W. Risk assessment for the harmful effects of UVB radiation on the immunological resistance to infectious diseases. *Environmental Health Perspectives* 1998; 106: 71-77.

Hu H. Effects of ultraviolet radiation. *Medical Clinics of North America* 1990; 74: 509-514.

NIOSH. *A Recommended Standard for Occupational Exposure to Ultraviolet Radiation.* Atlanta: NIOSH, 1977.

Infrared Radiation (IRR)

Like UVR, IRR causes damage primarily to the skin and eye, although IRR-induced injury in both tissues is primarily thermal in nature. Exposure to IRR is perceived as warmth to the skin, but eye damage can occur at energy levels less than the skin warmth threshold. Chronic IRR exposure leads to repeated cornea injury, which ultimately may lead to cataract formation. The skin surface responds to IRR by vasodilation of the capillary beds, leading to acute erythema (redness) at the exposed sites. Significant burn injury does not typically occur as the vasodilation increases blood flow and thus heat transfer, which along with the sweating process successfully dissipates the heat energy absorbed. IRR exposure can, however, be a significant risk for heat stress injury where increased temperature is an issue.

Occurrence

Significant IRR exposure is nearly always associated with the occupational or laboratory setting, either directly through lamp sources or indirectly through heat sources. Occupations at highest risk for cataract formation are glassblowers and furnace workers. Workers who work with molten metal, drying or dehydrating processes, or thermal conditioning of surfaces are also at significant risk. The proportion of adverse effects that are occupationally related is not known.

Causes

IRR injury is caused by excess absorbed energy from electromagnetic radiation in the IRR region (700 nm to 1 mm). Wavelengths less than 2000 nm are responsible for most of the injurious effects of IRR.

Pathophysiology

IRR with wavelengths less than 2000 nm cause molecular excitation and vibration, which is converted to thermal energy in tissue, with subsequent localized injury.

In the natural environment, IRR exposure is typically accompanied by a large amount visible light, which triggers the eyes' innate protective mechanisms. Conversely, occupational sources of IRR are often relatively void of visible light, which, coupled with the poor heat sensation and dissipation mechanism of the eye, increase the risk of injury.

Prevention

Engineering controls are the mainstay of IRR protection, with source shielding an effective control. The IRR source should be properly shielded, and eyes should be protected with IRR filters. There is a TLV set by ACGIH for the biologically-active IRR spectrum from 750 to 2000 nm. There is no OSHA exposure guideline.

Further Reading

Oriowo OM. Eye exposure to optical radiation in the glassblowing industry: An investigation in southern Ontario. *Canadian Journal of Public Health* 2000; 91: 471-474.

National Institute for Occupational Safety and Health. *Research Report: Determination of Ocular Threshold Levels for Infrared Radiation Cataractogenesis. (DHHS publication #80-121)*. Atlanta: NIOSH, 1980.

Microwave/Radiofrequency Radiation (MW/RFR)

MW/RFR causes tissue injury, at sufficient power intensity, through thermal effects. Microwave radiation is the portion of the radiofrequency between 10 MHz and 300 GHz (with wavelengths typically in the 0.1 to 10 cm range), and is the focus of public concern over health effects of radiofrequency radiation. The thermal effects caused by MW/RFR exposure can cause cataract formation, testicular degeneration (and decreased sperm counts), focal tissue burns (including keratitis), and, with extreme acute exposures, death. A phenomenon of "microwave hearing" has been reported by some workers as a clicking or buzzing associated with exposure to pulsed MW radiation, thought to be the result of thermoelastic expansion, a thermal effect. Tissue injuries through non-thermal mechanisms, below the MW/RFR thermal injury threshold are a much more controversial topic. Animal studies have shown chronic exposure to MW/RFR at intensity levels insufficient to cause thermal injury has been associated with alterations in neurological function (including acute behavior changes), immune cell dysfunction, and possibly cancer. Human studies regarding the health effects of chronic low-intensity MW/RFR have shown mixed results and are difficult to interpret. Associations between MW/RFR exposure and human behavior, immune system dysfunction, reproductive dysfunction, blood-tissue dysfunction, headache, depression, fatigue, endocrine dysfunction, and cardiovascular-system dysfunction have been shown in some studies. The significance of these associations is a subject of investigation and debate.

MW/RFR, including that produced by cellular telephone use, may interfere with implantable medical electronics, in particular pacemakers. Modern pacemaker leads are shielded to prevent MW/RFR interference, although consultation regarding specific risk should be sought from the pacemaker manufacturer and a cardiologist. Use of microwave ovens is not typically considered a hazard with respect to pacemakers due to the shielding of both the oven and the leads.

Occurrence

MW/RFR exposure is prevalent in many occupational settings, especially telecommunications, industries that utilize sealing and heating equipment, RF welding, surgical cautery, and medical diathermy applications. Currently, the greatest MW/RFR exposure to the general public and the typical worker is through cellular telephone use. The proportion of adverse effects that are occupationally related is not known.

Causes

Radiofrequency radiation ranges in frequency from 3 KHz to 300 GHz, with microwave radiation representing the subset of this range from 10 MHz to 300 GHz.

Pathophysiology

The MW portion of the RFR spectrum causes predominantly thermal injury via movement and realignment of molecules within tissues, especially those with asymmetrical charge density or polarity. Friction between molecules generates heat, which is the primary source of tissue injury at MW power densities above 10 mW/cm^2. The potential mechanism of injury in subthermal-intensity RF/MW is unclear and an area of significant interest.

Prevention

MW/RFR sources must be regularly monitored to detect and control "leakage" of the radiation. Screening should be done to appropriately counsel those with medical devices, which may be affected by MW/RFR. Appropriate shielding and engineering controls should be employed to reduce exposure at the MW/RFR source. Personal protective equipment

(PPE) in the form of protective clothing and eyewear is not recommended as a strategy to reduce MW/RFR, as reflection can cause increased risk to others and open circuits on the wearer may actually increase risk. OSHA has established PELs (29 CFR 1910.97), which are based on the thermal power density threshold.

Other Issues

There is much concern over the potential risk of brain cancer from cellular telephone use. Given the relatively short time that these phones have been in use, studies regarding the association between brain cancer and cellular phone use have not been conclusive; however, to this point, there has been no convincing evidence of such a link.

Video display terminals (VDTs) can emit x-rays, UVR, and MW/RFR, although modern terminals do not emit above background levels. Increased rates of spontaneous abortions and birth defects are the major concern with respect to VDTs, but most studies have shown no association. Any health effects associated with VDTs are thought to be associated with body posture or ergonomic in nature.

Further Reading

Elwood M. A critical review of epidemiologicial studies of radiofrequency exposure and human cancers. *Environmental Health Perspectives* 1999; 107 (Suppl 1): 155-168.

Goldsmith J. Epidemiologic evidence relevant to radar (microwave) effects. *Environmental Health Perspectives* 1997; 105 (Suppl 6): 1579-1587.

Wilkening GM, Sutton CH. Health effects of nonionizing radiation. *Medical Clinics of North America* 1990; 74: 489-507.

Laser Radiation

"Laser" is an acronym of Light Amplification by the Stimulated Emission of Radiation, an artificially generated nonionizing radiation. Laser radiation is a focused beam of monochromatic (single-wavelength) photons, from the ultraviolet, visible, or infrared portion of the electromagnetic spectrum that have a high energy density. Laser radiation exhibits a low level of divergence, which allows energy transfer over great distances. The tissues of the eyes are particularly susceptible to laser injury, either by direct exposure or diffuse reflection of lasers. Skin burns can occur with direct exposure to high-energy lasers. The mechanism of injury in laser injury can be photomechanical, thermal, or photochemical.

Occurrence

The use of lasers has increased dramatically in recent years, particularly in such areas as construction, industry, medicine, communications, entertainment, education, and the military. In the United States, all products containing lasers must be certified under the Federal Laser Product Performance Standard. Part of this standard requires that all laser products be classified from Class 1 to Class 4 in increasing order of power output, and thus injury risk. Any laser greater than Class 1 can cause eye injury by direct exposure; Class 4 lasers are high-power devices, which can also cause ocular injury through diffuse reflection of the laser beam. The proportion of adverse effects that are occupationally related is not known.

Causes

Laser radiation is unique among forms of radiant energy in its low divergence and highly directional nature. Whether UV, visible or infrared in spectrum, laser radiation is able to cause tissue injury over great distances when the beam path is crossed. Nonbeam hazards are created when the laser creates a potential health hazard through interaction of the beam with another medium. These can range from metal fumes, toxic gases/vapors, electricity, plasma radiation, fires, and explosions.

Pathophysiology

Thermal injury from laser exposure is caused by the direct absorption of the radiant energy creating movement of molecules within eye or skin tissue, which is dissipated as heat to the exposed tissue. Photochemical injury to tissues is typically caused by UV lasers, although photochemical eye injury is also caused by visible light lasers up to 550 nm (blue through green). The products of light-induced chemical reactions within the tissues are responsible for the photochemical injury.

Photomechanical injury can be caused by Q-switched or mode-locked lasers, which create a very short pulse, with high peak irradiance. This type of exposure can create a plasma in the tissues, which form shock waves and resultant destruction through propagation. Permanent eye injury can be sustained from pulses in the nanosecond range through this process.

Ocular damage can also occur indirectly from exposure to diffuse reflections from high-power (Class 4) lasers. Lasers producing nonvisible radiation can be particularly hazardous, given that exposure may not be readily apparent.

Prevention

Laser installations should be well marked and isolated, and should utilize attenuating or terminating interlocks when possible. Laser beams must be terminated by a nonreflective, fireproof material. Goggles can help only if they afford protection specific to the wavelength of the laser being used, and must not be utilized as a substitute for engineering controls. Proper worker education and surveillance eye exams are important components of the safety program. OSHA standard 29 CF 1926.54 provides general requirements for laser use on construction sites. ACGIH has published guidelines for the safe use of lasers, and exposure limits have been developed by ANSI (ANSI Z

136.1), and the FDA (for medical applications of laser). Awareness of the potential nonbeam hazards of a particular process is necessary to fully protect the worker.

Further Reading

American National Standards Institute. *American National Standard for the Safe Use of Lasers: ANSI Z 136.1.* New York: Laser Institute of America, 1993.

DiNardi SR. The *Occupational Environment-Its Evaluation and Control.* Fairfax, VA: American Industrial Hygiene Association Press, 1997, pp. 505-519.

Occupational Safety and Health Administration. *Laser Hazards (OSHA Technical Manual TED 1-0.15A,* Section III - Chapter 6). Washington, DC: OSHA, 1999.

Thach AB. Laser injuries of the eye. *International Ophthalmology Clinics* 1999; 39: 13-27.

Reactive Airways Dysfunction Syndrome (RADS)

ICD-10 J68.3

David C. Christiani

Reactive airways dysfunction syndrome (RADS), or more appropriately irritant asthma, is a form of asthma described following acute high-dose inhalation, or chronic moderate-dose inhalation, of an irritant substance. Once established, irritant asthma has similar or identical clinical manifestations of other forms of occupational asthma (reversible airway obstruction).

Symptoms typically begin within minutes to hours after acute inhalation, and persist for up to 1 year, or in many instances, permanently. Any agent with sufficient toxic potential that is inhaled in sufficient quantity may induce asthma. Agents inducing RADS are generally not sensitizers.

Airway obstruction is commonly a feature of the initial toxic response, especially to compounds with high or moderate water solubility, such as hydrofluoric acid and hydrochloric acid, which affect the upper airways and bronchial mucosa more than the lung parenchyma.

Clinical features of irritant asthma resulting from inhalation injury are similar to other forms of asthma. (See Asthma.) The patient may complain of the abrupt onset of dry cough, shortness of breath, chest tightness, or wheezing. The degree of bronchoconstriction depends upon the potency of the agent to induce airway hyperresponsiveness.

Occurrence

Work-related irritant asthma is not uncommon. A NIOSH study reported that a state-based surveillance system (SENSOR) described 14% of all new-onset occupational asthma as RADS, with prevalence ranging from 6% in California to 31% in New Jersey. The adverse health and economic consequences of RADS were similar to other forms of occupational asthma.

The SWORD (Surveillance of Work-Related Occupational Respiratory Diseases) surveillance system in the United Kingdom (UK) reported that about 9% of cases of occupational asthma reported were inhalation injury-related. A UK study of cases from 1990 to 1993 showed that occupations with both the largest proportion of cases and the highest rates were chemical processors and engineers and metalworkers. This study also found that inhalation injury events were associated with considerable morbidity. RADS resulting from multiple inhalation events of toxic substances in lower concentrations has also been described but is less well understood and less frequently recognized.

Causes

There is no comprehensive list of causes. Any compound or combination of compounds that are highly irritating is capable of inducing RADS. Acute insults to the airways may occur after exposure, such as to smoke from fires, chlorine, ammonia, and acid aerosols.

Pathophysiology

The precise mechanism(s) of asthma following inhalation of toxins remains undefined. However, it is clear that the airways become injured and, as a result, alter airway responsiveness. Bronchial biopsies of RADS patients demonstrate an inflammatory response characterized by epithelial desquamation and mucous cell hyperplasia. The likely pathophysiologic mechanisms include altered neural tone and vagal reflexes, modified beta-adrenergic sympathic tone, and the influence of a number of pro-inflammatory mediators. The direct irritant injury may expose and damage subepithelial irritant receptors. Subsequent repair and remodeling may result in alteration of neuromuscular receptors leading to altered airway responsiveness and altered airway compliance. Changes in epithelial permeability may also contribute to the airway hyperresponsiveness.

Prevention

The most important component of prevention of RADS is exposure control. Enclosure and exhaust of hazardous processes that generate or release gases, vapors, fumes, or particulates is essential. In addition, safety measures aimed at prevention of spills, splash, explosion, and fire are also essential in preventing exposure to airway toxins. Worker training is important, such as avoidance of creation of hazardous conditions when mixing ammonia and chlorine-containing compounds by building maintenance crews. Worker selection or placement based on history of allergy or by immunologic testing is not warranted; however, placement of individuals with a history of asthma in areas where moderate release of irritants may occur should be done cautiously with careful attention to exposure control and periodic monitoring of symptoms and serial lung function.

Further Reading
Bernstein IL, Chan-Yeung M, Malo JL, et al. *Asthma in the Workplace* (2nd ed.). Marcel Dekker, 1999.

Tarlo SM. Workplace respiratory irritants and asthma. *Occupational Medicine* 2000; 15: 471-484.

Renal Failure, Chronic: Tubular and Interstitial Kidney Disease
ICD-10 N10-N15, N16, N18

Carl-G. Elinder and C. Michael Fored

A variety of renal diseases may progress to end-stage renal disease (ESRD), including chronic glomerulonephritis, diabetic nephropathy, and polycystic kidney disease. All forms of chronic renal failure (CRF) are associated with marked tubulointerstitial injury (tubular dilatation and interstitial fibrosis), even if the primary process is a glomerulopathy. The degree of tubulointerstitial disease is a better predictor of the glomerular filtration rate and long-term prognosis than is the severity of glomerular damage in almost all progressive glomerular diseases. Thus, tubulointerstitial changes represent one component of CRF, but they may also occur in acute or chronic tubulointerstitial nephritis (TIN).

Multiple etiological factors throughout an individual's life, including a genetic predisposition, determine the patterns of CRF incidence.

Most known occupational nephrotoxins affect the renal tubule or the supporting interstitium. The manifestations of toxicity vary with the agent, the chronicity and intensity of exposure, and the age of the affected person. For example, both lead and cadmium primarily damage the renal tubules and interstitium (as opposed to the glomerulus), yet their clinical manifestations are markedly different: cadmium causes tubular dysfunction, whereas lead has an insidious effect on overall renal function.

Acute TIN typically presents with an acute rise in serum creatinine, usually temporally related to taking a prescribed medication or having a recent infection. Urine analysis reveals proteinuria (up to a few grams per day), often with hematuria and the presence of a few casts in the sediment. In allergic TIN due to drug reactions eosinophils may be increased in blood and present in urine. After the causative factor is removed or resolved, the acute TIN usually resolves and kidney function normalizes.

Chronic TIN may be asymptomatic until renal failure is relatively advanced. Creatinine clearance declines, although serum creatinine may remain normal until approximately 50% of renal function has been lost. Reduced renal function from chronic TIN will likely render the person more vulnerable to other medical or environmental risk factors for renal disease.

Several types of long-term toxic exposures, including to cadmium and lead, may cause chronic TIN. Inorganic mercury compounds accumulate in the proximal tubule and heavy exposure may cause acute tubular necrosis. The tubular proteinuria that may result from organic mercury compounds mostly regress with the discontinuance of exposure. Findings of less severe tubular effects include; slightly elevated urinary excretion of albumin, transferrin, and retinol binding protein (RBP) and the tubular enzymes β-galactosidase and N-acetyl-β-D-glucosaminidase (NAG). Both elemental mercury and organic mercury compounds have been associated with glomerular disease. This rare illness is believed to be immunologically mediated.

In addition, several non-occupational exposures may cause TIN and should be considered in the differential diagnosis: analgesic nephropathy from heavy and prolonged exposure to phenacetin, acetaminophen, and/or aspirin; and herbal nephropathy from aristolochic acid in certain Chinese tea.

Occurrence

Chronic TIN is the cause of a small percentage of ESRD cases. However, chronic TIN contributes to the development of ESRD that is due to many different causes. For most types of renal disease, a number of factors, such as hypertension, hyperlipidemia, and use of analgesics and prescription medications, have been reported to influence progression from early kidney failure to ESRD. In epidemiological studies, one should focus on ESRD, regardless of the presumed underlying disease, because of difficulties involved in making a correct diagnosis.

Regional epidemics of TIN have occurred in Bulgaria, Yugoslavia, Romania (Balkan nephritis), and in Japan (itai-itai disease). Balkan nephropathy afflicts approximately 25,000 persons in Eastern Europe and is suspected to be caused by an environmental toxin. Itai-itai disease caused by renal tubular calcium wasting, has affected approximately 100 elderly women in Japan, in areas where rice is contaminated with cadmium. More than 10,000 farmers have developed tubular damage from cadmium exposure. Cadmium-induced tubular dysfunction has also been reported in several areas in Europe and China. Occupational groups at risk of cadmium nephropathy include workers in brazing, nickel-cadmium battery manufacturing, primary and secondary smelting, scrap metal recovery, electroplating, and pigment manufacturing.

Populations exposed to lead include workers in lead battery plants, soldering, radiator repair, primary smelting, glass and ceramic manufacturing, and paint stripping. Non-occupational exposure to lead includes lead pollution from lead smelters, paint flakes and dust, leaded gasoline, lead-containing water pipes, and "moonshine alcohol."

Causes

Cadmium Nephropathy

The first sign of a cadmium-induced renal lesion is tubular dysfunction, characterized by an increased excretion of proteins with low molecular weight, such as β_2-microglobulin. In severe cases of cadmium-induced nephrotoxicity several signs of tubular damage are evident: renal glucosuria, aminoaciduria, hyperphosphaturia, hypercalcuria, polyuria, and deteriorated ability to handle an acid (NH_4Cl) load. In recent years, several highly sensitive indicators of the tubular damage has been developed and used in cross-sectional and epidemiological studies on cadmium-exposed groups. The urinary excretion of β_2-microglobulin has frequently been used. Other equally sensitive indicators of cadmium-induced tubular damage are retinol binding protein (RBP), apolipoprotein, and human complex-forming glycoprotein (protein HC or alpha$_1$-microglobulin), all of which are plasma proteins filtered through the glomerulus. Another urinary marker indicative of a subtle tubulotoxic effect is N-acetyl-β-D-glucosaminidase (NAG), an enzyme localized in lysosomes of the tubular cells.

The tubular dysfunction or damage caused by cadmium is often irreversible, even if exposure has ceased. Glomerular damage, with a decreased filtration rate (GFR), may develop in occupationally or environmentally exposed individuals. There is a significant association between tubular damage and decrease in GFR. Uremia was found to be a common cause of death among Japanese farmers suffering from itai-itai disease, and among a few former workers who had been heavily exposed to cadmium. In addition, in Japan cardiovascular and renal disease mortality is 40% to 100% higher in persons who have resided in cadmium-polluted areas and have shown signs of cadmium-induced renal damage or elevated urinary cadmium excretion. A cohort study of cadmium-exposed workers has demonstrated a dose-response relationship between occupational cadmium exposure and ESRD necessitating renal transplantation.

Cadmium-exposed workers also have an increased risk of renal stones and a decrease in bone mineral density.

Lead Nephropathy

The acute renal effects of high-dose lead exposure includes an impaired proximal tubular function with aminoaciduria, glycosuria, and hyperphosphaturia. These effects are reversible. Continued lead exposure over 5 to 30 years can induce progressive tubular atrophy and interstitial fibrosis. Affected patients typically present at this later stage with hypertension and chronic renal insufficiency. Chronic lead nephropathy usually affects proximal tubular function only to a limited extent, if at all. In adults, the clinical manifestations may include impaired tubular function with hyperuricemia (saturnine gout) due to diminished urate secretion, and lead nephropathy can be confused with hypertensive nephrosclerosis and primary chronic urate nephropathy, although the latter is rare. No one clinical test proves that nephropathy is due to lead; renal biopsy typically shows tubulointerstitial nephritis with relatively few inflammatory cells. Lead-inclusion bodies in the nuclei of proximal tubule cells are present infrequently in chronic lead nephropathy or after chelation. The diagnosis of lead nephropathy is thus based on documentation of high lead exposure and ruling out other explanatory causes.

Evidence for chronic lead nephropathy is not consistent. In some reports, workers with previous heavy and long-term lead exposure display little, if any, evidence of adverse renal effects. However, several extensive investigations of populations with lead exposure at a much lower intensity than among lead-exposed workers have shown significant relationships between blood lead levels, mean blood pressure, and elevated serum creatinine or BUN (blood urea nitrogen). Moreover, a recent report indicates that patients with CRF not primarily related to lead progressed more rapidly if they had an elevated lead-EDTA chelation test, and that those with a positive lead-EDTA test improved their renal function when treated with EDTA.

Pathophysiology

After ingestion or inhalation, cadmium is transported to the liver where metallothionein, a cadmium-binding protein, is synthesized. Metallothionein serves as a detoxifying protein by binding cadmium and preventing free cadmium ions from affecting normal cellular function. It is also responsible for pronounced accumulation of cadmium in the liver and kidneys. Metallothionein migrates slowly from liver cells into the blood, in which it is transported to the kidneys. It is freely filtered across the glomerulus and then reabsorbed by the proximal tubular cells via pinocytosis. In the tubular cells, metallothionein enters the lysosomes, where it is degraded by lysozymes, releasing free cadmium ions into the tubular cell cytoplasm. The renal tubular cells have a considerable capacity to synthesize metallothionein, thereby binding the toxic cadmium ions. At some point, however, the detoxifying capacity of the kidneys is surpassed and tubular damage can ensue. Later changes include interstitial inflammation and fibrosis, as well as glomerular injury.

Less is known about the mechanism by which continued lead exposure results in chronic interstitial disease. The pathogenesis of the renal disease possibly is related to the proximal tubule reabsorption of filtered lead and subsequent accumulation in intranuclear inclusion bodies. If high exposure persists, interstitial changes develop.

Prevention

Prevention by minimizing human exposure to cadmium is of utmost importance. Occupational exposure should be kept as low as technically feasible, preferably below 5 μg/m³, to protect workers over a 40-year working lifetime. Daily exposure via food should be kept below 30 to 50 μg of cadmium. It was previously suggested that 10 μg of cadmium per gram of creatinine (10 nmol of cadmium per mmol of creatinine) would be a safe level, below which kidney dysfunction would rarely develop. A urinary cadmium of 10 μg per gram of creatinine roughly corresponds to a kidney-cortex concentration of cadmium of 200 mg/kg, which was long considered to be the critical level above which kidney damage might first be seen. More recent reports however indicate that this is a serious underestimation of the risk and an overestimate of the critical concentration of cadmium in the kidney. Up to 10% of environmentally exposed individuals display evidence of cadmium-induced renal dysfunction already at urine cadmium concentrations of 2 to 3 nmol per mmol of creatinine.

The OSHA lead standard has substantially reduced lead exposure in large industries; small companies and secondary smelters, however, still have potentially high exposures. In adults, lead nephropathy may not occur if the blood lead level is below 40 μg/dL; however, exposure to lead should be kept as low as possible and well below this level. Recent studies have reported adverse lead effects on the kidneys and blood pressure at much lower levels.

For secondary prevention, urinary β_2-microglobulin or other early indicators of tubular dysfunction can be measured to determine whether cadmium exposure is adequately controlled, or to identify workers with early tubular nephropathy. Medical removal of such workers may prevent progression of injury, but will not correct the underlying problem of overexposure.

No good early detector for lead nephropathy exists; serum creatinine, BUN, and blood pressure are all insensitive to early disease.

Further Reading

Alfven T, Elinder CG, Carlsson MD, et al. Low-level cadmium exposure and osteoporosis. *Journal of Bone and Mineral Research* 2000; 15: 1579-1586.

Hellstrom L, Elinder CG, Dahlberg B, et al. Cadmium exposure and end-stage renal disease. *American Journal of Kidney Disease* 2001; 38: 1001-1008.

Jarup L, Berglund M, Elinder CG, et al. Health effects of cadmium exposure: a review of the literature and a risk estimate [published erratum appears in Scand J Work Environ Health 1998 June; 24 (3):240]. *Scandinavian Journal of Work, Environment and Health* 1998; 24 (Suppl. 1): 1-51.

Lin JL, Lin-Tan DT, Hsu KH, et al. Environmental lead exposure and progression of chronic renal diseases in patients without diabetes. *New England Journal of Medicine* 2003; 348: 277-286.

Loghman-Adham M. Renal effects of environmental and occupational lead exposure. *Environmental Health Perspectives* 1997; 105:928-939.

Nuyts GD, Daelemans RA, Jorens PG, et al. Does lead play a role in the development of chronic renal disease? *Nephrology, Dialysis, Transplantation* 1991; 6: 307-315.

Staessen JA, Lauwerys RR, Buchet JP, et al. Impairment of renal function with increasing blood lead concentrations in the general population. The Cadmibel Study Group. *New England Journal of Medicine* 1992; 327: 151-156.

Wedeen RP, De Broe M. Heavy metals and the kidney. In Davison AM, Cameron JS, Grunfeld JP, et al. (Eds.). *Oxford Textbook of Clinical Nephrology* (2nd Edition). Oxford: Oxford University Press, 1998, pp. 1175-1189.

Renal Failure, Acute

ICD-10 N17

Carl-G. Elinder and C. Michael Fored

Acute renal failure (ARF) can be caused by certain toxic chemicals [by accident or intent (suicidal)] and by severe trauma with crush injuries. A proper history is crucial to elucidating occupational exposures that may cause ARF.

It is important to distinguish between ARF and chronic renal failure with end-stage renal disease (ESRD) presenting with a sudden onset of symptoms of renal failure. It is not uncommon for patients who have slowly developed chronic renal failure over many years to present with symptoms due to renal failure, such as anorexia, abdominal discomfort, and weight loss (but usually not decreased urine production). ARF is characterized by reduced or absent urine production (oliguria or anuria) as well as rapidly rising serum creatinine (increasing >0.5 mg/dL/day) and blood urea nitrogen (BUN). Blood hemoglobin is often only marginally low. Ultrasound demonstrates normal- or large-sized kidneys (in contrast to chronic renal failure, where the kidneys often are reduced in size). Urinalysis is helpful: gross proteinuria (>1 g/L) or red cells casts indicate glomerulonephritis, and myoglobinuria indicates crush injury with muscle destruction. Microscopic hematuria is typical for glomerular and acute interstitial kidney disease, whereas macroscopic hematuria indicates injury to the urinary tract (by cancer, infection, or renal stone). Massive calcium oxalate crystaluria suggests ethylene glycol poisoning.

Occurrence

Occupationally induced parenchymal ARF is rare. It occurs after accidental or unintentional exposures to certain chemical substances, often by ingestion. ARF from rhabdomyolysis after severe injuries is common. Apart from acute tubular necrosis (ATN), intrarenal obstruction from hemoglobin or myoglobin may contribute to ARF after crush injuries.

ARF is a common complication of other diseases, such as infections, major surgery, medications, and fluid depletion and/or hypotension, especially among elderly patients and those with cardiovascular disease.

Causes

The most common cause of renal parenchymal ARF is ATN. This is caused by ischemia of the tubular cells, leading to tubular cell death. Certain chemicals, including frequently used medications, may also exert direct toxic effects on the tubule, causing ATN. Poor renal blood flow, however, is the most common cause of ATN, typically occurring in patients with markedly low blood pressure from major bleeding, surgery, or severe infections. Fluid loss from excessive sweating, gastrointestinal disorders with diarrhea and vomiting, and overuse of diuretic drugs may also cause prerenal ATN. Commonly used drugs such as ACE-inhibitors and NSAIDs increase the risk for reduced renal blood flow during volume depletion, and thereby increase the risk for prerenal ATN. Drug toxicity from aminoglycoside antibiotics and radiographic contrast media may induce parenchymal ATN. Isolated ATN, without severe interstitial or glomerular damage or inflammation, is reversible and has a good prognosis. However, it is sometimes necessary to perform dialysis for days or weeks in order for the tubular cells to recover and the patient to survive.

Occupational exposures (apart from physical injuries and volume depletion) that may cause ARF include pesticides (in particular paraquat and other bipyridyl herbicides), organic solvents (certain chlorinated hydrocarbons and glycols), and some metals and metallic compounds (arsenic, bismuth, cadmium, mercury, and lead).

Paraquat, which is used in large quantities as a herbicide in the tropics, can also cause ARF. Although it is often sprayed without proper PPE, hazardous exposure is usually from accidental or intentional (suicidal) ingestion. Ingestion of more than 4 ml per kg of body weight is immediately threatening to life. Pulmonary damage is the main adverse effect, but, if the patient survives for a few days, ARF often develops and dialysis treatment is needed.

Other potential occupational causes of parenchymal ARF include high concentrations of certain hydrocarbon solvents, such as chloroform, and inadvertent intense exposure to toxic metals, such as uranium hexafluoride gas, mercury vapor from heated elemental mercury, and cadmium oxide fumes from soldering cadmium.

Ethylene glycol, which is commonly used as an antifreeze agent, causes severe intoxications in people who drink it accidentally or as a substitute for ethanol. The lethal dose is approximately 100 ml. Symptoms and signs of ethylene glycol intoxication include unconsciousness or somnolence, severe metabolic acidosis, and deteriorating renal function. The solvent dioxane, which is metabolized to diethylene glycol, has caused renal failure and death in workers following skin and inhalation exposure.

Occupational causes should be differentiated from ARF caused by certain medications, drug abuse, some herbal medicines, and certain types of wild-growing mushrooms.

Pathophysiology

ARF often involves a combination of ischemia and direct toxicity to the renal tubules. Paraquat, after absorption from the GI tract, destroys the lipid cell membranes in various organs, causing severe damage to the lungs and ATN. Ethylene glycol is metabolized into many toxic compounds, including glycolic acid, which is toxic to the renal tubules, and oxalic acid, which may precipitate within the tubules. Mercuric chloride (sublimate), certain uranium salts, and arsine produce acute tubular necrosis.

Prevention

Banning, or at least strict regulation, of the use and availability of paraquat is the most important measure to prevent intoxication. If ingestion of paraquat is confirmed, for example, by measuring the paraquat concentration in serum or urine, intensive treatment, including prevention of further GI absorption, hemoperfusion, and hemodialysis to enhance elimination, and possibly administration of steroids and immunosuppressive drugs (cyclophosphamide) should be considered. Ethylene glycol poisoning could be reduced if antifreeze compounds were made unpalatable. In cases with confirmed or highly suspected intoxication, hemodialysis should be initiated as soon as possible. Alternatively, an effective antidote (Fomepizole), which rapidly and competitively inhibits alcohol dehydrogenase more potently than ethanol, can be administrated to slow down and delay the endogenous production of severely toxic metabolites of ethylene glycol.

Other workers at risk of ARF include workers in enclosed spaces; arsine-exposed workers in the metal smelting, refining, or etching industries; workers who handle uranium hexafluoride gas; and workers who solder or weld with cadmium alloys. A combination of engineering controls, information to workers, and good work practices is necessary to ensure safety in high-risk settings.

High-risk groups for rhabdomyolysis, such as police, firefighters, and military recruits and reservists, should review their training procedures and eliminate practices likely to cause severe myoglobinuria, hypovolumia, or heat stroke. Practices to avoid include severe unaccustomed exercise, insufficient hydration, insufficient acclimatization, and controlled exposure to heat.

Further Reading

Choi YH, Kim N, et al. ARF requiring hemodialysis after accidental perchloroethylene ingestion. *American Journal of Kidney Disease* 2003; 41: E11.

Post TW, Rose BD. Approach to the patient with renal disease including ARF. *UpToDate in Nephrology and Hypertension.* 2003, Version 11.2 Rose BD and Post TW. Boston, UpToDate® www.uptodate.com.

Schrier RW. Nephropathy associated with heat stroke and exercise. *Annals of Internal Medicine* 1967; 67: 356-376.

Winchester JF. Management of paraquat intoxication. *UpToDate in Nephrology and Hypertension.* 2003, Version 11.2 Rose BD and Post TW. Boston, UpToDate® www.uptodate.com.

Respiratory Tract Irritation
ICD-10 J60-J70, J80-J86, J96-J99
SHE/O

Barry S. Levy and David H. Wegman

Many agents cause primary respiratory tract irritation. There is no unique syndrome associated with these agents. The symptoms associated with the agents depend on locus of action. For agents acting in the upper respiratory tract, the symptoms can include sneezing, nasal discharge, nosebleeds, sore throat, hoarseness, cough, phlegm, and wheezing, as well as non-respiratory effects, such as excessive tearing of the eyes. For agents acting in the lower respiratory tract, the predominant symptoms are cough, chest tightness, and progressive shortness of breath. The most extreme reactions of chemical pneumonitis and pulmonary edema can progress to respiratory failure and death. Severe overexposure can also result in life-threatening laryngeal edema. In general, mild irritant responses are reversible with removal from exposure, but recovery from severe reactions is prolonged by the need to recover from the acute inflammatory response. Chronic sequelae of edema and pneumonitis, such as lung fibrosis, bronchiolitis obliterans (resulting in chronic airways obstruction), and airways hyperreactivity (asthma) have been reported.

Occurrence

There are no reliable data on the incidence or prevalence of respiratory tract irritation. Generally, any environment where such agents are used will, if uncontrolled, result in widespread reports of symptoms. Even in settings where controls are being implemented, accidents may occur and overexposures may result in symptoms. As a rule, every exposed person will react to irritants if there is a sufficient airborne concentration and length of exposure. Some people with asthma or other forms of airways hyperresponsiveness may react to lower concentrations of irritants than others. Increased rate and depth of breathing, such as in an emergency evacuation, may increase the depth of penetration in the respiratory tract of an inhaled gas or vapor.

Causes

Apart from their irritant reactions, many of these agents have disparate characteristics. Unfortunately, no common features separate respiratory irritants from agents that do not cause irritant reactions. Examples of agents that act primarily as irritants are chlorine and derivatives (hydrochloric acid, phosgene, and chlorine dioxide), fluorine and derivatives (hydrofluoric acid and silicon tetrafluoride), bromine and iodine, sulfur dioxide and sulfuric acid, nitrogen oxides (nitrogen dioxide being the most hazardous), ozone, isocyanates, ammonia, acetic acid, acrolein, formaldehyde, and cadmium oxide.

Pathophysiology

The important features to distinguish irritants are their solubility in water and, for particulates, the particle size. Irritants that are highly water-soluble, such as ammonia, sulfur dioxide, and acidic gases, dissolve readily in the mucosa of the upper airways; those that are relatively less soluble, such as oxides of nitrogen, phosgene, and ozone, will dissolve in the lower airways and parenchyma. Agents that usually dissolve in the upper airways will, under conditions of severe overexposure, overwhelm the defenses and can penetrate to the lower airways as well. Larger particles (greater than 10 µm in diameter) deposit in the upper airways; those between 2.5 and 10 µm deposit in the middle airways; and those less than 2.5 µm generally reach the air exchange units.

The irritant effects are generally attributed to the direct excitation of neural receptors in the mucous membranes of the respiratory tract. The associated neural responses include sensation of pain and reflex responses such as cough and reflex bronchoconstriction. In several circumstances, acute pulmonary edema results from a change in the permeability of the pulmonary vasculature and histamine release, which leads to bronchoconstriction, increased capillary pressure, and extravasation of fluids. Following episodes of high-intensity inhalation of toxic agents or multiple episodes of lower intensity inhalation, individuals may be left with persistent airways hyperresponsiveness. This condition has been labeled reactive airways dysfunction syndrome, or RADS. (See chapter on RADS.)

Prevention

There are no unique approaches to the control of respiratory irritants. General control of gas and vapor exposure through proper maintenance and local exhaust ventilation is the first line of defense. Canisters or lines containing or transporting toxic gases should be appropriately labeled. Probably the most problematic situation is accidental overexposure. Although such accidents cannot be entirely prevented, proper education of workers on risks and their avoidance is highly desirable. Furthermore, each worker should be carefully trained to avoid risks associated with improper attempts to rescue a co-worker from an overexposure environment. Rescue attempts should be made only with proper (and therefore readily available) rescue equipment, including air-supplied respirators or self-contained breathing apparatus.

Surveillance is unlikely to provide early warning, but investigation and analysis of overexposure events may provide information on why primary prevention systems are failing.

Other Issues

Exposure to relatively low levels of some irritants will acclimatize individuals to a higher odor threshold, thereby reducing awareness of potential overexposure.

Further Reading

Christiani D, Wegman DH. Respiratory disorders. In Levy BS, Wegman DH (Eds.). *Occupational Health: Recognizing and Preventing Work-Related Disease and Injury* (4th ed.). Philadelphia: Lippincott Williams & Wilkins, 2000, pp. 477-501.

Dalton P. Evaluating the human response to sensory irritation: implications for setting occupational exposure limits. *AIHAJ* 2001; 62: 723-729.

Dalton P. Upper airway irritation, odor perception and health risk due to airborne chemicals. *Toxicology Letters* 2003; 140-141: 239-248.

Delclos GL, Carson AI. Acute gaseous exposure. In Harber P, Schenker MB, Balmes JR (Eds.). *Occupational and Environmental Respiratory Disease*. St. Louis: Mosby, 1996, pp. 514-534.

Nowak D. Chemosensory irritation and the lung. *International Archives of Occupational and Environmental Health* 2002; 75: 326-331.

Shusterman D. Toxicology of nasal irritants. *Current Allergy and Asthma Reports* 2003; 3: 258-265.

Silicosis

ICD-10 J62

David Weissman and Gregory R. Wagner

Silicosis is an interstitial lung disease caused by inhalation of free crystalline silica (silicon dioxide). The surveillance case definition used by the Sentinel Event Notification System for Occupational Risks (SENSOR) program of NIOSH requires exposure to silica and x-ray or pathological changes consistent with silicosis. Clinically, the diagnosis of silicosis is based on history of inhalation exposure, presence of opacities on chest x-ray that are consistent with the diagnosis, and exclusion of other possible diagnoses.

Silicosis is associated with several patterns of chest x-ray abnormality. Simple silicosis is a profusion of small (<10 mm in diameter) opacities. These are generally rounded, but may be irregular, and occur predominantly in the upper lung zones. Conglomerate silicosis or progressive massive fibrosis (PMF) occurs when these small opacities enlarge and coalesce to form larger, upper or mid-zone opacities (>10 mm in diameter). The chest x-ray in acute silicosis, described below, is characterized by lower-zone alveolar filling and resembles conditions such as pneumonia or alveolar proteinosis.

Silicosis can be clinically and temporally classified into several different patterns. Chronic silicosis develops slowly, usually appearing 10 to 30 years after first exposure. It may be identified many years after leaving the job associated with exposure. The chest x-ray of most people with chronic silicosis shows diffuse nodular changes, but in a minority of people with chronic silicosis, nodules coalesce to become PMF. Accelerated silicosis develops less than 10 years after first exposure. It has the same radiographic appearance as chronic silicosis and is differentiated only by its more rapid development. Acute silicosis results from overwhelming exposure to fine respirable particles of crystalline silica over a short time, typically less than several years. The exposure results in symptoms within a few weeks to 4 or 5 years after the initial exposure.

Occurrence

Although preventable, silicosis continues to occur. It is an especially important problem in developing countries, but also still occurs in the United States and the European Union. A broad range of industries and occupations are associated with inhalation exposure to crystalline silica. These include any occupation that disturbs the earth's crust or involves processing of silica-containing rock or ores. Occupations classically associated with high risk have included miners, granite workers, sandblasters, ground silica (silica flour) workers, and foundry workers. Many other work settings are associated with exposures and new ones continue to be discovered. In the early 1990s, 200 000 miners and 1.7 million non-mining workers in the United States were estimated to have occupational exposures to inhaled silica. Silicosis was reported in 577 workers in Michigan between 1987 and 1995 through a state surveillance system. In the United States between 1987 and 1996, a total of 2787 deaths were attributed to silicosis. Morbidity and mortality figures are inaccurate because of both underreporting and under-recognition.

Silicosis is associated with significant morbidity and mortality. Silica exposure alone is associated with excessive decline in lung function. The development of radiographic findings of chronic or accelerated silicosis is associated, on average, with greater abnormality in pulmonary function. Chronic cough and dyspnea are common and become more frequent with worsening radiographic appearance. PMF is generally associated with more severe symptoms than the simple pattern of x-ray involvement and leads, in extreme cases, to respiratory failure and death. Acute silicosis is a particularly grim condition. Symptoms include cough, weight loss, fatigue, and sometimes pleuritic pain. Patients rapidly develop cyanosis, cor pulmonale, and respiratory failure. Survival after onset of symptoms is typically less than 4 years, with mycobacterial and fungal infections frequently complicating the clinical course.

Causes

Silicosis is caused by inhalation of dust containing particles of crystalline silica in the respirable size range. Although silica is the most abundant mineral on earth, not all silica is hazardous. Non-crystalline (amorphous) forms, such as diatomite (skeletons of prehistoric marine organisms) and vitreous silica, are relatively nontoxic after inhalation. In contrast, crystalline silica is very hazardous to inhale when aerosolized particles are in the respirable size range. Crystalline forms of silica include quartz, cristobalite, and tridymite. Quartz is the most common type and is a component of rocks such as granite, slate, and sandstone. Granite contains about 30% free silica and slate about 40%; sandstone is almost pure silica. Cristobalite and tridymite occur naturally in volcanic ash and are formed when quartz or amorphous silica is subjected to very high temperatures. Generation of cristobalite from amorphous silica or quartz can also occur during high-temperature industrial processes, such as in foundries when silica-containing clay molds are filled with molten metal to make castings. Processes such as hammering, crushing, pulverizing, grinding, and drilling through crystalline silica can generate silica aerosols with the potential to cause silicosis. Crystalline silica that has been freshly fractured through blasting or grinding may pose greater risk than aged silica.

"Free" crystalline silica is uncombined with other chemicals. "Combined" forms of silica, called silicates, are compounds in which silica is bound to other metal elements. Some silicates also have the potential to cause interstitial lung disease. Examples of silicates used in industry include asbestos, talc $[Mg_3Si_4O_{10}(OH)_2]$, and kaolinite $[Al_2Si_2O_5(OH)_4]$, a major component of china clay, or kaolin.

Pathophysiology

After inhalation, crystalline silica particles exert a direct cytotoxic effect on pulmonary target cells such as alveolar macrophages. Cytotoxicity is related to the ability of the crystalline surface to generate oxygen radicals when suspended in water. Freshly fractured silica particles generate oxygen radicals in greater amounts and are more cytotoxic than aged particles, perhaps accounting for the great risk associated with activities such as sandblasting. After exposure, pulmonary target cells, such as alveolar macrophages and epithelial cells, engage in cytokine networking with other resident cells, such as fibroblasts and endothelial cells. These, in turn, engage in cytokine networking with various circulating inflammatory cells leading to an influx of these cells into the lung. Depending on level of exposure, co-exposures, and individual susceptibility factors, such as genetic polymorphisms, these processes can eventually lead to inflammatory damage, fibrosis, and/or alveolar flooding with surfactant-like material.

Silicotic nodules are the pathologic hallmark of chronic and accelerated silicosis. These develop over a period of years. They have a central zone that is hyalinized and composed of concentrically arranged collagen fibers and a peripheral zone that is whorled and contains macrophages, lymphocytes, and lesser amounts of loosely formed collagen. Coalescence of silicotic nodules forms the PMF lesion. Chronic and accelerated silicosis of advancing severity is associated with progressively worse emphysematous changes detectable by pathology and computed tomography. The degree of emphysema correlates with severity of obstructive impairment. PMF further worsens obstructive ventilatory impairment through distortion of airways. It can also be associated with restriction and, in advanced cases, respiratory failure.

In acute silicosis, the lung interstitium is thickened with inflammatory cells. There is alveolar filling with proteinaceous material. The histologic appearance resembles idiopathic alveolar proteinosis, so acute silicosis has also been called silicoproteinosis. Silicotic nodules are only rarely seen and there is minimal pulmonary fibrosis. Acute silicosis is associated with profound respiratory impairment and death.

Prevention

Silicosis can be prevented. Since the mid-1990s, the ILO and WHO have joined in a 20-year effort for silicosis elimination. In both developed and developing countries, preventive interventions have successfully reduced the burden of silicosis. These interventions have focused on primary prevention through the hierarchy of substitution, engineering controls, administrative controls, and use of respirators to reduce respirable crystalline silica exposure. OSHA currently enforces a PEL for respirable dust based on a formula that accounts for the silica content. The respirable dust PEL is 10 mg/m^3 divided by (%SiO$_2$ + 2), or 250 million particles per cubic foot divided by (%SiO$_2$ + 5). This is equal to approximately 0.1 mg/m^3 of respirable quartz. This exposure limit is not protective for many workers with exposure over an entire working lifetime. For example, a recent study of retired Vermont granite workers showed that strict adherence to a 0.1 mg/m^3 respirable quartz exposure limit in the period since 1955 still resulted in radiologic abnormalities consistent with silicosis in 3.5% of exposed workers. It is likely that, for many of these and other workers, radiologic abnormalities may not develop until after retirement. In view of concerns that the current PEL is not entirely protective, NIOSH recommends a lower REL of 0.05 mg/m^3. Noncompliance with the PEL appears to be common in construction work, foundry work, mining, sandblasting, and in the cut-stone and stone-products industries.

Secondary prevention can be accomplished by monitoring exposed workers with serial chest radiographs and spirometry. Such health surveillance is mandated in many industrialized countries and may assist in early identification of silicosis. Ideally, individual workers found to have silicosis will be removed from further inhalation exposure. Identification of sentinel cases should result in efforts to determine what went wrong and improved interventions to reduce exposure for all of those comparably exposed. Tertiary prevention is a poor option in silicosis, as only symptomatic treatment is currently available.

Restriction or elimination of abrasive blasting with sand has resulted in sharp decreases in silicosis in many countries. Substitutes for abrasive agents are available; however, their use may also result in hazardous exposures that need to be controlled.

Other Issues

Silica exposure has a number of health effects other than pulmonary silicosis. As already noted, increasing silica exposure, even without radiologic silicosis, has been associated with accelerated decline in pulmonary function. Silica exposure and silicosis have been associated with increased risk for tuberculous and non-tuberculous pulmonary mycobacterial infections and systemic complications, such as increased risk for end-stage renal failure. Silicosis has also been associated with increased risk for collagen vascular diseases. Although the collagen vascular disease previously felt to be most strongly associated with silicosis was scleroderma, recent publications have suggested that, in the United States, there is a stronger association with rheumatoid arthritis. Finally, silicosis has been identified as a risk factor for lung cancer. Although some controversy still exists, many also consider silica exposure alone to be a risk factor for lung cancer. As is the case for silicosis, these related conditions should be preventable by limiting exposure to inhaled crystalline silica.

A number of NIOSH publications are available through the NIOSH website (http://www.cdc.gov/niosh/topics/ silica/default.html), addressing issues such as awareness of silica as a workplace hazard, environmental controls, personal protection, and medical monitoring.

Further Reading

American Thoracic Society. Adverse effects of crystalline silica exposure. *American Journal of Respiratory and Critical Care Medicine* 1997; 155: 761-765.

International Labour Organization. *Guidelines for the Use of the ILO International Classification of Radiographs of Pneumoconioses* (Revised edition, 2000). Geneva: International Labor Organization, 2002.

Kissell FN (Ed.). *Handbook for Dust Control in Mining.* IC 9465. (DHHS [NIOSH] Publication No. 2003-147). Washington, DC: NIOSH, 2003.

National Institute for Occupational Safety and Health. NIOSH *Hazard Review: Health Effects of Occupational Exposure to Respirable Crystalline Silica.* (DHHS [NIOSH] Publication No. 2002-129. Cincinnati: NIOSH-Publications Dissemination, 2002.

Wagner GR. *Screening and Surveillance of Workers Exposed to Mineral Dusts.* Geneva: World Health Organization, 1996.

World Health Organization. *Concise International Chemical Assessment Document 24: Crystalline Silica, Quartz.* Stuttgart: Wissenschaftliche Verlagsgesellschaft mbH, 2000.

Skin Cancer

ICD-10 C43-C44

Michael E. Bigby

Cutaneous neoplasms may be benign, premalignant, or malignant. Malignant cutaneous tumors include basal and squamous cell carcinomas as well as melanomas. Basal cell carcinomas occur most commonly on sun-exposed areas, such as the face, neck, and hands, and typically appear as translucent papules with pearly borders, telangiectasia, and central ulceration. Squamous cell carcinomas may be recognized by their rough, irregular, scaly surface; indistinct borders; and, frequently, ulceration. Melanoma is the only common cutaneous neoplasm that often metastasizes. Melanomas appear as pigmented macules, papules, or nodules that have irregular, often-notched borders, irregular contours, and variegations in color, including red, white, blue, black, or gray.

Occurrence

In the United States in 2000, there were over 1.3 million new cases of basal and squamous cell carcinoma; and 54 200 new cases of and 7600 deaths due to cutaneous melanoma. The percentage of cases of skin cancer that are occupationally related is not known.

Causes

Workers exposed to ionizing radiation, non-ionizing radiation (ultraviolet [UV] light), polycyclic hydrocarbons, arsenic, and tar are at highest risk for developing basal and squamous cell carcinomas of the skin. These skin cancers occur more often in latitudes where there is more intense sunlight.

Pathophysiology

Cutaneous neoplasms are induced by agents that can directly damage the DNA of epidermal cells, inhibit DNA repair, or possibly decrease immune surveillance of the epidermis. Basal cell carcinomas are the most common type of cancer. They arise from mutations in the genetic material of basal-layer keratinocytes induced primarily by ultraviolet radiation. A mutation in the patched gene was identified as the cause of some cases of basal cell nevus syndrome, a syndrome in which patients develop many basal cell carcinomas and have palmar pits and skeletal defects.

The pathogenesis of melanoma involves activation of genes that stimulate melanocytes to divide uncontrollably. Patients with defects in DNA repair, such as xeroderma pigmentosum, have rates of melanoma more than 1000 times that of the normal population. Melanoma cell lines express genes not expressed by normal melanocyte cell lines. These genes include those that regulate the cell cycle: CDKN2 (encoding p16), CDKN2B (p15), CDKN2C (p18), CDK4, and p53. Only the p16 mutation has been confirmed to be of etiologic importance, having been found in a number of kindreds with familial melanoma.

Prevention

Outdoor workers in sunny climates should always be provided with wide-brimmed hats, topical sunscreens, and, if practical, long-sleeved shirts. Wearing sun-protective clothing and avoiding unnecessary sun exposure during the peak hours of UV radiation (10:00 a.m. to 3:00 p.m.) are the most effective measures to prevent skin cancer. Sunscreen of at least an SPF of 15 should be applied 15 to 30 minutes before exposure at $2gm/cm^2$ and reapplied at the interval recommended in the product label (usually every 2 to 3 hours). It should also be reapplied after swimming or sweating. A global solar ultraviolet index (UVI) has been developed by WHO. Its values range from 1 to 11+ and are grouped into low, moderate, high, very high, and extreme exposure categories with corresponding color codes and pictograms, indicating recommended sun protection strategies. Easy-to-understand information on the global solar UVI and recommended sun protection measures are available from the WHO World Wide Web page (www.who.int/uv). Use of arsenical insecticides should be discontinued. Exposure to other chemical carcinogens can be reduced by using protective clothing, gloves, adequate ventilation, and closed systems when carcinogenic agents are being used.

Other Issues

Exposure to UV radiation in sunlight is currently the leading cause of occupational skin cancer. Workers exposed to UV light may develop actinic keratoses, which are premalignant lesions, or basal and squamous cell carcinomas. Oncogenic polycyclic hydrocarbons, which are found in coal tar and petroleum products, may cause skin precancers and cancers in workers who come in contact with tar, pitch, soot, crude paraffin, or asphalt.

Further Reading

Adams RM. *Occupational Skin Disease* (3rd ed.). Philadelphia: W.B. Saunders, 1999.

American Academy of Dermatology: http://www.aad.org/pamphlets/

Centers for Disease Control and Prevention. Preventing skin cancer. *MMWR* Vol. 52, No. RR-15, 1-18.

Meves A, Repacholi MH, Rehfuess EA. Promoting safe and effective sun protection strategies. *Journal of the American Academy of Dermatology* 2003; 49: 1203-1204.

Meves A, Repacholi MH, Rehfuess EA. Global solar UV index: A physician's tool for fighting the skin cancer epidemic. *International Journal of Dermatology* 2003; 42: 846-849.

The Skin Cancer Foundation: http://www.skincancer.org/prevention/index.php

Skin Infections

ICD-10 L00-L08

Michael E. Bigby

Diagnoses of skin infections are made by clinical appearance, Gram stain, potassium hydroxide preparation, Tzanck smear, direct fluorescence antibody staining (Herpes simplex and Varicella zoster), and culture. Lesions vary and are determined by etiologic agents. Bacterial infections, most commonly caused by streptococci or staphylococci, may appear as superficial crusted plaques and papules (impetigo); follicular papules and pustules (folliculitis); or deep, tender, erythematous nodules or plaques, with or without lymphatic streaks (furuncles, carbuncles, or cellulitis). Fungal (dermatophyte) infections commonly appear as annular plaques with central clearing and an erythematous, raised, scaly border. Viral infections vary tremendously in clinical appearance; for example, herpes simplex infections most commonly present as grouped vesicles on an erythematous base.

Occurrence

No reliable population-based data for the United States are known to exist.

Causes

Pyodermas, induced by streptococci or staphylococci, are the most common bacterial skin infections. These infections may occur as a result of trauma or as a complication of other occupational dermatoses. Fungal infection with dermatophytes (ringworm) or *Candida albicans* are often found in a local environment of moisture, warmth, and maceration. They therefore occur frequently in body folds and in warm climates. Herpes simplex is the most frequently identified work-related viral skin infection encountered in the U.S.

Pathophysiology

Bacteria and viruses gain access through disruptions in the normal epidermal barrier caused by trauma or underlying skin diseases, such as atopic dermatitis and irritant or allergic contact dermatitis. Yeast and fungi can proliferate within the epidermis in susceptible individuals, especially in areas that are moist and warm. Infectious agents are acquired from contact with infected customers, animals, carcasses, hides, or materials found in the workplace. After gaining access, organisms are able to proliferate intracellularly or extracellularly. Clinical manifestations are caused by multiple factors, including direct cellular injury or necrosis, elaboration of toxins, release of inflammatory mediators, and induction of an immune response.

Prevention

Preventive measures depend on the etiologic agents and include identifying workers at risk, using properly selected protective clothing, washing hands, identifying and avoiding sources of infectious agents in the workplace, and reducing environmental factors associated with cutaneous infections (excess heat, high humidity, and frequent exposure to water). This presumes that facilities are available for hand washing.

Other Issues

Fungal and yeast infections often occur in workers, such as factory workers, dishwashers, and canning industry workers whose local environment is humid or hot or whose work involves frequent exposure to water. Health care workers-especially those who are exposed to oral secretions, such as dentists, dental technicians, nurses, respiratory therapists, and anesthetists-are at high risk for herpetic infections of the hands. Daycare workers are also at increased risk. Sheep handlers are at high risk for orf (contagious ecthyma). Barbers and cosmeticians who contact customers suffering from contagious skin diseases are particularly at risk for bacterial and fungal infections.

Further Reading

Adams RM. *Occupational Skin Disease* (3rd ed.). Philadelphia: W.B. Saunders, 1999, pp. 86-110.

Bigby M, Arndt KA, Coopman SA. Skin disorders. In Levy BS, Wegman DH (Eds.). *Occupational Health: Recognizing and Preventing Work-Related Disease and Injury* (4th ed.). Philadelphia: Lippincott Williams & Wilkins, 2000, pp. 537-552.

Lushniak, BD. Occupational infectious diseases with dermatologic features. In Couturier AJ (Ed.). *Occupational and Environmental Infectious Diseases*. Beverly Farms, MA: OEM Press 2000, pp. 389-404.

Skin Injuries

ICD-10 T14.0, T14.1, T14.7-T14.9

Michael E. Bigby

Occupational skin injuries result from acute trauma or a brief exposure to an injurious chemical. The clinical appearance is variable, depending on the cause.

Occurrence

During 2001, skin injuries accounted for 18.5% of the total of 1.5 million reported injuries and illnesses in private industry that required recuperation away from work beyond the day of the incident, according to data reported to the Bureau of Labor Statistics. (See Table 1.)

Burns and contusions occurred most commonly in services, retail trade, manufacturing, and transportation and public utilities industries. Cuts and lacerations were most common in manufacturing, retail, construction, and services industries. Heat burns were most common in the retail trades. Chemical burns were most common in manufacturing.

Table 1. Number of occupational injuries and illnesses involving time away from work, by selected nature of injury and illness, United States, 2001

Type of Injury	Total Number of Cases (in thousands)
Bruises and contusions	136.4
Cuts and lacerations	114.8
Heat burns	25.1
Chemical burns	9.5

Causes

There are many causes of occupational skin injuries, including sharp, rough-surfaced, and hard objects that workers either hit or are hit by; caustics, acids, and other corrosive chemicals; hot objects, liquids, gases, or flames; electricity; and ionizing and nonionizing radiation.

Pathophysiology

Acute trauma or exposure to strong irritants or to physical agents causes direct injury to cells of the epidermis and, occasionally, the underlying tissues. Both the type and extent of damage vary considerably according to etiology.

Prevention

The wide variety of types of skin injuries make a broad-based effort to prevent skin injuries necessary. (See Chapter 1 in Part I and chapter on Burn Injuries in Part II.)

Other Issues

Hydrofluoric acid (HF) is capable of producing rapidly progressive and deeply destructive burns. HF burns cause severe pain and may be accompanied by systemic effects, such as hypocalcemia. Initial management of these burns includes removing all clothing contaminated with HF and washing affected areas copiously with cold water. Further therapy is directed at removing remaining fluoride ions to prevent further tissue destruction and systemic absorption. Acceptable measures include benzethonium chloride or benzalkonium chloride soaks, calcium gluconate gel application, and calcium gluconate injection.

Further Reading

Bureau of Labor Statistics. Lost-worktime injuries and illnesses: characteristics and resulting days away from work, 2001 (USDL 03-138). Washington, DC: BLS, 2003. Available at: http://www.bls.gov/iif/home.htm (Document osnr0017.pdf).

Smell and Taste Disorders

ICD-10 R43.0

Marcia H. Ratner, Seth Nelson, and Robert G. Feldman

Intact smell and taste function is essential to food selection and appreciation, as well as to the recognition of exposure to potential airborne toxic substances. Olfaction can be an important asset in the prevention of work-related exposures to toxic chemicals. The olfactory threshold may be below, equal to, or greater than the ambient air permissible exposure limit for a chemical. A worker may be exposed to a concentration of a chemical that is less than its exposure limits, but may nevertheless develop resultant symptoms of olfactory dysfunction.

Physical examination of the nose and mouth can be used to screen for overt underlying pathology. Computed tomography (CT) scans, magnetic resonance imaging (MRI), as well as commercially available standardized tests may be also useful. MRI can be used to ascertain if there has been damage to the olfactory bulbs and tracts, frontal lobes, and temporal lobes. Tests such as smell discrimination and nasal challenges generally help to assess impairment of the sense of smell.

Anosmia, a complete loss of the sense of smell, is usually due to prolonged, chronic exposure to chemicals. Hyposmia, a diminished sense of smell, is a more common disorder that often occurs in workers chronically exposed to low levels of olfactory irritants. Disturbances of odor distortion are dysosmia, in which there is a distorted quality of an odorant stimulation, and *troposmia*, in which an odor is perceived when no odorant is present.

Occurrence

At least 2 million Americans suffer from smell and taste disorders. Smell disorders are more common than taste disturbances because of (a) the vulnerability and anatomical distinctiveness of the olfactory system, and (b) a decline in olfactory function that is part of the normal aging process. Smoking plays a key role in the onset of hyposmia. Loss of smell progresses with age due to normal aging processes. Evidence of a relationship between the duration and concentration of exposure to olfactory irritants and the severity of hyposmia has been seen in individual workers. An increase in the number of workers with olfactory deficits is considered indicative of a greater level of exposure as well as the pervasiveness of the chemical throughout the workplace.

Causes

Smell and taste disturbances may be due to head trauma, medications, upper respiratory infections, nasal and paranasal sinus diseases, and damage to the olfactory, facial, and trigeminal nerves that supply taste and smell. Taste disorders can also result from trauma to the tongue and oropharynx, and lesions of peripheral receptors, taste control pathways, and cortical areas.

Blunt trauma to the head can temporally and/or permanently disrupt olfactory function. Among the causes of olfactory dysfunction, head trauma is associated the greatest loss of function. Prognosis among patients with olfactory dysfunction due to head trauma depends on the severity of the injury and the specific structure involved. Patients with trauma-related anosmia rarely regain normal olfactory ability.

Traumatic injury to the facial nerve is a common occurrence during wartime combat, but may also be seen among civilians. The clinical picture may be confusing because partial damage to the peripheral nerve may mimic impairment of the central facial motor mechanism.

Repeated or chronic exposure to industrial solvents can lead to olfactory adaptation before nerve damage occurs. Chronic exposures to common solvents, including benzene, toluene, trichloroethylene, and xylene, can result in hyposmia of varying degrees that at first may be completely reversible with cessation of exposure. If exposure continues, ability to recognize exposure is further diminished as persistent deficits emerge insidiously.

The chemical composition of solvents plays a role in smell and taste disturbances and the ability of the chemical to stimulate the olfactory nerve as compared with the trigeminal nerve. The amount of pain and irritation associated with an odorant chemical is dependent on its volatility and its irritant and aromatic properties. The vapors of acidic compounds, such as hydrochloric acid and hydrogen sulfide, tend to irritate nasal mucosa and damage olfactory neurons more than do pH-neutral compounds, such as *n*-hexane and trichloroethylene.

Heavy metals are another cause of work-related olfactory disorders. Occupational exposures to manganese, nickel, and cadmium have been associated with olfactory dysfunction. It is not yet known whether mercury is toxic to the olfactory system in mammals, but this metal is known to alter olfaction and olfactory-related behavior in fish. In a study of workers exposed to cadmium, results showed (a) a distinct olfactory deficit in most of the workers exposed to cadmium; and (b) a relationship between olfactory impairment and the concentration of cadmium in blood, urine, and workplace air.

Pathophysiology

Damage to the olfactory system can occur in many ways. First, there is a possibility that chemosensory structures such as the olfactory receptors in the regio olfactoria can be damaged due to nasal inhalation of irritating chemicals since this is the first structure that chemically interacts with the odorant molecule. The nasal epithelium and mucosa can be damaged from chronic exposure to olfactory-irritating chemicals. Nasal epithelium can regenerate if exposure to the irritating chemicals cease.

Workers who are exposed to low levels of irritating toxins, such as organic solvents and heavy metals, begin to lose their sense of smell gradually in a process known as smell fatigue or olfactory adaptation, due in part to the physiology and life cycle of olfactory receptor cells. Receptor cells are turned over approximately every 40 days. If the receptors are being irritated by chemicals daily, a worker's sense of smell will be reduced.

Marked loss of olfactory neurons is associated with anosmia. The pathophysiology underlying olfactory distortions (troposmia) is typically a decreased number of functioning olfactory primary neurons so that an incomplete characterization of the odorant is made. By contrast, phantosmia may be due to an abnormal signal or inhibition from the primary olfactory neurons or peripheral olfactory or trigeminal signals that "trigger" a central process.

The interaction between the olfactory and gustatory systems and the trigeminal nerve and facial nerve and their central projections is an important determinant of sensations of odor and taste that appears to change as a result of aging and disease. Interruption of the trigeminal-olfactory-gustatory pathway occurs with damage to the trigeminal nucleus and its related structures and therefore, may occur with exposures to certain industrial solvents, such as trichloroethylene and its environmental breakdown product dichloroacetylene.

Prevention

Treatment of these disorders is still limited to conditions with discernible and reversible causes. Therefore, prevention of injuries that cause permanent damage to olfactory structures is essential.

There is a marked difference in the concentration of odor that can be detected, and the concentration of odor that will be detected. Those who are distracted while experiencing an odor tend to detect the smell at higher levels than those who are focusing their attention on the act of smelling the odor.

Early detection is key to keeping workers safe. Workers are encouraged to report any suspect odors or nasal irritation immediately so that appropriate investigation can occur.

Preventing head trauma significantly reduces the risk for olfactory and taste disorders among workers. Workers who are at risk for head injury should wear hardhats at all times.

Prevention of acute and chronic exposures to those chemicals that may specifically cause injury to the trigeminal nerve and its pathways, such as trichloroethylene and dichloroacetylene, minimizes the risk of associated smell and taste disturbances.

Further Reading

Bromley SM. Smell and taste disorders: a primary care approach. *American Family Physician* 2000; 61: 427-436, 438.

Doty RL, Yousem DM, Pham LT, et al. Olfactory dysfunction in patients with head trauma. *Archives of Neurology* 1997; 54: 1131-1140.

Leopold D. Distortion of olfactory perception: diagnosis and treatment. *Chemical Senses* 2002; 27: 611-615.

Ship JA, Weiffenbach JM. Age, gender, medical treatment, and medication effects on smell identification. *Journal of Gerontology* 1993; 48: M26-32.

Spielman AI. Chemosensory function and dysfunction. *Critical Reviews in Oral Biology and Medicine* 1998; 9: 267-291.

Soft Tissue Sarcoma

ICD-10 C49

Richard W. Clapp

Soft tissue sarcomas comprise a set of individually rare cancers of muscle, adipose tissue, blood vessels, cartilage, and other mesenchymal tissues. Some types, such as Kaposi's sarcoma, are malignancies of blood vessels that are clinically obvious, pigmented, patchy lesions on the skin. Other types, such as liposarcomas and fibrosarcomas, may present as isolated lumps or multicentric lesions in fat or muscle. Rhabdomyosarcomas, neurosarcomas, and angiosarcomas can occur in a variety of locations and organs in the body. Leiomyosarcomas typically occur in uterine smooth muscle, although they can also occur throughout the body. There are myriad subclassifications and varying degrees of aggressiveness of these cancers, and a wide variety of surgical and combined therapies are available, but the overall prognosis is poor. Angiosarcoma of the liver is considered a "sentinel" cancer of occupational origin.

Occurrence

Each year, approximately 8000 cases of soft tissue sarcoma are diagnosed in the United States, and approximately 4000 people die of this disease. It is not known how many cases and deaths are occupationally related. Male rates are generally higher than female rates, and African-Americans typically have higher rates than whites in the U.S. One exception is the much higher rate of Kaposi's sarcoma in young white males, which has paralleled the HIV/AIDS epidemic of the past two decades. There are somewhat higher rates in young children, including cases found in genetic syndromes, such as the Li-Fraumeni syndrome. Incidence rates decline in the teenage years and then increase sharply starting from about age 30. Prevalence was estimated at about 65 000 cases in the U.S. population in early 2000. There have been improvements in 5-year survival for specific types of sarcomas in recent decades, but the overall percent survival is about 35% in white patients and 45% in African-American patients.

Causes

Exposure to ionizing radiation, including Thorotrast, is known to cause soft tissue sarcoma. Vinyl chloride monomer is an established cause of angiosarcoma of the liver. Phenoxyacetic acid herbicides contaminated with dioxin have been associated with soft tissue sarcomas among (a) Americans who are Vietnam veterans; (b) residents exposed to dioxin near a chemical plant explosion in Seveso, Italy; and (c) those exposed to these chemicals in farming and forestry work. A number of studies have shown increased risk of soft tissue sarcoma in chemical production workers, pulp and paper workers, and herbicide sprayers who may be exposed to dioxins as well as other chemicals. Vietnam veterans are compensated if they have been diagnosed with soft tissue sarcoma since time of service in Vietnam. Studies of cohorts of dioxin workers in the U.S. and European Union countries have also shown excess soft tissue sarcoma, especially among those most heavily exposed. Some studies have suggested increased risk of soft tissue sarcoma in patients exposed to arsenical medications. There are also numerous reports of "wound sarcomas" at the location of surgical scars and burns. Kaposi's sarcoma in young males in the U.S. is now thought to be caused by human herpes simplex virus, type 8, which is often found in patients with AIDS.

Pathophysiology

Soft tissue sarcoma may be initiated by some exposures, such as ionizing radiation, and promoted by others. For example, some exposures may increase cell division or inhibit the p53 gene and reduce apoptosis. The precise mechanisms are not well understood and probably differ with individual types of soft tissue sarcoma.

Prevention

Primary prevention of soft tissue sarcoma involves avoiding exposure to ionizing radiation, arsenical medicines, vinyl chloride, and dioxin. Patient education about risk factors, such as family history of these tumors and pesticides used in farming and forestry, is also warranted. Screening for early detection of this malignancy has not been definitively shown to be effective.

Diagnosis of soft tissue sarcoma should prompt a review of potential dioxin exposure. Evidence of dioxin exposure should alert health care providers to seek signs and symptoms of soft tissue sarcoma.

Further Reading

Bailar J. How dangerous is dioxin? *New England Journal of Medicine* 1991; 241-242.

Bertazzi PA, Pesatori AC, Consonni D, Tironi A, Landi MT, Zocchetti C. Cancer incidence in a population accidentally exposed to 2,3,7.8-tetrachlorodibenzo-para-dioxin. *Epidemiology* 1993; 4: 398-406.

Fingerhut MA, Halperin WE, Marlow DA, et al. Cancer mortality in workers exposed to 2,3,7,8-tetrachlorodibenzo-p-dioxin. *New England Journal of Medicine* 1991; 324: 212-218.

Zahm SH, Fraumeni JF Jr. The epidemiology of soft tissue sarcoma. *Seminars in Oncology* 1997; 24: 504-514.

Solvents, Organic, Adverse Effects

ICD-10 T52

Barry S. Levy

Organic solvents are volatile compounds or mixtures that are used for extracting, dissolving, or suspending materials such as fats, waxes, and resins that are not soluble in water. Some categories of organic solvents and illustrative associated health effects are shown in Table 1. Diagnosis of solvent-related health effects and determination of their association with organic solvent exposure is often challenging. Critical aspects of identification are signs and symptoms suggestive of or consistent with these adverse health effects, an occupational and environmental exposure history, and laboratory tests indicating (a) dysfunction associated with these compounds; and/or (b) the presence of these substances or their metabolites in blood, urine, and exhaled air.

Occurrence

The magnitude of adverse health effects due to organic solvents is not known. Approximately 10 million workers in the United States are potentially exposed to organic solvents at work. At high risk for adverse health effects are workers who are chronically exposed to high concentrations of airborne organic solvents, including workers in the plastics, printing, graphics, metal, and dry-cleaning industries, and workers who manufacture or use adhesives, lacquers, and paints.

Causes

Illustrative adverse health effects that organic solvents have been shown to cause are shown in Table 1.

Pathophysiology

Virtually all organic solvents depress the central nervous system with acute exposure, causing symptoms ranging, with degree of exposure, from lethargy to unconsciousness and death. With chronic exposure, organic solvents affect both the central and the peripheral nervous systems. Studies of solvent-exposed workers have demonstrated (a) neurobehavioral effects, including disorders with reversible subjective symptoms, such as irritability and memory impairment; sustained changes in personality or mood, such as emotional instability and diminished impulse control and motivation; and impaired intellectual function, with learning disability as well as decreased memory and concentration ability; and (b) chronic changes in peripheral nerve function, with severe peripheral neuropathy occurring with exposure

Table 1. Major Categories of Organic Solvents and Illustrative Adverse Health Effects

Category	Examples	Illustrative Adverse Health Effects
Aliphatic hydrocarbons	*n*-Hexane	Peripheral neuropathy
Aromatic hydrocarbons	Benzene	Aplastic anemia, leukemia
	Styrene	Loss of color vision
Halogenated hydrocarbons	Carbon tetrachloride	Liver and kidney damage, cardiac sensitization
	Methylene chloride	Carboxyhemoglobin formation (with the same adverse health effects as carbon monoxide)
	Freons	Arrhythmias
Alcohols	Methyl alcohol	Optic atrophy, metabolic acidosis, respiratory depression
Glycol ethers	Ethylene glycol monomethyl ether (EGME)	Toxic encephalopathy, macrocytic anemia, abnormal pregnancy outcome
Ketones	Methyl butyl ketone (MBK)	Peripheral neuropathy
Miscellaneous	Carbon disulfide	Psychosis, suicide, peripheral neuropathy, parkinson-like syndrome, coronary artery disease

to some solvents, such as *n*-hexane, methyl butyl ketone, and trichloroethylene. A number of organic solvents are definite human carcinogens, such as benzene, or probable human carcinogens, such as trichloroethylene and perchloroethylene. Organic solvents also affect the skin, causing contact dermatitis and other disorders. Most organic solvents, except for aliphatic hydrocarbons, irritate the eyes, nose, and throat. There is increasing evidence that exposure to organic solvents can cause spontaneous abortions (miscarriages) and congenital malformations. In addition, a number of organic solvents cause cardiovascular, hepatic, hematologic, and renal disorders.

Prevention

Efforts should focus on primary prevention by reducing or eliminating exposure. Exposed workers should be educated about the solvents with which they work, the hazards these solvents can cause, and ways to prevent adverse health effects. Air monitoring for solvents, by personal and area measurements, should be performed periodically in all workplaces where there is solvent exposure. The most effective primary prevention measures are engineering measures. These include (a) effective design, installation, operation, and maintenance of closed systems, and, where they cannot be used, exhaust ventilation systems; (b) isolation of hazardous processes or procedures, which can often be an effective approach if the process lends itself to this type of measure; and (c) substitution of highly toxic or carcinogenic solvents with safer solvents, such as toluene instead of carbon tetrachloride, and use of alternatives—even soap and water in some circumstances. Substitution for toxic solvents has been the most effective preventive measure and is the preferred public health approach.

Gloves, aprons, and other personal protective equipment should be used to prevent direct skin contact. Chemical safety goggles and face shields should be used if there is a potential for splashing. Use of respirators should not be routine and should be reserved only for short-term maintenance situations, emergencies, implementation of engineering measures, and similar situations. Workers who may be, or are being, exposed should have preplacement and periodic examinations that focus on those organ systems most likely to be adversely affected by organic solvents.

OSHA has promulgated PELs and NIOSH has established RELs for many organic solvents.

Further Reading

Darcey D, Langley R. Solvents. In Frazier LM, Hage ML (Eds.). *Reproductive Hazards of the Workplace*. New York: John Wiley & Sons, Inc., 1998, pp. 162-191.

Feldman RG. *Occupational and Environmental Neurotoxicity*. Philadelphia: Lippincott-Raven Publishers, 1999.

Snyder R, Andrews LS. Toxic effects of solvents and vapors. In: Klaassen CD (Ed.). *Casarett & Doull's Toxicology: The Basic Science of Poisons* (5th Edition). New York: McGraw-Hill, 1996, pp. 737-771.

Wahlberg JE, Adams RM. Solvents. In Adams RM (Ed.). *Occupational Skin Disease* (3rd Edition). Philadelphia: W.B. Saunders Company, 1999, pp. 484-500.

Zimmerman HJ. Occupational toxicity. *Hepatotoxicity: The Adverse Effects of Drugs and Other Chemicals on the Liver* (2nd ed.). Philadelphia: Lippincott Williams & Wilkins, 1999, pp. 365-390.

Stomach Cancer

ICD 10 C16

Elizabeth Ward

Gastric cancer often arises in patients with extensive premalignant changes of atopic gastritis and progressive dysplasia, which may cause non-specific symptoms. When symptoms do occur, they include indigestion or a burning sensation (heartburn), discomfort or pain in the abdomen, nausea and vomiting, diarrhea or constipation, bloating of the stomach after meals, loss of appetite, weakness, fatigue, and bleeding (vomiting of blood or blood being present in the stool). In addition to history and physical exam, x-ray diagnosis is commonly used but is less sensitive and specific than gastroscopy, with biopsy. Exfoliative cytology should be included if no clear lesions are seen.

Occurrence

An estimated 22 400 cases and 12 100 deaths occurred in the United States in 2003, with a male-to-female incidence ratio of 2:1 and a black-to-white incidence ratio of 2:1. The disease is strongly influenced by environmental risk factors, as evidenced by geographic variations in incidence and changes in incidence after migration. Environmental factors include consumption of salted or smoked foods, diets deficient in fresh fruits and vegetables, *Helicobacter pylori* infection, and tobacco use. Incidence rates have declined dramatically over the past 50 years in many countries, presumably related to dietary changes resulting from refrigeration and from reduced prevalence of *H. pylori* infection. There have been no formal estimates of the proportion of stomach cancers that are due to occupation.

Causes

There is substantial evidence for an association of stomach cancer with high levels of asbestos exposure. Evidence is also strong for increased risk of stomach cancer among coal miners, who are potentially exposed to dusts containing coal, metals, polycyclic aromatic hydrocarbons, and silica as well as nitrogen oxides. Excess stomach cancer has also been reported for underground iron, lead, zinc, and gold miners; it is unclear to what extent this is associated with silica exposure. Consistent associations have been reported for workers in other dusty trades, including brick and stonemasons, construction iron workers, boilermakers, filers, grinders and polishers, metal molders, cement plant and quarry workers, welders and cutters, millwrights and blacksmiths, and furnace and furnace repair workers in foundries. Several studies have reported excess stomach cancer among jewelry manufacturers, among workers with wood dust exposure, and among seamen and fishermen. Stomach cancer has been associated with ionizing radiation exposure in atomic-bomb survivors, radium-dial painters, and radiologists. Farmers have been found to have an excess of stomach cancer, possibly related to pesticides, fertilizers, and waterborne nitrate. Occupational exposure to nitrosamines or precursors of nitrosamines occurs in many industries and occupations where excess stomach cancer is found, including metalworking using synthetic and soluble cutting fluids, foundries, leather tanning, and some chemical manufacturing processes. There is some evidence for increased stomach cancer risk associated with ethylene oxide production, with exposure to sulfites and sulfates in the pulping process in the paper industry and among painters. Among non-occupational factors, diet, cigarette smoking, and *H. pylori* are important risk factors. Pernicious anemia and blood group A also increase the risk of gastric cancer.

Pathophysiology

H. pylori infection seems to act synergistically with exogenous carcinogens. In one study, heavy smokers were found to have a 2.5-fold increased risk of stomach cancer, *H. pylori* infection alone carried a 5-fold increased risk, while among *H. pylori* infected current smokers an 11-fold increased risk was observed. *H. pylori is* assumed to have an indirect action because it provokes gastritis, which is a precursor of gastric atrophy, metaplasia, and dysplasia. Inhaled coal and other dusts are thought to reach the stomach by being swallowed after clearance from the bronchial tree. Impairment of pulmonary clearance appears to protect against stomach cancer risk in coal miners.

Prevention

Primary prevention entails eliminating and reducing exposure to carcinogenic agents, such as asbestos, metal and mineral dusts, and nitrosamines and nitrosamine precursors. Secondary prevention includes early detection; however, there have been no reports of early detection programs in high-risk occupational groups. In Japan, where baseline incidence is much greater than in the United States, 5-year survival increased from 34% to 60% after introduction of mass screening programs. Because of the demonstrated efficacy of early detection, a low threshold for testing symptomatic individuals is recommended in the U.S. The 5-year survival rate in the United States for the period 1992-1998 was 22%; among the 22% of cases diagnosed at a localized stage, the - year survival rate was 58.9%.

Further Reading

American Cancer Society. All about stomach cancer: detailed guide. Learn about cancer [serial online]. American Cancer Society, Atlanta, GA. [updated 01/04/2001]. Available from URL:
http://www.cancer.org/docroot/CRI/CRI_2_3x.asp?dt=40

Bruckner HW, Morris JC, Mansfield P. Neoplasms of the stomach. In Bast RC, Kufe DW, Pollock RE, et al. (Eds.). *Cancer Medicine* (5th ed.). Ontario: B.C. Decker, Inc., 2000, pp. 1355-1390.

Cocco P, Ward MH, Buiatti E. Occupational risk factors for gastric cancer: An overview. *Epidemiologic Reviews* 1996; 18: 218-234.

Gamble JF, Ames RG. The role of the lung in stomach carcinogenesis: a revision of the Meyer hypothesis. *Medical Hypotheses* 1983; 11: 359-364.

Neugut AI, Wylie P. Occupational cancers of the gastrointestinal tract. In Brandt-Rauf PW (Ed.). *Occupational Cancer and Carcinogenesis: State of the Art Reviews in Occupational Medicine* 1987; 2: 109-135.

Nomura A. Stomach cancer. In: Schottenfeld D, Fraumeni JF, Jr. (eds.). *Cancer Epidemiology and Prevention* (2nd ed.). New York: Oxford University Press, 1996, pp. 707-724.

Siman JH, Forsgren A, Berglund G, et al. Tobacco smoking increases the risk for gastric adenocarcinoma among Helicobacter pylori-infected individuals. *Scandinavian Journal of Gastroenterology* 2001; 36: 208-213.

Stress

ICD-10 F43

Jeffrey V. Johnson

There is increasing consensus in the scientific community that occupational stress is a growing threat to human health. Engineers originally defined stress as a forcewhich deforms bodies. The biomedical understanding of stress began with the observation that the human body works to maintain an internal steady state (homeostasis) in the face of sometimes dramatic alterations in the external environment. Research demonstrated that exposure to physical stimuli such as extremes in temperature or noise would trigger an adaptation response—a "hard-wired" physiological program designed to return the body's internal environment to a homeostatic equilibrium as quickly as possible. Stress—according to Hans Selye, one of the originators of the stress concept—was the nonspecific component, the common denominator, of all the body's physiological adaptation responses to environmental demands. More recently, the term allostatic load has been developed to describe the long-term cumulative costs to the body of these repeated and sustained adaptation efforts which may eventually lead to functional and physiological changes.

Occurrence

Nationally representative surveys of the United States, Canadian, British and European Union populations generally agree that from 20% to 30% or more of the workforce reports that they are often or always stressed on their jobs. Recent prevalence estimates in the United States range from 36% who report being frequently stressed in the 1998 General Social Survey to 26% in the 1997 Family and Work Institute's National Study of the Changing Workforce. The Canadian Community Health Survey performed by Statistics Canada in 2001 found that 30% of the labor force reported being quite a bit or extremely stressed on their jobs. The Working Conditions Survey performed in 2000 by the European Foundation for the Improvement of Living and Working Conditions found that 28% of workers in the 15 European Union member states experience work-related stress as a major health problem, making it the second most common health problem, second only to back pain. In addition, the European Union's Agency for Safety and Health at Work estimated in 2002 that 50% to 60% of all lost working days are related to stress.

Studies performed by and for the British Health and Safety Executive provides some of the strongest population-based estimates of work stress prevalence and incidence as well as temporal trends. The Stress and Health at Work Survey performed in the United Kingdom (UK) in 2000 reported that 20% of workers rated their jobs as very or extremely stressful. Based on large population surveys of work-related illness performed in 2001 and 2002, the British Health and Safety Executive estimates that 560,000 workers in the UK are actively suffering from work-related stress, depression, or anxiety, with an additional 80,000 reporting work-related heart disease. British data from 2001 and 2002 provide estimates of the incidence of occupational stress: 265,000 persons per year report initial onset of work-related stress conditions. Self-reported prevalence of occupational stress has doubled between 1990 and 1999, and has remained stable from 1999 to 2001. In 2001, an estimated 13.4 million working days were lost in the UK due to work-related stress, depression, or anxiety. On average, each worker lost 29 days due to an occupational stress condition in 2001. This is a substantial increase from 1995, when 6.5 million working days were lost with an average of 16 days per affected person. Although equivalent data do not exist in the United States, the Bureau of Labor Statistics has shown that the lost-time cases due to occupational stress are among the most serious in terms of the length of the disability period, with the median absence period for work stress cases being 23 days—more than four times the median for all other injuries and illnesses.

Occupational stress also contributes to the risk of developing chronic diseases, most notably hypertension and coronary heart disease. In order to determine the degree to which work stress contributes to cardiovascular disorders, the population attributable risk (PAR%) has been estimated in a number of studies. (PAR% is the reduction of incidence in disease if the population is entirely unexposed to any given risk factor, such as stress.) Investigators have estimated that 29% of hypertension among working men in New York could be attributed to exposure to stressful working conditions, such as job strain or high demands with low control. European studies report that 6% to 20% of coronary heart disease mortality may be attributed to stressful working conditions, depending on the sex and age group under investigation.

Causes and Pathophysiology

Although there continues to be a lack of consensus on the exact definition of the term "stress," there is general agreement that it is best understood as a process originating in (1) environmental demands (stressors), which (2) if appraised as threatening will trigger (3) acute emotional and physiological reactions, that if repeated and prolonged will give rise to (4) biological and behavioral effects, which may, in turn, lead to (5) long-term health consequences, such as chronic disease, and, eventually, death. Throughout the stress process individual or environmental moderators (6) may protect or buffer those exposed from the potentially adverse impact of stress. A brief discussion of each of these stages follows.

(1) Environmental Demands (Stressors)

Although many aspects of the work environment may induce physiological adaptation responses, stress research has largely focused on threatening and challenging stimuli that promote the activation of a particular class of emergency reactions known as the "fight-or-flight" response. According to the Demand–Control Model—one of the most widely accepted conceptual models of work stress—these physiological reactions are most likely to be triggered in work environments where high demands for performance are coupled with low levels of control over task-related decision latitude. Other researchers have shown that the capacity to exercise environmental control, and the potential or actual loss of that control may be key to predicting which environments will be stressful.

Other types of work environments place chronic demands on individuals requiring constant effort and concentration. Jobs like bus driving and assembly-line work, although not constantly triggering the classic fight-or-flight response, are nevertheless associated with an increased activation of the sympathetic nervous system. Monotonous and repetitive jobs-underload-also trigger neurohormonally mediated adaptation responses.

Life events can also be significant "stressors." For example, nonoccupational stressful events include death of a spouse or child, significant financial or legal problems, and serious health events. In modern industrial societies, one's position in the labor market provides access to or "control over" many of the most valued aspects of life itself, such as social position, human relationships, financial security, temporal structure, and individual identity. This explains why unemployment or the threat of job loss is one of life's most stressful experiences.

(2) Cognitive Appraisal

Some researchers have suggested that environmental demands must be perceived by the individual as stressful in order for neurohormonal arousal to occur. However, more recent research suggests that neurohormonal activation may occur even when individuals do not report being anxious or feeling "stressed." This may be the case when demanding and potentially stressful environmental conditions are a normal and routine part of an individual's daily experience at work. Moreover, certain work stressors, such as high-demand/low-control jobs, are likely to be stressful to most people much of the time.

(3) Acute Reactions

Two distinct patterns of acute response have been identified:

Active Distress: The "fight-or-flight" response is an adaptation syndrome from our hunting and gathering past that prepares us for vigorous, survival-related activity. It involves "effortful" or active adaptation to those environmental challenges that threaten our control over valued environmental resources. This reaction pattern involves the activation of the sympathetic adrenal medullary system by the amygdala in the brain's limbic system. Marked elevations in catecholamine excretion (adrenaline and noradrenaline) are triggered, with concomitant increases in testosterone, blood pressure, and heart rate. If this acute reaction pattern is frequently triggered, if it is prolonged, and if return to baseline (recovery) is not rapid, then adverse pathophysiological changes are more likely to occur.

Passive Distress: The "conservation-withdrawal" response is activated by environmental events or situations that involve loss or threats of loss-as in unemployment, chronic job insecurity, or contingent working arrangements. It can also be associated with the exhaustion and emotional depletion that occurs after active efforts to adapt to environmental challenges have been unsuccessful in preventing a loss of control. This acute reaction pattern is based in the hippocampus and involves the activation of the pituitary-adrenal-cortical system with increases in cortisol, decreases in testosterone, and increases in depressive symptoms.

Research on the Effort/Distress Model—another conceptual model of work stress—suggests that the most "toxic" jobs are those where both of these types of acute reactions occur simultaneously. For example, workers with high demands have high adrenaline levels, while those with low control have high levels of cortisol. Workers exposed to both high demands and low control have high levels of adrenaline and cortisol. Researchers who have developed the Effort/Distress Model report that the presence of cortisol makes the cardiovascular system more vulnerable to damage from the other stress hormones.

Cognitive changes can also occur along with these physiological reactions and may include alterations in emotions, in the appraisal of risk, in motivation, in vigilance behavior, and in memory. These may also be accompanied by behavioral changes, such as increases in absenteeism, sleep disturbances, increased fatigue, sexual dysfunction, smoking intensity, and alcohol and substance use.

(4) Biological and Behavioral Effects

In the presence of acute or chronic stressors, the body struggles to adapt. Over time, this process causes the wear and tear that researchers now refer to as allostatic load. As stress increases in frequency, intensity, and duration, the efficiency in coordinating the body's adaptive responses begins to break down. Allostatic responses are no longer terminated appropriately, thereby maintaining a level of neurohormonal activation within the body long after the stress in the environment has ceased. This may explain why, after exposure to years of job stress, blood pressure may become permanently elevated above earlier baseline levels. The impact of allostatic load has been most clearly demonstrated in the cardiovascular system where chronic stress has been shown to be linked to increases in cholesterol, elevations in plasma fibrinogen, increases in heart muscle mass, disturbances in cardiac rhythm, and increases in heart rate. More recently, researchers have examined the impact of stress-related muscle tension on the musculoskeletal system and have linked the stress-related modulation of the immune system to inflammatory, infectious, and autoimmune diseases.

For many people, the workplace is also the major context in which adult socialization and learning takes place. Work may have a permanent impact on the development of the adult personality through a "learning generalization" effect, whereby the types of activities that are present in the work setting are reproduced by the individual during their leisure time. There is considerable evidence that workers exposed to passive jobs tend to engage in passive leisure time pursuits. Jobs with higher levels of control over task content and decision-making are much more likely to provide the ongoing experience of occupational self-determination that has been identified as the major predictor of cognitive flexibility and active styles of behavior. Through this process of occupational socialization and learning, work environments may increase or decrease the willingness and ability of individual workers to undertake efforts to stop smoking and to adopt healthy lifestyle behaviors.

(5) Long-term Health Consequences

Workers in jobs with low control are at significantly increased risk of premature death from cardiovascular disease, even after adjusting for all known biomedical and behavioral risk factors. High-demand, low-control work environments are associated with increased risk of cardiovascular disease (CVD) and premature CVD mortality. Workers in passive (low-demand, low-control) jobs are also at increased risk of premature all-cause mortality. Social support at work has also been shown to be related to CVD morbidity and mortality, particularly when it is linked with high levels of work control.

Work stress has also been linked to an increased risk of musculoskeletal disorders (MSDs). Monotony, time pressure, and high workload as well as low levels of work control and work-related social support have all been found to increase the risk for MSDs. Low levels of social support and work control have also been shown to increase the risk of back pain. The health impact of work stress on MSDs is much greater among workers who are also exposed to heavy physical demands.

Work stress also has an impact on the risk for chronic mental illness. High demands and low control are related to increased risks of depression and alcohol dependence. Lack of control at work and low levels of work-related social support are also important factors in burnout.

Work stress does not seem to have a substantial impact on gastrointestinal disorders, such as ulcers and irritable bowel syndrome. Nor is there yet any strong scientific evidence that work stress is a risk factor for cancer, although there remains comparatively little research in this area.

(6) Stress Moderators

Individual or situational stress moderators provide ways of coping with stress or even changing the sources of stress. Research findings suggest that for individuals hopeful optimism combined with realism is the most effective personal coping strategy. One's coping style may be influenced by work experience; for example, workers in jobs that lack occupational self-determination are more likely to be socialized into relatively passive and fatalistic coping styles.

Researchers using the Effort/Reward Imbalance Model have focused on the expectations workers develop based on their previous employment experience. Stress is most likely to occur when there is a sustained imbalance among the efforts individuals expend at work, the rewards they expect to get, and the rewards they actually receive. Individuals also differ in their degree of intrinsic commitment to work with some being much more immersed in the pursuit of work-related goals. This degree of "immersion" places overly committed workers at greater risk for CVD when their efforts are not rewarded with anticipated results. This suggests that a "healthy detachment" may also serve to moderate the impact of stressful work situations, particularly when there is a marked imbalance between effort and reward.

The presence of social support at work is one of the most important situational moderators. Support from co-workers buffers (or dampens) the physiological stress response. It assists workers in understanding and reinterpreting the meaning of potentially stressful job experiences. In addition it may provide a collective means by which the working conditions that cause stress may be identified and changed.

Another important moderator is work control. Workers exposed to high levels of demands for performance and productivity, yet who have high levels of control over how the work is to be done, are much less likely to experience stress than those without control. Workers with high demands and high control experience a condition of challenge and active learning.

Further Development of Conceptual Models

The three conceptual models reviewed here—The Demand/Control Model, the Effort/Distress Model, and the Effort/Reward Imbalance Model—have contributed much to our understanding of the potentially hazardous impact occupational stress may have on the working population. However, these models examine only a small range of potential stressors in the work environment. Following these models has encouraged researchers to focus considerable attention on an important set of core work exposures. This has provided a much stronger base of research findings than existed 10 years ago. A number of relatively new areas are now being addressed:

(1) The three models discussed here are in the process of being integrated into one overarching model.
(2) Much more attention is being given to human service work and to the stress of emotional labor, particularly among women workers.
(3) The employment relationship has become increasingly unstable and "contingent work arrangements" are being studied as a source of stress.
(4) Researchers are examining if occupational stress is increasing with the intensification of the work process and the introduction of lean production and "just-in-time" methods.
(5) Many of the social security and health-care provisions that were part of the labor-management social contract during the post-World War II period have been diminished and this may be adding to the potential stress burden of the workforce.

The current guiding models of work stress will eventually be modified, or new models will be created to more adequately address these fundamental historical shifts in the nature of the labor process and the labor market. However, it is likely that we will continue to be guided by some of the essential insights derived from our current conceptual models. We can predict with some confidence that the biological aspects of the stress process are most likely to occur in situations where the overall environmental and survival related demands that occur within any given historical period overwhelm the resources available to individuals and groups to meet these demands.

Prevention

Primary prevention efforts have largely focused on increasing the levels of job control and social support available to workers. Participatory intervention strategies have been designed to mobilize workers to participate in identifying the most significant stressors, as well as redesigning work organizations. The results of these studies are mixed. In some studies, these types of efforts have been effective in reducing or eliminating occupational stressors and improving both health and productivity outcomes. Interventions often focus on changing very specific, concrete aspects of the job. Stressful demands may be decreased by reducing time pressure and by modifying deadlines and reducing caseloads; or by decreasing the length of the workweek and the workday and by increasing work breaks. Work control might be enhanced by increasing skill levels through training, and by giving workers more authority over specific decisions regarding products, clients, or customers. Social support at work can be improved by training supervisors to be less hostile and more cooperative, and by providing opportunities for workers to develop a greater sense of collectivity through an increase in interaction opportunities.

Secondary prevention interventions, such as cognitive behavioral interventions, have been shown to be particularly effective among those who already have relatively high levels of control in their workplace. Interventions focused on increasing the coping abilities of these high-control workers have been shown to be effective in reducing stress levels. Stress management interventions using relaxation techniques have also been shown to be moderately effective in reducing stress in some studies.

At an organizational level, workers and their union organizations can use the collective bargaining process to reduce occupational stress. Collective bargaining provides employees with a means to improve stressful conditions and a protective system from the potentially arbitrary actions of the employer. Contracts have been used to moderate demands, increase levels of control, and create more social support. According to Landsbergis, specific contract language should focus on "job security, work standards, assignments, performance evaluations, technological change, harassment, discrimination, staffing, comparable worth, skills training, and career development." Unions can also anticipate major organizational and work process changes that might increase occupational stress and intervene before they are introduced.

In some societies, broader multilevel strategies have been undertaken to regulate stressful work organization. This approach has been most successful in the Nordic countries where work environment laws have been enacted such as the Norwegian Worker Protection Act of 1977, which calls for "the avoidance of undiversified, repetitive work and work that is governed by machine or conveyor belt in such a manner that the employees themselves are prevented from varying the speed of the work...effort shall be made to arrange the work so as to provide possibilities for variation and for contact with others...."

More recently, there has been a call for reference values or threshold limit values (TLVs) for psychosocial work environment exposures. Researchers suggest that using benchmarks or reference values will stimulate both employers and employees to take preventive measures of various kinds. For example, the Health and Safety Executive in the United Kingdom is currently developing and piloting "standards of good management practice" that will provide a benchmark that employers can use to assess their performance in creating healthier work organizations. The Health and Safety Executive has established a standard cut-off point for demands, control, and support. Firms will achieve the standard only if at least 85% of employees indicate that they are satisfied with how much of each of these core dimensions are available on their jobs.

Historically, the most successful prevention strategies have used a multilevel approach consisting of participatory interventions, local organizational change efforts involving unions and other grassroots organizations, and national level legislation that provides both the legal authority for worker empowerment and protective environmental regulations. By the late 1990s, political and economic changes associated with globalization and the increasing international competition among firms had made the climate for environmentally-oriented stress prevention activities less favorable. Yet these same societal forces contribute to making work environments even more stressful. This seeming paradox suggests the need for a renewed multilevel prevention strategy that combines scientific investigation, local democratization, and political mobilization at societal and global levels.

Further Reading

Committee on Behavior, Institute of Medicine. *Health and Behavior: The Interplay of Biological, Behavioral, and Societal Influences.* Washington, DC: National Academy of Sciences Press, 2001.

European Agency for Safety and Health at Work. Research on Work-related Stress. Luxembourg, 2000. Available as a pdf file for no charge: http://agency.osha.eu.int/publications/reports/104/en/stress.PDF (Accessed February 21, 2004).

European Agency for Safety and Health at work. Working on Stress: Prevention of Psychosocial Risks and Stress at Work in Practice. Luxembourg, 2002. Available as a pdf file at no charge: http://agency.osha.eu.int/publications/reports/203/en/stress.pdf (Accessed February 21, 2004).

Health and Safety Executive of the United Kingdom. Stress Homepage Website: http://www.hse.gov.uk/stress/index.htm (Accessed on February 21, 2004)

Job Stress Network Web Site: http://www.workhealth.org/ (Accessed on February 21, 2004).

Johnson JV, Johansson G (Eds.). *The Psychosocial Work Environment: Work Organization, Democratization and Health.* Amityville, NY: Baywood Publishing Co., 1991.

Karasek R, Theorell T. Healthy Work: Stress, *Productivity, and the Reconstruction of Working Life.* New York: Basic Books, Inc., 1990.

National Institute for Occupational Safety and Health. The Changing Organization of Work and the Safety and Health of Working People: Knowledge Gaps and Research Directions, 2002. Available as a pdf file at no charge: http://www.cdc.gov/niosh/02-116pd.html (Accessed February 21, 2004).

National Institute for Occupational Safety and Health. Stress at Work, 1999. Available as a pdf file at no charge:
http://www.cdc.gov/niosh/pdfs/stress.pdf

National Institute for Occupational Safety and Health. Stress at Work Web Site. http://www.cdc.gov/niosh/topics/stress/ (Accessed February 21, 2004)

Sauter SL, Murphy LR, Hurrell JJ, et al. Psychosocial and organizational factors. In Stellman JM (Ed.). ILO *Encyclopedia of Occupational Health and Safety* (4th ed.). Geneva: International Labour Office, 1998, pp 34.1-34.77.

Siegrist J. Adverse health effects of high-effort/low-reward conditions. *Journal of Occupational Health Psychology* 1996; 1: 27-41.

Schnall PL, Belkic K, Landsbergis P, et al. (Eds.). Why the workplace and cardiovascular disease? *Occup Medicine: State Art Reviews* 2000; 15: 1-6.

Theorell T. Working conditions and health. In Berkman LF, Kawachi I. *Social Epidemiology.* New York: Oxford University Press, 2000, pp. 95-117.

Suicide

ICD-10 X60-X84

E. Lynn Jenkins

The relationship between work and suicide is complex. The fundamental issue researchers and practitioners face is simply defining what is meant by the term "occupational or work-related suicide." Most occupational injury fatality surveillance systems only capture incidents that occur on an employers' premises or while a worker is engaged in work activities. In the case of suicide, deaths that occur at a work site may not have any relationship to the work itself, but rather to the availability of a means for committing suicide or simply being the place at which the decision to take one's life is ultimately made. Conversely, suicides that are indeed driven by work-related factors may occur at a worker's home or other place and the work-related component is unclear or unknown. For this reason, existing statistics on workplace suicide should be interpreted with caution. The vast majority of the literature on occupation and suicide is based on mortality studies in various occupational groups.

There are other dimensions to the relationship between work and suicide that must also be considered, such as the notion of meaningful work being a potential deterrent or protective factor for suicide. However, if work or the stress that comes from work is or becomes negative, it may become a risk factor.

Occurrence

In 2000, suicide was the fourth leading cause of death for persons of "working age," that is, aged 16 to 64 years old, accounting for 23 547 deaths in the United States. For all ages combined, males are more than four times more likely to die from suicide than females. Data from the Census of Fatal Occupational Injuries surveillance system indicate there were an average of 221 suicides identified as occupationally related annually during the 5-year period from 1997 through 2001. In 2002, there were 199 such incidents.

Causes

Occupational mortality studies have suggested a range of potential causes for excess suicide risk, including exposure to pesticides, solvents, or electromagnetic fields; work overload; stress; depression; alcohol use; isolation; economic downturns or threats to job security; seasonal variations for outdoor workers; shift work; and socioeconomic status. In addition, a number of studies discuss the role of the availability of a means for committing suicide at work such as chemicals for chemists, drugs for physicians, nurses, pharmacists or other health care workers, and firearms for law enforcement and military personnel. These mortality studies rarely, however, have robust methods or data that allow specification of causal factors for suicide.

Some studies have examined suicide for female workers specifically and suggest that there may be particularly elevated risks for females in balancing home and work responsibilities and stressors related to child-bearing and child-rearing responsibilities and their relationship to participation in the labor force and career advancement.

The only clear message from the existing data and literature is that the relationship between work and suicide is complex. Published findings are sometimes conflicting and the reality is that the causes are likely multifactorial and will be difficult to ascertain with administrative or other existing data.

Pathophysiology

Suicide may be committed by a number of means with various outcomes including firearms, causing a range of trauma to tissues and internal organs; chemicals or drugs, causing various reactions within the nervous, respiratory, or cardiovascular systems; hanging, causing potential trauma to the neck as well as suffocation; exposure to toxic or oxygen-deficient environments; jumping from a height, causing multiple trauma to tissues and internal organs; and traffic-related incidents either as a pedestrian or driver, again causing multiple trauma to tissues and internal organs.

Prevention

Potential avenues for the prevention of suicide among working age persons, and in the workplace specifically, may include screening and appropriate treatment for depression, alcohol abuse, and stress. Managers, labor unions, co-workers, employee assistance professionals, and others may have roles to play in raising awareness and creating an environment in which treatment for stress, depression, or substance abuse can be sought without fear of retribution or stigmatization. Some studies, particularly in law enforcement and military settings, have identified these as major barriers to proper identification and treatment of psychological and substance abuse issues.

In 1999, the U.S. Department of Health and Human Services published the Surgeon General's Call to Action to Prevent Suicide. This document provides a framework for addressing suicide and calls for an approach termed AIM—Awareness, Intervention, and Methodology. This includes 15 recommendations based on "consensus and evidence-based findings" from a workshop of experts. While a specific role is not specified for occupational safety and health, industry, or labor groups, there is clearly a call for multidisciplinary action in raising awareness, reducing stigma, improving treatment, and improved research to understand the risk and protective factors for suicide.

Other Issues

In recent years, the Japanese term "Karoshi"—death by overwork—has come into the common vernacular. The Labor Ministry of Japan has drawn up guidelines defining karoshi and the compensation to which workers and their families are entitled. The organization of work and long hours of work are of particular interest for researchers, labor unions, and industry-with a particular focus on the impact of the organization of work and changes in the organization of work on occupational safety and health outcomes. The role of public health and occupational safety and health professionals in better understanding these issues is multidimensional, including both research and practice.

Further Reading

Boxer PA, Burnett C, Swanson N. Suicide and occupation: a review of the literature. *Journal of Occupational and Environmental Medicine* 1995; 37: 442-452.

Feskanich D, Hastrup JL, Marshall JR, et al. Stress and suicide in the nurses' health study. *Journal of Epidemiology and Community Health* 2002; 56: 95-98.

Hawton K, Clements A, Sakarovitch C, et al. Suicide in doctors: a study of risk according to gender, seniority and specialty in medical practitioners in England and Wales, 1979-1995. *Journal of Epidemiology and Community Health* 2001; 55: 296-300.

Helmkamp JC. Occupation and suicide among males in the US armed forces. *Annals of Epidemiology* 1996; 6: 83-88.

Kposowa A. Suicide mortality in the United States: differentials by industrial and occupational groups. *American Journal of Industrial Medicine* 1999; 36: 645-652.

National Institute for Occupational Safety and Health. *Stress at Work* (DHHS [NIOSH] Publication No. 99-101). Washington, DC: NIOSH, 1999.

Peipins LA, Burnett C, Alterman T, et al. Mortality patterns among female nurses: A 27-state study, 1984-1990. *American Journal of Public Health* 1997; 87: 1539-1543.

U.S. Public Health Service. *The Surgeon General's Call to Action to Prevent Suicide.* Washington, DC: USPHS, 1999.

Violanti JM, Vena JE, Marshall JR. Suicides, homicides, and accidental death: A comparative risk assessment of police officers and municipal workers. *American Journal of Industrial Medicine* 1996; 30: 99-104.

Sunburn

ICD-10 L55

Michael E. Bigby

Sunburn is the most common adverse reaction to sun exposure. It is easily recognized in lightly-pigmented individuals as erythema in sun-exposed areas. Erythema appears within about 4 hours after sufficient exposure, peaks in 8 to 12 hours, and fades in about 2 days. It is commonly accompanied by edema and tenderness and is followed in a few days by desquamation. Severe sunburn may cause blistering and be accompanied by systemic symptoms, such as fever and malaise. In darker individuals, erythema may not be perceptible and sunburn can only be recognized by desquamation several days after heavy sun exposure.

Occurrence

Based on a telephone survey of more than 156,000 adults in the 50 United States, the District of Columbia, and Puerto Rico, 32% of adults reported having a sunburn in the previous year. Risk factors associated with a higher prevalence were age between 18 and 29 years and "white" skin color. A similar prevalence (31%) was found among adolescents in New Zealand and a higher prevalence (43%) was found among white children aged 6 months to 11 years in the United States. The proportion of sunburn cases that is occupationally related is not known.

Causes

Sunburn is caused by exposure to ultraviolet (UV) light. The UV spectrum is divided into three bands: UVA—wavelength 400 to 320 nm, UVB—320 to 290 nm, and UVC—wavelength 290 to 200 nm. Sunburn is caused predominantly by exposure to UVB. UVA irradiation contributes about 15 to 20% to the sunburn reaction. Very little UVC reaches the surface of the earth. A number of phototoxic and photoallergic reactions occur, including to outdoor workers taking tetracycline and other drugs and those exposed to lime or celery, resulting in skin reactions.

Pathophysiology

The biological effects that result from ultraviolet radiation (UVR) are due to absorption of UVR predominantly by nucleic acids and to a lesser extent by proteins in the skin. Nucleic acid absorption of UVR results in the formation of pyrimidine dimers. These dimers are repaired by several mechanisms, including photo-reactivation, excision repair, post-replication repair, and sos repair. If exposure is sufficient, UVR is associated with apoptosis of keratinocytes, melanocytes, and Langerhans cells, and the initiation of a cascade of cytokine release that results in skin inflammation. The erythema seen with sunburn is due to vasodilatation of the superficial plexus of the skin's blood supply.

Epidermal thickening and tanning also result from UVB. These compensatory mechanisms increase the skin protection against future UVR damage.

Prevention

Sunburn will result from summer UVR exposure in 20 to 96 minutes, depending on latitude and skin color in lightly-pigmented, unprotected individuals. Decreasing exposure to UVR can prevent sunburn. This goal can be achieved by avoiding exposure during peak hours of UVR (10:00 a.m. to 4:00 p.m.), and adequate use of sunscreens and protective clothing (tightly-woven, wide-brimmed hats and long-sleeved shirts and pants).

The sun protection factor (SPF) of sunscreens is determined by photo-testing in an in vivo assay, based on an internationally agreed upon application density of 2 mg/cm^2 of sunscreen. A SPF of 15 indicates that 15 times as much UVR is required to induce minimal erythema compared to unprotected skin. Application of a sunscreen at 2 mg/cm^2 (30 gm to cover the entire cutaneous surface) is adequate to protect most individuals from sunburn even for a full day of sun exposure, unless it washed off by sweat or water. Sunscreen failure is most commonly due to application of inadequate amounts of sunscreen (most studies indicate that, in general use, sunscreens are applied at 1 mg/cm^2 or less), or failure to reapply sunscreens at the frequency of manufacturer recommendations (most sunscreens should be reapplied every 2 to 4 hours).

Other Issues

UVR-induced damage to epidermal-cell nucleic acids with inadequate repair are thought to be the causes of the development of basal cell carcinoma, squamous cell carcinoma, and melanoma. Exposure to UVA and UVB are also responsible for the development of photoaging, which is recognized as dryness, wrinkling, sagging, loss of elasticity, mottled pigmentation, and telangiectasia. Sunburns in childhood are linked directly to an increase in the number of melanocytic nevi and indirectly to the development of melanoma.

Further Reading

Diffey BL. Solar ultraviolet radiation effects on biological systems. *Review in Physics in Medicine and Biology* 1991, 36: 299-328. Available at:
   http://www.ciesin.org/docs/001-503/001-503.html
Diffey B. Has the sun protection factor had its day? *British Medical Journal* 2000; 320: 176-177. Available at:
   http://bmj.com/cgi/content/full/320/7228/176
Hall HI, McDavid K, Jorgensen CM, et al. Factors associated with sunburn in white

children aged 6 months to 11 years. *American Journal of Preventive Medicine* 2001; 20: 9-14.

Kang K, Stevens S, Cooper K. Pathophysiology of Ultraviolet Irradiation. Available at: http://www.aad.org/education/uvirradiation.htm

Richards R, McGee R, Knight RG. Sunburn and sun protection among New Zealand adolescents over a summer weekend. *Australia and New Zealand Journal of Public Health* 2001; 25: 352-354.

Saraiya M, Hall HI, Uhler RJ. Sunburn prevalence among adults in the United States, 1999. *American Journal of Preventive Medicine* 2002; 23: 91-97.

Synthetic Vitreous Fiber (SVF)-Related Diseases

ICD-10 J30.3

David C. Christiani

Commonly referred to as "man-made mineral fibers (MMMFs)," but more appropriately called "synthetic vitreous fibers (SVFs)," these fibers have been historically classified into three main groups: (1) glass fiber, such as glass wool, continuous glass filament, and special purpose glass fiber; (2) mineral wool, such as rock and slag wool; and (3) refractory ceramic fiber. Most SVFs are manufactured through a melting and fiber-forming (fiberization) process. Fiber diameter is a critical factor determining the potential for SVF penetration into mucous membranes (and skin) when their surfaces are exposed to fiber-contaminated air. Fibers of 4 to 5 μm diameter can cause intense skin itching and rash. Factors related to pulmonary toxicity are fiber, dose delivered to the lung, and fiber durability to the tissues.

Occurrence

Clinical conditions related to SVF exposures vary by type of material. Skin irritation is the most common health complaint. It is estimated the 5% of new workers involved with SVF leave employment because of skin irritation. Irritation of the eyes and upper and lower airways may also occur.

Small opacities on chest x-rays consistent with pneumoconiosis were observed in 1% to 6% of workers in five fibrous glass and two mineral wool manufacturing plants, especially among those exposed to fine diameter (1 to 3 μm) fibers.

Refractory ceramic fiber (RCF) exposure can cause parenchymal and pleural changes. Findings include pleural plaques and, less commonly, linear interstitial fibrosis. Studies have shown plaque prevalence of up to 11% in workers with more than 20 years since their first RCF exposure, and small linear opacities in up to 13%. However, some studies involve populations that have also been exposed to asbestos.

Respiratory malignancy has been evaluated among workers exposed to various forms of SVFs. The preponderance of evidence suggests that workers exposed to rock or slag wool have a significantly elevated risk of lung cancer, while glass filament workers had a non-significantly elevated lung cancer risk, and glass workers had no elevated risk.

NTP has determined that for ceramic fibers and glass wool fibers of respirable size there is sufficient evidence for carcinogenicity in animals. No data are available on the carcinogenicity of ceramic fibers in humans.

Causes

The respiratory health effects are caused by exposure to airborne fibers in the manufacture and use of the products noted above.

Pathophysiology

The most recent understanding of toxicologic mechanisms for SVF-induced lung disease is biopersistence as the primary determinant of fiber toxicity. Pulmonary biopersistence refers to the time that an intact fiber remains in the lung, including its movement and residence in target cells and tissues. Special-purpose glass fibers and refractory ceramic fibers have the greatest durability, followed by rock and slag wool, then fibrous glass. Additional determinants of toxicity include fiber chemical composition, surface charge, and site of deposition.

Prevention

Prevention of respiratory and skin effects requires minimizing exposure. Exposures can be controlled by engineering (exhausts and ventilation), proper work practices, and worker training. Where skin contact or airborne exposures cannot be controlled fully with these measures, personal protection equipment should be used, such as long-sleeved shirts, eye protection, and respirators.

Regular air sampling for fibers should be done in all relevant workplaces.

Currently, there are no specific regulations governing exposure to SVFs. In 1998, ACGIH adopted a TLV of 1 f/mL for conventional glass fibers, rock, and slag work. It also recommended a lower permissible exposure limit for RCFs, but provided no specific standard. In other countries, regulatory agencies have adopted exposure levels in the range of 0.5 to 1.0 f/mL.

Other Issues

New SVFs are being developed for high-temperature insulation with high stability and low biopersistence. Such product development will lead to the primary preventive strategy of substitution of higher toxicity materials with lower toxicity fibers in the future.

Further Reading

Boffetta P, Saracci R, Anderson A, et al. Cancer mortality among man-made vitreous fiber production workers. *Epidemiology* 1997; 8: 259-268.

Cowie HA, Beck J, Wild P, et al. Epidemiological Research in European Ceramic Fibre Industry 1994-98 *(IOM Report TM/99/01)*. Institute of Occupational Medicine, June 1999.

Lockey JE, Levin LS, Lemasters GK, et al. Longitudinal estimates of pulmonary function in refractory ceramic fiber manufacturing workers. *American Journal of Respiratory and Critical Care Medicine* 1998; 157: 1226-1233.

World Health Organization European Program for Occupational Health. *Validity of Methods for Assessing the Carcinogenicity of Man-made Fibres. Executive Summary, 19-20 May 1992.* Copenhagen: WHO Regional Office for Europe.

Tendonitis and Tenosynovitis

ICD-10 M65, M68, M70.8, M70.9, M75

Laura Punnett and Judith E. Gold

Tendonitis is an inflammation of a tendon, the ropelike structure that attaches each end of a muscle to bone. Tenosynovitis is a tendon inflammation that also involves the synovium, the tendon's protective and lubricating sheath. However, the underlying primary pathology in chronic tendon disorders may be collagen degeneration rather than inflammation. Tendinopathy and tendinosis (implying a degenerative process) have been suggested as terms more appropriate to the pathologic findings in surgical specimens. (See Pathophysiology.) Here, the terms tendon disorder, tendonitis, and tendinosis are used interchangeably.

Many forms of tendonitis have names that indicate the specific location of the disorder, such as lateral or medial epicondylitis ("tennis elbow" and "golfer's elbow"), bicipital tenosynovitis (above the biceps muscle), and rotator cuff tendonitis (specifically affecting shoulder tendons).

De Quervain's disease is a tendon disorder that affects the thumb extensor or abductor tendons—those on the back of the hand at the base of the thumb (the "anatomical snuffbox").

If inflammation involves the bursa—a fluid-filled pocket near a joint that cushions the joint structures, it is termed "bursitis."

"Trigger finger" or "trigger thumb" occurs when the inflammation in a finger tendon causes the development of an enlarged nodule on the tendon or a narrowing of the tendon sheath. This results in the finger locking into position when bent because the nodule interferes with normal motion. It is often necessary to use the other hand to straighten out the finger.

A ganglion cyst is a fluid-filled swelling or lump that develops at the site of an inflammation. It is not always painful, but, like trigger finger, it may interfere with the smooth motion of the tendon. The fluid may be withdrawn from the ganglion with a needle, both to assist in diagnosis and to attempt to reduce or reverse it.

A related condition that may occur due to overuse is Dupuytren's contracture, a permanent shortening of the finger flexor tendon, usually in the fourth finger.

These conditions are characterized by localized pain, especially during active motion. Redness, warmth or swelling may also be present. Symptoms may progress to burning, aching, or swelling while the body part is at rest. Pain may also spread along the involved muscle. The onset of symptoms is usually gradual; sometimes, however, rapid onset may accompany intensive work at unaccustomed tasks or motions, such as after transfer to a new job, return to work following a vacation, or increase in work pace.

Physical examination methods are not well standardized, and the prevalence of findings may vary greatly among examiners. On physical examination, pain at the site of inflammation can be elicited by putting pressure on the tendon or its point of attachment to the bone. Usually, symptoms are precipitated or exacerbated by resisted motion of the muscle; fine crepitus (a grating sensation) may also be present. Resisted motion of the antagonist muscle, in general, should not produce pain. In de Quervain's disease, sharp pain is produced by Finkelstein's test, in which the subject holds the thumb inside the fingers of the same hand, and the examiner stabilizes the forearm and bends the wrist, moving the hand toward the little finger. In rotator cuff tendonitis, there may be a painful arc within the range of motion-that is, pain between 70 and 120 degrees of shoulder abduction (sideways elevation of arm). In epicondylitis or in wrist and finger flexor tendon disorders, there may be objective loss of strength in the muscle, resulting from pain that interferes with active or resisted motions in normal use.

Occurrence

Utilizing 1988 National Health Interview Survey data, the prevalence in the U.S. working population of all hand/wrist and elbow/forearm tendon disorders combined was estimated at 4.6 per 1000. Ganglion cysts comprised about one-third of the total, while trigger finger a much smaller proportion. Tendon disorders were more prevalent in women than in men. Twenty-eight percent of all tendon disorder cases were self-reported as work-related. A significant problem with these data is the tendency of individuals with nonspecific upper extremity pain and their physicians to overuse the term tendonitis when the diagnosis is not confirmed.

In occupational groups substantially exposed to ergonomic stressors, the prevalence rate of hand/wrist tendonitis has been estimated to range from 4% to 56%. This rate varies according to the ergonomic features of work, the diagnostic criteria used, and body location.

Work-related tendon disorders are more common at some body locations than at others, depending on ergonomic features of the work. The wrist and shoulder tend to be affected more often than in the middle of the arm. In one study, distal upper extremity symptoms were 4 times more common among assembly-line food packers (with rapid, repetitive hand movements and static muscle loading) than among sales clerks. Elsewhere, women employed in the fish processing industry with highly repetitive work tasks (few pauses and high wrist extension/flexion velocity) were 3.5 times as likely to have shoulder tendonitis as female municipal employees in the same town. The risk of hand and wrist tendonitis associated with highly forceful manual exertions has been estimated to be 6 times higher than in low-force jobs, and the risk associated with highly repetitive exertions 3 times higher than in less repetitive work. Many manual-intensive jobs involve simultaneous exposure to more than one of the factors listed above. Such combined exposures appear to have a greater than additive effect on risk. For example, work that is both forceful and repetitive has been estimated to increase the risk of hand and wrist tendonitis by 29 times.

Among those with hand/wrist tendon disorders, 30% have reported sleep disturbances and 22% have missed work. A major change in work activities was necessary for 13% of patients.

Causes

In general, high-risk occupational activities include any rapid, repetitive, and forceful manual tasks in service or manufacturing, especially in assembly-line or packing jobs. Sustained contractions of any muscle in a fixed position ("static loading"), such as needed to hold a tool or pairs or to maintain a body posture, may strain the muscle-tendon system. Working with the wrist bent, the arm raised at the shoulder, or with any other joint in a non-neutral posture puts the muscle(s) at a biomechanical disadvantage, requiring more forceful muscular contractions to perform the task and increasing the internal forces exerted by the tendon. Using a pinch grip to hold parts or tools similarly creates a biomechanical disadvantage for the working muscles. Mechanical stress concentrations on top of a tendon may interfere with its movement and lead to injury. This may be caused by scissors, other tools, or controls with sharp edges that press on the sides or backs of the fingers.

Based on laboratory evidence, exposure to vibration while performing manual work may be another occupational risk factor; however, the epidemiologic data supporting this as a risk factor are sparse. Vibration interferes with reflexes that enable the nervous system and brain to control the precise amount of force exerted by a muscle. The force exerted is higher than necessary, causing additional strain on tendons. Because the higher grip forces result in increased transmission of vibration to the hand, a vicious cycle may ensue in which damage to both tendons and nerves is progressively more severe. (See Hand-Arm Vibration Syndrome.)

The specific cause of any tendon disorder depends on the particular muscle-tendon system and the motion(s) that it controls. For example, repetitive or prolonged work with the arms at or above shoulder height may cause rotator cuff tendonitis and eventually chronic bursitis. With downward rotation of the forearm, especially with repetitive or forceful wrist flexion, the tendon attachment at the outside of the elbow may become inflamed. Repeated or forceful bending of the wrist toward the little finger and repeated or forceful thumb motion in opposition to the fingers or to operate a pull-trigger or push-button tool have been associated with de Quervain's disease. Dorsiflexion of the wrist, especially in combination with rapid or forceful finger extension, such as when typing on a high keyboard placed too high, may injure the extensor tendons on the back of the hands.

Based on its review of the epidemiologic literature, NIOSH has concluded that there are associations between ergonomic exposures and hand/wrist tendonitis: the range of odds ratios for repetition, 1.4 to 17.0; for force, 2.5 to 17.0; and for postural factors, 1.4 to 6.2. There is also strong evidence for a combined effect of highly repetitive and forceful exertions.

Epicondylitis is also associated with forceful work, and there is strong evidence for the combined effect of force and posture as well as force and repetition. In addition, risk factors for shoulder disorders, although not exclusively rotator cuff tendonitis, include repetitive work and shoulder postures greater than 60 degrees of flexion or abduction. Odds ratios ranged from 1.6 to 5.0 in the four highest quality studies examining repetition, and from 3.5 to 5.0 in the four highest quality studies examining posture.

Nonoccupational causes of tendon disorders may include amateur sports and recreational and household activities, especially when high forces are generated, when the activities are performed repeatedly, or when there is not adequate training or a "break-in" period. In addition, tendon disorders are more prevalent among current smokers.

Predisposing conditions for tendonitis include diabetes mellitus, rheumatoid arthritis, and, less commonly, gout, hypothyroidism, psoriatic arthritis, osteoarthritis, systemic sclerosis, Reiter's syndrome, hypercholesteremia, calcium pyrophosphate deposition, collagen vascular disease, and indolent infection with mycobacteria or fungi.

Pathophysiology

There are gaps in the understanding of the pathophysiology underlying tendonitis and tendinosis, particularly with regards to early responses to stressors. Chronic tendon disorders must be distinguished from acute tendon injuries. Histological studies of affected tissue from those with chronic epicondylitis, Achilles tendinopathy, patellar tendinosis, and rotator cuff tendonitis have exhibited disorganization (microtears and separation of fibers) of collagen, the primary material of tendons. An increase in blood vessels and fibroblasts (the cells capable of forming collagen) has also been seen. There is a lack of inflammatory cells in these samples.

Thickened areas of the tendons develop where they are compressed, change direction, or where they attach to bone. These regions have fewer blood vessels than other sections of the tendon, and, hence, localized repair processes are poor. Tendons appear to increase these fibrocartilagenous characteristics under compressive loading, such as may be encountered in repetitive or forceful movements.

Each time a muscle contracts, it causes a slight strain or stretching of the tendon; if the next contraction occurs before the tendon has fully recovered, the strain accumulates. Cell and tissue injury eventually result. The time required for recovery is a function of the force and duration of each exertion, the frequency of rest periods, and the total work time with exposure to repetitive or forceful exertions. Incomplete recovery may disrupt normal regeneration of the tendon matrix, which is also subject to influences from hormones and neurotransmitters.

When the nature of the work involves prolonged static postures and muscle contraction rather than rapid movements, the nature of the injury may be more complicated. The relationship between this type of work and tendonitis is less well established. In addition to the chronic strain produced by prolonged contractions, static loading reduces blood supply to the muscle and tendon, which may be an additional cause of cell injury and death and of eventual tendon degeneration.

Prevention

Primary prevention involves reducing the exposure to occupational ergonomic risk factors, especially forceful manual exertions, awkward or non-neutral body positions during work, static muscle loads, highly repetitive motions, and a rapid and unvarying work pace, such as with piece-rate wages or machine-paced work. Whenever possible, engineering controls (appropriate ergonomic design, selection, and use of tools, tasks, and workstations) should be used instead of administrative controls. (See Upper Extremity Musculoskeletal Disorders for a detailed description of primary preventive measures.)

The most important secondary preventive measure is training health care providers in the appropriate interview and clinical examination procedures necessary to identify work-related tendon disorders. Workers who report symptoms should receive immediate attention, as these conditions may appear deceptively minor in the early stages. However, care is needed to avoid overdiagnosis of nonspecific pain as tendonitis.

Once reported, cases should be treated conservatively, with total rest for the affected tendon(s), and jobs should be analyzed for ergonomic features that may be modified. Use of a splint or elastic bandage on the job should be considered only if it does not interfere with work or require the worker to exert more force or stress on another joint to perform the task. Return to work should be gradual to allow for the reconditioning of the muscle-tendon group. Follow-up is important to ensure that job modifications have been correctly selected to avoid continuing stress to the wrist and to ensure that symptoms and signs have not progressed.

For more severe cases or when the above measures fail, physical or occupational therapy is an option. Treatments, such as local hot and cold contrast baths, myofascial release, deep tissue massage, ultrasound, TENS (transcutaneous electrical nerve stimulation), and iotophoresis are of unknown efficacy.

Concerning tertiary measures, steroid injections and surgery for tendons not prone to rupture may be considered. Surgery is generally a last resort; it may be only temporarily effective if the worker is returned to an ergonomically stressful job that has not been modified. Possible loss of muscle strength and build-up of scar tissue following surgery make it imperative that job assignments be selected carefully to avoid recurrence.

Further Reading

Astrom M, Rausing A. Chronic Achilles tendinopathy: A survey of surgical and histopathologic findings. *Clinical Orthopaedics and Related Research* 1995; 316: 151-164.

Bernard BP. *Musculoskeletal Disorders and Workplace Factors: A Critical Review Of Epidemiologic Evidence for Work-Related Musculoskeletal Disorders of the Neck, Upper Extremity, and Low Back.* Washington, DC: NIOSH, 1997.

Kahn K, Cook J, Kannus P, et al. Time to abandon the "tendinitis" myth. *British Medical Journal* 2002; 324: 626-627.

Leadbetter WB. Cell-matrix response in tendon injury. *Clinical Sports Medicine* 1992; 11: 533-578.

Moore JS. Biomechanical models for the pathogenesis of specific distal upper extremity disorders. *American Journal of Industrial Medicine* 2002; 41: 353-369.

Piligian G, Herbert R, Hearns M, et al. Evaluation and management of chronic work-related musculoskeletal disorders of the distal upper extremity. *American Journal of Industrial Medicine* 2000; 37: 75-93.

Potter HG, Hannafin JA, Morwessel RM, et al. Lateral epicondylitis: Correlation and MR imaging, surgical and histopathologic findings. *Radiology* 1995; 196: 43-46.

Tanaka S, Petersen M, Cameron L. Prevalence and risk factors of tendinitis and related disorders of the distal upper extremity among U.S. workers: Comparison to carpal tunnel syndrome. *American Journal of Industrial Medicine* 2001; 39: 328-35.

Thyroid Cancer

ICD-10 C73

Laura Welch

Most thyroid cancer presents as a hard, asymptomatic nodule in the thyroid gland. The thyroid scan shows diminished function in the nodule. There are four histological types: papillary, follicular, anaplastic, and medullary.

Occurrence

Papillary thyroid cancer is the most common form. It occurs with a bimodal age distribution with one peak in the 30s and 40s and a second peak later in life. The proportion of cases that are occupationally related is not known.

Causes

Exposure to ionizing radiation is a well-documented cause of thyroid cancer both from gamma or x-radiation of the thyroid and from exposure to radioactive isotopes of iodine. Many children exposed to fallout from the Chernobyl nuclear reactor incident in 1986 developed thyroid cancer. There may be a risk from exposure during adulthood, although the risk may be limited to those exposed before age 20, with the highest risk clearly apparent in children exposed at a younger age. A high incidence of thyroid neoplasms and hypothyroidism occurred among inhabitants of the Marshall Islands following above-ground detonation of an atomic bomb in 1954.

Pathophysiology

Ionizing radiation is a well-known carcinogen. Iodine is selectively absorbed by the thyroid so that exposure to radioactive iodine can result in high localized exposure of the thyroid to ionizing radiation.

Prevention

In the event of acute radiation exposure, oral intake of iodine blocks the uptake of radioactive iodine from inhalation. After radiation exposure, periodic examination of the thyroid for nodules is imperative.

Further Reading

Mazzaferri EL. Thyroid cancer. In Becker KL, Bilezikian JP, Bremner WJ (Eds.). *Principles and Practice of Endocrinology and Metabolism.* Philadelphia: Lippincott Williams & Wilkins, 2001, pp. 382-402.

Welch LS. Environmental toxins and endocrine function. In Becker KL, Bilezikian JP, Bremner WJ (Eds.). *Principles and Practice of Endocrinology and Metabolism.* Philadelphia: Lippincott Williams & Wilkins, 2001, pp. 2134-2140.

Thyroid Disorders (Hypothyroidism)

ICD-10 E03

Laura Welch

Hypothyroidism is a decrease in thyroid hormone production. It has a varied presentation, depending on its severity. It can range from an asymptomatic decrease in serum thyroxine to a life-threatening condition of myxedema. Usual symptoms include lethargy, cold intolerance, constipation, weight gain, skin changes, and slowed mental functioning. A wide range of chemicals affect thyroid function in humans and animals; most of them cause goiter (a nonspecific enlargement of the thyroid) and/or hypothyroidism.

Occurrence

Clinically apparent hypothyroidism is present in about 2% of the population with a mean age of 57; the disease was 10 times more common in women than in men. The percentage of uses that are occupational in origin is not known.

Causes

Lead has been reported to affect the pituitary-thyroid axis. Lead exposure in a battery plant resulted in decreased serum thyrotropin levels and low serum thyroxine, with flat or delayed response to thyrotropin-releasing hormone stimulation, suggesting anterior pituitary involvement. Cessation of exposure was followed by improvement over months to years. The affected humans had nonspecific symptoms such as fatigue, muscle pains, and impotence, which may have been independently caused by lead exposure.

Thiouracil, thiocyanates, and thiourea are goiter-producing agents. Ethylene thiourea is used as an accelerator in the manufacture of rubber and as a chemical intermediate in the manufacture of pesticides, fungicides, dyes, and chemicals. It is in the same chemical class as propylthiouracil, which is used to treat hypothyroidism. Thiocyanate occurs naturally in plants in the cabbage family and causes goiter in animals. Consumption of milk from cows that feed on these plants has been associated with goiter in humans.

Halogenated aromatic hydrocarbons can also affect thyroid function in humans and animals. Polybrominated biphenyls (PBBs) caused primary hypothyroidism in 11% of men working in a PBB production facility. The pesticide hexachlorobenzene (HCB) caused goiter in humans. In the 1950s, a hexachlorobenzene epidemic occurred in Turkey: 40% of those exposed—accounting for 60% of the women and 27% of the men—had thyroid enlargement.

Pathophysiology

Thiocyanates and thiouracil interfere with organic iodination in the thyroid, an essential step in thyroid hormone synthesis. The mechanism of action of lead on the thyroid is not known. Animal experiments suggest that the halogenated aromatic hydrocarbons impair the release of thyroid hormone. There is a specific receptor for this family of chemicals in the thyroid, the Ah receptor. This receptor is similar to the receptor for steroid hormones.

Prevention

Occupational endocrine disorders can be prevented using standard industrial hygiene control of exposure: Substitute less toxic materials, use either positive engineering to eliminate release into the work environment or environmental controls such as local exhaust ventilation, or, if these are not feasible, use personal protective equipment. On-site changing rooms, washrooms, and separate eating facilities also help reduce exposure to toxins. Exposure and biological monitoring are useful in controlling episodes.

Further Reading

Shapiro L, Surks M. Hypothryoidism. In Becker KL, Bilezikian JP, Bremner WJ (Eds.). *Principles and Practice of Endocrinology and Metabolism.* Philadelphia: Lippincott Williams & Wilkins, 2001, pp. 445-453.

Welch LS. Environmental toxins and endocrine function. In Becker KL, Bilezikian JP, Bremner WJ (Eds.). *Principles and Practice of Endocrinology and Metabolism.* Philadelphia: Lippincott Williams & Wilkins, 2001, pp. 2134-2140.

Tremor

ICD-10 G25.0

Robert G. Feldman and Marcia H. Ratner

A tremor is a rhythmic alteration in movement of an extremity or the head with a consistent pattern, amplitude, and frequency. It is due to the reciprocal contraction of a muscle group and its antagonist muscle group. Tremors are classified by location, speed, amplitude, relationship to rest position and movement, etiology, and suspected underlying pathological process. A physiological tremor, normally low in amplitude and invisible, occurs in all individuals when a muscle is under tension.

An abnormal tremor is associated with disturbed motor control; it has lower frequency of movement than a physiological tremor and occurs only when the person is awake. Abnormal tremors can be further categorized as: (a) intention or action tremors that appear only with deliberate movement,

are coarse and irregular, and associated with cerebellar dysfunction and ataxia; and (b) static or nonintention tremors present at rest and associated with
dysfunction of the basal ganglia and extrapyramidal pathways, such as in
Parkinson's disease. Non-intention tremors are more rhythmic and often
localized in the hands, feet, jaw, lips, or tongue. Electromyography, psychomotor tests of eye-hand coordination, finger tapping performance tests,
and neurological tests of cerebellar functioning are used to objectively assess
tremor. The frequency and amplitude of tremor may be measured by an
accelerometer connected to a tape recorder, through a precision sound-level
meter or electromyographic device.

Occurrence

No reliable occurrence data exist concerning intention tremor. Certain
occupational tasks bring out these tremors while performing tasks. A fast
intention tremor occurs as a result of exposure to toxic chemicals, such as
mercury. Of 220 chemicals studied, 149 have caused neurotoxicity in humans.

Causes

Exposure to certain neurotoxic chemicals in the workplace can cause
tremor (Table 1). Tremor has frequently been observed in workers exposed to
heavy metals including inorganic and organic mercury, and manganese.
Acute exposures to certain gases including carbon monoxide may result in
tremors similar to those of extrapyramidal syndromes or "parkinsonisms."
Chronic exposure to methyl bromide, a colorless gas used as a fumigant,
refrigerant, fire extinguisher, and insecticide, can cause tremor, mixed with
cerebellar symptoms and peripheral neuropathy, whereas acute exposure
causes delirium and visual and speech disturbances. Chronic exposures to
solvents, including toluene and trichloroethylene, can also produce tremor.
Paraquat (1,1-dimethyl-4,4-bipyridylim dichloride), a herbicide, can produce
hand tremor 4 to 5 days after exposure, as well as renal damage and mental
disturbance. Acute exposure to DDT can cause tremors and seizures.
Chlordecone (Kepone) exposure can cause an intention tremor, although a
static tremor can occur in those severely affected. High-level chronic exposure
to the insect repellent DEET has been associated with damage to cerebellar
neurons and tremor similar to that seen among patients exposed to toluene.

Nonoccupational causes of tremor include hyperthyroidism, nonoccupational parkinsonism, tardive dyskinesia, Tourette's syndrome, epilepsy,
Wilson's disease, and dystonia. Tremor may also be due to alcoholic withdrawal, nicotine, caffeine, amphetamines, and barbiturates as well as other
medications such as asthma medications. Hepatic and other
encephalopathies may cause asterixis, a flapping of outstretched, extended
hands. Patients with thyroid disease may have tremor. Lithium may cause
tremor, an indication of possible overdose. Tremors of any etiology must be
differentiated from familial tremors, which by definition are reduced by alcohol. This is often an historical pearl.

Table 1. Exposures Associated with Tremors

Neurotoxin	Major Uses or Sources of Exposure
Metals	
Manganese	Iron, steel industry
	Welding operations
	Metal-finishing operations
	Fertilizers
	Manufacture of fireworks, matches
	Manufacture using oxidation catalysts
	Manufacture of dry cell batteries
Mercury	Scientific instruments
	Electrical equipment
	Amalgams
	Electroplating industry
	Photography
	Felt making
	Taxidermy
	Textiles
	Pigments
Solvents	
Carbon disulfide	Manufacture of viscose rayon
	Paints
	Preservatives
	Textiles
	Rubber cement
	Electroplating industry
Trichloroethylene	Degreasers
	Painting industry
	Paints
	Lacquers
	Varnishes
	Dry-cleaning industry
	Rubber solvents
	Adhesive in shoe and boot industry
	Process of extracting caffeine from coffee
Gases	
Methyl bromide	Insect fumigant
	Fire extinguisher
	Refrigerant
	Degreasing agent for wool
Carbon monoxide	Exhaust fumes of internal combustion engines, incomplete combustion
	Acetylene welding
Insecticides/Herbicides	
Chlorinated hydrocarbons (DDT, Chlordecone)	Agricultural industry
Paraquat	Agricultural industry
	Field workers

Pathophysiology

Intention tremor is generally a manifestation of impaired cerebellar functions. Degeneration of the basal ganglia and extrapyramidal pathways is seen among patients with resting or nonintention tremor, such as in Parkinson's disease. This is often accompanied by rigidity. The pathology of resting tremor involves disturbances of neurotransmission and/or loss of neurons within the substantia nigra and basal ganglia. A specific pathology to explain essential intention tremor reflects an imbalance in the stretch receptor-spinal cord-extrapyramidal system that maintains muscle tone.

Prevention

Exposure to neurotoxic chemicals should be minimized. Industrial hygiene monitoring and biological monitoring may be helpful. Removal from exposure and/or chelation of mercury, if performed early enough, may reverse the symptoms of tremor. N-acetyl-penicillamine is the preferred chelating agent for elemental mercury exposure. L-dopa therapy is sometimes useful in treating the parkinsonian-like tremor due to manganese poisoning.

Other Issues

Some non-workplace factors may confound the identification of a work-related cause of tremor. Familial tremors, medium to coarse in amplitude and absent at rest, tend to progress in severity as a person ages and are often accentuated by physical stress, stimulants, and excitement. They tend to lessen upon ingestion of alcohol and sedatives.

Further Reading

Biernat H, Ellias SA, Wermuth L, et al. Tremor frequency patterns in mercury vapor exposure, compared with early Parkinson's disease and essential tremor. *Neurotoxicology* 1999; 20:945-952.

Feldman RG. *Occupational and Environmental Neurotoxicology*. Philadelphia: Lippincott-Raven, 1999.

Grandjean P, Sandoe SH, Kimbrough RD. Non-specificity of clinical signs and symptoms caused by environmental chemicals. *Human and Experimental Toxicology* 1991; 10: 167-173.

Peters HA, Levine RL, Matthews CG, et al. Carbon disulfide-induced neuropsychiatric changes in grain storage workers. *American Journal of Industrial Medicine* 1982; 3: 373-391.

Ratner M, Cabello D, Thaler D, et al. Movement disorder in an adult following exposure to DEET. *Journal of Toxicology. Clinical Toxicology* 2001; 39: 477.

Roels H, Malchaire J, Van Wamkeke JP, et al. Development of a quantitative test for hand tremor measurement. *Journal of Occupational Medicine* 1983; 25: 481-487.

Tuberculosis

ICD-10 A15-A19

Yvonne Boudreau

Tuberculosis (TB) is a bacterial disease caused by *Mycobacterium tuberculosis*. It is usually transmitted from person to person through the air when a person with TB in the lungs or throat coughs, sneezes, speaks, or sings. TB most often affects the lungs, but it can affect other parts of the body, such as the brain, kidneys, or spine. The initial infection rarely produces noticeable symptoms, but the infected person usually develops a delayed hypersensitivity immune response within 2 to 12 weeks after exposure, and may develop x-ray evidence of infection in the lung.

Some people infected with TB progress from the initial infection to develop TB disease. General symptoms of TB disease include fatigue, fever, night sweats, and weight loss; lung symptoms include cough, chest pain, and hemoptysis (coughing up blood). Treatment with standardized regimens can cure TB disease and can prevent infected individuals from developing TB disease.

The Mantoux tuberculin skin test (TST) is used to determine whether a person is infected with *M. tuberculosis*. This test is performed by injecting 0.1 milliliter of purified protein derivative tuberculin into the skin (intradermally) on the inner surface of the forearm. The site is evaluated for induration (swelling) 48 to 72 hours after administration. The size of the induration that is considered to represent a "positive" reaction varies according to the individual's risk factors, such as occupation, recent exposure history, and competency of the immune system.

False-positive reactions may occur due to infection with mycobacteria other than *M. tuberculosis* or immunization with bacille Calmette-Guérin (BCG) vaccine. BCG is a strain of *M. bovis*, the organism that causes TB in cattle, which has been modified to produce immunity against TB without causing disease. BCG is the most widely used vaccine in the world, but it is generally not recommended in the United States because of the low risk of infection with *M. tuberculosis*, the variable effectiveness of the vaccine against pulmonary TB in adults, and interference by the vaccine with tuberculin reactivity.

False-negative reactions can occur if (a) the skin test is performed too soon after exposure, prior to the development of an immune response; (b) skin testing is done soon after administration of a live-virus vaccination, such as measles vaccine; or (c) the person being tested is unable to show an immune response because of an immune system that is not intact.

In addition, in some infected persons, the ability to react to tuberculin may wane over time. When given a skin test years after infection, these persons may exhibit a negative reaction. However, the administration of the skin test itself may stimulate the immune system, causing a positive reaction on subsequent tests—the "booster phenomenon." Two-step testing can be used to distinguish between the booster phenomenon and a new infection. If the reaction to the first test is negative, a second test is given 1 to 3 weeks later. If the reaction to the second test is positive, this could be considered a boosted reaction and not a reaction due to a new infection. Two-step testing can be useful for the initial skin testing of adults who are going to be retested periodically, such as workers in occupations involving exposure to TB.

In 2001, the FDA approved another test for detecting latent *M. tuberculosis* infection, the QuantiFERON®-TB test (QFT). This is a blood test that assesses cell-mediated immune response to *M. tuberculosis*. Unlike the TST, this test requires only a single visit to a healthcare provider. In addition, this test does not cause the booster phenomenon, is less subject to reader bias and error than the TST, and is more specific than the TST (is less affected by prior BCG vaccination and infection with nontuberculous mycobacteria).

The CDC recommends that the QFT be considered for initial and serial testing of persons with an increased risk of latent TB infection, such as recent immigrants, injection-drug users, and residents and employees of prisons, jails, and homeless shelters; for initial and serial testing of persons who are by history at low risk for latent TB infection, but whose future activity may place them at increased risk of exposure, such as healthcare workers and military personnel; and for testing of persons for whom screening is performed for reasons such as entrance requirements for certain schools and workplaces. Currently, the test is not recommended for evaluation of persons with suspected TB disease; assessment of contacts of persons with infectious TB disease; confirmation of TST results; diagnosis of *M. avium* complex disease; or for screening children under 17 years of age, pregnant women, or people with medical conditions that increase the risk of progression from infection to disease.

A positive TST or QFT only indicates infection. To determine whether people have TB disease, they should receive additional evaluation, including a medical history, physical examination, chest x-ray and, if indicated, a sputum smear and culture. The sputum smear is examined for the presence of acid-fast bacilli (AFB), the bacteria that cause TB. However, although AFB on a sputum smear often indicates TB disease, it does not confirm the diagnosis because not all AFB are *M. tuberculosis*; therefore, culture examinations should be done on all specimens from TB suspects. A positive culture for *M. tuberculosis* confirms the diagnosis of TB disease; however, TB may also be diagnosed on the basis of clinical symptoms and signs in the absence of a positive culture. Recently developed systems that use a liquid environment for growth, such as the Mycobacteria Growth Indicator Tube (MGIT), allow detection of mycobacteria in 4 to 14 days. Approved nucleic acid (RNA or DNA) amplification tests may enhance diagnostic certainty, but should be interpreted in a clinical context and on the basis of local laboratory performance. The initial *M. tuberculosis* isolate should be tested for drug resistance in order to ensure appropriate treatment. Drug susceptibility patterns should be repeated for patients who do not respond adequately to therapy.

Follow-up bacteriologic examinations (sputum smears and cultures) are important for assessing a patient's infectiousness and response to therapy. Specimens should be obtained at monthly intervals until the culture results convert to negative. Culture conversion is the most important objective measure of response to treatment. Conversion is documented by the first negative culture in a series of cultures (subsequent culture results must remain negative).

TB reporting is required by law in every state in the United States. Clinicians should report all new TB cases and suspect cases to health departments. Cases may also be reported by infection control nurses or by pharmacies when TB drugs are dispensed. In addition, all positive TB smear and culture results should be reported promptly by laboratories. Early reporting is important for the control of TB and gives clinicians access to the health department resources for assistance in case management and contact investigation.

Occurrence

Worldwide, an estimated 2 billion persons (one-third of the world population) are infected with TB, and approximately 2 million TB-related deaths occur annually. In the United States, an estimated 15 million people are infected. The global incidence rate of TB is growing at approximately 0.4% per year, but much faster in sub-Saharan Africa and in countries of the former Soviet Union.

After years of decline in the United States, the number of reported TB cases increased 20% from 1985 to 1992. This increase was associated with deterioration of the infrastructure for TB services, the human immunodeficiency virus (HIV) epidemic, immigration of persons from countries where TB is endemic, TB transmission in congregate settings (such as hospitals and prisons), and development of multidrug-resistant TB. However, between 1992 and 2002, the number of TB cases in the United States decreased by 43.5% due to a renewed emphasis on TB control and prevention. Despite this progress, the 2002 rate in the United States (5.2 per 100,000 population) remained higher than the 2000 interim goal of 3.5, set as part of the national strategic plan for TB elimination—less than 1 case per 1 000 000 by 2010.

Groups of persons known to have high rates of *M. tuberculosis* transmission and increased risk for being recently infected include contacts of persons with TB disease, foreign born persons from geographic areas with a high prevalence of TB who are within 5 years of their arrival into the United States, persons who are medically underserved and of low social and economic status, homeless persons, current or former inmates of correctional facilities, persons who abuse alcohol, and injection drug users. Groups with a high risk for progression from TB infection to disease include recently infected persons, young children, persons with fibrotic lesions on chest x-ray, and persons with certain medical conditions (HIV infection, silicosis, gastrectomy or jejunoileal bypass, weight more than 10% below ideal, chronic renal failure with renal dialysis, diabetes mellitus, immunosuppression resulting from organ transplantation, receipt of high dose corticosteroid or other immunosuppressive therapy, and some malignancies).

Some occupational exposures, such as silica dust, increase the risk of TB disease if infection occurs. Silica particles appear to exert their effect by impairing the ability of macrophages to inhibit the growth of, and kill, TB bacteria. Occupational groups at risk for silicosis include miners, quarry workers, stonemasons, sandblasters, and foundry and pottery workers.

Persons in occupations in which low socioeconomic groups are disproportionately represented have higher TB rates than those of the general population. These occupations include unskilled laborers, food handlers, migrant farm workers, and lower-paid health-care workers.

In institutions that serve high-risk populations, the workplace risk may be substantial. These venues include inpatient settings, outpatient settings, TB clinics, correctional facilities, agencies that deliver home-based care, nursing homes, shelters for the homeless, extended care facilities, mental hospitals, drug addiction treatment centers, and emergency medical services. Health-care workers who may be at increased risk include, but are not limited to, physicians, nurses, aides, dental workers, technicians, workers in laboratories and morgues, emergency medical personnel, students, part-time workers, temporary and contract workers, and persons not involved directly in patient care but potentially at risk for occupational exposure, such as volunteers, outreach workers, and dietary, housekeeping, maintenance, clerical, and janitorial workers. People who are stationed abroad in areas of high TB prevalence may be at high risk for TB infection, especially if they are health-care providers or are in prolonged contact with members of the local population in TB endemic areas. Social workers and others who work with the poor may also be at higher risk.

Other occupational environments where workers may be exposed to TB include autopsy rooms, mycobacteriology laboratories, medical waste-treatment facilities, funeral homes, zoos, primate research centers, and laboratories that handle clinical specimens that may contain *M. tuberculosis*.

Causes

In the United States, the vast majority of TB cases are caused by *M. tuberculosis*, sometimes referred to as the tubercle bacillus. *M. tuberculosis* and three very closely related mycobacterial species (*M. bovis*, *M. africanum*, and *M. microti*) can cause tuberculous disease, and they compose what is known as the *M. tuberculosis* complex. *M. microti* does not cause disease in humans; *M. bovis* and *M. africanum* are less usual causes of disease in the United States. *M. bovis* primarily affects cattle, but may occur in humans where disease in cattle has not been adequately controlled. Human infection with *M. bovis* usually results from ingestion of raw, unpasteurized milk or dairy products; however, it occasionally results from airborne spread to farmers and animal handlers, and may be transmitted from person to person. Mycobacteria other than those comprising the *M. tuberculosis* complex are called nontuberculous mycobacteria. Nontuberculous mycobacteria may cause pulmonary disease resembling TB.

Pathophysiology

Tubercle bacilli are carried in airborne particles, called droplet nuclei, which can be generated when persons who have TB of the lungs or throat sneeze, cough, speak, or sing. The particles are estimated to be 1 to 5 µm in size. Normal air currents can keep them airborne for prolonged periods and spread them throughout a room or building.

Infection occurs when a susceptible person inhales droplet nuclei containing *M. tuberculosis* bacilli, and the droplet nuclei reach the alveoli (air sacs) of the lungs. Once in the alveoli, the organisms are engulfed by macrophages, but they resist killing by macrophages and continue to multiply. The bacilli spread to more distant sites via the lymphatic system and bloodstream. Usually within 2 to 12 weeks, cellular immunity develops and this limits further multiplication and spread of the bacilli. However, some bacilli remain dormant, but viable, for many years, a condition referred to as latent TB infection. The infected individual usually remains asymptomatic, the lesions heal, and the immune response can be demonstrated by testing for hypersensitivity to tuberculin.

The probability that a person who is exposed to *M. tuberculosis* will become infected depends on the concentration of infectious droplet nuclei in the air, the number of bacteria, and the duration of exposure. Characteristics that enhance the likelihood of infectious transmission include presence of cough; failure to cover the mouth and nose when coughing; presence of respiratory tract disease with involvement of lungs, airways, or larynx; presence of AFB in the sputum; cavitation on chest x-ray; lack of, inappropriate, or short duration of chemotherapy; exposure in relatively small, enclosed spaces; inadequate local or general ventilation that results in insufficient dilution or removal of infectious droplet nuclei; and recirculation of air containing infectious droplet nuclei.

In some people, infection may progress to clinical illness (TB disease). The risk of this is greatest during the first several years after infection and in immunocompromised persons, especially those infected with HIV.

Prevention

Primary prevention measures include maintaining vigilance for the disease, especially in high-risk populations and environments. The rapid identification, isolation, and treatment of someone with TB is the most important infection control measure. Patients should be taught how to cover the mouth and nose when coughing or sneezing. Infectivity should be controlled by prompt, specific drug therapy, which usually results in sputum conversion within a few weeks. Hospitalized patients with sputum-positive pulmonary TB should be placed in respiratory isolation.

Additional primary prevention measures include providing adequate ventilation, directional airflow, air filtration, and selective use of ultraviolet germicidal irradiation (UVGI) in areas where there is a high risk of transmission; using approved personal respiratory protective equipment, when indicated; ensuring proper cleaning, sterilization, or disinfection of potentially contaminated equipment; improving social conditions that increase the risk of infection, such as overcrowding, and of disease, such as undernutrition; applying measures to prevent silicosis in industrial settings; administering BCG vaccination to people in countries with a high risk of infection; eliminating TB among dairy cattle by tuberculin testing and slaughter of reactors; and pasteurizing or boiling milk.

Secondary prevention measures include performing sputum smear examination of symptomatic people in high-incidence areas, performing tuberculin testing of household members and close nonhousehold contacts of people with infectious cases, providing chest x-ray evaluations on positive tuberculin reactors and symptomatic individuals (regardless of tuberculin test reaction), and performing targeted tuberculin testing on groups at high risk for infection and disease.

Other Issues

There must be strong collaboration among national and international health partners to reach those at highest risk for TB. Innovative strategies to improve testing and treatment among high-risk populations should be identified. One such effort involves the CDC's establishment of a binational TB referral system for TB patients who cross the United States-Mexico border. In another effort, the United States Advisory Council for the Elimination of Tuberculosis has made recommendations to resolve issues concerning the post-detention treatment of persons with TB who are released or deported from custody after detention by the Immigration and Naturalization Service.

Another well-established strategy is the use of directly observed therapy (DOT) throughout the world. This involves the required presence of a health-care representative each time persons being treated for TB take their prescribed medications. This leads to reductions in treatment failure, relapse, and drug resistance. Joint participation by health departments and community healthcare providers is essential to ensure successful implementation of efforts to prevent TB in high-risk groups. Community sites where persons at high risk may be accessed and where targeted testing programs have been evaluated include neighborhood health centers, jails, homeless shelters, inner-city sites, and methadone and syringe/needle-exchange programs.

In the United States, TB control among groups of low socioeconomic status, minorities, and people who are foreign-born presents special problems because these groups often suffer from economic hardships, coexisting diseases, lack of shelter and transportation, chemical dependencies, population mobility, noncompliance with long-term therapy, language and cultural barriers, or fear of deportation. People in these groups should be promptly evaluated when TB is suspected. If limited access to transportation is an issue, evaluation at the worksite, community site, or living area should be considered. Those with significant tuberculin reactions should be placed on therapy for latent TB infection in accordance with American Thoracic Society/CDC guidelines. DOT should be considered if self-administered therapy is not likely to be successful. In cases in which an individual is moving and requires continued treatment or other follow-up, health care providers should route pertinent information to the destination local or state health department. If there is an issue of frequent relocation, individuals on treatment should be given records that they can take with them to indicate their current treatment status. Those with a significant tuberculin reaction who are not prescribed therapy should be counseled about the significance of the skin test reaction and instructed to seek medical attention should they develop symptoms suggesting TB.

Prevention of exposure to hazardous levels of silica dust (and, thus, of silicotuberculosis) can be achieved by engineering controls, such as improved ventilation, process enclosure, wet mining techniques, and use of substitutes for some industrial silica in manufacturing processes. Approved respiratory protection devices should be considered if these methods fail. Periodic tuberculin testing of workers exposed to silica dust is indicated and therapy for latent TB infection should be considered for those who have significant reactions. Workers who have symptoms and signs suggestive of current TB disease should receive a chest x-ray and sputum cultures. Although modern treatment regimens are usually effective in curing TB, treatment failures and relapses are more frequent in patients with silicosis.

Currently, under the general duty clause of the OSHAct, OSHA can require employers to comply with guidelines for protecting employees from TB exposure. Targeted workplaces include health care settings, correctional institutions, homeless shelters, long-term care facilities for the elderly, and drug treatment centers. There are many resources available to assist employers and employees in preventing TB exposures in their facilities. The CDC and state and local health departments can provide more information about these resources.

Further Reading

American Thoracic Society/Centers for Disease Control and Prevention. Diagnostic standards and classification of tuberculosis in adults and children. *American Journal of Respiratory and Critical Care Medicine* 2000; 161: 1376-1395.

American Thoracic Society/Centers for Disease Control and Prevention/Infectious Diseases Society of America. Treatment of tuberculosis. *American Journal of Respiratory and Critical Care Medicine* 2003; 167: 603-662.

Institute for Medicine. *Tuberculosis in the Workplace.* Washington, D.C.: National Academy Press, 2001.

Snider DE. The relationship between tuberculosis and silicosis. *American Review of Respiratory Disease* 1978; 118: 455-460.

US Centers for Disease Control and Prevention. Adverse event data and revised American Thoracic Society/CDC recommendations against the use of rifampin and pyrazinamide for treatment of latent tuberculosis infection-United States, 2003. *MMWR* 2003; 31: 735-739.

US Centers for Disease Control and Prevention. Guidelines for preventing the transmission of Mycobacterium tuberculosis in health-care facilities, 1994. *MMWR* 1994: 43(No.RR-13).

US Centers for Disease Control and Prevention. Guidelines for using the QuantiFERON®-TB test for diagnosing latent Mycobacterium tuberculosis infection. *MMWR* 2003; 52(No. RR-02): 15-18.

US Centers for Disease Control and Prevention. Targeted tuberculin testing and treatment of latent tuberculosis infection. *MMWR* 2000; 49(No. RR-6): 1-54.

US Centers for Disease Control and Prevention. Treatment of tuberculosis. *MMWR* 2003; 52: 1-77.

World Health Organization. *Global Tuberculosis Control: Surveillance, Planning, Financing.* Geneva: WHO/CDS/TB, 2003, p. 316.

Note: The CDC plans to publish revised guidelines for tuberculosis in 2005.

Upper-Extremity Musculoskeletal Disorders
ICD-10 M60-M63

Laura Punnett and Judith E. Gold

Many different conditions may affect the muscle, tendons, ligaments, and joints of the musculoskeletal system, as well as the associated peripheral nerves that control and regulate muscular movement. Work-related musculoskeletal disorders (MSDs) have been known to affect virtually every part of the body: back, neck, shoulders, hands, arms, hips, legs, and feet.

These disorders are known as soft tissue disorders, cumulative trauma disorders, repetitive strain injuries, repetitive motion disorders, occupational cervicobrachial disorders, or overuse syndromes. Such terminology reflects the reality that various clinical disorders may occur in a group of workers with a common set of occupational ergonomic exposures. These conditions are typically chronic or subchronic rather than acute in onset. Furthermore, the available clinical diagnoses do not cover all work-related conditions. It may thus be useful to consider upper-extremity MSDs collectively in order to evaluate the total morbidity burden and the effectiveness of intervention efforts. Nevertheless, for an affected individual, if diagnosis of the specific disorder is possible, it will likely facilitate the choice of appropriate treatment, rehabilitation, and prevention measures.

The most important upper-extremity disorders associated with risk factors in the workplace are hand-arm vibration syndrome; tendinitis, tenosynovitis, and other conditions affecting muscle-tendon systems; peripheral nerve disorders, especially carpal tunnel syndrome (CTS); and degenerative joint disease (osteoarthritis) affecting any of the joints in the upper extremity. Specific conditions are discussed in separate entries on these topics. Tension neck syndrome involves chronic pain in the back of the neck, head, and shoulders, thought to be produced by spasm of neck and upper back muscles. Other, less well-defined conditions include myositis and fibromyalgia, as well as regional and sometimes poorly localized pain and paresthesia not attributable to specific pathologies.

Diagnosis of upper extremity MSDs typically relies on the description of symptoms by the affected worker as well as a physical examination and sometimes a laboratory test. The location and quality of symptoms can sometimes differentiate one type of problem from another. For example, muscle-tendon disorders may be distinguished from joint problems on physical examination by their greater discomfort on active and resisted maneuvers rather than on passive motion. However, diagnostic criteria are not standardized and are often inconsistent from one examiner to another, even though there are consensus documents for several upper-extremity MSDs.

In addition, signs and symptoms of some MSD patients do not conform to specific diagnostic entities, such as CTS, rotator cuff tendonitis, and de Quervain's disease. Some clinicians have characterized a diffuse nonspecific upper-extremity pain syndrome, upon which specific MSD symptoms may be superimposed. Others report that patients frequently present with multiple MSDs simultaneously. For example, 71% of 537 workers' compensation upper extremity claimants in Maryland had more than one diagnosis, including a variety of tendinoses as well as CTS.

Occurrence

MSDs are widespread in the general population and account for much pain and disability. In both the United States and Canada, more people are disabled from working as a result of MSDs than from any other group of diseases. MSDs are also among the top three leading causes of work absenteeism in Sweden and England. Numerous surveys of working populations have reported upper extremity symptom prevalences of 20% to 30% or higher.

One study estimated 40% of all upper extremity disorders in U.S. workers are attributable to occupational exposures, representing more than 500 000 people affected annually. In general, high-risk occupational activities are those characterized by rapid, repetitive, continuous, or forceful manual tasks, especially in manufacturing, service, construction, or agriculture. Industries with especially high rates include meat, poultry, and fish processing; and manufacturing of aircraft, motor vehicles, furniture, appliances, electrical equipment, electronic products, shoes, textiles, apparel, and upholstery. MSDs have also been found to occur frequently in clerical work, postal service, janitorial work, industrial inspection and packaging, and other occupations with labor-intensive or assembly-line work processes.

MSDs are consistently the largest category of work-related illness reported to the BLS by private employers and represent about one-third of all injuries and illnesses causing lost workdays. The incidence of all work-related "disorders associated with repeated trauma" in the private sector, excluding firms with fewer than 11 employees, is about 11 per 10 000 full-time workers per year. However, many industries and occupations have rates that are much higher. For example, one study found more than 50% of newly-hired computer operators report new musculoskeletal symptoms during the first year on the job.

Attempts to count cases of work-related MSDs are hampered by workplace under-reporting for a variety of reasons. Thus, most available prevalence and incidence data, such as those from the BLS surveys, probably underestimate the true magnitude. Diagnostic difficulties and inconsistencies compound this problem. The rates of MSDs reported to BLS increased in the early 1990s and then appeared to reach a plateau. It is not clear to what extent these variations reflect the true rates of occurrence as compared with trends in workplace reporting of these disorders or medical diagnosis and recognition of work-relatedness.

Recurrences of upper-extremity disorders are common. Economic and noneconomic outcomes of MSDs, including interference with both work and family activities, can be severe and sometimes persist even after retirement from the job at time of onset.

Among people with MSDs, correlates of work disability and other poor outcomes include high physical effort requirements as well as work that is monotonous, repetitive, and lacking autonomy over work pace and schedule. This indicates that appropriate job accommodation in return-to-work is especially important when the job may be physically difficult for a symptomatic individual to tolerate; such adjustments may be a particular concern for women, who are more likely than men to be employed in highly repetitive jobs.

Causes

As with most chronic diseases, MSDs have multiple occupational and non-occupational risk factors. In addition to work demands, other aspects of daily life, such as sports and housework, may present physical stresses to the musculoskeletal tissues. The musculoskeletal and peripheral nerve tissues are affected by rheumatoid arthritis, gout, systemic lupus erythematosus, kidney disease, diabetes mellitus, and other systemic endocrine, neurologic, and collagen vascular diseases. There is little evidence that physical work capacity, especially muscle strength, is a predictor of upper-extremity MSDs.

The physical ergonomic features of work that are frequently cited as risk factors for MSDs include rapid work pace and repetitive motion patterns; insufficient recovery time; forceful manual exertions; non neutral body postures (either dynamic or static); mechanical pressure concentrations; vibration; and local or environmental cold. Exertion to overcome external load, such as in lifting or carrying, involves muscle contraction and tendon loading; it may also cause pressure on adjacent nerves or blood vessels. With sufficient recovery time, relative to the total effort required, most such exertions are conditioning rather than injurious. However, excessive intensity or frequency of loading may cause tissue damage. Sustained contractions of any muscle in a fixed position (static loading), such as needed to hold a tool or pairs or to maintain a body posture, may strain the muscle-tendon system. Working with the wrist bent, the arm raised at the shoulder, or with any other joint in a non-neutral posture puts the muscle(s) at a biomechanical disadvantage, requiring more forceful muscular contractions to perform the task and increasing the internal forces exerted by the tendon. Using a pinch grip to hold parts or tools similarly creates a biomechanical disadvantage for the working muscles.

Another type of stressor that has received increasing interest with respect to MSDs is that of psychosocial features of the work environment that arise from the organization of production activities. Most forms of work organization-task structure, the division of labor, and skill utilization-constrain individual worker behavior and thus influence physical as well as psychological job features. The occupational psychosocial stressor most consistently associated with MSDs is low job control (few opportunities to exercise autonomy or decision-making).

Piece-rate wages and machine-paced work deserve particular attention. Machine pacing requires a constant work speed, regardless of whether the person is fatigued or in pain. Piecework (for which employees are paid per unit produced) induces workers to increase both their work pace and often the hand forces they exert, as well as to reduce their total rest time, which is required for tendons and other tissues to recover from mechanical strain. Both forms of work organization are usually found in a highly segmented work process, implying both stereotyped physical motion patterns and adverse psychosocial conditions, such as low decision latitude and skill utilization.

Prevention

Primary prevention should focus on engineering controls, especially the ergonomic design of workstations, equipment, tools, and work organization to fit the size, speed, and strength capabilities of the human body. Although a mix of specific conditions may be occurring in the affected population, the general goal is to reduce or eliminate the generic ergonomic risk factors cited above. In high-risk jobs, often two or more of these factors are present and synergistically exert adverse effects. Control measures, to be effective, usually need to address all existing exposures.

(1) Effects of rapid and repetitive work may be mitigated by slowing the overall work pace, providing longer and more frequent rest breaks, avoiding piece-rate wages in favor of hourly rates, and permitting operators to set their own work pace. Administrative controls, such as job rotation or work enlargement (adding more tasks to the job description), may reduce the duration of exposure or increase the variability of motions performed. However, care must be taken to ensure that additional tasks do not simply replicate exposures of the primary job. Performing many tasks that are all ergonomically stressful is not an effective control strategy and should not be allowed to divert attention from ergonomic redesign efforts.

(2) The force required to perform a job may be reduced by lighter loads or less frequent lifting, improved friction between hand and tool or part, and better quality control on parts. Slip clutches reduce the torque produced by powered hand tools. Tools may be installed on overhead balancers or articulating arms mounted on the workbench; this decreases both the weight that must be supported and the torque transmitted to the arm. Parts, tools, and equipment should be designed and installed so that awkward body postures or pinch grips do not have to be used, especially for repeated, prolonged, or forceful motions, because these postures result in biomechanical disadvantage and increase the force necessary to perform a task. Regularly scheduled, frequent maintenance of tools and equipment, including sharpening of knives or scissors, avoids unnecessary mechanical resistance. Gloves (whether to protect the hand from abrasion, mechanical stress, cold, vibration, or chemicals) should be selected with caution to avoid increasing the force required to grip tools and parts; such gloves should be available in a wide range of sizes to fit the entire working population.

(3) Mechanical stress or pressure concentrations refer to highly concentrated forces applied by physical contact between any object in the workplace and the skin, with pressure on the underlying tissues, such as blood vessels and tendons. These forces may be reduced by rounding and padding edges and corners (such as on tool handles, table edges, and chair seats) that come into contact with the body.

(4) Work postures may be kept close to anatomically neutral by avoiding work locations that are above shoulder height, farther forward than arm's length, or behind the midline of the body. Eliminating the need

for twisting motions, and designing and selecting tools that permit work to be performed with the wrist straight and the rest of the body in a comfortable position, will also contribute to keeping work postures neutral. Providing adequate illumination and locating work close to eye height will eliminate the need for bending and twisting the neck. Suspending tools overhead and providing seating wherever possible may minimize prolonged static muscular constructions. Chairs should be designed and selected for compatibility with the specific task being performed, taking into account functional reach distances and hand work heights. In general, equipment and furniture should be adjustable in height and other relevant dimensions so as to conform to the range of body sizes of workers.

(5) Transmission of vibration to the hand and arm may be reduced by engineering controls, such as mechanical isolation and damping in tool handles or vehicle seats. (See Hand-Arm Vibration Syndrome.) Alternatively, the duration of exposure may be reduced by administration measures, such as rotation or rest breaks.

(6) Insulating air-powered tools and directing exhaust air away from the tool handle may avoid local cold exposure to the hand. Parts stored in an unheated warehouse should be moved into the work area early enough so that they warm before use. Workers in cold environments need adequate insulating clothing and rest breaks in warm locations whenever possible.

Certain administrative controls have been proposed for primary prevention purposes, especially workplace exercise and stretching programs and medical screening of prospective employees to eliminate those with predisposing factors. Unfortunately, neither of these approaches has been shown to be effective for prevention. Exercise and stretching can logically be expected to increase an individual's strength, flexibility, and tolerance of the physical demands of work. However, only limited formal evaluations of these programs have been undertaken to date, and no benefits have been found for an entire working population either in prevention of new problems or in improvement of preexisting conditions. (In fact, in some cases, exercise might even aggravate or worsen a problem once it has begun, if inflamed tendons, strained muscles, or compressed nerves require rest in order to recover.)

There is also no evidence that medical evaluation of prospective workers to eliminate those at high risk reduces future occurrences, because no individual factors have been shown conclusively to predict which workers will or will not develop musculoskeletal disorders after exposure to ergonomically stressful working conditions.

Secondary prevention measures include reducing obstacles that (a) prevent or deter workers with symptoms from reporting them to their physicians or the workplace health service, (b) prevent affected workers from being allowed adequate recuperation time away from work, and (c) deter reassignment of affected workers to "light-duty" work. Initially, symptoms may disappear with cessation of work, only to recur after the employee resumes work. However, if the injury gets worse, pain may continue even when work ceases. The longer a person works in pain on the same job, the slower the recovery and the greater the risk of permanent disability.

Health care providers should be trained in the appropriate interview and clinical examination procedures to diagnose occupational musculoskeletal disorders. Exposed workers and their supervisors should be informed of the symptoms and signs associated with these syndromes. Because not all workers with symptoms will be likely to seek medical attention promptly, active medical surveillance (interviews and physical examinations of all workers) should be carried out to identify both those in need of treatment and the jobs or areas where ergonomic redesign should be undertaken. Once reported, cases should be treated conservatively, and jobs should be analyzed for ergonomic features that may be modified.

In the context of tertiary prevention, the best treatment for usage-related musculoskeletal disorders is rest for the affected tendon, joint, or nerve; this includes rest from work, all household duties, and other manual activities. Rest is sometimes combined with local heat or cold compresses for temporary pain relief, anti-inflammatory medication, physical or occupational therapy, or use of splints. However, none of these treatments is likely to be effective if the individual continues to work without appropriate ergonomic modification of the workstation, tools, equipment, and/or work pace. Splint use on the job should be considered only if it does not interfere with work or require the worker to exert more force or strain on another joint to perform the task. Follow-up is important to ensure that job modifications have been effective, that light-duty jobs have been correctly selected to avoid continuing stress on the affected part, and that symptoms and signs do not progress.

Surgical treatment is available for some disorders, such as carpal tunnel syndrome, cubital tunnel syndrome, and epicondylitis. However, the evidence base for selecting optimum treatment is very limited because most studies evaluating surgery, steroid injections, and other treatments suffer from inadequate design with regard to sample size, comparison groups and follow-up periods. Surgery may be only temporarily effective if the worker is returned to an ergonomically stressful job that has not been modified; possible loss of strength and flexibility and increased vulnerability to technical insult following surgery make it essential that job assignments be selected carefully to avoid recurrence. Use of transitional workshops for affected workers and graded retraining under the supervision of an experienced physical therapist may also reduce the risk of recurrence.

Further Reading

Bernard BP (Ed.). *Musculoskeletal Disorders and Workplace Factors: A Critical Review of Epidemiologic Evidence for Work-Related Musculoskeletal Disorders of the Neck, Upper*

Extremity, and Low Back. Cincinnati, OH: National Institute of Occupational Safety and Health, 1997.

Cailliet R. Soft Tissue Pain and Disability (3rd ed.). Philadelphia: F.A. Davis, 1996.

Gordon SL, Blair SJ, Fine LJ. *Repetitive Motion Disorders of the Upper Extremity.* Rosemont, IL: American Academy of Orthopaedic Surgeons, 1994.

Hagberg M, Silverstein BA, Wells RP, et al. (Eds.). *Work-Related Musculoskeletal Disorders (WMSD): A Handbook for Prevention.* London: Taylor and Francis, 1995.

ICOH Scientific Committee for Musculoskeletal Disorders. Musculoskeletal disorders: Work-related risk factors and prevention. *International Journal of Occupational and Environmental Health* 1996; 2: 239-246.

Mackinnon SE, Novak CB. Clinical commentary: Pathogenesis of cumulative trauma disorders. *Journal of Hand Surgery* 1994; 19A: 873-883.

Moore JS. Biomechanical models for the pathogenesis of specific distal upper extremity disorders. *American Journal of Industrial Medicine* 2002; 41: 353 369.

National Academy of Sciences. Panel on Musculoskeletal Disorders and the Workplace Commission on Behavioral and Social Sciences and Education, National Research Council. *Musculoskeletal Disorders and the Workplace: Low Back and Upper Extremities.* Washington, DC: National Academy Press, 2001.

Punnett L, Herbert R. Work-related musculoskeletal disorders: Is there a gender differential, and if so, what does it mean? In: MB Goldman, MC Hatch (Eds.). *Women Health.* San Diego, CA: Academic Press, 2000.

Sluiter JK, Rest KM, Frings-Dresen M Criteria document for evaluation of the work-relatedness of upper extremity musculoskeletal disorders. *Scandinavian Journal of Work, Envirnoment, and Health* 2001; 27 (Suppl 1): 1-102.

Zoonoses*

David J. Weber and William A. Rutala

Zoonoses are infectious diseases transmitted from animals to humans. Contact with animals is frequent among Americans, even in urban areas, for a number of reasons. First, Americans keep a wide variety of animals as household pets, including cats, dogs, birds, and fish, and, increasingly, more exotic animals, such as ferrets, monkeys, reptiles, rodents, wolves, and prairie dogs. In addition, a variety of farm animals may be kept as pets, such as cattle, chickens,

---

This chapter is derived, in part, from a review chapter, Zoonotic Infections, that was published in RL Lange (ed.). Animal Handlers. *Occupational Medicine* 1999;14:247-284.
*For ICD-10 codes, see Heymann DL (Ed.). *Control of Communicable Diseases Manual* (18th ed.). Washington, DC: American Public Health Association, 2004.

horses, pigs, and sheep. Second, leisure pursuits in rural areas, such as hunting, camping, spelunking, and hiking, are increasingly more popular and bring people into close contact with wild animals, arthropods, and potentially contaminated water supplies. Third, increasing numbers of Americans travel to remote areas of the world where they may be exposed to zoonotic diseases rarely seen in the United States. Fourth, immigrants from other countries may introduce zoonoses into the United States. Fifth, many Americans work in occupations involving direct contact with animals, such as farming, fishing, and animal management. Persons who handle agricultural products, animal hides, and other products of animal origin are also at increased risk of zoonoses. Finally, scientists using animals for research, health care workers caring for patients with zoonotic diseases, and laboratory personnel involved in isolating or culturing zoonotic pathogens may also acquire a zoonosis.

In recent years, scientists and public health officials have expressed concern regarding the potential threat from biological, chemical, and nuclear terrorism. This concern was realized in late 2001, when letters containing the bacteria that cause anthrax led to 22 cases of clinical anthrax in the United States. Many of the potential agents of highest concern for bioterrorism are zoonotic pathogens, including *Bacillus anthracis* (anthrax), *Yersinia pestis* (plague), *Francisella tularensis* (tularemia), filoviruses that cause Ebola and Marbury hemorrhagic fever, and arenaviruses that cause Lassa fever and Argentine hemorrhagic fever. All physicians and public health personnel should be knowledgeable regarding the clinical characteristics, isolation precautions, decontamination methods, diagnosis, pre- and post-exposure prophylaxis for exposures, and therapy of these agents.

This chapter reviews human infection acquired from animals, focusing on the most common occupational hazards. Other occupational hazards related to animal contact include allergic reactions, envenomation injuries or stings, and animal bites. Many of the occupations that bring people into contact with animals also feature contact with environmental reservoirs of infectious agents that may lead to disease from direct injury, such as tetanus and sporotrichosis, or inhalation, such as cryptococcosis and coccidioidomycosis.

Occurrence

More than 300 zoonotic pathogens have been described and new pathogens continue to be recognized because a microbial pathogen is newly isolated, its potential to cause human disease is newly recognized, or the pathogen has recently jumped species lines and now causes disease in humans. Zoonotic agents may be transferred to humans by contact, arthropod vectors, airborne, or common-vehicle routes. Contact includes direct contact, indirect contact, and droplet transmission. Direct contact includes animal bites, scratches, or direct contact with infected animal products, such as animal skins or fur. Droplet transmission is by large aerosolized particles that are infectious over a distance of less than 3 feet. Vector-borne diseases are transmitted by arthropod vectors, such as mosquitoes, ticks, flies, and other insects. Airborne diseases are transmitted by aerosols over long distances. Common-vehicle transmission is by food or water.

Understanding the epidemiology of zoonoses requires a detailed understanding of the reservoir and source of microbial agents. The reservoir is where the organism normally lives and reproduces. In the case of zoonotic diseases the reservoir is most commonly the animal host. The source is the means by which the infectious agent reaches the human. For most zoonotic infections, the normal life cycle does not involve humans. Rather, humans are accidental hosts, such as in rabies and toxoplasmosis, representing dead-end vectors. Some pathogens share maintenance of their life cycle with both animals and humans, such as those that cause schistosomiasis.

Factors affecting transmission of zoonotic diseases to humans include the geographical range of animal hosts that serve as reservoirs, the number and type of animal hosts, frequency of activities that bring people into contact with infected animals, prevalence of infection in host animals, the means by which infection can be transmitted from an infected animal to humans, and susceptibility of humans to infection. For vector-borne zoonoses, the epidemiology is critically dependent on the biology of the vector.

Zoonoses may be classified by phylogeny, mode of transmission, and occupations at risk of infection (Table 1). Given the many zoonoses, it is impossible to comprehensively review all of them. For this reason, diseases are grouped below by major mode of transmission.

Animal Bites

Animal bites represent an important hazard for workers in many occupations, including animal control officers, mail carriers, and veterinarians.

The infection rate from penetrating dog bites has generally been reported in the range of 5% to 15%. Cat bites are more likely to become infected than dog bites. About 1% of rat bites become infected. The incidence of infection following the bite of other animals is unknown. Factors that increase the risk of infection include full-thickness puncture; wounds involving the joints, tendons, ligaments, or fractures; and wounds in patients with impaired host defenses. Many pathogens may infect animal bites. However, a few generalizations can be made:

- Dog bites most commonly result in infections due to *Pasteurella spp.*, *Streptococcus spp.*, and *Staphylococcus spp.* Less commonly, a dog bite may cause an infection due to *Capnocytophaga canimorsus*. Infections with *C. canimorsus* generally have generally occurred in adults with impaired host defenses, most commonly splenectomy; the case fatality rate has been approximately 25%. The clinical syndrome is characterized by fever, disseminated intravascular coagulation, thrombocytopenia, hypotension, and renal failure.

Table 1. Selected Medically Important Diseases Acquired From Animals

Disease	Pathogen	Transmission	Persons At Risk
Viral Diseases			
Bovine popular stomatitis	Parapoxvirus	DC	I, II
California/La Crosse encephalitis	Bunyavirus	V	I, III, rural public
Herpesvirus simiae	Vessiculovirus	DC	IV, V
Colorado tick fever	Orbivirus	V	I, III
Eastern equine encephalitis	Alphavirus	V	III, V, VI, public
Lymphocytic choriomeningitis	Arenavirus	I, R	I, IV, V, public
Milker's nodule	Parapoxvirus	DC	I, II
Newcastle disease	Paramyxovirus	DC, R	I, II, IV, V
Orf (contagious ecthyma)	Parapoxvirus	DC	I, II
Rabies	Lyssavirus	DC, R (rare)	III, VI, public
Rotavirus	Rotavirus	I	I, III, IV, IX, public
Sin Nombre virus	Hantavirus	R	I, III, IV, VI, rural public
St. Louis encephalitis	Flavivirus	V	I, public
Vesicular stomatitis virus	Vesiculovirus	DC, R	I, V
Venezuelan equine encephalitis	Alphavirus	V	III, V, VI, public
West Nile encephalitis	Flavivirus	V	III, V, VI, public
Western equine encephalitis	Alphavirus	V	III, V, VI, public
Bacterial Diseases			
Aeromonas	*Aeromonas* spp.	DC, R	III, IV, VIII, public
Anthrax (Woolsorter's diseases)	*Bacillus anthracis*	DC, IC, R	I, II, V
Brucellosis	*Brucella* spp.	I, R	I, II, III, V
C. canimorsus sepsis	*Capnocytophaga canimorsus*	DC	III, IV, IX, public
Campylobacteriosis	*Campylobacter jejuni*	I	I-VI
Cat scratch fever	*Bartonella henselae*	DC	IV, IX, public
Edwardsiella infection	*Edwardsiella tarda*	DC, I	IV, VIII
Haverhill or rat bite fever	*Streptobacillus moniliformia*	DC	Public
Leptospirosis (Weil's disease)	*Leptospira interrogans* spp.	DC, I, R (rare)	I-V
Listeriosis	*Listeria monocytogenes*	I	IX, public
Lyme disease	*Borrelia burgdorferi*	V	I, III, IV, VI, public
Pasteurellosis	*Pasteurella multocida*	DC, R	III, IV, IX, public
Plague	*Yersinia pestis*	R, V	III, IV, V, VII
Plesiomonas infection	*Plesiomonas shigelloides*	I	VIII
Psittacosis	*Chlamydia psittaci*	R	I-VI
Q fever	*Coxiella burnetii*	R	I, II, V
Relapsing fever	*Borrelia* spp.	V	I, II, IV
Rocky Mountain Spotted Fever	*Rickettsia Rickettsii*	V	I, III, IV, VI, IX, public
Salmonellosis	*Salmonella enteriditis*	I	I-VI, VIII, IX
Staphylococcal infection	*Staphylococcus aureus*	DC	I, II, IV, IX
Streptococcal infection	Streptococcus, group A	R	I, II, IV, public
Tuberculosis	*M. bovis/M. tuberculosis*	I, R	I, II, IV, V, IX
Tularemia	*Francisella tularensis*	DC, I, R, V	I, II, V
Vibriosis	*Vibrio parahemolyticus*	I	III, VIII

(continued)

Table 1. *(cont.)* Selected Medically Important Diseases Acquired From Animals

Disease	Pathogen	Transmission	Persons At Risk
Bacterial Diseases (cont.)			
V. vulnificus infection	*Vibrio vulnificus*	DC, I	VIII, IX
Yersiniosis	*Yersinia enterocolitica*	I	I-V, VIII
Fungal Disease			
Ringworm	*Dermatophytes*	DC	I-VI
Parasitic Diseases			
Acanthamoebiasis	*Acanthamoeba* spp.	R	III
Babesiosis	*Babesia microti*	V	III, IX
Cryptosporidiosis	*Cryptosporidia* spp.	I	I-IV, VI, IX, public
Dipylidiasis	*Dipylidium caninum*	I	IV
Dirofilariasis	*Dirofilaria immitis*	V	III
Echinococcosis	*Echinococcus granulosus*	I	I
Giardiasis	*Giardia lamblia*	I	I, III, IV, public
Malaria	*Plasmodium* spp.	V	I, III, IV, VI, VII, public
Naegleriasis	*Naegleria fowleri*	R	III
Toxocariasis	*Toxocara canis, T. cati*	I	IV
Toxoplasmosis	*Toxoplasma gondii*	I	IV, IX

Transmission: DC = infection acquired by direct contact (animal bite, animal scratch, contact with live animal); IC = infection acquired via indirect contact (hides, wool, hair); I = infection acquired via ingestion; R = infection acquired via inhalation (droplet or airborne transmission); V = infection acquired via bite of arthropod vector.

Persons At Risk: Group I (Agriculture) = Farmers or other people in close contact with livestock and their products; Group II (Animal-product processing and manufacture) = All personnel of abattoirs and of plants processing animal products or byproducts; Group III (Forestry, outdoors) = Persons frequenting wild habitats for professional or recreational reasons; Group IV (Recreation) = Persons in contact with pets or wild animals in the urban environment; Group V (Clinics, laboratories) = Health care personnel who attend patients, and health workers (including laboratory personnel) who handle specimens, corpses, or organs; Group VI (Epidemiology) = Public health professionals who do field research; Group VII (Emergency) = Public affected by catastrophes, refugees, or people temporarily living in crowded or highly stressful situations; Group VIII (Fishermen) = People catching or cleaning fish or engaging in recreational activities in the water; Group IX (Immunocompromised hosts) = People who are immunocompromised because of acquired immunodeficiency, cancer chemotherapy, organ transplants, immunosuppressive medication, liver and/or renal disease.

- Cat bites most commonly result in infections due to *P. multocida.*
- Small rodents may transmit the agents of rat bite fever, *Streptobacillus moniliformis*, and *Spirillum minor.*
- Aquatic animals most commonly transmit *Aeromonas hydrophila* or *Edwardsiella tarda.*
- Sea mammals may transmit the agent of "seal bite finger," *Mycoplasma phocacerebrale.*

- Nonhuman primate bites may result in infections due to herpes B virus, monkeypox, Ebola virus, Marburg virus, *Eikenella corrodens*, *Corynebacterium spp.* and α-hemolytic streptococci. More than 25 cases of herpes B virus have been reported with a case-fatality rate greater than 50%.

Rabies

Rabies ranks 10th worldwide as a cause of mortality, with an estimated 50,000 to 60,000 deaths annually, despite the availability of effective vaccines for post-exposure treatment. In addition, millions of persons, primarily in developing countries, undergo costly post-exposure treatment. In the United States, rabies continues to be endemic in many wild animals. Although human rabies is rare in the United States, post-exposure rabies prophylaxis is likely provided to more than 35,000 people per year at a cost of between $230 million and $1 billion. Rabies is primarily a disease of animals. The epidemiology of human rabies is a reflection of both the distribution of the disease in animals and the degree of human contact with these animals. The following animals very rarely have been found to be rabid and hence their bites almost never require post-exposure prophylaxis for rabies: squirrels, hamsters, guinea pigs, gerbils, chipmunks, rats, mice, rabbits, hares, and other small rodents. During the past half-century, the number of humans dying of rabies in the United States has dramatically decreased to an average of 2 to 3 per year. In the United States, the number of reports of wildlife rabies far exceed those of rabies in domestic animals; rabies variants in bats are associated with a disproportionate percentage (about 90%) of infections in humans, although bats constitute only about 15% of all reported rabies cases in animals. Most other human rabies cases diagnosed in the United States are attributable to infections acquired in areas of enzootic canine rabies outside the United States; most persons with a case of rabies that originated in the United States have no history of an animal bite; rabies is diagnosed after death in more than one-third of these individuals.

Direct Contact with Animal Products

Direct contact with animals or animal products (hides, hair, or bone) may lead to transmission of a variety of diseases if contaminated animal products come into contact with nonintact skin or mucous membranes. Diseases transmitted by direct contact include cutaneous anthrax, brucellosis, erysipeloid, leptospirosis, listeriosis, melioidosis, orf, plague, and tularemia. Direct contact may also lead to infection with ectoparasites and fungi.

Vector-Borne Infections

In recent years, vector-borne infections have increasingly burdened the health of the residents of North America. For example, West Nile virus was first detected in the New York area in 1999. By 2002, this virus had spread across the continental United States and was responsible in that year for 4,156 human clinical infections leading to 284 deaths. Changes in recreational activity and residential placement have exposed people to a diverse array of novel enzootic infectious agents. In addition, increases in foreign travel have led to isolated clinical cases of diseases due to exotic "tropical" infectious agents and vectors into North America.

Vector-borne diseases may be classified by vector (mosquito or tick), phylogeny (viruses, bacteria, or parasites), or clinical illnesses (encephalitis or hemorrhagic fever).

Arthropod-borne viruses (arboviruses), a subset of vector-transmitted zoonoses, are an ecologically defined set of viruses that have in common (a) replication in both arthropods and vertebrate hosts; and (b) transmission between vertebrate animals by the bite of mosquitoes, ticks, sandflies, and midges. Most arboviruses belong to Togaviridae, Flaviviridae, Bunyaviridae, Reoviridae, or Rhaboviridae. Arboviruses are usually maintained in a reservoir cycle that consists of both arthropod and vertebrate hosts. Some arboviruses are transmitted vertically through the egg of the arthropod and, in these cases of transovarial transmission, the arthropod alone may be the reservoir of the virus and may maintain the virus in the absence of a vertebrate animal. More than 500 arboviruses and related zoonotic agents transmitted directly from vertebrates have been recognized worldwide. More than 50 have been isolated in North America, of which more than 15 are known to cause human illness.

Mosquito-Borne Diseases

Mosquito-borne diseases in the United States include a variety of arboviral diseases, malaria, and dirofilariasis. In the past, mosquito-borne dengue and yellow fever were major problems.

In the United States, arbovirus encephalitis viruses of concern include Eastern equine encephalitis (EEE), Western equine encephalitis (WEE), California encephalitis (CE), St. Louis encephalitis (SLE), West Nile encephalitis (WNE), and Venezuelan equine encephalitis (VEE) viruses (Table 2). These diseases differ in their geographical localization, frequency, symptomatology, and mortality. The clinical illnesses resulting from these infections has been summarized as follows: asymptomatic infection (WEE, SLE, CE, and WNE); mild febrile illness (WEE, SLE, WNE, CE, and yellow fever); influenza-like illness with aching and joint pains (dengue); mild encephalitis (CE, WEE, and SLE); severe encephalitis (SLE, EEE, WEE, WNE, and CE); acute ascending paralysis (WNE); jaundice and proteinuria (yellow fever); rash, sometimes with hemorrhagic manifestations (dengue); and shock syndrome (dengue, following a second infection with a different dengue serotype).

Table 2. Important Arboviral Encephalitis Viruses of North America

Genus	Species	Peak Age Infected	Principal Distribution	Mortality
Flavivirus	St. Louis	>50	Entire US	~8%*
	West Nile	>50	Entire US	~10%*
Bunyavirus	California#	<15	Eastern and central US	~1%
	LaCrossse	Children	Ohio, Indiana, Minnesota, Wisconsin, southern US	<1%
Alphavirus	Western equine	Infants, young children	Western and central US	2-5%
	Eastern equine	Young children	Eastern US and Gulf coast	30-60%
	Venezuelan equine	Children	Florida, Louisiana, Texas	<1%

* Higher mortality in the elderly.
# The California group of viral encephalitis that occurs in the US includes Jamestown Canyon, Jerry Slough, Keystone, San Angelo, Trivittatus, and others.

Tick-Borne Disease

Ticks can transmit infections due to viruses, bacteria, rickettsia, and parasites, and can also account for toxic reactions (Table 3). The incidence of tick-borne diseases, especially Lyme disease, has been rising as a result of reforestation, movement of humans into rural areas, and increased recreational outdoor activities.

Ticks (phylum Arthropoda, class Arachnida, order Acarina) are obligate, blood-sucking parasites found worldwide. There are three families of ticks: two of these, the Ixodidae (hard ticks) and Argasidae (soft ticks), are known to cause human and animal disease. Starting as an egg, all ticks go through three stages during their growth: larva, nymph, and adult. Hard ticks, which require a blood meal at each stage for morphogenesis, generally remain attached to the host for hours or days at a time. Three genera of hard ticks are known to transmit disease to humans in the United States: Ixodes, Dermacentor, and Amblyomma. Soft ticks, which may have several nymphal stages, may take multiple blood meals, each meal usually lasting less than 30 minutes.

Table 3: Important Tick-borne Diseases of North America

Disease	Causative Agent	Principal Distribution
Babesiosis	*Babesia* spp. (protozoa)	Northeast US
Colorado tick fever	Family *Reoviridae* (virus)	Western US and Canada above 5000 feet
Human granulocytic ehrlichiosis	HME agent (related to veterinary pathogens, *E. equi* and *E. phagocytophila*)	Eastern, Upper Midwestern, South-central, West coast of US
Human monocytic ehrlichiosis	*Ehrlichia chaffeenis*	South-central and Southeast US (concentrated in Oklahoma, Texax, Arkansas, Missouri)
Lyme disease	*Borrelia burgdorferi* (bacteria)	Northeast, Mid-Atlantic, and North-central US
Powassan encephalitis	Flavivirus	Northern US, Southern Canada
Relapsing fever	*Borrelia* sp. (bacteria)	Western US and Canada
Rocky Mountain Spotted Fever	*Rickettsia rickettsii*	Entire US (concentrated in Mid-Atlantic, Southeast, and South-central states)
Tick paralysis	Toxin	Pacific Northwest, Rocky Mountain states, occasionally Southeastern states
Tularemia	*Francisella tularensis* (bacteria)	US (concentrated in Arkansas, Missouri, Oklahoma)

Respiratory Tract Infections Transmitted from Animals

A number of zoonotic agents transmitted from animals may produce respiratory infections (Table 4). Diseases caused by infections that are transmitted by the aerosol route include anthrax, brucellosis, plague, psittacosis, Q fever, Sin Nombre virus infection, tularemia, and severe acute respiratory syndrome (SARS). Other zoonotic agents that may involve the lungs include cat scratch fever, dirofilariasis, echinococcosis, ehrlichiosis, leptospirosis, melioidosis, pasteurellosis, Rocky Mountain spotted fever, toxocariasis, and toxoplasmosis. Few zoonotic diseases causing pneumonia are transmitted person-to-person. These include pneumonic plague and SARS.

Animals commonly kept as pets may transmit disease via the aerosol route, including *Chlamydia psittaci* from infected birds, *Coxiella burnetii* from infected cats, and *P. multocida* from close contact with cats or dogs. Although animal-to-human transmission of the environmental fungi *Histoplasma capsulatum* and *Cryptococcus neoformans* does not occur, common-source outbreaks involving both dogs and humans have been described. Although birds do not seem to be infected, *H. capsulatum* tends to concentrate in sites where bird and bat wastes accumulate; there is great human risk of acquiring infection by disturbing materials contaminated with bird or bat wastes, so animals are indirectly responsible for cases of histoplasmosis.

Table 4. Characteristics of Zoonotic Pathogens That Cause Pneumonia

Agent	Disease	Epidemiologic Associations	Distribution
B. anthracis	Inhalation anthrax	Imported animal hides, raw wool, sick domestic animals	Worldwide, warmer latitudes of US
Brucella sp.	Brucellosis	Food animals and product handling, ingestion, inhalation	Worldwide, especially where animals are not immunized
C. psittaci	Psittacosis	Bird exposure—pet and pet shop, veterinarians, turkey farmers	Worldwide in parrots, song birds, pigeons, and poultry, especially turkeys and ducks
C. burnetti	Q fever	Inhaled endospores from animal contaminated soil; cat afterbirth, ticks	Worldwide, especially where domestic animals are raised
F. tularensis	Tularemia	Aerosol from dead birds, animals; bacteremia spead from bubo; ticks and biting flies	North America, Europe, Asia in wild animals and birds, occasionally in domestic animals
L. interrogans ss.	Leptospirosis	Domestic and wild animals, farmers, veterinarian, contaminated water	Worldwide in many wild and domestic animals
P. multocida	Pasteurellosis	Underlying respiratory disease; contact with cat, dog in home	Worldwide in felines, birds, other mammals
P. pseudo-mallei	Melioidosis	Penetrating injury in endemic area in rodent-contaminated soil or water	Latitude 20°N to 20°S, especially Asia
R. rickettsii	Rocky Mountain spotted fever	Tick associated, typical rash	North America in local areas, especially mid-Atlantic states
T. gondii	Toxoplasmosis	Contact with domestic food animals and pets, ingestion cysts, pneumonia in immuno-compromised persons	Worldwide in cats and other mammals
Y. pestis	Plague	Contact with mammals and fleas; veterinarians; and outdoor activities in endemic area; cats	Worldwide including Asia, Southwest United States

The etiologic agents of human tuberculosis include *M. tuberculosis, M. bovis, M. africanum*, and *M. microti*. The main reservoir for *M. tuberculosis* is the human, but other animals can become infected, including monkeys, large apes, dogs, and cats. Rarely animal-to-human transmission of *M. tuberculosis* has been reported. The main reservoir of *M. bovis* infection in mammals is cattle, but badgers, bison, camels, deer, ferrets, foxes, nonhuman primates, opossums, rats, and other animals have been found to be infected. Transmission to humans is usually via milk from infected cattle. However, pulmonary tuberculosis due to *M. bovis* acquired via airborne transmission is an occupational zoonosis in abattoir and farm workers. An outbreak of *M. bovis* has been reported in which zoological workers caring for an infected rhinoceros developed infection (as measured by PPD conversion), but not disease with *M. bovis*. Pasteurization and destruction of infected cattle have eliminated *M. bovis* infection in the United States, but worldwide *M. bovis* infection remains a major public health problem. Although cats and dogs can be infected with *M. bovis*, these species do not represent an epidemiologically significant source of infection.

Gastrointestinal Infections Transmitted from Animals

Many gastrointestinal pathogens of humans have animal reservoirs. These pathogens may be acquired via ingestion of contaminated surface waters, raw milk, and uncooked or undercooked foods, such as shellfish, poultry, and meat. They may also be acquired by bathing in contaminated waters and with ingestion of water. Enteric bacterial pathogens acquired from animals include: *Aeromonas hydrophila, Plesiomonas shigelloides, Campylobacter jejuni, Edwardsiella tarda, Salmonella spp., Yersinia enterocolitica*, and *Yersinia pseudotuberculosis*. Parasitic pathogens include *Cryptosporidium parvum* and *Dipylidium caninum* (humans acquire infection by ingesting fleas, which are the intermediate hosts).

Prevention

Preventing acquisition of zoonoses relies on several general preventive measures including appropriate immunizations; use of personal protective equipment to prevent diseases transmitted by direct contact (gloves, gowns, boots) or aerosols (mask, goggles); and proper clothing, bed netting, and repellents for arthropods. Post-exposure prophylaxis is available for several zoonoses, including animal bites (rabies vaccine, tetanus toxoid vaccine, antibiotics), anthrax (doxycycline, ciprofloxacin), herpes B (antiviral agents), monkeypox (vaccinia vaccine), plague (doxycycline, ciprofloxacin), and rabies (HRIG and rabies vaccine).

Immunizations

Immunizations are available to prevent several zoonotic infections including anthrax, Japanese encephalitis, rabies, typhoid, and yellow fever.

The Lyme disease and plague vaccines are no longer available in the U.S. Cholera vaccine is not recommended for persons in the U.S. since there are no endemic foci of infection here; it is not recommended for travelers because the vaccine is only marginally effective. Japanese encephalitis and yellow fever vaccines are recommended only for certain travelers to endemic countries and to laboratory personnel who might be exposed to wild-type and vaccine strains of either Japanese encephalitis or yellow fever. Although tetanus is not a zoonosis, tetanus has followed animal bites or scratches and environmental injury. In addition to being a "universally recommended" vaccine, all persons working outdoors or with animals should receive appropriate tetanus immunizations. (General guidelines regarding the indications, contraindications, administration, and side effects of immunizations are available at the CDC web page from the Advisory Committee on Immunization Practices [ACIP] http://www.cdc.gov/nip/publications/ACIP-list.htm.)

Pre-exposure prophylaxis with the rabies vaccine is recommended for persons at high risk of exposure as defined by occupation and level of endemicity in the animal population (see CDC web page). Local health authorities should be consulted regarding the risk of acquiring rabies by persons with exposure to wild or domestic animals. Pre-exposure prophylaxis is given because (a) it may provide protection to persons with inapparent exposure to rabies; (b) it may protect persons whose post-exposure therapy might be delayed; and (c) although pre-exposure vaccination does not eliminate the need for additional therapy after a rabies exposure, it simplifies therapy by eliminating the need for HRIG and decreasing the number of doses of vaccine needed. The frequency of booster doses and need for post-immunization titers to assure protection depends on the risk of acquiring rabies. Pre-exposure prophylaxis against rabies does not alleviate the need for post-exposure prophylaxis following potential exposure to the virus or an infected animal.

Prevention of Laboratory-Acquired Disease

The key principle of biosafety is "containment," a collection of engineering controls designed to allow the safe handling of infectious materials in the laboratory environment. Primary containment, the protection of personnel and the immediate laboratory environment from exposure to infectious agents is provided by good microbiologic techniques and the use of appropriate safety equipment (such as biological safety cabinets and enclosed containers). Pre-exposure immunization may be available and recommended (such as with vaccinia vaccine for people working with monkeypox). Secondary containment, the protection of the environment external to the laboratory from exposure to infectious material, is provided by a combination of facility design and operational practices (such as specialized ventilation systems to ensure directional airflow and controlled access zones).

The U.S. Public Health Service has provided excellent guidelines for assuring the safe handling of microbes (see Further Reading list). These guidelines can be customized for each individual laboratory and utilized with other scientific information to help prevent laboratory-associated infections. Practices, procedures, and facilities differ, depending on how an agent is being manipulated. These guidelines group microbes into four categories, depending on several factors, including pathogen virulence, modes of transmission, and availability of vaccine and treatment. The pathogen group then defines four levels of recommended biosafety (BSL-1 to BSL-4) that require increasingly elaborate primary and secondary containment.

- BSL-1 practices, safety equipment, and facility design and construction are appropriate for undergraduate and secondary educational training and teaching laboratories, and for other laboratories in which work is done with defined and characterized strains of viable microorganisms not known to consistently cause disease in health adult humans, such as infectious canine hepatitis virus.

- BSL-2 practices, safety equipment, and facility design and construction are applicable to clinical, diagnostic, teaching, research, or production facilities in which work is done with the broad spectrum of indigenous moderate-risk agents that are present in the community and associated human disease of varying severity, such as *Salmonella* and *Toxoplasma* species. Primary hazards to personnel working with these agents relate to accidental percutaneous or mucous membrane exposures, and ingestion of contaminated materials.

- BSL-3 practices, safety equipment, and facility design and construction are applicable to clinical, diagnostic, teaching, research, or production facilities in which work is done with indigenous or exotic agents with a potential for respiratory transmission, and which may cause serious and potentially lethal infection, such as St. Louis encephalitis virus and *Coxiella burnetii*. Primary hazards to personnel working with these agents relate to autoinoculation, ingestion, and exposure to infectious aerosols.

- BSL-4 practices, safety equipment, and facility design and construction are applicable for work with dangerous and exotic agents that pose a high individual risk of life-threatening disease, which may be transmitted via the aerosol route and for which there is no available vaccine or therapy, such as the Marburg and Ebola viruses. The primary hazards to personnel working with BSL-4 agents are respiratory exposure to infectious aerosols, mucous membranes or non-intact skin exposure to infectious droplets, and autoinoculation.

Prevention Against Contact-Transmitted Diseases

Animal bites frequently lead to infection and efforts should be made to avoid the bites or scratches of pets, farm animals, and wild animals. Understanding the psychology of cats and dogs and avoiding provocative behavior, such as interference with feeding, interactions with young animals being cared for by their mother, intrusions into "home" territory, petting, and threatening behaviors, is important in preventing attacks. Appropriate herding and capture techniques should be applied with farm and wild animals.

Persons dealing with farm or wild animals or their products, such as hides or wool, should wear appropriate protective garments including gloves and clothes, and, when appropriate, masks to prevent exposure to potential pathogens.

Prevention Against Mosquitoes and Other Arthropod Vectors

Although vaccines or chemotherapeutic drugs are available against some important vector-borne diseases, such as Japanese encephalitis, malaria, and yellow fever, there are none against most mosquito-borne diseases, such as Eastern equine encephalitis, and most tick-borne diseases, such as Rocky Mountain spotted fever and ehrlichiosis. In addition, the effectiveness of malaria chemoprophylaxis is variable, depending on patterns of resistance and compliance with medication. Therefore, travelers and persons working outdoors in locations where transmission of arthropod-borne diseases may occur should avail themselves of repellents and other general protective measures against arthropods.

General Preventive Measures

Excellent general guidelines for the prevention of arthropod-borne diseases are available from the CDC. Although designed for the traveler, they also provide useful guidance for persons whose occupations require outdoor work, especially in wooded areas or forests. The CDC guidelines state that the principal approach to prevention of vector-borne diseases is avoidance. Tick- and mite-borne infections characteristically are diseases of "place"; whenever possible, known foci of disease transmission should be avoided. Although many vector-borne infections can be prevented by avoiding rural locations, certain mosquito- and midge-borne arboviral and parasitic infections are transmitted seasonally, and simple changes in itinerary can greatly reduce risk for acquiring them. However, certain mosquito- and midge-borne arboviral and parasitic infections are preferentially transmitted in urban areas and around human residences. In the United States, most arthropod-borne illnesses are transmitted in late spring through early fall, and changes in work or leisure activities in these time periods may reduce exposure.

Prevention of Mosquito-Borne Diseases

Exposure to arthropod bites can be minimized by modifying patterns of activity or behavior. Some vector mosquitoes are most active in twilight periods at dawn and dusk or in the evening. Avoidance of outdoor activity during these periods may reduce the risk of exposure. Wearing long-sleeved shirts, long pants, and hats will minimize areas of exposed skin. Shirts should be tucked in. Repellents applied to clothing, shoes, tents, mosquito nets, and other gear will enhance protection.

When accommodations are not adequately screened or air-conditioned, bed nets are essential to provide protection and comfort. Bed nets should be tucked under mattresses and can be sprayed with a repellent, such as permethrin. Aerosol insecticides and mosquito coils may help clear rooms of mosquitoes.

Prevention of Tick-Borne Diseases

Individuals living in endemic areas should attempt to minimize exposure to ticks by avoiding tick-infested areas, using tick repellents, and keeping pets tick-free. When exposure to mites or ticks is a possibility, pants should be tucked into socks and boots should be worn; sandals should be avoided. However, protective clothing is useful, but may be impractical during hot weather. An effective preventive measure is careful examination twice per day of the body, including scalp, pubic area, axillary hair, and other anatomic crevices, with prompt removal of ticks. Removal is best accomplished by grasping the head of the tick with a forceps and gently pulling until removed. Care should be taken to avoid crushing attached ticks, spraying blood from engorged ticks, and excoriating the area. Following removal, the area should be cleaned with soap and water or a disinfectant.

Repellents

Most authorities recommend repellents containing DEET (*N,N*-diethylmetatoluamide) as the active ingredient. DEET repels mosquitoes, biting fleas, gnats (black flies or midges), chiggers, ticks, and other insects. DEET may be applied to skin or clothing. Although DEET is safe for cotton, wool, and nylon fabrics, it may damage spandex, rayon, acetate, and pigmented leathers. In addition, DEET can dissolve plastic and vinyl, such as in eyeglass frames. In flammability tests, polyester acrylic cloth treated with some DEET products ignited immediately when near a flame. Formulations containing less than 50% DEET are recommended because the additional gain in repellent effect with higher concentrations is not significant, when weighed against the potential for toxicity. In general, the more DEET a repellent contains, the longer time it can protect against mosquito bites. A microencapsulated, sustained-release formulation can have a longer period of activity than liquid formulations at the same concentrations. Length of protection also varies with ambient temperature, amount of perspiration, any water exposure, abrasive removal, and other factors. Overall, DEET appears to be remarkably free of serious toxicity when applied to the skin. It is toxic when ingested and high concentrations applied to the skin may cause blistering. Consumers who apply both a DEET-based repellent and a sunscreen should be aware that the repellent might reduce the effectiveness of the sunscreen. Users should adhere to precautions and recommendations for use available in the package inserts.

Permethrin-containing repellents are recommended for use on clothing, shoes, bed nets, and camping gear. Permethrin is highly effective as an insecticide/acaricide and as a repellent. Permethrin-treated clothing repels and kills ticks, mosquitoes, and other arthropods, and retains this effect after repeated laundering. However, the insecticide should be reapplied after every five washings. There appears to be little potential for toxicity from permethrin-treated clothing. While permethrin-containing shampoo and cream are available for treatment of head lice and scabies infestations and potentially could be useful as repellents when applied to skin or hair, the CDC and most authorities recommend the use of repellents containing DEET for this purpose.

Prevention Against Contaminated Food and Water

Ingestion of contaminated food and water are common sources for the introduction of infection into the body by international travelers and persons working in or enjoying rural areas, such as hunters, hikers, and campers. Persons who may become contaminated with potential zoonotic pathogens as a result of their work, such as diary farmers and abattoir workers, should carefully wash their hands, preferably with an antibacterial agent before eating or drinking.

Water

Water that has been adequately chlorinated, using minimum recommended waterworks standards as practices in the United States, will afford significant protection against viral and bacterial waterborne diseases. However, chlorine treatment alone, as used in the routine disinfection of water, may not kill some enteric viruses and the parasitic organisms that cause giardiasis, amebiasis, and cryptosporidiosis. Where water may be contaminated, ice should also be considered contaminated and should not be used in beverages.

Boiling is the most reliable method to make water of uncertain purity safe for drinking. Water should be brought to a vigorous rolling boil for 1 minute and allowed to cool to room temperature (without adding ice). At altitudes above 2,000 meters (about 6,500 feet), for an extra margin of safety, one should boil water for 3 minutes or use chemical disinfection after the water has boiled for 1 minute. Chemical disinfection with iodine is an alternative method of water treatment when it is not feasible to boil water. Disinfection may be achieved by use of tincture of iodine (2% solution; add 0.25 mL [5 drops] to clear water or 0.5 mL [10 drops] to cold or cloudy water) and letting clear water stand for 30 minutes, or very turbid or very cold water stand for up to several hours prior to use. Alternatively chemical disinfection may be achieved by the use of tetraglycine hydroperiodide tablets (following manu-facturer's recommendations). A variety of commercially available portable filters will provide various degrees of protection against microbes. Persons using filters should follow manufacturer's recommendations. Proper selec-tion, operation, care, and maintenance of water filters are essential to pro-ducing safe water.

Food

Contaminated food may also serve as a source for infection. Raw food is especially vulnerable to contamination. In areas where hygiene and sanita-tion are inadequate, persons should be advised to avoid salads, uncooked vegetables, and unpasteurized milk and milk products such as cheese, and to eat only food that has been cooked and is still hot, or fruit that has been just peeled. Some species of fish and shellfish can contain poisonous biotoxins, even when well cooked; for example, ciguatera fish poisoning can be due to barracuda, red snapper, grouper, amberjack, sea bass, and a wide range of tropical reef fish.

In order to prevent acquisition of gastrointestinal pathogens from farm and wild animals at risk, persons should adhere to the above recommenda-tions when applicable. Water obtained from potentially contaminated sites should also be treated, either by boiling or adding appropriate levels of chem-icals. Persons handling animals should always wash their hands before and after animal contact. Human food should always be stored to prevent con-tamination by animals or their excreta. All food should be appropriately stored, refrigerated, and cooked. When possible, persons should only swim in water evaluated by local public health authorities for evidence of contam-ination.

Prevention Against Aerosol-Transmitted Infections

Persons in environments in which inhalation of infected aerosols is pos-sible should wear appropriate respiratory protective devices. Examples include spelunkers entering caves containing bats, and persons working with wool or hides that might be infected with anthrax. Of the zoonotic agents that cause pneumonia, only *M. bovis*, plague, and Q fever may be transmitted between humans. It is recommended that persons with pneumonic plague be placed on droplet precautions when being cared for in healthcare facilities.

Cases of brucellosis, glanders, tularemia, plague, Q fever, and those due to many other zoonotic agents now need to be reported to appropriate state health departments under the Select Agent rule.

Further Reading

Acha PN, Szygres B. *Zoonoses and Communicable Diseases Common to Man and Animals* (2nd ed.). Washington, DC: Pan American Health Organization, 1989.

Bell JC, Palmer SR, Payne JM. *The Zoonoses.* London: Edward Arnold, 1988.

Brown C, Bolin C. *Emerging Infectious Diseases of Animals.* Washington, D.C: American Society of Microbiology, 2000.

Centers for Disease Control and Prevention. *Health Information for International Travel, 2003-2004.* Atlanta: US Department of Health and Human Services, Public Health Service, 2003.

Heymann DL (Ed.). *Control of Communicable Diseases Manual* (18th Ed.). Washington, DC: American Public Health Association, 2004.

Langley RL. Animal handlers. *Occupational Medicine* 1999; 14: 1-478.

Palmer SR, Soulsby L, Simpson DIH. *Zoonoses.* Oxford: Oxford Press, 1998.

Schlossberg D. *Infections of Leisure.* Washington, DC: American Society of Microbiology, 1999.

U.S. Department of Health and Human Services. *Biosafety in Microbiological and Biomedical Laboratories* (4th Ed.). Washington, DC: US Government Printing Office, 1999.

PART III

Special Topics

Chapter 6
Taking an Occupational History

Barry S. Levy and Rose H. Goldman

Recognition of occupational diseases and injuries depends on obtaining an occupational history. An occupational history is taken in the context of a complete medical history, which includes a description of the patient's chief complaint and other current medical problems, past medical history, review of systems, family history, residential history, current medications, and personal habits, including use of tobacco products, alcohol, and recreational drugs, as well as hobbies and other nonoccupational activities. The minimal components of a screening occupational history could include attention to the chief complaint (or diagnosis) for clues suggesting a relationship to activities at work or at home; a list of current and longest held jobs; a current job description; and questions about symptoms during or after exposures to fumes, dusts, chemicals, loud noise, radiation, or musculoskeletal stresses. Findings from the clinical encounter may prompt the clinician to take a more thorough occupational history. Such triggers could include a suggestive temporal relationship between an exposure and symptoms; a "sentinel health event" or diagnosis that has been strongly linked to exposures, such as pulmonary tuberculosis, asthma, contact dermatitis, bladder cancer, peripheral neuropathy, or pulmonary fibrosis; the occurrence of an illness in an unexpected person, such as lung cancer in a nonsmoker; and symptoms without any other clear etiology or diagnosis.

A detailed occupational history contains the following information. A standardized form may be useful in recording this information. (See an example of a standard form at the end of this chapter.)

Description of All Jobs Held

Starting with the current or last (previous) job, determine for each job: name of employer and type of industry, title and description of job and tasks performed, and starting and ending dates. Obtain information on routine tasks as well as overtime and unusual tasks. If workers have multiple tasks, have them describe a typical workshift and have them simulate the motions of complex tasks. Since the physician or other health care provider is often not knowledgeable about technical details of the person's work, it is important to have the person explain work tasks in detail. Visiting the current workplace may be helpful. In the list of past jobs, include military service and part-time jobs, even part-time and summer jobs while in school, and work as a homemaker and parent. Also include jobs in the informal sector of the economy. Some workers, such as workers in the construction trades and temporary office workers, work at many different workplaces and therefore one may be able to obtain only an overview of their jobs and information on the most hazardous workplaces that they can specifically recall.

Review of Occupational Exposures

Obtain information on known or suspected hazardous exposures in all jobs held. Include chemical exposures, physical exposures (such as loud noise, vibration, radiation, and temperature extremes), biologic exposures (to infectious agents), and psychosocial exposures (to working conditions and situations that provoke increased stress). Obtain information on both routine exposures as well as accidents and unusual incidents, such as spills and machinery malfunctions.

To what was the worker exposed? Often the worker may know the substance only by a brand name, slang term, or code number, or may only be able to describe its color, odor, and other characteristics. If certain exposures are suspected by the interviewer, he or she should specifically ask the worker if those exposures were present. Sometimes it is helpful to ask questions from a list of general or industry-specific chemicals that has been prepared in advance. It is typical for workers to be exposed to multiple hazards at work. The availability of material safety data sheets (MSDSs), along with reliance on workers' rights guaranteed by the federal Hazard Communication Standard, can often facilitate access to detailed information on hazards to which a worker has been or is being exposed.

When, for how long, and in what manner was the worker exposed? Describe worker inhalation, ingestion, and/or skin contact with the hazard that occurs routinely as well as unusual exposures, such as when hazardous liquids may have been splashed into the eyes or onto the skin or mucous membranes.

What was the level of exposure? Workers are unlikely to be able to exactly quantitate exposures, but usually can provide descriptions of the intensity of the exposure, such as, "It was so dusty that I couldn't see the wall at the other end of the building 100 feet away." In addition, workers may be aware that OSHA or other governmental agencies have obtained quantitative measurements of air contaminants or other hazards in the workplace.

Was personal protective equipment (PPE), such as gloves, masks and respirators, safety glasses and goggles, and protective clothing and footwear, available and used? If it was, determine what types of PPE were used, and under what circumstances and for how long they were used. Determine if PPE fit properly. Determine if respirator cartridges were changed at appropriate intervals.

Also determine if the worker washed hands, showered, and changed workclothes before going home.

Did the worker eat, drink, or smoke in his or her work area? If so, for how long and how frequently?

Timing of Symptoms in Relation to Exposures

Determine any temporal relationship between symptoms and work exposures. Do the symptoms decrease or clear entirely during weekends and vacation periods only to recur on return to work? This pattern often occurs when symptoms are due to chemicals, such as organic solvents, that are easily absorbed and have short biologic half-lives.

Presence of Similar Illness Among Co-workers

Determine if there may be similar illnesses or symptoms among other workers at the same workplace or at other workplaces of the same employer. If similar illnesses or symptoms are found among co-workers, what work tasks and what work exposures do they share in common? A number of occupational diseases have been discovered by clinicians who recognized such clusters or illness and undertook or facilitated further investigations to elucidate their causes.

Nonoccupational Exposures

It is important to question the worker about nonoccupational exposures, such as those that may occur in the home environment, while engaged in hobby activities at home or elsewhere, and use of medications, tobacco, alcohol, recreational drugs, and other substances. As with other elements of the occupational history, one should obtain information on the amount, frequency, and timing of these other exposures.

Synthesis and Use of Information

Once having obtained an occupational history, it is important to synthesize this information and correlate it with the individual's health problem, additional information concerning work history and workplace exposures, and relevant scientific and medical literature in order to determine (a) whether or not the individual's illness or injury has been initiated or promoted by his or her work, and, if so, (b) what further investigation should be performed and what intervention, and/or preventive measures should be implemented.

Further Reading

Levy BS, Wegman DH, Halperin WE. Recognizing occupational disease and injury. In Levy BS, Wegman DH (Eds.) *Occupational Health: Recognizing and Preventing Work-Related Disease and Injury* (4th ed.). Philadelphia: Lippincott Williams & Wilkins, 2000, pp. 99–122.

Goldman RH. Suspecting occupational disease: The clinician's role. In McCunney RJ (Ed.), *A Practical Approach to Occupational and Environmental Medicine*. Philadelphia: Lippincott Williams & Wilkins, 2003, pp. 279–294.

OCCUPATIONAL HISTORY FORM

I. IDENTIFICATION

Name: _____ Soc. Sec. No.: _____

Address: _____ Sex: M F

_____ Birthdate: _____

Telephone: Home _____ Work _____

II. OCCUPATIONAL PROFILE

Fill in the table below listing all jobs at which you have worked, including short-term, seasonal, and part-time employment, and including military service. Start with your present job and go back to the first. Use additional paper if necessary.

Workplace (Employer name and city)	Dates Worked From	To	Average hours per week	Type of Industry	Describe your job duties	Known health hazards in workplace dusts, solvents, etc.)	Protective equipment used (gloves, masks, etc.)

Adapted from a form developed by the American Lung Association of San Diego and Imperial Counties, CA.

III.   OCCUPATIONAL EXPOSURE INVENTORY

1. Please describe any health problems or injuries you have experienced connected
    with your present or past jobs:

2. Have any of your co-workers also experienced health problems
    or injuries connected with the same jobs?   No   Yes
    If yes, please describe:

3. Do you or have you ever smoked cigarettes, cigars, or pipes?   No   Yes
    If so, which and how many per day? _____
    Are you still smoking?   No   Yes
    If no, when did you stop? _____

4. Do you smoke while on the job, as a general rule?   No   Yes

5. Do you have any allergies or allergic conditions?   No   Yes
    If so, please describe:

6. Have you ever worked with any substance which caused you
    to break out in a rash?   No   Yes
    If so, please describe your reaction and name the substance.

7. Have you ever been off work for more than a day because of an illness or
    injury related to work?   No   Yes
    If so, please describe:

8. Have you ever worked at a job which caused you trouble breathing,
    such as cough, shortness of wind, or wheezing?   No   Yes
    If so, please describe:

9. Have you ever changed jobs or work assignments because
    of any health problems or injuries?   No   Yes
    If so, please describe:

10. Do you frequently experience pain or discomfort in your lower back
    or have you been under a doctor's care for back problems?   No   Yes
    If so, please describe:

11. Have you ever worked at a job or hobby in which you came into direct contact
    with any of the following substances by breathing, touching, or direct exposure?
    If so, please check the box beside the substance:

❏	Acids	❏	Lead
❏	Alcohols (industrial)	❏	Manganese
❏	Alkalis	❏	Mercury
❏	Ammonia	❏	Methylene chloride
❏	Arsenic	❏	Nickel
❏	Asbestos	❏	Noise (loud)
❏	Benzene	❏	PCBs
❏	Beryllium	❏	Perchloroethylene
❏	Cadmium	❏	Pesticides
❏	Carbon tetracholoride	❏	Phenol
❏	Chlorinated naphthalenes	❏	Phosgene
❏	Chloroform	❏	Radiation
❏	Chloroprene	❏	Rock dust
❏	Chromates	❏	Silica dust
❏	Coal dust	❏	Solvents
❏	Cold (severe)	❏	Styrene
❏	Cotton dust	❏	Talc
❏	Dichlorobenzene	❏	Toluene
❏	Ethylene dibromide	❏	TDI or MDI
❏	Ethylene oxide	❏	Trichloroethylene
❏	Fibrous glass	❏	Trinitrotoluene
❏	Formaldehyde	❏	Vibration
❏	Heat (severe)	❏	Vinyl chloride
❏	Isocyanates	❏	Welding fumes
❏	Ketones	❏	X-rays

If you have answered "yes" to any of the above, please describe your exposure on a separate sheet of paper.

IV. ENVIRONMENTAL HISTORY

1. Have you ever changed your residence or home because of a health problem?

No   Yes

  If so, please describe:

2. Do you live next door to or very near an industrial plant?   No   Yes
  If so, please describe:

3. Do you have a hobby or craft which you do at home?   No   Yes
  If so, please describe:

4. Does your spouse or any other household member have contact
  with dusts or chemicals at work or during leisure activities?   No   Yes
  If so, please describe.

5. Do you use pesticides around your home or garden?   No   Yes
  If so, please describe:

6. Which of the following do you have in your home?
   (Please check those that apply.)

   ❏   Air conditioner
   ❏   Air purifier
   ❏   Humidifier
   ❏   Gas stove
   ❏   Electric stove
   ❏   Fireplace
   ❏   Central heating

Additional Questions for the Evaluation of Occupational Associations to the Present Illness or Injury[1]

Questions	Interpretation
Is your condition better or worse when you are off work for a few days or on vacation?	Identify patterns suggesting either improvement or exacerbation on withdrawal from exposure.
Is your condition better or worse when you return to work after a weekend or vacation?	Identify patterns suggesting return of the condition on reexposure in the workplace.
Does your condition get worse or better after you have been back at work for several days or shifts?	Identify patterns suggesting either tolerance or cumulative effects with multiple exposure.
Describe your workplace. (Please draw a diagram and indicate your work station.)	Evaluate the proximity to exposure, protection available (ventilation or barriers), mobility within the workplace, and location of co-workers who may also be affected.
What ventilation systems are used in your work space? Do they seem to work?	Obtain a general impression of adequacy of ventilation by air movement and odors.
Does the protective equipment you are issued fit properly? Do you receive instructions in its proper use? Do you ever fix or make changes in the equipment to make it more comfortable?	Consider the possibility that protective equipment is not fully effective. In the case of respirators (masks), ask if they are "fit-tested" to comply with Occupational Safety and Health Administration (OSHA) regulations.
Where do you eat, smoke, and take your breaks when you are on the job?	Identify opportunities for food- and cigarette-borne intake, and evaluate the adequacy of rest stations (isolation from heat, noise, fumes).

Question	Purpose
Where are your (and your spouse's or partner's) work clothes laundered?	Identify possibility of passive exposure at home or of prolonged skin contact.
How often do you wash your hands at work, and how do you wash them?	Identify the potential for contamination of hands or contact with solvents or drying agents.
What is your spouse's or partner's occupation?	Identify the potentials for passive exposure (an occupational history for the spouse or partner may be indicated.).
Have any of your fellow workers experienced similar conditions?	Identifying others who may have been affected may lead to inquiries that clarify the individual patient's problem. Prevention-oriented interventions or requests for investigation by the state or federal enforcement agency may be required.
Do you recall a specific incident or accident that occurred on the job? Were others also affected?	Identify unusual or transient conditions that may have resulted in an exposure not reflected in the occupational history, such as leaks, fires, or uncontrolled exothermic chemical reactions.
Are animals (pets, livestock, birds, or pests such as mice) present in the vicinity of the workplace? Has there been a change in their health, appearance, or behavior?	Animals (and especially animal wastes) may be a source of infectious or allergic hazards. Animals may also respond to toxic exposures that affect humans.

[1] Modified and reproduced, with permission, from Occupational and Environmental Health Committee of the American Lung Association: Taking the occupational history. *Ann Intern Med* 1983; 99:641.

Chapter 7
Work Organization

Bill Kojola and Nancy Lessin

W ork itself is inherently an organized activity. Decisions about the organization of work are generally made by management. Reorganization of work often occurs in response to such things as demand for products, availability of raw materials, and changes in production methods. However, contemporary changes in the organization of work are revolutionary and amount to a second industrial revolution made possible, in part, by innovations in information and communications technologies. Often rationalized as a response to intense pressure to increase competitiveness and productivity in the global economy, change in the organization of work, promoted by advanced technological societies, also creates intense pressures on workers. The goals of many of the work reorganization schemes employers have introduced are to increase output (in part through increased demands on the workforce and the steady removal of any downtime or micro-breaks from the work process) and to decrease localized control over the work.

Like the first industrial revolution, new forms of work organization are being introduced with scant attention paid to the potential adverse health and safety consequences for workers. However, we are discovering that some work organization changes contribute significantly to the risk of occupational injuries and illnesses. Both NIOSH and the European Agency for Safety and Health at Work have made changing organization of work one of their top research priorities.

Forms of Work Organization

Work organization is fundamentally about control of work and the division of labor. It is about tasks that are performed, who performs them and how they are performed in the process of making a product or providing a service. Included in the scope of work organization are the pace of work (such as the speed of an assembly line), work load, number of people performing a job (staffing levels), hours and days on the job, length and amount of rest breaks and days away from work, layout of work (such as cellular manufacturing or "pods" as opposed to assembly–line production), skill mix of those on the job, and training for the tasks being performed. These parameters are measured and analyzed in detail in order to improve productivity, control, and efficiency. When work is restructured, these and other factors can undergo significant changes.

In general, work is being restructured to achieve goals of standardization (often referred to as "quality") and efficiency. Some common systems of work organization include the following:

Continuous Improvement/Kaizen: process of continually improving productivity and efficiency by "leaning out" the work process (eliminating "nonvalue–added" activity, such as micro-breaks or any time spent not producing/serving) while increasing production goals or reducing staff.

Just-in-Time Production: limiting or eliminating inventories, including work–in–progress inventories, using single piece production techniques to highlight and eliminate "waste" in the production process, including any activity that does not add value to the product.

Total Quality Management (or its newer versions, such as "Six Sigma"): process of statistical analysis of variation in outcomes combined with intimate employee knowledge of the work process (often obtained through work teams) to standardize work process and results.

Total Productive Maintenance: designed to eliminate all nonstandard, nonplanned maintenance with the goal of eliminating unscheduled disruptions, simplifying (de-skilling) maintenance procedures, and reducing the need for "just-in-case" maintenance employees.

Work Teams: one type of work team is used in a production process, where teams of workers are responsible for completing whole segments of work product; another type of work team meets separately from the production process and provides ideas on how to improve quality, production, and efficiency.

Outsourcing/Contracting Out: transfer of work formerly done by employees to outside firms.

Many restructured workplaces are touted as being facilities where workers become an integral part of decision-making and do not "check their brains at the door." The reality is very different. Workers' knowledge about the production process is harvested through "employee involvement" groups such as "process improvement teams," and, coupled with information gathered via computers that are ubiquitous in most workplaces, is then used to enhance the "leaning out" and standardization process, thereby actually reducing the reliance on employee skill and creativity.

This "leaning out" of work and these restructuring schemes have resulted in job loss for some workers, while increasing the work load and work pace of those who remain on the job. Employers—through the systems of work created—put pressure on the remaining workforce to increase the quantity and quality of production. Workers are forced to adjust continuously to new work methods and management practices, producing (or serving) greater numbers, in less time, with greater efficiency and quality, under tighter deadlines. The result of these changes in work organization is that it is no longer just machines that are wearing down and wearing out—it is workers themselves.

Occurrence

The extent to which these new forms of work organization have been implemented across workplaces is difficult to determine, in part because there is such a wide diversity of changes. Surveys have indicated that 30% to 50% of employers with more than 50 employees have implemented some aspects of lean production and "high performance work organization" practices.

One measure of change is changes in the hours that workers spend on their jobs. While the annual hours worked in European countries and Japan have been decreasing, in the United States the number of hours worked annually has been steadily increasing over the past couple of decades to the point where American workers work more hours than workers in any other industrialized country. Overtime hours, some of which are mandatory for workers to work, have also risen within the last decade in the United States.

Health and Safety Outcomes

What is known about the adverse effects of the new forms of work organization is giving rise to a heightened concern about their impacts on the health and safety of workers. The organization of work itself can influence the level of psychological stress and exposure to physical hazards, either of which can lead to injuries or illnesses. The new forms of work organization can result in work intensification, leading to working faster and harder. This work intensification may be increasing stress on the job, with higher demand and low control.

Various changes in work organization systems have been associated with the development of musculoskeletal disorders in health care, automobile manufacturing, meatpacking, and contingent work. Work-related musculoskeletal disorders associated with organization of work changes have been linked to exposure to physical hazards and psychologically stressful attributes of these changes, including machine-paced work, inadequate work-rest cycles, wage incentives, time pressure, low job control, low social support, and repetitive work.

In the health care industry, organizational changes associated with a shortage of nurses and high hospital patient–to–nurse staffing ratios, as well as poor working climates, have been linked to increases in needlestick injuries and near misses. Hospital nurse understaffing has also been shown to lead to increases in nurse burnout, job dissatisfaction, and greater surgical patient mortality.

A considerable literature exists concerned with the effects of work stress on the cardiovascular system. High job strain (jobs with high work demands and low job control) is associated with increases in blood pressure and increased risk of cardiovascular mortality. (See chapter in Part II on Stress.)

Long hours of work also appear to have an acute impact on the cardiovascular system. Overtime work has been shown to increase 24-hour average blood pressure in workers who worked 60 or more hours of overtime per month compared to those working 30 or fewer hours of monthly overtime. A recent study demonstrated an increased risk of having an acute myocardial infarction (AMI) associated with working greater than 11 hours per day in the month prior to the event while another study reported a twofold increased risk of an AMI when working 61 or more hours on a weekly basis.

An association between long hours of work and worker injuries on the job has been reported. The risk of experiencing a workplace injury increases with long work hours and increases exponentially beyond the ninth hour of work.

Working long hours increases the exposure period to physical hazards and psychological stresses, thus in effect delivering a greater dose of the hazards and stresses to the worker. This may account for some of the reported increases in injuries among workers who work extended hours on the job. Increased levels of fatigue may also play a significant role with the increases in injury rates for workers working long hours.

While some research has suggested that the current employer trend toward "employee involvement" is an antidote to the problem of low worker control over job tasks, employer-introduced involvement strategies may be more ceremonial than substantive and may even erode workers' means to influence job conditions through more traditional labor-management mechanisms (such as collective bargaining). Cross-functional teamwork and job enlargement strategies may in some instances multiply the number of tasks workers perform.

Prevention

Given the scope and complexity of work organization, there is no single prevention strategy that would improve workers' health and safety. In addition, health and safety are only one concern among many that arise from work organization. Among the others are job security, job satisfaction, wage levels, and means of settling disputes. Workers in today's workplaces do not need new research and studies to know that they are working faster, harder and longer, and that they are suffering from increased stress, injuries, and illnesses as a result. Top occupational safety and health concerns of workers and unions today include downsizing/understaffing, mandatory overtime, 12-hour shifts, outsourcing, lack of training for added job duties, untrained/undertrained subcontractors in their workplaces, increased workload, increased work pace, and pressures pushing them to cut corners to get the job done (often under the threat that work will be contracted out or moved to a different location if tighter deadlines or increased production goals are not met). It is well understood by workers and their unions that these features of work contribute to the risk of injuries, musculoskeletal disorders, stress, and other effects. The responses for addressing these concerns and problems are inherent in their definition: job security, voluntary overtime, shorter work shifts, job training, reasonable workloads and pace of work, and support for performing jobs well.

While employers may have taken notice of increased worker injury and illness in today's workplaces, few, if any, are looking at their core work processes and work restructuring as the problem that needs to be addressed. Instead, many have introduced safety programs that focus on individual workers' "unsafe behaviors" and "unsafe acts." These "behavioral safety" schemes ignore unsafe workplace conditions—especially production pressures, lack of staff, work overload, and long work hours and weeks—and blame workers for being inattentive or working carelessly if they suffer injuries.

This leaves two major problems to address: eliminating or reducing hazards created or intensified by work restructuring schemes, and eliminating "blame-the-worker" approaches to safety and health.

The solutions to both lie in workers and unions having a greater say in how work is organized and restructured, how technology is used, and the policies and practices employers seek to impose. Strategies for correcting these problems emphasize both a collective process and collective solutions to what are systemic—not individual—problems.

Unions have been more aggressive in taking on management's "rights to manage" when those rights have brought about conditions that cause or contribute to worker injury, illness, and death. Collective bargaining agreements, legislative campaigns, and job actions have been vehicles for changing harmful work restructuring practices and eliminating or preventing hazards associated with them.

Unions have successfully negotiated language in collective bargaining agreements that has instituted minimum staffing requirements, limited or prohibited mandatory overtime, reduced production quotas, put limits on the pace of work, mandated rest breaks, and ensured a focus of worksite safety programs on finding and fixing hazards rather than blaming workers.

Workers and unions have also engaged in legislative campaigns creating laws that mandate safe staffing levels and limit mandatory overtime.

Training and education have been indispensable in building successful campaigns to address hazards created by work restructuring. The use of surveys, body mapping, and hazards mapping (collectively identifying injuries, illnesses, and stresses suffered by workers in a particular department or workplace; and locating the hazards in a workplace that caused or contributed to those problems) has been critical in illustrating the extent of harm being done, and identifying the hazards—including the hazards of work restructuring—responsible for that harm.

Because of the turbulence in today's workplaces created by continuous changes in technologies, work organization, and management policies, unions need to push even further into the decision-making process around work restructuring and technological change, and should consider adopting a "continuous bargaining" approach to addressing these changes. Using such an approach, unions will research new forms of work organization and their health and safety impacts prior to their implementation in the workplace, communicate with their members about health, safety, and other concerns posed by the restructuring of work, and identify intervention strategies that will be successful in preventing harmful changes in work organization and promoting real—not illusory—worker involvement and control over how work is organized and restructured.

Further Reading

Aiken LH, Clarke SP, Sloane DM, et al. Hospital nurse staffing and patient mortality, nurse burnout, and job dissatisfaction. *Journal of the American Medical Association* 2002; 288: 1987–1993.

Bluestone B, Rose S. *Public Policy Brief. The Unmeasured Labor Force: The Growth in Work Hours.* Annandale–on–Hudson, NY: The Jerome Levy Economics Institute of Bard College, 1998.

Braverman H. *Labor and Monopoly Capital. The Degradation of Work in the Twentieth Century.* New York: Monthly Review Press, 1974.

Clarke SP, Sloane DM, Aiken LH. Effects of hospital staffing and organizational climate on needlestick injuries to nurses. *American Journal of Public Health* 2002; 92: 1115–1119.

European Agency for Safety and Health at Work. *Research on Changing World of Work.* Luxembourg: The European Agency, 2002.

Golden L, Jorgensen H. *Time After Time: Mandatory Overtime in the U.S. Economy.* Washington, DC: Economic Policy Institute, 2001.

Graham, L. *On the Line at Subaru-Izusu.* Ithaca, NY: ILR/Cornell University Press, 1995.

Hammer M, Champy J. *Reengineering the Corporation: A Manifesto for Business Revolution.* New York: Harper Business Books, 1993.

Hanecke K, Tiedemann S, Nachreiner F, et al. Accident risk as a function of hour at work and time of day as determined from accident data and exposure models for the German working population. *Scandinavian Journal of Work, Environment & Health* 1998; 24 (Suppl. 3): 43–48.

Hayashi T, Kobayashi Y, Yamaoka K, et al. Effect of overtime work on 24-hour ambulatory blood pressure. *Journal of Occupational and Environmental Medicine* 1996; 38: 1007–1011.

Howard R. *Brave New Work–Place.* New York: Viking Penguin Books, 1985.

International Labour Organization. *Key Indicators of the Labour Market.* Geneva: ILO, 1999.

Kamata S. *Japan in the Passing Lane: An Insider's Account of Life in a Japanese Auto*

Factory. New York: Pantheon Books, 1982.

Kivimaki M, Leino-Arjas P, Luukkonen R, et al. Work stress and risk of cardiovascular mortality: prospective cohort study of industrial employees. *British Medical Journal* 2002; 325: 857–861.

Landsbergis PA. The changing organization of work and the safety and health of working people: A commentary. *Journal of Occupational and Environmental Medicine* 2003; 45: 61–72.

Landsbergis PA, Cahill J, Schnall P. The impact of lean production and related new systems of work organization on worker health. *Journal of Occupational Health and Psychology* 1999; 4: 108–130.

Landsbergis PA, Schnall PL, Pickering TG, et al. Life-course exposure to job strain and ambulatory blood pressure in men. *American Journal of Epidemiology* 2003; 157: 998–1006.

Liu Y, Tanaka H. Overtime work, insufficient sleep, and risk of non–fatal acute myocardial infarction in Japanese men. *Occupational and Environmental Medicine* 2002; 59: 447–451.

Mureau M. *Workplace Stress: A Collective Bargaining Issue.* Geneva: International Labor Organization. Available at:
www.ilo.org/public/english/dialogue/actrav/publ/126/mureau.pdf

National Institute for Occupational Safety and Health. *The Changing Organization of Work and the Safety and Health of Working People* (DHHS [NIOSH] Publication No. 2002–116). Cincinnati, OH: NIOSH, 2002.

Noble DF. *Forces of Production: A Social History of Industrial Automation.* New York: Alfred Knopf, 1984.

Rones PL, Ilg RE, Gardner JM. Trends in hours of work since the mid–1970s. *Monthly Labor Review* 1997: 3–14.

Rosa RR. Extended workshifts and excessive fatigue. *Journal of Sleep Research* 1995; 4 (Suppl. 2): 51–56.

Schenk C, Anderson J (Eds.). *Reshaping Work 2: Labour, the Workplace and Technological Change.* Ontario: Canadian Centre for Policy Alternatives and Garamond Press, 1999.

Sokejima S, Kagamimori S. Working hours as a risk factor for acute myocardial infarction in Japan: case–control study. *British Medical Journal* 1998; 317: 775–780.

Spurgeon A, Harrington JM, Cooper CL. Health and safety problems associated with long working hours: A review of the current position.

Chapter 8
Populations at High Risk

Sherry Baron

Over the last half of the 20th century the size of the working population in the United States doubled, increasing by 79 million workers between 1950 and 2000. The composition of this new workforce reflects the changing social, political, and demographic characteristics of the United States. Many more women and immigrant workers have entered the workforce, changes in laws and technology have increased job opportunities for disabled workers, and the aging of the baby boom generation has increased the median age of the workforce. At the same time, the economy of the United States has been transformed, as traditional permanent, full–time, often unionized manufacturing jobs have moved to Latin America and Asia and new nonunion, temporary and contract, and service–sector jobs have taken their place. Although there have been many significant advances in civil rights, the African American population is still disproportionately employed in high–hazard jobs, while racism and other forms of discrimination, both in the community and the workplace, contribute to additional health risks.

Innovative programs in the workplace and society have not kept pace with the demands of this changing workforce, leaving many workers increasingly vulnerable to the various forms of occupational injury, illness, and work stress described in this book. For example, working parents have few low–cost, high–quality options for child care; training and education programs are not prepared to serve the linguistic and cultural needs of the new immigrant workforce; few new programs exist to help those in temporary jobs find stable and safe employment; and almost half of all low–wage, full–time workers lack health care coverage. Although a variety of workers are affected by these social and economic trends, this chapter will highlight the particular issues facing some of the most important sectors. The statistics presented here are primarily generated by either the Census Bureau or the Bureau of Labor Statistics, and can be found at the Internet sites listed under Further Reading.

Women Workers

In 1950, fewer than one–third of workers in the United States were female, while today they make up close to half of the workforce; in 2000, 61% of the female population over age 16 was part of the labor force. Overall, the entry of women into the workforce has created increased economic and social opportunities. However, 59% of women workers are considered low wage, meaning they earn a wage that places a family of four below the poverty level, and for one out of six the entire family income is less than $25,000. Low–wage women workers are also more likely to be single mothers, have less than a high school education, and are either an immigrant worker or African American. More than two–thirds of low–wage women workers do not receive health insurance coverage through their work. One out of five single gle mothers work evening, night, or rotating shifts and one in three work weekends. The combination of low wages and benefits combined with non-standard work hours creates added stress, as women struggle to find reasonable and safe day care options and to balance their roles as wage earners and mothers. Women workers also are more likely to work part–time or part–year, which, while increasing their flexibility to balance work and home responsibilities, also means that they are more likely to work in temporary jobs that lack job security and health benefits.

Although the number of women workers has increased, the job market remains highly segregated by sex. In 2000, over half of all women worked in just three occupational categories: administrative support, such as secretaries and other clerical support jobs (24%); professional specialty jobs, such as nurses and primary and secondary school teachers (18%); and services workers, such as cashiers, restaurant workers, and hairdressers. Although overall women are underrepresented in the manufacturing sector, they are the majority in certain sectors, such as textile and garment manufacturing. Not surprisingly, occupational injuries and illnesses for women workers are concentrated in the occupations where low–wage workers are employed. About one–third of the work–related injuries and illnesses in women occur among service–sector occupations and about one-fourth to laborers, fabricators, or operators; yet those occupations only account for about one–fourth of all women workers.

African American Workers

Despite many changes in civil rights legislation, and increases in educational and employment opportunities for African Americans since the 1960s, many economic and social disparities remain, as compared to the majority non–Hispanic white population. Most (55%) African Americans still live in the South, and over half live in the center city of a large metropolitan area, compared to only 21% of the non–Hispanic white population. Twice as many African Americans over age 25 have not completed a high school education (21% for both males and females) compared to the non–Hispanic white population. About one-fourth (23%) of the African American population lives below the poverty level, three times the rate for non–Hispanic whites. While African American men are less likely to be part of the civilian labor force than non–Hispanic whites (68% v. 74% workforce participation rate), African American women are more likely (62% v. 60%). African American males are twice as likely to work in service occupations and as laborers, fabricators, and operators; yet they are half as likely to be in managerial or professional specialty occupations, as compared to non–Hispanic whites. Yet these occupations that are disproportionately African American are also some of the occupations with the highest concentration of occupational injuries and illnesses. Operators, fabricators, and laborers—occupations where African Americans are overrepresented—comprise 13% of the workforce, but they account for 39% of occupational injuries and illnesses.

Many recent health reports have focused on the disproportionate burden of diseases, such as cancer, cardiovascular disease, and asthma, in the African American community. Scientific discussion and debate has attempted to disentangle the many factors potentially contributing to these disparities, including access to health care, nutritional factors, environmental exposures, and genetic factors. One component of this debate has focused on understanding the role of racism in the causation of disease, and one of the pathways by which racism may contribute to increased health burden is through the disproportionate placement of minorities in the most dangerous industries and occupations. Although it is hard to systematically examine the impact of racial discrimination as a risk factor for occupational injury or illness, historically there are several clear examples. For example, one landmark study in the coke oven industry showed that the increased risk for lung cancer among African American workers was not due as originally thought to genetic factors, but rather to those workers disproportionately being assigned to the job categories with the highest exposure to carcinogenic emissions.

African Americans are also disproportionately affected by economic recession. Much of the rising social and economic status of the African American population can be attributed to increases in employment in the relatively well–paid manufacturing sector between 1939 and 1959. As the economy was restructured from manufacturing to service, particularly since 1970, African Americans have suffered a disproportionate share of the rise in unemployment; between 2001 and 2003, the increase in the African American unemployment rate was twice that of whites. In addition to the economic effects of unemployment, studies have also shown adverse effects on health and well-being associated with the stress of unemployment. The unemployment rate is especially high for African American youth, which may have long–term social and economic impacts as youth become disillusioned and leave the job market.

Immigrant Workers

During the 1990s, more immigrants entered the United States than during any other period of history, with close to 1 million immigrants arriving each year. Foreign–born workers now make up 12% of the entire workforce; they accounted for about 50% of the net increase in the labor force during the second half of the 1990s. As with the African American population, for-eign–born workers are more likely to work in service occupations and as operators, fabricators, and laborers, and also are twice as likely to work in farming, forestry, and fishing, as compared to native–born workers. Central Americans and Mexicans, especially those who have had less than 10 years of residence in the United States and are not citizens, are especially likely to work in these sectors. In 1999, the median earnings for all foreign–born work-ers with less than 10 years of residence in the United States was $21,600. The median earnings for Mexican and Central–American women was less than $16,000 and for Mexican males was only slightly more than $19,000.

Since foreign–born workers, especially new immigrants, often have few geographic ties in this country, they often travel in search of jobs. This dis-persion of new immigrants has led to a rapid shift in the demographics of the workforce in certain areas. While states, such as California, Texas, New York, and Florida, continue to have the most foreign–born workers, during the 1990s many states, especially in the South and the Midwest, more than dou-bled their foreign-born population. This movement was initially fueled by employment in the low–wage and high–hazard meat– and poultry–process-ing industry and in agriculture. However, once immigrants began to settle in these areas, they not only found that jobs were available, but also many found the lifestyle preferable to the congestion, high cost, and crime of the tradi-tional inner–city immigrant communities. Employment opportunities soon expanded to include construction, services, and other manufacturing jobs.

Although systematically collected data on nonfatal occupational injuries for foreign–born workers are not available, government statistics do show that foreign–born workers have a higher occupational fatality rate compared to other workers. Between 1995 and 2000, the occupational fatality rate for all foreign–born Hispanic workers was 50% higher (6.1 per 100,000 workers) than the rates for all workers and for native–born Hispanic workers. The cause of this disparity is, in large part, due to the disproportionate distribu-tion of foreign–born workers in high–risk industries, such as construction, agriculture, and manufacturing. Additionally, even within these high–risk industries, foreign–born workers may face the highest risk for injury. A recent analysis found that Hispanic construction workers had an 80% greater fatali-ty rate compared to non–Hispanic construction workers. However, concerns have also been raised regarding many of the linguistic, cultural, and legal bar-riers that foreign–born workers face. Most foreign–born workers, especially from Mexico and Central America, have less than a high school education and, for many, raised in rural indigenous communities, Spanish is not their first language. Immigrants, especially new immigrants, may be unfamiliar with local laws regarding safety and health protection or workers' compen-sation. Although systematic statistics are not available on the proportion of foreign–born workers who lack legal documentation, a Department of Labor survey in the agricultural industry found that over half of crop–farm workers are undocumented. Systemic programs to improve the safety and health conditions for foreign–born workers must not only address the industry– or occupation–specific hazards, but must also develop linguistically and culturally appropriate training and education programs, and address the legal barriers resulting from their immigration status.

Youth Employment

Since the passage of strong federal child labor laws in the 1930s, exploitative child labor has been rare in the United States. Youth employment, however, is extremely common; according to official statistics, between 1996 and 1998, 2.9 million youth aged 15–17 worked during the school year, 4 million worked during the summer, and there was no difference between male and female youth employment rates. Youth in higher–income families are more likely to work than those in lower–income families, and white youth are almost twice as likely as black and Hispanic youth to work. This may be a function of the greater availability of jobs in higher–income than lower–income communities, as well as increased access to transportation to find jobs outside of the community. It also may be a result of lower–income youth needing to provide more assistance at home and providing childcare for younger siblings. While most youth work part–time, 6% of employed youth work full–time during the school year and 20% during the summer months. Young workers are largely low–wage workers, most earning less than $7 per hour. The most common jobs for youth, both male and female, are in the retail trade sector, with about one–third employed in eating and drinking establishments and 30% in other retail establishments. During the summer months, there is some shift in employment as youth, especially males, are also employed in manufacturing, construction, and agriculture, and females in the service sector.

While employment provides many benefits to youth, including increased self–confidence, job skills, and income, it also poses potential hazards. There are a number of characteristics of young workers that raise particular concerns regarding their safety and health. Like all new workers, young workers are at increased risk for injury. In many surveys of working youth, almost half report they did not receive health and safety training on the job. Developmental characteristics may also place youth at risk, as the level of physical and cognitive development in teens is variable. Smaller teens may have a harder time reaching machines and also may not have the physical strength required for certain tasks. Even when youth have reached adult stature, their psychological and cognitive maturity may lag. Employers may assign them tasks to which they are not yet cognitively prepared. Youth also approach work with extreme enthusiasm and a desire to do well, and although these may be very positive attributes, they may also make them less comfortable asking questions or expressing concerns about their ability to perform a challenging task. Additionally, studies of youth working more than 20 hours a week during the school year show important effects, such as increased daytime fatigue and substance abuse.

An average of 68 fatalities occur at work in youth every year, and severe work–related injuries in youth can have long–term effects on future job possibilities and on the quality of their lives in general. Occupational fatalities occur most commonly during agricultural work (42%), construction work (14%), and in retail trade (19%). Tragically, 28% of fatalities occur in children less than 15 and 30% to those working in a family business, mostly agriculture. Among fatalities in youth working in the retail sector, two–thirds are homicides, many occurring as the result of a robbery. In 1996, about 15,000 young workers sustained a work–related injury that resulted in lost workdays. These injuries occurred most commonly in those industries where young workers are most commonly employed; about 60% of injuries were in retail–trade workers and about 20% in service workers. The most typical workplaces where injuries occurred were restaurants, grocery stores, health services, and amusement and recreation services.

Older Workers

The dominant factors affecting the changing demographics of the United States since the 1950s have been the "baby boom" following World War II (1946–1964), combined with the continued increase in life expectancy. Together, these factors have dramatically changed the shape of the population pyramid, such that between 1990 and 2000, the number of workers aged 25–44 did not change, while the number of workers aged 45–65 increased by more than 12 million. This phenomenon has caused a reevaluation of conventional ideas of retirement, and has raised questions about the positive and negative associations between aging and work.

As health researchers and policy experts explore the issues raised by the increasing age of the population, one major question is the impact of aging on health and working capacity and the impact of working on the aging process. Although the answer may, in part, be job–dependent, important questions have been raised regarding the relative importance of the physiological and cognitive deterioration associated with the aging process versus the positive attributes of experience and expertise associated with being a longer–term worker. With age, occupational injury rates decrease, but are associated with greater severity, as measured in the number of lost workdays. In 2000, the work-related fatality rate for workers aged 55–64 was 50% greater than for workers aged 45–54.

Although it appears that age at retirement is not increasing, more older workers are returning to the workforce in part–time and temporary jobs, either for economic or social reasons. For example, the proportion of those over 55 who are collecting a pension, but are still working, has increased. These older workers are employed in occupations with a higher rate of part–time work, such as sales and service occupations. Also, as life expectancy increases and many families try to keep older individuals out of institutional care settings, many older individuals, especially women, either as paid or unpaid workers, are providing home-services for their friends and family members.

Workers with Disabilities

The category of adults with disabilities includes a broad range of individuals, from those with severe physical limitations that preclude participation in the workforce to those who, either with or without accommodations, can have productive working careers. Beginning in the 1970s, there has been an increasing social movement of those with disabilities who have fought for improved programs that would allow disabled individuals to live more independently and to have more work opportunities. One result was the passage of the Americans with Disabilities Act (ADA) of 1990 that included provisions to promote economic equality in employment by prohibiting discrimination based on disability status, and by requiring workplaces to make reasonable accommodations for qualified applicants with disabilities.

In 1997, it was estimated that one in five adults in the United States had some type of disability. While the disability rate increases with age, almost one in four adults (23%) aged 45–54 and almost one in eight (13%) of those 25–44 have some disability. Of the more than 17 million adults between the ages of 21 and 64 who had a severe disability in 1997, about one–fourth were employed. Despite the passage of the ADA, it appears that the proportion of adults with disabilities who are working has not increased. Almost no research has been done to explore why this is true. The ability of disabled adults to find appropriate jobs that they can perform safely throughout a working career may be one explanation; the lack of enforcement of the ADA may be another.

Impact of Globalization on Workers in the United States and Abroad

During the past 25 years, the process known as globalization—the reorganization of world production by multinational corporations and governments—has led to the relocation of many manufacturing plants, especially from the advanced industrial economies, such as the United States, Germany, and Japan, to export–processing zones (EPZs) in developing nations. Typically these host countries have established EPZs that offer tax incentives, infrastructure, and low–labor costs to corporations in order to attract capital and to provide industrial development and jobs. In Mexico, for example, EPZs were first established in 1965; as of June 2002, there were 3164 EPZs (called maquiladoras) in Mexico, employing 850 000 hourly workers, who were paid, on average, about $300 per month. More recently, many manufacturing jobs have moved to Vietnam and China, where wages are even lower.

In 1994, the North American Free Trade Agreement (NAFTA) went into effect, allowing for the free movement of goods and capital among Mexico, Canada, and the United States, but not permitting free movement of workers. In response to concerns from communities, labor unions, and advocacy organizations, NAFTA included special side agreements designed to protect the environment and to assure that labor rights, including occupational health protections, were harmonized among the countries. Based on the official findings from governmental investigations of complaints that have been filed under these side agreements, while Mexican occupational health legislation is equivalent to that of the United States and Canada, it is rarely enforced. Similarly, recent nongovernmental investigations into working conditions in EPZs producing athletic equipment, fueled by concerns of university students and administrators who purchase their products, have found violations of many international labor laws. The continued growth in globalization, accompanied by the passage of new international trade agreements, raises new challenges, responsibilities, and opportunities for occupational safety and health professionals throughout the world to collaborate to ensure that workers are protected against preventable injuries and illnesses at work.

The Rise of Temporary Work

The increasing globalization of the world economy and the pressure to compete with the lower cost of production in the developing world have led many U.S. workplaces to shift from full–time permanent, and often unionized, workers to more temporary or contract and nonunion employees. While unionized workers' wages, on average, are 10% higher than nonunion workers, the overall unionization rate in the United States has dropped steadily from about one–third of all workers in the 1950s to less than 14% in 2001. In addition to increasing wages, labor unions have also provided improved safety and health protection through collective bargaining agreements, advocated for legislation such as the Occupational Safety and Health Act, provided protection for workers against retaliation for refusing unsafe work, and, through joint labor-management safety and health committees, developed training and intervention programs.

Many employers have turned to the use of more temporary workers and contracted services, where the number and hours of work are more flexible, in order to control production costs. Some of the shift in the size of the labor force away from the manufacturing sector and into the service sector can be attributed to manufacturing corporations employing workers through temporary services agencies. Twenty–two percent of the over 14 million jobs added to the economy between 1988 and 1996 were in the business, engineering, and management services industry. The most rapidly growing occupations in this sector are in personnel supply services, commonly known as "temp agencies," where there has been a large increase in both administrative support (clerical) workers and in helpers, laborers, and material movers. In 1996, one out of four new administrative support workers and 44% of "helpers" were employed by a personnel supply service company.

BLS statistics have used the term "contingent worker" to refer to any worker who does not have an explicit or implicit contract for long–term employment. A study of contingent workers between 1997 and 1999 found that, while the overall economy expanded and unemployment decreased, the percentage of all workers who were contingent remained the same (about 4% of all workers). It also found that contingent workers, whether part–time or full–time, earned less, on average, than noncontingent workers; except for those under age 25, most would prefer to work in a more permanent job, if one were available. It also found that only one in five contingent workers had employer-provided health insurance, compared to three of five noncontingent workers. Although theoretically covered by the same occupational safety and health protections as permanent workers, workers who move among different workplaces and different jobs have reduced ability to receive comprehensive training and health and safety protection. Often the temporary services agency is responsible for training, although it may be unfamiliar with the specific workplace. In addition, all workers, even after adequate training, require experience at a specific workplace before they can comfortably and safely perform a given job; contingent workers who have frequent job changes may never become accustomed to any workplace.

Further Reading

Frumkin H, Pransky G (Eds.). Special populations. *Occupational Medicine: State of the Art Reviews* 1999; 14: 479–484.

Hipple S. Contingent work in the late 1990s. *Monthly Labor Review*, March 2001.

Levy B, Wegman D (Eds.). *Occupational Health: Recognizing and Preventing Work-Related Disease and Injury* (4th ed.). Philadelphia: Lippincott Williams & Wilkins, 2000.

Kim M. Women paid low wages: Who they are and where they work. *Monthly Labor Review*, September 2000.

National Research Council. *Protecting Youth at Work*. Washington, DC: National Academies Press, 1998.

National Research Council. *Safety is Seguridad*. Washington, DC: National Academies Press, 2003.

Solomon J. Trading away rights: The unfulfilled promise of NAFTA's labor side agreement. *Human Rights Watch* 2001; 12 (2B).

U.S. Department of Labor. *Report on the Youth Labor Force, November 2000*. Available at http://www.bls.gov/opub/rylf/rylfhome.htm, accessed 12/04/03.

U.S. Census Bureau. *Population Profile of the United States, 2000* (internet release). Available at http://www.census.gov/population/www/pop-profile/profile2000.html, accessed 12/04/03.

Chapter 9
Preventing Occupational Disease and Injury in Developing Countries

Barry S. Levy, Kathleen M. Rest, and Marilyn A. Fingerhut

The following represents additional considerations for preventing occupational disease and injury in developing countries, where average annual incomes are low, the informal sector of the economy is generally large, most people are subsistence farmers, and work tends to be labor-intensive.

Occupational Health and Safety Is Often a Low Priority: Occupational health and safety is likely to receive relatively little attention due to high rates of endemic infectious diseases affecting the population, as well as understandable governmental and public focus on critical economic, social, and health issues. Infectious diseases of concern include diarrheal diseases (due largely to contaminated water supplies), respiratory diseases (often worsened by high levels of outdoor and indoor air pollution), parasitic diseases (including malaria, which accounts for much morbidity and mortality), and HIV/AIDS and other sexually transmitted diseases. Life expectancy is often markedly less than in developed (more-developed), industrialized countries, and infant and child mortality rates are likely to be high.

Intense In-Country Occupational Health and Safety Problems: Workers in developing countries often face intense exposure to highly toxic pesticides, organic solvents, heavy metals, and many other health and safety hazards. Because of the proximity between homes and workplaces, children and other family members may be exposed to hazardous materials in addition to toxic substances being brought home on parents' work clothes, skin, or hair. In addition, workers in developing countries are often at greater risk because of inadequate information about hazards, inadequate engineering and other workplace-based prevention and control measures, and inadequate personal protective equipment.

Child Labor: Although child labor violates international conventions, it is widespread in many countries. The ILO estimates that almost 250 million children worldwide (about one in every six children aged 5 to 17) are involved in child labor. Of these children, 179 million (one in eight) are exposed to child labor that endangers their physical, mental, or moral well-being.

The Export of Hazard: Developed, industrialized countries, including the United States, export large amounts of restricted and banned pesticides and other hazardous substances to developing countries, where there is inadequate awareness of these hazards and little or no regulation of them. For example, after the pesticide dibromochloropropane (DBCP) was banned in the United States in the late 1970s because it was found to cause sterility in men, the two U.S. companies that produced it sent their remaining stocks to many developing countries, where many male banana and pineapple plantation workers were exposed and became sterile.[1]

Workers Less Likely to Report Problems: Unemployment and underemployment are likely to be high, making people who get jobs less likely to complain about hazardous working conditions for fear of losing them. In addition, employers are aware that, for every employee who cannot work because of an illness or injury sustained on the job, many people may be willing and eager to work and to do the same job. Also, for workers who complain about or report workplace hazards, the potential for being fired, demoted, or otherwise discriminated against may be higher in developing countries, where employment protections may be lacking.

Inadequate Adoption and Enforcement of Laws and Regulations: In most developing countries, existing occupational health and safety laws and regulations are weak, incomplete, and not strongly enforced. For example, in many developing countries laws do not adequately address farm hazards and agricultural workers, and fines and other sanctions are minimal for violation of laws and regulations.

Inadequate Numbers of Trained Health and Safety Professionals: Most developing countries have very few adequately trained occupational health and safety professionals. Physicians, nurses, epidemiologists, industrial hygienists, safety engineers, and other professionals in this field are small in number. Some who obtain further education and training in developed countries do not return home. In addition, there are often too few adequately trained professionals in related fields, such as clinical pathology, toxicology, laboratory sciences, pulmonary medicine, neurology, and dermatology.

Inadequate Health Care Facilities: In general, health care facilities are inadequate in number, quality, and accessibility in many developing countries. These problems are usually far more extreme for health care facilities that specialize in occupational medicine. As a result, workers who become injured or ill as a result of workplace hazards are less likely to receive adequate treatment and rehabilitation, and their health problems are less likely to trigger investigations and interventions to prevent these problems from occurring in other workers.

---

[1] Levy BS, Levin JL, Teitelbaum DT (Eds). Symposium: DBCP-Induced Sterility and Reduced Fertility among Men in Developing Countries: A Case Study of the Export of a Known Hazard. *International Journal of Occupational and Environmental Health* 1999; 5: 115-153.

Opportunities to Facilitate Improvements

In most countries, the health and labor ministries have responsibility for occupational health. WHO and ILO, which have global responsibility for occupational health, have joint programs to provide assistance to countries in need. These include the WHO/ILO Joint Effort on Occupational Health and Safety in Africa and the ILO/WHO Global Program to Eliminate Silicosis. These programs rely on involvement of occupational health experts and institutions from many countries. WHO has a global network of 70 collaborating centers in occupational health located on all continents. The global network currently is chaired by NIOSH and includes 17 collaborating centers in the WHO Region of the Americas. These collaborating centers work with WHO, ILO, and nongovernmental organizations for the benefit of developing nations through a 5-year global work plan involving more than 300 projects in 15 priority areas.

The Fogarty International Center (FIC) at NIH conducts the International Training and Research in Occupational and Environmental Health (ITREOH) program that offers competitive grants for U.S. universities to work with universities in developing nations in research and capacity building. This program currently supports 17 U.S. universities that are working with institutions in more than 25 developing nations. Partnerships exist among these universities, WHO, ILO, and the WHO collaborating centers, and provide a sustainable framework for involvement of additional occupational professionals and institutions.

Individual and institutional opportunities exist for occupational health and safety professionals at home and abroad. Advances in distance-learning technologies make possible the inclusion of students from developing nations in courses at U.S. universities. Sabbatical programs can be arranged through WHO, ILO, the WHO Collaborating Center Network, and the Fogarty program's universities. Global work is facilitated by professional associations, including the International Commission on Occupational Health (ICOH), the International Ergonomics Association (IEA), the International Council of Nurses (ICN), and the International Occupational Hygiene Association (IOHA). National professional associations also contribute to global efforts of capacity-building, information-sharing, and training.

In summary, occupational health and safety professionals in the United States and other developed countries can help improve the prevention of occupational disease and injury in developing countries by working with existing global networks or independently to (a) learn more about global problems and issues; (b) advocate for an effective public health infrastructure and stronger laws and regulations to benefit developing countries; (c) work in developing countries; and (d) establish educational, research, and consulting partnerships with counterparts in developing countries.

Further Reading and Useful Websites

European Agency for Safety and Health at Work. <www.agency.osha.eu.int>

Fleming LE, Herzstein JA, Bunn WB (Eds.). *Issues in International Occupational and Environmental Medicine.* Beverly, MA: OEM Press, 1997.

International Commission on Occupational Health (ICOH). <www.icoh.org.sg>

International Council of Nurses. <www.ich.ch>

International Ergonomics Association. <www.iea.cc>

International Journal of Occupational and Environmental Health, published quarterly. <www.ijoeh.org>

International Labor Organization (ILO). <www.ilo.org>

International Occupational Hygiene Association. <www.ioha.com>

NIH Fogarty International Center. <www.nih.gov/fic>

WHO/ILO Joint Effort on Occupational Health and Safety in Africa. <www.sheafrica.info>

World Health Organization (WHO) Global Occupational Health Program. <www.who.int/oeh>

World Health Organization (WHO). *World Health Report.* Geneva: WHO, October 2002. <www.who.int/whr/2002>

Chapter 10
Resources for Disaster Preparedness

Barry S. Levy

Preparedness for disasters affecting the workplace, ranging from fires and explosions to natural disasters to terrorist attacks, is increasingly important. Preparedness encompasses a wide range of activities from evacuation drills to ensuring the availability of emergency services during a disaster. It is beyond the scope of this book to provide detailed information on this subject. The following websites and other sources of information can provide relevant information and guidance.

Planning Guides

Cetaruk EW. *Worker Preparedness and Response to Bioterrorism Core Program*. Available from: Association of Occupational and Environmental Clinics, 1010 Vermont Avenue NW, #513, Washington, DC, 20006, (202) 347–4976, <www.aoec.org>.

Emergency Management Guide for Business and Industry (FEMA).
External Link:
<www.fema.gov/library/bizindex.shtm>
pdf version available on this site—229 KB (67 pages).
Provides information on how to create and maintain a comprehensive emergency management program. It can be used by manufacturers, corporate offices, retailers, utilities, or any organization where a sizable number of people work or gather.

Critical Incident Protocol (Michigan State University)
External Link:
<www.cj.msu.edu/~outreach/CIP/CIP.pdf>
Provides information about the public and private sectors working together to plan for emergencies. Elements include planning, mitigation, business recovery, lessons learned, best practices, and plan exercising.

OSHA Evacuation Plans and Procedures eTool
External Link:
<www.osha.gov/SLTC/etools/evacuation/index.html>
Guidance for retail businesses on implementing an emergency action

plan. Also includes information on workplace evaluation, education, and training.

Small Business Disaster Planning Guide (Small Business Association/Institute for Business & Home Safety)
External Link:
<www.ibhs.org/docs/openforbusiness.pdf>
Disaster planning toolkit that enables small businesses to identify hazards, as well as plan for and reduce the impact of disasters. Also provides advice on insurance, disaster supplies, and other things that make a small business more disaster resistant.

Developing a Preparedness Plan and Conducting Emergency Evacuation Drills (National Fire Protection Association)
External Link:
<www.nfpa.org/Research/nfpafactsheets/emergency/emergency.asp>
Fact Sheet provides information about developing an emergency action plan, including fire prevention plans.

Model Shelter-in-Place Plan for Businesses (National Institute for Chemical Studies)
External Link:
<www.nicsinfo.org/SIP%20plan%20for%20offices%20NICS%20feb2003.pdf>
Provides information about establishing a shelter–in–place program for your office building.

Shelter–in–Place in an Emergency (American Red Cross)
External Link:
<www.redcross.org/services/disaster/beprepared/shelterinplace.html>
Includes information about shelter–in–place at home, work, school, and in a vehicle.

Business and Industry Preparedness Guide (American Red Cross)
External Link:
<www.redcross.org/services/disaster/beprepared/busi_industry.html#fema>
Guidance about planning for disasters, reducing potential damage, and protecting employees, customers, and business.

Guidance for Filtration and Air-Cleaning Systems to Protect Building Environments from Airborne Chemical, Biological, or Radiological Attacks

Guidance for Protecting Building Environments from Airborne Chemical, Biological, or Radiological Attacks (NIOSH).
Both are available at
<www.cdc.gov/niosh/topics/prepared/prepared facility.html>

Other Publications

Bevelacqua A, Stilp R. *Terrorism Handbook for Operational Responders*. Albany, NY: Delmar Publishers, 1998.

Henderson DA, Inglesby TV, O'Toole T. *Bioterrorism: Guidelines for Medical and Public Health Management*. Chicago: AMA Press, 2002.

Institute of Medicine, National Research Council. *Chemical and Biological Terrorism: Research and Development to Improve Civilian Medical Response*. Washington, DC: National Academy Press, 1999.

Landesman LY. *Public Health Management of Disasters: The Practice Guide* (with an Epilogue on 9/11). Washington, DC: American Public Health Association, 2001.

Levy BS, Sidel VW. *Terrorism and Public Health*. New York: Oxford University Press, 2003.

Maniscalco PM, Christen HT. *Understanding Terrorism and Managing the Consequences*. Upper Saddle River, NJ: Prentice Hall, 2002.

Pan American Health Organization. *Disasters: Preparedness and Mitigation in the Americas*. Monthly Newsletter available from: PAHO, 525 Twenty-third Street NW, Washington, DC 20037–2895.

The Second National Symposium on Medical and Public Health Response to Bioterrorism: Public Health Emergency and National Security Threat. *Public Health Reports* 2001; 116 (Suppl. 2): 1–118.

Sifton DW (Ed.). *PDR Guide to Biological and Chemical Warfare Response*. Montvale, NJ: Thomson/Physicians' Desk Reference, 2002.

Websites

CDC Bioterrorism Website: <www.bt.cdc.gov>

NIOSH Emergency Response Resources:
<www.cdc.gov/niosh/topics/emres>

OSHA Emergency Preparedness and Response Website:
<www.osha.gov/SLTC/emergencypreparedness/index.html>

Department of Homeland Security Website:<www.dhs.gov/dhspublic/>
Its materials on preparedness are at <www.ready.gov>

NIOSH references for businesses to prepare for terrorism and other disasters:
<www.cdc.gov/niosh/topics/prepared/>

Chapter 11
Other Sources of Information: Organizations and Websites

Barry S. Levy

General Resources

Emergency Services

Centers for Disease Control and Prevention
CDC Emergency Response
<www.cdc.gov>

Chemical Transportation Emergency Center
CHEMTREC® (Chemical Transportation Emergency Center), established by the chemical industry as a public service hotline for fire fighters, law enforcement, and other emergency responders to obtain information and assistance for emergency incidents involving chemicals and hazardous materials; offers a broad range of services. In the event of an emergency or spill, CHEMTREC provides information on hazardous materials to local emergency responders and links them with the shipper, product manufacturer, or other expert resources that can supply technical information, product disposition guidance, or other assistance on how to successfully handle the incident.
<www.chemtrec.org>

National Pesticide Information Center
A source of factual chemical, health, and environmental information about more than 600 pesticide active ingredients incorporated into over 50,000 different products registered for use in the United States since 1947.
<www.npic.orst.edu>

National Response Center
Serves as the sole national point of contact for reporting all oil, chemical, radiological, biological, and other discharges into the environment anywhere in the United States and its territories.

Occupational Safety and Health Administration (OSHA)

OSHA Emergency Hotline: 800-321-OSHA (800-321-6742)
This is a toll-free number for occupational safety and health emergencies that relate to a fatality or imminent threat to life.

Poison Control Centers

Each state or region has a 24-hour poison control center offering expert emergency medical information and referrals. Centers are listed on the inside cover of local phone directories. They can also be reached by dialing 911 and asking for the Poison Control Center or by calling the National Response Center Hotline listed earlier.

Guides, Catalogs, and Directories

National Institute for Occupational Safety and Health (NIOSH) Publications
<www.cdc.gov/niosh/homepage.html>

Occupational Safety and Health Administration

OSHA Publications

<www.osha.gov>
OSHA Computerized Information System (OCIS) on CD-ROM
OSHA Regulations, Documents, and Technical Information on CD-ROM

ORGANIZATIONS

Federal Government

Hazardous Materials Information Center

The Office of Hazardous Materials Safety, which is within the United States Department of Transportation's Research and Special Programs Administration, is responsible for coordinating a national safety program for the transportation of hazardous materials by air, rail, highway, and water.
<www.hazmat.dot.gov>

U.S. Department of Health and Human Services (DHHS)

The Department of Health and Human Services is the United States government's principal agency for protecting the health of all Americans and providing essential human services, especially for those who are least able to help themselves.
<www.hhs.gov>

Centers for Disease Control and Prevention (CDC)

The Centers for Disease Control and Prevention is recognized as the lead fed-

eral agency for protecting the health and safety of people-at home and abroad, providing credible information to enhance health decisions, and promoting health through strong partnerships. The CDC serves as the national focus for developing and applying disease prevention and control, environmental health, and health promotion and education activities designed to improve the health of the people of the United States.
<www.cdc.gov>

National Institute for Occupational Safety and Health (NIOSH)
NIOSH is the federal agency responsible for conducting research and making recommendations for the prevention of work-related injury and illness. NIOSH is part of the CDC in the Department of Health and Human Services.
<www.cdc.gov/niosh/homepage.html>

NIOSH Technical Information Service (800-356-4674)
A toll-free service that provides convenient public access to NIOSH and its information resources.

NIOSH Research Laboratories and Field Offices
NIOSH is headquartered in Washington, DC, with research laboratories and field offices in Cincinnati, Morgantown, Pittsburgh, Spokane, and Atlanta. NIOSH is a professionally diverse organization with a staff of over 1400 people representing a wide range of disciplines including epidemiology, medicine, industrial hygiene, safety, psychology, engineering, chemistry, and statistics.

Occupational Safety and Health Administration (OSHA)
Its mission is to save lives, prevent injuries, and protect the health of America's workers.
<www.osha.gov>

OSHA Directorate of Science, Technology, and Medicine (DSTM)
The DSTM includes offices located in the National Office along with the Salt Lake and Cincinnati Technical Centers. DSTM has developed special expertise in a number of areas to ensure that OSHA's capabilities are state-of-the-art with regard to occupational safety and health, including industrial hygiene, ergonomics, occupational health nursing and medicine, and safety engineering, along with sample analysis and equipment calibration and repair.

Experts in these areas are used by OSHA to supplement other program staff by providing specialized technical expertise and advice. The wide range of expertise and services available in DSTM results in its being involved in every major program area of OSHA, as well as in compliance assistance and outreach activities. Implementation of DSTM's programs provides support, service, and solutions for the rest of OSHA.

Other Federal Agencies

Agency for Toxic Substances and Disease Registry (ATSDR)

Its mission, as an agency of the U.S. Department of Health and Human Services, is to serve the public by using the best science, taking responsive public health actions, and providing trusted health information to prevent harmful exposures and disease related to toxic substances.
<www.atsdr.cdc.gov>

U.S. Department of Labor Bureau of Labor Statistics (BLS)

Principal fact-finding agency for the federal government in labor economics and statistics.
<www.bls.gov>

Environmental Protection Agency (EPA)

Its mission is to protect human health and to safeguard the natural environment—air, water, and land—upon which life depends.
<www.epa.gov>

Mine Safety and Health Administration (MSHA)

Its mission is to administer the provisions of the Federal Mine Safety and Health Act of 1977 (Mine Act) and to enforce compliance with mandatory safety and health standards as a means to eliminate fatal accidents; to reduce the frequency and severity of nonfatal accidents; to minimize health hazards; and to promote improved safety and health conditions in the nation's mines.
<www.msha.gov>

National Institute of Environmental Health Sciences (NIEHS)

Its mission is to reduce the burden of human illness and dysfunction from environmental causes by understanding each of these elements and how they interrelate.
<www.niehs.nih.gov>

National Toxicology Program (NTP)

Coordinates toxicological testing programs within the Department of Health and Human Services, strengthens the science base in toxicology, develops and validates improved testing methods, and provides information about potentially toxic chemicals to health regulatory and research agencies, the scientific and medical communities, and the public.
<www.ntp-server.niehs.nih.gov>

U.S. Nuclear Regulatory Commission (NRC)

An independent agency established by the Energy Reorganization Act of 1974 to regulate civilian use of nuclear materials.

Professional Organizations in the United States

American Association of Occupational Health Nurses (AAOHN)
Ensures occupational and environmental nurses (OHNs) are the authorities on health, safety, productivity and disability management for worker populations. Furthers the profession of occupational and environmental health nursing by advancing, protecting, guiding, and promoting the profession.
<www.aaohn.org>

American College of Occupational and Environmental Medicine (ACOEM)
Represents thousands of physicians and other health care professionals specializing in the field of occupational and environmental medicine. ACOEM is the nation's largest medical society dedicated to promoting the health of workers through preventive medicine, clinical care, research, and education. ACOEM develops positions and policies on vital issues relevant to the practice of preventive medicine both within and outside of the workplace.
<www.acoem.org>

American Conference of Governmental Industrial Hygienists (ACGIH®)
A member-based organization and community of professionals that advances worker health and safety through education and the development and dissemination of scientific and technical knowledge.
<www.acgih.org>

American Industrial Hygiene Association (AIHA)
Promotes, protects, and enhances industrial hygienists and other occupational health, safety and environmental professionals in their efforts to improve the health and well-being of workers, the community, and the environment.
<www.aiha.org>

American Medical Student Association (AMSA)
The oldest and largest independent association of physicians-in-training in the United States. Focuses on problems of the the medically underserved, inequities in U.S. health care system, and related issues in medical education. A fully independent student organization. Represents concerns of physicians-in-training. With a membership of medical students, premedical students, interns, and residents, AMSA is committed to improving medical training and the nation's health.
<www.amsa.org>

American National Standards Institute (ANSI)
A private, nonprofit organization that administers and coordinates the U.S. voluntary standardization and conformity assessment system. The Institute's mission is to enhance both the global competitiveness of U.S. business and the U.S.

quality of life by promoting and facilitating voluntary consensus standards and conformity assessment systems, and safeguarding their integrity.
<www.ansi.org>

American Public Health Association (APHA)
The oldest and largest organization of public health professionals in the world. APHA brings together researchers, health service providers, administrators, teachers, and other health workers in a multidisciplinary environment of professional exchange, study, and action. APHA is concerned with a broad set of issues affecting personal and environmental health, including federal and state funding for health programs, pollution control, programs and policies related to chronic and infectious diseases, a smoke-free society, and professional education in public health.
<www.apha.org>

American Society of Safety Engineers (ASSE)
The oldest and largest professional safety organization. Its members manage, supervise, and consult on safety, health, and environmental issues in industry, insurance, government, and education.
<www.asse.org>

Human Factors and Ergonomics Society
Promotes discovery and exchange of knowledge concerning the characteristics of people that are applicable to design of systems and devices of all kinds. Encourages education and training for those entering the human factors and ergonomics profession and for those who conceive, design, develop, manufacture, test, manage, and participate in systems.
<www.hfes.org>

National Safety Council (NSC)
Leading advocate for safety and health. Educates and influences society to adopt safety, health, and environmental policies, practices, and procedures that prevent and mitigate human suffering and economic losses arising from preventable causes.
<www.nsc.org>

Society for Occupational and Environmental Health (SOEH)
Brings together professionals in government, industry, labor, and academia in order to reduce occupational and environmental health hazards through the presentation of scientific data and dynamic exchange of information across institutions and disciplines.

Association of Occupational and Environmental Clinics (AOEC)

Improves practice of occupational and environmental health through information sharing and collaborative research. Aids in identifying, reporting, and preventing occupational and environmental health hazards and their effects; encourages provision of high quality clinical services for people with occupationally or environmentally related health problems; provides means for occupational/environmental health clinics to share information that will better enable them to diagnose and treat occupational/environmental diseases; increases communication among clinics concerning issues related to patient care; facilitates liaison between clinics and agencies responsible for workplace/environmental monitoring; and provides a source of data for research projects related to occupational/ environmental health.
<www.aoec.org>

Committees on Occupational Safety and Health (COSH Groups)

The National COSH Network is a collection of loosely affiliated local and statewide COSH groups—Committees/Coalitions on Occupational Safety and Health. COSH groups are private, nonprofit coalitions of labor unions, health and technical professionals, and others interested in promoting and advocating for worker health and safety.
<www.coshnetwork.org>

COSH Groups

ALASKA
Alaska Health Project
907-276-2864

ARKANSAS
Arkansas Coalition on Safety and Health (ARCOSH)
501-569-8483

CALIFORNIA
Santa Clara Center for Occupational Safety and Health (SCCOSH)
408-998-4050

Southern California COSH (SoCalCOSH-formerly Los Angeles COSH)
310-206-0860

WORKSAFE California (Bay Area)

CONNECTICUT
Connecticut Council on Occupational Safety and Health (ConnectiCOSH)
860-953-2674

GEORGIA
Georgia Committee for Occupational Safety and Health (GCOSH)
404-245-7040

ILLINOIS
Chicago Area Committee on Occupational Safety and Health (CACOSH)
<www.cacosh.org>

MAINE
Maine Labor Group on Health
<www.mlgh.org>

MASSACHUSETTS
Massachusetts Coalition for Occupational Safety and Health (MassCOSH)
<www.masscosh.org>

Western Massachusetts Coalition for Occupational Safety and Health
(WesternMassCOSH) Springfield
<www.westernmasscosh.org>

MICHIGAN
Southeast Michigan Coalition on Occupational Safety and Health (SEM-
COSH)
<www.semcosh.org>

NEW HAMPSHIRE
New Hampshire Coalition for Occupational Safety and Health (NHCOSH)
603-226-0516

NEW JERSEY
New Jersey Work Environment Council (WEC)
<www.njwec.org>

NEW YORK
Central New York Committee on Occupational Safety and Health (CNY-
COSH), Syracuse Area
315-471-6187

Mid-State Education and Service Foundation, Ithaca Area
607-277-5670

New York Committee for Occupational Safety and Health (NYCOSH), New York City Area
<www.nycosh.org>

Western New York Council on Occupational Safety and Health (WNYCOSH), Buffalo
716-833-5416

NORTH CAROLINA
North Carolina Occupational Safety and Health Project (NCOSH)
<www.ncosh.org>

PENNSYLVANIA
Philadelphia Project on Occupational Safety and Health (Philaposh)
<www.philaposh.org>

RHODE ISLAND
Rhode Island Committee on Occupational Safety and Health (RICOSH)
401-751-2015

WISCONSIN
Wisconsin Committee on Occupational Safety and Health (WISCOSH), Milwaukee
<www.wiscosh.org>

South Central Wisconsin Committee on Occupational Safety and Health (SCWCOSH), Madison
603-345-1111

Educational Resource Centers

NIOSH funds 15 university programs that train occupational health professionals, conduct research, and provide services in this area. A number of other programs have been awarded NIOSH training funds to support graduate studies in the various disciplines (occupational medicine, occupational nursing, industrial hygiene, occupational ergonomics, and occupational safety). A complete list of these programs can be obtained from NIOSH.

ALABAMA
Deep South Center for Occupational Health and Safety
     University of Alabama at Birmingham School of Public Health
     Birmingham, AL 35294
     <www.uab.edu/dsc>

CALIFORNIA-NORTHERN
Northern California Center for Occupational and Environmental Health
   University of California, Berkeley, Richmond Field Station
   Richmond, CA 94804
   <www.coeh.berkeley.edu>

CALIFORNIA-SOUTHERN
   Southern California Education and Research Center
   University of Southern California
   Los Angeles, CA 90033

FLORIDA
   Sunshine Education and Research Center
   University of South Florida
   College of Public Health
   Department of Environmental and Occupational Health
   Tampa, FL 33612-3805
   <www.publichealth.usf.edu/erc>

ILLINOIS
   Great Lakes Center for Occupational and Environmental
    Safety and Health
   University of Illinois at Chicago
   School of Public Health
   Chicago, IL 60612-7260
   <www.uic.edu/spha/glakes>

MARYLAND
Johns Hopkins Education and Research Center for Occupational Safety and
Health
   Johns Hopkins University
   School of Hygiene and Public Health
   Baltimore, MD 21205
   <www.jhsph.edu/erc>

MASSACHUSETTS
   Harvard Education and Research Center
   Center for Continuing Professional Education
   Harvard School of Public Health
   Boston, MA 02115-6023
   <www.hsph.harvard.edu/erc>

MICHIGAN
   Michigan Education and Research Center
   University of Michigan

Center for Occupational Health and Safety Engineering
Ann Arbor, MI 48109-2117
<www.engin.umich.edu/dept/ioe/COHSE>

MINNESOTA
Minnesota Education and Research Center
Midwest Center for Occupational Safety and Health
St. Paul, MN 55101
<www1.umn.edu/mcohs>

NEW YORK/NEW JERSEY
New York/New Jersey Occupational Safety & Health Education and
Research Center
EOHSI Centers for Education and Training
Piscataway, NJ 08854-3923
<www.nynjerc.org>

NORTH CAROLINA
North Carolina Occupational Safety and Health Education and Research
Center
North Carolina Institute for Public Health
University of North Carolina at Chapel Hill
Chapel Hill, NC 27514
<www.sph.unc.edu/osherc>

OHIO
NIOSH Education and Research Center
University of Cincinnati
Cincinnati, OH 45267-0056
<www.uc.edu/erc>

TEXAS
Southwest Center for Occupational and Environmental Health
The University of Texas Health Science Center at Houston
School of Public Health
Houston, TX 77030
<www.sph.uth.tmc.edu/swcoeh>

UTAH
University of Utah Education and Research Center
Rocky Mountain Center for Occupational and Environmental Health
University of Utah
Salt Lake City, UT 84112

WASHINGTON
> Northwest Center for Occupational Health and Safety
> Department of Environmental Health
> University of Washington
> Seattle, WA 98105
> <www.depts.washington.edu/ehce/NWcenter/index.html>

Agricultural Research Centers

Great Lakes Center for Agricultural Safety & Health
<www.ag.ohio-state.edu/~agsafety>
States served: IL, IN, KY, MI, OH, PA, WI, WV

Great Plains Center for Agricultural Health
<www.public-health.uiowa.edu/gpcah>
States served: IA, KS, MS, NE

High Plains Intermountain Center for Agricultural Health and Safety
<www.hicahs.colostate.edu>
States served: CO, MO, ND, SD, UT, WY

Midwest Center for Agricultural Disease and Injury Research, Education, and Prevention
<www.marshfieldclinic.org/nfmc/projects/default.htm>
States served: IL, IN, MI, MN, OH, WI

Northeast Center for Agricultural Safety and Occupational Health
<www.nycamh.com>
States served: CT, DE, MA, MD, ME, NH, NJ, NY, PA, RI, VA, VT, WV

Pacific Northwest Agricultural Safety and Health Center
<www.depts.washington.edu/pnash>
States served: AL, ID, OR, WA

Southeast Center for Agricultural Health & Injury Prevention
<www.mc.uky.edu/scahip>
States served: GA, KY, NC, SC, TN, VA

Southern Coastal Agromedicine Center
<www.ncagromedicine.org/scac.htm>
States served: AL, FL, GA, MS, NC, SC, VA, Puerto Rico, Virgin Islands

Southwest Center for Agricultural Health, Injury Prevention, and and Education

States served: AR, LA, NM, OK, TX

Western Center for Agricultural Health and Safety
<www.agcenter.ucdavis.edu>
states served: AZ, CA, HA, NV

Labor-Oriented Advisory and Research Groups

Center to Protect Workers' Rights (CPWR)
Develops practical ways to improve safety and health for construction workers and their families. As the research, development, and training arm of the
Building and Construction Trades Department and the construction unions in
the AFL-CIO, CPWR works with more than 30 organizations nationwide.
<www.cpwr.com>

Environmental Defense
Protects environmental rights of people, including future generations, including rights to clean air and water, healthy and nourishing food, and a flour-
ishing ecosystem. Guided by science, it evaluates environmental problems
and works to create and advocate solutions that win lasting political, economic, and social support.
<www.edf.org>

Highlander Research and Education Center
Gathers workers, grassroots leaders, community organizers, educators, and
researchers to address the most pressing social, environmental, and economic problems facing people of the South. Sponsors educational programs and
research into community problems, as well as a residential Workshop Center
for social change organizations and workers active in the South and internationally.
<www.highlandercenter.org>

Migrant Legal Action Program
Provides legal representation and a national voice for migrant and seasonal
farm workers. Works to enforce rights and to improve public policies affecting farm workers' working and housing conditions, education, health, nutrition, and general welfare.
<www.mlap.org>

9to5 National Association of Working Women
Strengthens women's ability to work for economic justice.

Public Citizen

National, nonprofit consumer advocacy organization that represents consumer interests in Congress, the executive branch, and the courts.
<www.citizen.org>

International Organizations

International Commission on Occupational Health (ICOH)

An international nongovernmental professional society that fosters scientific progress, knowledge, and development of occupational health and safety.
<www.icoh.org.sg>

International Labor Organization (ILO)

UN specialized agency that promotes social justice and internationally recognized human and labor rights.
<www.ilo.org>

Pan American Health Organization (PAHO)

The Pan American Health Organization (PAHO), an international public health agency with 100 years of experience working to improve health and living standards of the people of the Americas, has an essential mission to strengthen national and local health systems and improve the health of the peoples of the Americas, in collaboration with ministries of health, other government and international agencies, nongovernmental organizations, universities, social security agencies, community groups, and many others. PAHO promotes primary health care strategies, which reach people in their communities, to extend health services to all and to increase efficiency in the use of scarce resources. It addresses a wide range of disorders, including those due to occupational factors.
<www.paho.org>

World Health Organization (WHO)

UN specialized agency for health that promotes attainment by all people of the highest possible level of health. Health is defined in the WHO constitution as a state of complete physical, mental, and social well-being and not merely the absence of disease or infirmity.
<www.who.int>

Labor Organizations

American Federation of Labor-Congress of Industrial Organizations (AFL-CIO)

Works to improve the lives of America's working families, bring fairness and dignity to the workplace, and secure social and economic equity in the United States. Strengthens working families by enabling more workers to join together in unions, builds a stronger political voice for working families, provides a

new voice for workers in the global economy, and creates a more effective voice for working families in our communities.
<www.aflcio.org>

Building and Construction Trades Department, AFL-CIO
Helps its 15 affiliated building trades unions to make job sites safer, delivers apprenticeship and journey-level training, organizes new workers, supports legislation that affects working families, and assists in securing improved wages, hours, and working conditions through collective bargaining.
<www.buildingtrades.org>

AFL-CIO Affiliates

American Federation of Government Employees (AFGE)
<www.afge.org>

American Federation of State, County and Municipal Employees (AFSCME)
<www.afscme.org>

Communications Workers of America (CWA)
<www.cwa-union.org>

International Chemical Workers Union (ICWU)
<www.icwuc.org>

United Food and Commercial Workers International Union (UFCW)
<www.ufcw.org>
International Union of Automobile, Aerospace and Agricultural Implement Workers of America (UAW)
<www.uaw.org>

Paper, Allied-Industrial, Chemical & Energy Workers International Union (PACE)
<www.paceunion.org>

Service Employees International Union (SEIU)
<www.seiu.org>

United Mine Workers of America (UMWA)
<www.umwa.org>

Union of Needletrades, Industrial and Textile Employees (UNITE!)
<www.uniteunion.org>

Index